SCOTT M. STARRATT

D0621844

How will the world's vegetation, from 'natural' ecosystems to intensively managed agricultural systems, be affected by changes in land use, the composition of gases in the atmosphere, and climate? This major new book presents a collection of essays by leading authorities who address the current state of knowledge. The chapters bring together the early results of an international scientific research programme designed to answer the following questions: what will happen to our ability to produce food and fibre, and what effects will there be on biological diversity under rapid environmental change? How will these changes to terrestrial ecosystems feed back to further environmental change?

Among the special features of the book are descriptions of a dynamic global vegetation model, developing generic crop models and a special section on the emerging discipline of global ecology.

Global Change and Terrestrial Ecosystems

The International Geosphere-Biosphere
Programme was established in 1986 by the
International Council of Scientific Unions,
with the stated aim

*To describe and understand the interactive
physical, chemical and biological processes that
regulate the total Earth system, the unique
environment that it provides for life, the
changes that are occurring in this system, and
the manner in which they are influenced by
human activities.*

A wide-ranging and multi-disciplinary project
of this kind is unlikely to be effective unless it
identifies priorities and goals, and the IGBP
defined six key questions that it seeks to
answer. These are;

- How is the chemistry of the global atmosphere regulated, and what is the role of
  biological processes in producing and
  consuming trace gases?

- How will global changes affect terrestrial
  ecosystems?

- How does vegetation interact with physical
  processes of the hydrological cycle?

- How will changes in land-use, sea level and
  climate alter coastal ecosystems, and what
  are the wider consequences?

- How do ocean biogeochemical processes
  influence and respond to climate change?

- What significant climatic and environmental
  changes occurred in the past, and what
  were their causes?

The **International Geosphere-Biosphere
Programme Book Series** brings new work
on topics within these themes to the attention
of the wider scientific audience.

INTERNATIONAL GEOSPHERE–BIOSPHERE PROGRAMME BOOK SERIES

Global Change and Terrestrial Ecosystems

Edited by

Brian Walker
CSIRO Division of Wildlife & Ecology

Will Steffen
GCTE Core Project Office
CSIRO Division of Wildlife & Ecology

CAMBRIDGE
UNIVERSITY PRESS

Published by the Press Syndicate of the University of Cambridge
The Pitt Building, Trumpington Street, Cambridge CB2 1RP
40 West 20th Street, New York, NY 10011-4211, USA
10 Stamford Road, Oakleigh, Melbourne 3166, Australia

First published 1996

Printed in Great Britain at
the University Press, Cambridge

A catalogue record for this book is available from the British Library

Library of Congress cataloguing in publication data available

ISBN 0 521 57094 8 hardback
ISBN 0 521 57810 8 paperback

RO

Contents

Contributors

D. J. Barrett
CSIRO Division of Plant Industry, GPO Box 1600, Canberra ACT 2601, Australia

S. L. Bassow
Department of Organismic and Evolutionary Biology, Harvard University,
Cambridge MA 02138, USA

F. A. Bazzaz
Department of Organismic and Evolutionary Biology, Harvard University,
Cambridge MA 02138, USA

G. M. Berntson
Department of Organismic and Evolutionary Biology, Harvard University,
Cambridge MA 02138, USA

L. B. Brubaker
College of Forest Resources, University of Washington, Seattle WA 98195, USA

R. Bunce
Institute of Terrestrial Ecology, Merlewood Research Station, Grange over Sands,
Cumbria LA11 6JU, United Kingdom

F. S. Chapin III
Department of Integrative Biology, University of California, Berkeley CA 94720-3140,
USA

M. B. Coughenour
Natural Resource Ecology Laboratory, Colorado State University, Fort Collins
CO 80523, USA

W. Cramer
Department of Global Change and Natural Systems, Potsdam Institute for Climate
Impact Research (PIK), P.O. BOX 60 12 03, D-14412 Potsdam, Germany

C. M. D'Antonio
Department of Integrative Biology, University of California, Berkeley CA 94720-3140,
USA

G. Dedieu
Centre d'Etudes Spatiales de la Biosphere (CESBIO), 18 Av E Belin, bpi 2801, F-31055 Toulouse Cedex, France

V. M. Eckhart
Department of Integrative Biology, University of California, Berkeley CA 94720-3140, USA

A. Fischer
Centre d'Etudes Spatiales de la Biosphere (CESBIO), 18 Av E Belin, bpi 2801, F-31055 Toulouse Cedex, France
Current address: Department of Global Change and Natural Systems, Potsdam Institute for Climate Impact research (PIK), P.O. Box 60 12 03, D-14412 Potsdam, Germany

G. Fischer
International Institute for Applied System Analysis, Schlossplatz 1, A-2361 Laxenburg, Austria

R. H. Gardner
Appalachian Environmental Laboratory, Center for Environmental and Estuarine Studies, University of Maryland, Frostburg MD 21532, USA

R. M. Gifford
CSIRO Division of Plant Industry, GPO Box 1600, Canberra ACT 2601, Australia

J. Goudriaan
Department of Theoretical Production Ecology, Agricultural University, Wageningen, The Netherlands

L. Gunderson
Arthur R. Marshall Laboratory, Department of Zoology, University of Florida, Gainesville FL 32611, USA

W. W. Hargrove
Oak Ridge National Laboratory, Environmental Division, PO Box 2008, Oak Ridge TN 37831, USA

O. W. Heal
Institute of Terrestrial Ecology, Bush Estate, Penicuik, Edinburgh EH26 0QB, United Kingdom

K. L. Heong
IRRI, P.O. Box 933, Manila, Philippines

C. S. Holling
Arthur R. Marshall Laboratory, Department of Zoology, University of Florida, Gainesville FL 32611, USA

D. Y. Hollinger
USDA Forest Service, P.O. Box 640, Durham NH 03824, USA

J. E. Hossell
Department of Geography, University of Birmingham, Birmingham B15 2TT, United Kingdom

J. S. I. Ingram
GCTE Focus 3 Office, Centre for Ecology and Hydrology, Maclean Building, Crowmarsh Gifford, Wallingford, Oxfordshire OX10 8BB, United Kingdom

P. J. Jones
Centre for Agricultural Strategy, University of Reading, 1 Earley Gate, Reading RG6 2AT, United Kingdom

F. M. Kelliher
Manaaki Whenua – Landcare Research, P.O. Box 31-011, Christchurch, New Zealand

L. Kergoat
Centre d'Etudes Spatiales de la Biosphere (CESBIO), 18 Av E Belin, bpi 2801, F-31055 Toulouse Cedex, France

A. Kjøller
Department of General Microbiology, University of Copenhagen, Solvgade 83H, DK-1307 Copenhagen K, Denmark

Ch. Körner
Institute of Botany, University of Basel, Schönbeinstr. 6, CH-4056 Basel, Switzerland

M. J. Kropff
IRRI, P.O. Box 933, Manila, Philippines

D. Lambert
Arthur R. Marshall Laboratory, Department of Zoology, University of Florida, Gainesville FL 32611, USA

J. J. Landsberg
Centre for Environmental Mechanics, CSIRO, GPO Box 821, Canberra ACT 2601, Australia

R. Leemans
Department of Terrestrial Ecology and Global Change, National Institute of Public Health and Environmental Protection, PO Box 1, 3720 BA Bilthoven, The Netherlands

S. Linder
Swedish University of Agricultural Sciences, Department of Ecology and Environmental Research, P.O. Box 7072, S-750 07 Uppsala, Sweden

J. Lloyd
Research School of Biological Sciences, Australian National University, GPO Box 475, Canberra ACT 2601, Australia

S. Louahala
Media Science International, Av. Pre des Agneaux 83, B-1160 Bruxelles, Belgium

J. L. Lutze
CSIRO Division of Plant Industry, GPO Box 1600, Canberra ACT 2601, Australia

P. Maisongrande
Centre d'Etudes Spatiales de la Biosphere (CESBIO), 18 Av E Belin, bpi 2801, F-31055 Toulouse Cedex, France

P. Marples
Arthur R. Marshall Laboratory, Department of Zoology, University of Florida, Gainsville, FL 32611, USA

J. S. Marsh
Centre for Agricultural Strategy, University of Reading, 1 Earley Gate, Reading, RG6 2AT, United Kingdom

J. S. McLachlan
Harvard Forest, Harvard University, Petersham MA 01366, USA

R. E. McMurtrie
School of Biological Science, P.O. Box 1, Kensington NSW 2033, Australia

J. M. Melillo
The Ecosystems Center, Marine Biological Laboratory, Woods Hole MA 02543, USA

H. A. Mooney
Department of Biological Sciences, Stanford University, Stanford CA 94305, USA

R. P. Neilson
USDA Forest Service, 3200 S.W. Jefferson Way, Corvallis OR 97331, USA

I. R. Noble
Ecosystem Dynamics, Research School of Biological Sciences, Australian National University, Canberra ACT 0200, Australia

F. W. Nutter
Department of Plant Pathology, Iowa State University, Ames IA 50111, USA

M. L. Parry
Department of Geography, University College London, 26 Bedford Way, London WC1H 0AP, United Kingdom

W. J. Parton
Natural Resource Ecology Laboratory, Colorado State University, Fort Collins CO 80523, USA

G. Peterson
Arthur R. Marshall Laboratory, Department of Zoology, University of Florida, Gainesville FL 32611, USA

K. Redford
The Nature Conservancy, International Headquarters, 1815 North Lynn Street, Arlington VA 22209, USA

T. Rehman
Department of Agriculture, University of Reading, P.O. Box 236, Reading RG6 2AT, United Kingdom

H. L. Reynolds
Department of Integrative Biology, University of California, Berkeley CA 94720-3140, USA

W. H. Romme
Department of Biology, Fort Lewis College, Durango, CO 81301, USA

C. Rozenzweig
Columbia University / Goddard Institute for Space Studies, 2880 Broadway, New York NY 10025, USA

S. W. Running
School of Forestry, University of Montana, Missoula MT 59812, USA

O. E. Sala
Departamento de Ecologia, Facultad de Agronomia, Universidad de Buenos Aires, Av San Martin 4453, Buenos Aires 1417, Argentina

A. B. Samarakoon
CSIRO Division of Plant Industry, GPO Box 1600, Canberra ACT 2601, Australia

E.-D. Schulze
Universität Bayreuth, Lehrstuhl Plazenokölogie, D-95440, Bayreuth, Germany

J. Sendzimir
Arthur R. Marshall Laboratory, Department of Zoology, University of Florida, Gainesville FL 32611, USA

H. H. Shugart
Department of Environmental Sciences, University of Virginia, Charlottesville VA 22901, USA

T. M. Smith
Department of Environmental Sciences, University of Virginia, Charlottesville VA 22901, USA

W. L. Steffen
GCTE Core Project Office, CSIRO Division of Wildlife & Ecology, PO Box 84, Lyneham ACT 2602, Australia

S. Struwe
Department of General Microbiology, University of Copenhagen, Solvgade 83H, DK 1307 Copenhagen K, Denmark

R. W. Sutherst
c/o CTPM, Gehrmann Laboratories, University of Queensland, St Lucia QLD 4072, Australia

P. S. Teng
IRRI, P.O. Box 933, Manila, Philippines

S. C. Thomas
Department of Organismic and Evolutionary Biology, Harvard University, Cambridge MA 02138, USA

P. B. Tinker
Department of Plant Sciences, University of Oxford, South Parks Road, Oxford OX1 3RB, United Kingdom

R. B. Tranter
Centre for Agricultural Strategy, University of Reading, Reading RG6 2AT, United Kingdom

M. G. Turner
Department of Zoology, University of Wisconsin, Madison WI 53706, USA

C. Valentin
ORSTOM, BP 11416 Niamey, Niger

N. N. Vygodskaya
Severtsov Institute of Animal Evolutionary Morphology and Ecology, Russian Academy of Sciences, Moscow, Russia

B. H. Walker
CSIRO Division of Wildlife & Ecology, PO Box 84, Lyneham ACT 2602, Australia

F. I. Woodward
Department of Animal and Plant Sciences, University of Sheffield, PO Box 601, Sheffield S10 2UQ, United Kingdom

Preface

The First Science Conference of the Global Change and Terrestrial Ecosystems (GCTE) Core Project of the International Geosphere-Biosphere Programme (IGBP) was held at the Marine Biological Laboratory, Woods Hole, Massachusetts, USA, in May 1994. The invited, plenary presentations at the Conference have been brought together in this book. Some 430 scientists participated in the Conference and the lively discussions following these plenary sessions served two important purposes. First, they advanced the understanding and general awareness of global change science in regard to terrestrial ecosystems. Second, they provided a valuable input to the leaders of the variouss Activities and Tasks of the GCTE Core Project. New directions were set and new projects resulted from the Conference. On both counts, the Conference was very successful.

Many people helped to make the Conference possible, but we would particularly like to acknowledge the following: Jerry Melillo, Director of the Ecosystems Centre, Marine Biological Laboratory, for organizing the Conference venue and overseeing the local arrangements. Susan Pennington, Suzanne Donovan and the rest of the MBL staff at Woods Hole, who worked tirelessly to ensure that the Conference was a success. Rowena Foster, of the GCTE Core Project Office, Canberra, Australia, and her assistant, Sheree Baker, for the enormous organizational effort they contributed before, during and after the Conference.

Brian Walker
Will Steffen

Canberra, May 1996

Part one

Introduction

GCTE science: objectives, structure and implementation

B. H. Walker and W. L. Steffen

Introduction

This volume represents a synthesis of our current understanding of global change interactions with terrestrial ecosystems, and includes the initial achievements of the Global Change and Terrestrial Ecosystems (GCTE) Core Project of the International Geosphere-Biosphere Programme (IGBP).

GCTE is one of six Core Projects of the IGBP. Its place in the overall structure of the Programme is shown in Fig. 1.1. The IGBP was established as an international research programme in 1986 by the International Council of Scientific Unions (ICSU). The role of terrestrial ecosystems in the functioning of the Earth system was one of the major themes arising from the first deliberations of the Special Committee, formed to plan the IGBP programme. The planning process involved input from a large number of scientists, in a number of workshops, and the deliberations and conclusions from these workshops are published in a series of IGBP Reports (Nos. 5, 9, 10, 11, 12 and 21).

In 1990 GCTE became an accepted Core Project within IGBP. In December of that year the Core Project Office was established in Canberra, Australia. By late 1991 planning for the GCTE research effort was largely completed, and the GCTE Operation Plan (IGBP Report No. 21) was published in April 1992.

Later in 1992 GCTE officially became operational with the acceptance of the first set of contributing projects into its Core Research Programme. Also at that time, the first of GCTE's coordinated research structures – the Elevated CO_2 Consortium and the Wheat Network – were launched. Since then the Core Research Programme has continued to grow and additional coordinated activities have coalesced around specific Tasks in GCTE (see IGBP Report No. 21 for a definition of Focus, Activity and Task within the context of GCTE's structure). Currently some 400 research scientists and technicians from 39 countries around the world are involved in GCTE Core Research projects, and the total annual expenditure on this research is over $US 20 million. GCTE Report No. 1 (GCTE 1994) lists the initial

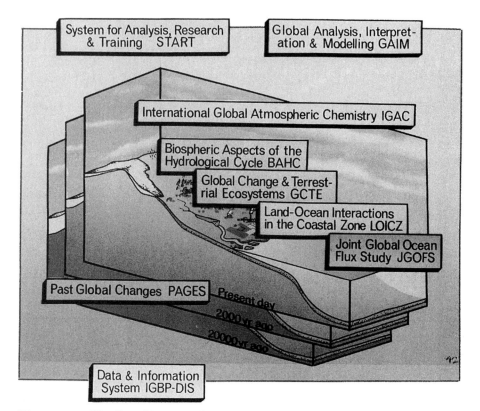

Figure 1.1 *The Core Projects (IGAC, BAHC, GCTE, LOICZ, JGOFS and PAGES) and Framework Activities (START, GAIM, IGBP-DIS) of the IGBP.*

Core Research projects and gives information on how to become formally involved with the Core Research Programme. Annual updates will be published in the GCTE Report series.

Objectives

GCTE has two overall objectives:

- To predict the effects of changes in climate, atmospheric composition and land use on terrestrial ecosystems, including (i) agriculture, forestry, soils, and (ii) ecological complexity.
- To determine how these effects lead to feedbacks to the atmosphere and the physical climate system.

These two objectives are closely linked throughout GCTE research, as demonstrated in many of the chapters in this volume.

Note also that global change, as defined operationally within IGBP, has three components, or drivers: change in (i) atmospheric composition, (ii) climate and (iii) land use. Included in (ii) are aspects of radiation, including UV-B. All three interact strongly in their effects on terrestrial ecosystems, and the challenge of developing a predictive understanding of ecosystem response to this suite of drivers is central to the GCTE effort.

Structure

The GCTE research programme is built around four major themes, or Foci. Fig. 1.2 shows the overall structure, with the four Foci represented by large boxes. The component Activities of each Focus are shown as bullets in the boxes.

The primary aim of Focus 1, Ecosystem Physiology, is to understand and model the effect of global change on primary ecosystem processes, such as the exchange of energy, water and trace gases with the atmosphere, element cycling and storage, and biomass accumulation or loss. A central thesis of Focus 1 is that the ways in which ecosystems function – their physiology – will be strongly affected by the combined and interactive suite of changes in atmospheric CO_2, land-use practices, and the likely changes in the means and extremes of temperature and rainfall.

The driving forces of global change will also lead to changes in the distributions of plant and animal species and the species composition of ecosystems. Changes in ecosystem composition will, in turn, lead to changes in ecosystem physiology, such as evapotranspiration and nutrient cycling. The goal of Focus 2, Change in Ecosystem Structure, is to model this complex suite of impacts and responses so that the pattern of change in ecosystem composition and structure can be predicted.

The world's terrestrial ecosystems constitute a continuum from virtually pristine to intensively managed and highly modified systems devoted to production of food and fibre. Many of these systems are already threatened by damage to soil and water resources through poor technology and management practices, and will be further impacted by global change. Through its Focus 3, Global Change Impact on Agriculture, Forestry and Soils, GCTE aims to improve our capability to predict global change impacts on key agricultural systems, livestock production, forestry and soils, and also to provide means for mitigating these impacts. Focus 3 interacts strongly with Foci 1 and 2 to share expertise gained from studying systems from the more applied and the more fundamental points of view respectively. It also works closely with Focus 4 to study the role of ecosystem complexity in multi-species agriculture, and how global change will affect it.

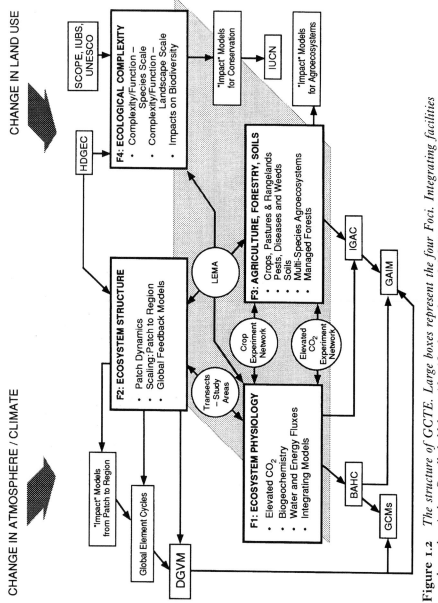

Figure 1.2 *The structure of GCTE. Large boxes represent the four Foci. Integrating facilities are shown by circles. Small, bold boxes identify the products of GCTE, and the others are some of the major groups with which GCTE interacts.*

CHANGE IN LAND USE

CHANGE IN ATMOSPHERE / CLIMATE

SCOPE, IUBS, UNESCO

F4: ECOLOGICAL COMPLEXITY
- Complexity/Function – Species Scale
- Complexity/Function – Landscape Scale
- Impacts on Biodiversity

"Impact" Models for Conservation

IUCN

"Impact" Models for Agroecosystems

HDGEC

F2: ECOSYSTEM STRUCTURE
- Patch Dynamics
- Scaling: Patch to Region
- Global Feedback Models

LEMA

F3: AGRICULTURE, FORESTRY, SOILS
- Crops, Pastures & Rangelands
- Pests, Diseases and Weeds
- Soils
- Multi-Species Agroecosystems
- Managed Forests

Crop Experiment Network

Elevated CO_2 Experiment Network

Transects – Study Areas

IGAC

GAIM

F1: ECOSYSTEM PHYSIOLOGY
- Elevated CO_2
- Biogeochemistry
- Water and Energy Fluxes
- Integrating Models

"Impact" Models from Patch to Region

Global Element Cycles

DGVM

BAHC

GCMs

Ecological complexity encompasses the spatial patterning between and within ecosystems, the numbers and relative abundances of species, and the structure of their trophic networks. The relationships between this complexity and the functioning of ecosystems, and how global change will affect these relationships, is the central theme of Focus 4.

The structure built around these four Foci is proving to be robust and effective in organizing the GCTE research effort. Thus, this volume is also built around that structure, with papers grouped by Focus, each addressing an important issue within that Focus. An overview chapter by the Focus leader pulls together the main results highlighted in the other papers and provides an integrated assessment of the current state of the science in each of GCTE's four major thematic areas.

The research undertaken in GCTE is not isolated within the four Foci, however. There is much collaboration amongst them, and a number of integrating mechanisms have been developed to foster this interaction, such as the Long-term Ecological Modelling Activity (LEMA) network and the IGBP/GCTE transects. These are shown as circles in Fig. 1.2. In addition, there are a number of Activities/Tasks that have major relevance for more than one Focus, and these are jointly 'owned' and managed by the two or more Foci involved.

Overviews of the status of each of the four Foci are given in this volume by Mooney, Shugart, Tinker & Ingram and Sala, respectively.

Cross-cutting themes are also emphasized in Part VI of this volume. A particularly important area is GCTE's contribution to earth system science, which relies on research throughout the programme to provide an integrated understanding of the terrestrial biosphere's role in global biogeochemical cycles. An example is the role of terrestrial ecosystems in the global carbon cycle (see chapter by Melillo), which draws on work from all four Foci.

Implementation

A phased approach to implementing the research programme has been adopted. Three broad stages of development can be identified, and good examples of all three are represented by chapters in this volume.

Some areas of research are already well developed, and GCTE is playing an international coordinating role, where appropriate. In these areas, there was a recognition that a large body of excellent work already existed; GCTE has further promoted and supported that work, sponsoring synthesis workshops and model intercomparisons, and has facilitated experimental cooperation. Examples include research on the effects of enhanced atmospheric CO_2 on terrestrial ecosystems (see chapters by Körner, Bazzaz *et al.* and Gifford *et al.*); the review and synthesis of stomatal and bulk surface conductance (see chapter by Schulze *et al.*); and the model intercomparisons of the GCTE wheat network (see chapter by Goudriaan). In these cases synthesis studies have already been undertaken, and some conclusions and generalizations are becoming available.

Other areas of GCTE-type science are still in a formative stage but much basic, underlying research has been done and they are poised for rapid progress. Here GCTE is taking a more proactive role in the initiation, organization and implementation of the work. A good example is global vegetation modelling, where just two years ago the only models were equilibrium models based on a broad correlation between biome type and climate. Recent advances, at least partly catalysed by GCTE's push for a dynamic global vegetation model, have been impressive. Global vegetation models are now more firmly based on plant and ecosystem physiology (see chapter by Woodward), have been linked to models of ecosystem function (see chapter by Neilson & Running), are using paleodata to reconstruct vegetation patterns of the past, and now are rapidly moving towards capturing much more realistic dynamics of vegetation change.

Other areas are still diffuse, full of good and stimulating ideas at the very forefront of scientific thought and primed for the initiation of an exciting and vigorous research effort. In many cases, these areas are strongly multidisciplinary, and require the active and close collaboration of scientific disciplines that have not often worked together in the past. GCTE is just initiating its research in these areas. The chapters in this volume describe the major scientific issues to be resolved, and point the way towards future development in these areas by outlining plans for coordinated GCTE research activities. Examples of these new areas include the links between agriculture, ecology and human social systems (e.g. complex agro-ecosystems); the intriguing question of ecological complexity, ecosystem function, and global change (see the chapters by Sala, Chapin *et al.* and Holling *et al.*); and the move toward closer interaction between the social and natural sciences in studying the human dimensions of land use/cover change (see the chapters by Noble, Parry *et al.* and Leemans).

What about the future? There are still a number of gaps where more research is clearly needed, and where GCTE and other groups interested in global change interactions with terrestrial ecology have much to do. The final chapter in the book provides a summary of these gaps, and poses challenges that promise plenty of interesting research for ecologists over the next 10–15 years.

The chapters in this book are based on the invited plenary presentations at the GCTE Science Conference. There were over 250 presented and poster papers in all, and many of the non-plenary papers are to be published as a special edition of the *Journal of Biogeography*.

References

GCTE Report No. 1: GCTE Core Research: 1993 Annual Report. Canberra, 1994, 135 pp.

IGBP Report No. 5: Effects of Atmospheric and Climate Change on Terrestrial Ecosystems. Report of a Workshop Organized by the IGBP Coordinating Panel on Effects of Climate Change on Terrestrial Ecosystems at CSIRO, Division of Wildlife and Ecology, Canberra, Australia, 29 February–2 March 1988. Compiled by B. H. Walker and R. D. Graetz (1989). 61 pp.

IGBP Report No. 9: Southern Hemisphere Perspectives of Global Change. Scientific Issues, Research Needs and Proposed Activities. Report from a Workshop held in Mbabane, Swaziland, 11–16 December 1988. Edited by B. H. Walker and R. G. Dickson (1989). 55 pp.

IGBP Report No. 10: The Land Atmosphere Interface. Report on a Combined Modelling Workshop of IGBP Coordinating Panels 3, 4 and 5. Brussels, Belgium, 8–11 June 1989. Edited by S. J. Turner and B. H. Walker (1990). 39 pp.

IGBP Report No. 11: Proceedings of the Workshops of the Coordinating Panel on Effects of Global Change on Terrestrial Ecosystems. I. A Framework for Modelling the Effects of Climate and Atmospheric Change on Terrestrial Ecosystems, Woods Hole, USA, 15–17 April 1989. Edited by B. H. Walker. II. Non-Modelling Research Requirements for Understanding, Predicting, and Monitoring Global Change, Canberra, 29–31 August 1989. Edited by B. H. Walker and S. J. Turner. III. The Impact of Global Change on Agriculture and Forestry, Yaounde, 27 November–1 December 1989. Edited by S. J. Turner, R. T. Prinsley, D. M. Stafford Smith, H. A. Nix and B. H. Walker (1990). 108 pp.

IGBP Report No. 12: The International Geosphere-Biosphere Programme: A Study of Global Change (IGBP). The Initial Core Projects (1990). 330 pp.

IGBP Report No. 21: Global Change and Terrestrial Ecosystems: The Operational Plan. Edited by W. L. Steffen, B. H. Walker, J. I. Ingram and G. W. Koch (1992). 97 pp.

Part two

Ecosystem physiology

Ecosystem physiology: overview and synthesis

H. A. Mooney

Introduction

The threat and reality of global change has had a major impact on the direction and tempo of scientific research. This is clearly evident from the material discussed in Part II of this volume. A whole new discipline, earth system science, has arisen in the past decade providing answers to questions about the potential impact of a suite of changing and interacting factors on the responses of entire ecosystems. Why this has become an important area of research is the growing knowledge that the exchanges of gases and materials from ecosystems interact with the atmosphere and drive, in part, the climate system, which in turn influences ecosystem processes.

The challenge has been to develop an understanding of what controls the exchanges of resources, such as radiation, energy, water, carbon and nitrogen, of entire ecosystems. Research on resource balance of individual organisms has had a long history. What is new is a focus on the physiology, or 'metabolism' of whole ecosystem units. This progression of research focus to a higher level of integration than has been utilized in the past has necessitated the development of new research approaches and paradigms.

This has been difficult for a number of reasons including the need to study larger biological units, over longer periods of time, than has been done in the past. The larger spatial domain of focus has meant the severing of the very strong experimental base that has been the hallmark of the study of physiological ecology of individual organisms. There have been moves to bring experimentation to the ecosystem level (Mooney *et al.*, 1991) but progress in this area has been slow because of the cost and complexity of this endeavour. There has been, however, the development of new tools for the study of large systems that have facilitated progress including methods for studying the fluxes of gases from large units, the improvement of remote sensing tools and algorithms, and the development of whole ecosystem models that incorporate feedbacks among vegetation, soil and the atmosphere.

Global models of ecosystem physiology

At the moment the development and utilization of models, of necessity, dominates the field of ecosystem physiology. These models have allowed us to integrate our knowledge of the interactions that drive the fluxes of energy, water and materials and increasingly have incorporated physiological mechanisms into model formulation. Examples of these are the ecosystem models of the Woods Hole group (Melillo *et al.*, 1993), the CASA model (Potter *et al.*, 1993) and the BGC models (Running & Coughlan, 1988; Running & Gower, 1991). These models operate on time scales from days to centuries. This in fact has been one of the challenges – how to capture the details of short-term atmospheric phenomena with the longer time constants embodied in soil processes. The ecosystem models can only couple to general circulation models (GCMs) by using the GCM outputs as inputs. More ecological models that directly couple to GCMs, and which operate on hourly time steps, are being developed but they are not yet able to incorporate soil biogeochemistry, which is a crucial global change feedback. The new class of models (e.g. SiB; Sellers *et al.*, in press), does utilize physiology, as embodied in gas fluxes and their controls, and vegetation features, such as canopy structure, carboxylating enzyme distribution and photosynthetic pathway. This is certainly an important beginning of marrying not only physiology and climate, in real time, but also biodiversity. The strength of the SiB-type models is coupling land surface with atmospheric features.

These then are clearly advances in the development of ecosystem physiology – the characterization of the system metabolism of biome types and of the entire globe. These models, in their various configurations, allow us to examine functional linkages among vegetation, soils and the atmosphere, and to predict the consequences of climate change on these linkages. Some are directly linked to GCMs, others can utilize remote sensing of land surface features as drivers, and others offer the capacity to look at long-term feedbacks of changes in biogeochemistry. None, however, can be directly tested with experimental data because of their global character except through the use of historical proxies. Another pathway for validation in a sense is inter-model comparisons. These are beginning to be done and have shown the considerable differences in model output given the same initial boundary conditions.

Direct experimentation

What about the experimental study of ecosystem metabolism? One long step down in spatial scale from the use of global models are the current attempts to examine directly the physiology of ecosystems. There have been decades of high quality research on the impact of elevated CO_2 on plant physiology (see overviews in Lemon, 1983; Strain & Cure, 1985). This area has received high research priority, since of all the elements of global change, CO_2 increase has been the best documented and is of greatest concern. At the beginnings of the IGBP programme on Global Change and Terrestrial Ecosystems it was realized that the data available were nevertheless insufficient for predicting the responses of whole ecosystems to CO_2 change since they did not provide crucial information on feedbacks of plants with the soil and the atmosphere. New, more integrated approaches were needed.

Two experimental approaches have been utilized, in addition to the continuation of greenhouse and growth chamber work on the specific process responses. These have been the use of open-top chambers and, the other, the use of so-called FACE (Free Air CO_2 Enrichment) arrays, where CO_2 is injected on demand within a ring encompassing the test area (Schulze & Mooney, 1993). Both of these techniques were borrowed from research on the effects of air pollutants on community responses.

These approaches, above all, have given us new understanding of the responses of whole ecosystems to increasing CO_2. The early results of a new network of CO_2 investigators sponsored by the Global Change and Terrestrial Ecosystem Programme (Steffen et al., 1992) are outlined in the chapter by Körner in this volume. He notes those results which are common to all experiments, such as the increase of carbon to nitrogen ratios of tissues, the accumulation of carbohydrates, and the differential responses of species within a community.

The complex web of interaction at the whole ecosystem level that was proposed to result from an increase in CO_2 (Fig. 2.1) (Steffen et al., 1992) has certainly been substantiated. In Fig. 2.1 it can be seen that in addition to direct effects, two feedback loops were predicted: one on the biogeochemistry of the system and the other on water and energy balance. Recent studies have shown that the water feedback may actually be more significant than the biogeochemical one, resulting in changes in community composition (unpublished data).

Although the results accumulating are giving us a much deeper understanding of the complex responses that can be anticipated in natural systems to our changing earth system, much more work needs to be done. Korner indicates how crop systems may be more responsive than natural

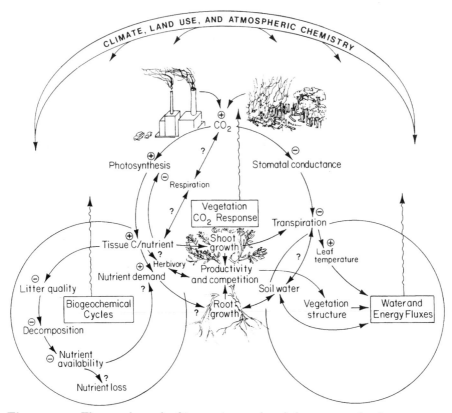

Figure 2.1   *The complex web of interaction at the whole ecosystem level resulting from an increase in CO_2.*

systems, but our sample of both is fairly small. More importantly, the protocols that have been utilized in these initial experiments will have to be amplified in the future since information on responses of systems to an instantaneous doubling of CO_2 is not totally relevant to what will occur, and may even be misleading when considering highly buffered woody ecosystems. It is heartening though that for herbaceous systems, responses to CO_2 have been successfully modelled and validated with direct field experiments (see Coughenour & Parton, this volume).

The time dimension of responses of ecosystems to enhanced CO_2 is a difficult and critical issue. In the past many investigators have extrapolated their results on short-term, and even instantaneous responses, to long-term phenomena. Gifford *et al.*, in this volume, and Oechel elsewhere, have correctly noted the many different response phenomena that occur through time, from seconds to centuries, that make linear time scaling problematic. At present, there is no easy way to abbreviate the learning of these many response phenomena in their own particular time domain.

A different sort of difficulty is outlined by Bazzaz and his co-workers, who have made substantial contributions to our understanding of the responses of biotic systems to elevated CO_2 (see Bazzaz, 1990). They have concentrated their efforts at the level of individual plant responses embedded within a community. They have provided abundant evidence, as reviewed in their chapter in this volume, of the individualistic response of individuals and species to increased CO_2. They correctly note that recent efforts in ecosystem physiology have concentrated on whole system responses, particularly on carbon balance, rather than on that of the responses of individuals. Thus we do not have guidelines, as of yet, for predicting species competitive outcomes under elevated CO_2. However, this situation is not unique to the CO_2 question. We are unable to predict outcomes under increasing temperature or under increasing pollutant loads for many species. The basic problem is that each species and population has unique responses to any resource or stress. This is the case for basic physiological responses as well as for reproductive development. The potential combination of stress factors that will change in the future are great, making this a daunting problem area. Fundamental advances in linking whole plant physiology, reproductive output and competitive capacity are needed.

Where are we? We have, in a relatively short time, explored the issue of the responses of whole ecosystems to enhanced CO_2 and have found very complex interactions that make simple extrapolations from short-term physiological measurements to predictions of long-term whole system responses untenable. However, we have uncovered some very powerful generalities of whole system responses, but we are just beginning the process.

Measuring fluxes

In addition to global modelling and direct experimentation, as tools for studying whole ecosystem physiology, there have been efforts to directly measure the metabolism of ecosystems and even of landscapes. Technology in this area is advancing rapidly and measurement networks are very likely to be in place soon that will give us continuous measurements of gas fluxes of the world's major ecosystem types. These measurements will help us pinpoint, for example, sources and sinks for CO_2, through time. Schulze *et al.*, this volume, demonstrate some of the breakthroughs that have occurred in this exciting area. Their work has set the boundary conditions for gas fluxes of the major biomes of the earth and has provided a context in which to interpret the emerging results of direct measurements of ecosystem gas fluxes.

Other approaches

There is one area of ecosystem physiology that is not treated in Part II but that nevertheless has made fundamental contributions to our understanding of ecosystem physiology. This is the whole area of remote sensing, particularly from satellites, that has, for example, enabled us to view the changing productive capacity of the earth through time (Mooney & Field, 1989). The remotely sensed normalized difference vegetation index, or greenness index, has proved to be a powerful analytical tool whose applications seem endless. The normalized difference vegetation index can be viewed globally and the results linked directly to global models of metabolism as described earlier.

Summary

The research community has certainly risen to the challenge of devising the means for making predictions of the responses of ecosystems to the new world of global change. In order to achieve this new understanding they have had to learn to experiment and model at much larger temporal and spatial scales than they have operated in the past. The resulting study of ecosystem physiology is a frontier area that will be providing major new insights into the functioning and responses of whole ecosystems in the years ahead.

References

Bazzaz, F. A. (1990). The response of natural ecosystems to the rising global CO_2 levels. *Annual Review of Ecology and Systematics*, **21**, 167–96.

Lemon, E. R. (eds.) (1983). *CO_2 and Plants. The Response of Plants to Rising Levels of Atmospheric Carbon Dioxide.* Boulder: Westview Press.

Melillo, J. M., McGuire, A. D., Kicklighter, D. W., Moore, B., Vorosmarty, C. J. & Grace, A. L. (1993). Global climate change and terrestrial net primary production. *Nature*, **363**, 234–40.

Mooney, H. A. & Field, C. B. (1989). Photosynthesis and plant productivity – scaling to the biosphere. In *Photosynthesis*, ed. W. R. Briggs, pp. 19–44. New York: Allan R. Liss.

Mooney, H. A., Medina, E., Schindler, D. W., Schulze, E.-D. & Walker, B. H. (eds.) (1991). *Ecosystem Experiments.* Berlin: Springer.

Potter, C. S., Randerson, J. T., Field, C. B., Matson, P. M., Vitousek, P. M., Mooney, H. A. & Klooster, S. A. (1993). Terrestrial ecosystem production: a process model based on global satellite and surface data. *Global Biogeochemical Cycles*, **7**, 811–41.

Running, S. W. & Coughlan, J. C. (1988). A general model of forest ecosystem processes for regional applications. I. Hydrologic balance, canopy gas exchange and primary production processes. *Ecological Applications*, **42**, 125–54.

Running, S. W. & Gower, S. T. (1991). FOREST-BGC, a general model of forest ecosystem processes for regional applications. II. Dynamic carbon allocation and nitrogen budgets. *Tree Physiology*, **9**, 147–60.

Schulze, E.-D. & Mooney, H. A. (ed.) (1993). *Design and Execution of Experiments on CO_2 Enrichment*. Brussels: Commission of the European Communities.

Sellers, P. J., Randall, D. A., Collatz, C. J., Berry, J. A., Field, C. B., Dazlich, D. A., Zhang, C. & Colello, G. D. (in press). A revised land surface parameterization (SiB2) for atmospheric GCMs. I. Model formulation. *Journal of Climate*.

Strain, B. R. & Cure, J. D. (1985). *Direct Effects of Increasing CO_2 on Vegetation*. Washington, DC: US Department of Energy.

Steffen, W. L., Walker, B. H., Ingram, J. S. I. & Koch, G. W. (1992). *Global Change and Terrestrial Ecosystems. The Operational Plan* (21). Stockholm: The International Geosphere-Biosphere Programme.

The response of complex multispecies systems to elevated CO_2

Ch. Körner

Our current base of information

The study of CO_2 effects on plants has a century-long history and research started to boom once it became evident that horticultural yields can be increased by CO_2 fertilization (Enoch, 1978; Allen, 1979; Kimball & Idso, 1983; Wittwer, 1984; Acock & Allen, 1985). The ecological interest arose when it became apparent that atmospheric CO_2 levels are increasing at unprecedented speed and possibly will influence plant life all over the world. Today, approximately 1500 publications on plant responses to elevated CO_2 levels are available, and have been summarized in several reviews (Hoffman, 1984; Strain & Cure, 1985; Sionit & Kramer, 1986; Eamus & Jarvis, 1989; Lawlor & Mitchell, 1991; Woodward et al., 1991; Hogan et al., 1991; Gifford, 1992; Bazzaz & Fajer, 1992; Körner, 1993; Amthor, 1995; Norby et al., 1995; Koch & Mooney, 1996; Körner & Bazzaz, 1996). While we can predict with confidence that doubling ambient CO_2 levels under horticultural or agro-industrial conditions will facilitate CO_2-stimulated enhancements of seasonal yields by about one-third compared with controls (Cure, 1985), very little is known about CO_2 responses of non-agroecosystems (Körner, 1993). For instance, all whole plant CO_2 enrichment experiments that have been conducted with forest tree species were, for practical reasons, done with seedlings, mostly growing in isolation and in highly disturbed or even horticultural substrates (Eamus & Jarvis, 1989), but trees form approximately 90% of the global biomass compared with 1% in crop biomass. The situation is better for short stature vegetation (e.g. tundra, grassland) which can more easily be handled as a whole by CO_2 enrichment technology, but the number of studies completed with natural communities in undisturbed natural soil can be counted on one hand.

Despite this extremely unbalanced and, therefore, limited basis of information, one conclusion is almost certain: once interspecific interactions and resource limitation of growth come into play, CO_2 responses do not fit the patterns observed in agrosystems, making sound predictions for the future development of the biosphere impossible. In order to change this situation,

GCTE is advocating CO$_2$ research with more complex systems, in particular natural ecosystems (Mooney, 1990; Mooney *et al.*, 1991; Walker, 1991). This chapter will (1) explain why a reductionistic approach to the CO$_2$ problem is inadequate; (2) discuss some fundamental misconceptions affecting predictive models; and (3) try to summarize briefly what we have learned from CO$_2$ research with complex systems so far.

From diversity to complexity

Our thinking in experimental biology is still strongly bound to the world of Linnaeus, where nature first had to be subdivided, classified and tagged before one could dare to think about its functioning. The 'classified unit', most commonly the species, became the self-evident object in research, including the study of CO$_2$ effects. At the species level we find an over-whelming diversity of responses not found at the ecosystem level. As an example, the maximum rate of photosynthesis or the maximum leaf diffusive conductance found in vascular plant species varies by almost two orders of magnitude, whereas the community means of these characteristics for biomes as different as tundra, mediterranean shrubland, boreal forests, temperate forest and tropical forests do not differ significantly (Körner, 1994; Schulze *et al.*, 1994; Fig. 3.1). A nice regional example of such 'homeostatic' trends at ecosystem level has recently been published for a forest transect in Oregon by Pierce *et al.* (1994). Hence, the variation found at species (leaf) level by far outranges the variation at the ecosystem level, strongly constraining up-scaling attempts.

Growth: more than just CO$_2$ assimilation

Growth at the level of the individual, as the ultimate net result of photosyn-thetic activity, cannot be predicted from photosynthesis, simply because most of the other growth determinants listed in Table 3.1 are quantitatively or even qualitatively unknown (cf. Strain, 1987). In particular, phenological controls are poorly understood, but they are of key importance for all other growth determinants (e.g. Coleman *et al.*, 1994). Yet, within a given isolated individual of a species, exposed to a fairly constant environment and over a relatively short period of time, with all other plant-bound growth determinants remaining constant, there may be a positive correlation between the rate of photosynthesis and the rate of growth. Since it is unlikely that all the parameters listed in Table 3.1, in particular phenorhythmics and the organismic interactions, can be quantified

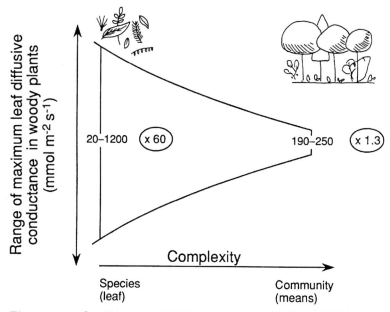

Figure 3.1 *Species versus community ranges of physiological key variables, exemplified here by maximum diffusive conductance of leaves of the main woody biota of the world (Körner 1994). Functional diversity is greatest at the species level and smaller at the ecosystem level.*

separately, the study of overall system behaviour (ecosystem physiology) is the more promising approach (Mooney, 1990). Models on CO_2 responses of ecosystem carbon pools in essence (given the many unknowns mentioned above, by necessity) still base their estimates on photosynthetic CO_2 response curves (e.g. Kohlmaier *et al.*, 1987; Thornley *et al.*, 1991; Melillo *et al.*, 1993). The following section will illustrate some difficulties encountered in scaling from physiological response curves at tissue level or single-factor focussed approaches to long-term responses of complex natural systems.

Scaling assumptions

In predicting global change of terrestrial ecosystems, including effects of climate warming and elevated CO_2, some assumptions are commonly made that have been given almost 'axiomatic' status. A few of these assumptions need challenging (Table 3.2).

The capacity of leaves of most terrestrial C_3 plants to fix CO_2 is far from being saturated at current ambient CO_2 levels. In the short term virtually all C_3 plants investigated showed increased rates of photosynthesis when exposed to elevated CO_2. Given the multitude of controls of growth shown

Table 3.1 *The control of plant growth by internal and external biological determinants (i.e. not considering additional influences by the physical and chemical environment)*

Recruitment and establishment	Seed quality Germination Stress resistance
Rate of photosynthesis	As a function of age and position of leaves
Carbon investment	Specific leaf area Leaf area ratio Stem mass ratio Root mass ratio Storage organs Reproduction Plasticity of resource foraging Exudation Mycorrhiza
Rate of respiration	Specific rates for each of the above plant compartments
Rate of biomass turnover	Leaf life span Root turnover Rate of plant development (maturation, senescence)
Morphological determinants	Leaf angle Self shading Branching patterns Maximum plant height/width Structural alteration of organs
Plant interactions	Competition for resources (in space and time) Other interactions
Plant – animal interactions	Herbivory Pollination
Microbial interactions	Symbiosis Parasitism Other pathogenic effects Biomass (nutrient) recycling

Table 3.2 *Predicting complex system responses to elevated CO_2 and climate warming*

Some incorrect assumptions

Long-term leaf photosynthetic responses: similar to initial responses

Faster growth rate: greater landscape C pools

Lower photosynthetic light compensation point (improved C balance of shade leaves): increased LAI

Warmer climate: proportional increase in tissue respiration

Some yet unproven assumptions

Increased C/N ratio in plant tissues: slow down of decomposition rates, increased soil C pools

Improved mineral nutrient acquisition as better soil exploration

Effects on N balance: respresentative of effects on other nutrient balances, e.g. P

Changes in food quality: compensatory feeding even when herbivores have a choice of food species

in Table 3.1, it is clear that it is not to be expected that all these components will respond in a concerted manner, if photosynthesis is stimulated by a step rise of CO_2. In addition, growth requires resources other than CO_2 and their availability, and in particular their pool size also cannot be expected to increase proportionally. Therefore, the most common initial response of plants to CO_2 fertilization is overshooting of assimilate levels in leaves (mostly starch), reflecting limitations to structural investment or export of these photosynthesis products from chloroplasts (e.g. Ehret & Jolliffe, 1985; Körner *et al.*, 1995a). Species specific and environmental characteristics will then determine the long-term response. Even well-fertilized agricultural plants exposed to elevated CO_2 may exhibit quite contrasting photosynthetic adjustment (Sage *et al.*, 1989). Fig. 3.2 illustrates that there is a wide spectrum of possible responses with the uppermost response (proportional stimulation of photosynthesis) to be the most unlikely (but the most often assumed). If – due to unequal initial rates of photosynthesis in CO_2-enriched and control plots – biomass will follow the dynamics illustrated in Fig. 3.3, the duration of an experiment (date of harvest) will determine the resultant 'CO_2 response' (Körner, 1995).

A second assumption commonly made is that elevated CO_2 will induce an increase of leaf area index (LAI), even under situations where canopy closure is achieved under current ambient CO_2 levels (e.g. Lieth *et al.*,1986; Long & Drake, 1992). The basis for this assumption is the reduction of the light compensation point of photosynthesis under elevated CO_2 (Valle *et al.*, 1985; Wong & Dunin, 1987; Long & Drake, 1991). Leaves living under limiting light conditions can be expected to profit relatively more from CO_2

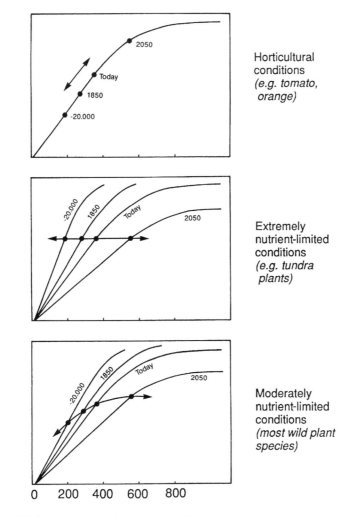

Figure 3.2 *Photosynthetic adjustment to changing CO₂ concentration: a multitude of possible long-term responses.*

enrichment, although the absolute carbon gain may still be small. However, when mineral nutrient availability is limiting growth, which may be the case even in moderately fertile soils, it appears that plants in dense communities could respond in a different way. Despite improved carbon balance in their shaded leaves, the nitrogen trapped in these leaves could be invested much more efficiently in the fully sunlit crown, where the relative stimulation by CO_2 enrichment may be smaller, but the absolute carbon gain due to elevated CO_2 is much greater. This is in line with the theory for optimal nitrogen distribution in dense canopies (Werger & Hirose, 1991; Chen *et al.*, 1993) but does not apply to understorey plants which are

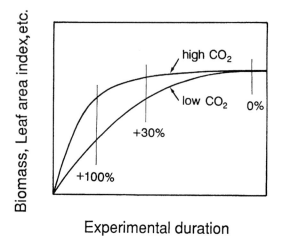

Experimental duration

Figure 3.3 *When CO₂ enrichment alters the rate of plant expansion (e.g. by faster growth), the experimental duration may determine results in cases of incomplete 'niche occupancy' at the beginning of the experiment.*

bound to the lowest canopy level. Indeed, closed canopies that have achieved a steady state LAI under experimental conditions, exhibit either similar or even a smaller LAI under elevated CO_2 (tundra: Oechel & Riechers, 1986; tropical model ecosystems: Körner & Arnone, 1992; Arnone & Körner, 1995; spruce model ecosystems: Hättenschwiler & Körner, 1996; calcareous grassland: Leadley & Körner, 1996; alpine grassland: Schäppi & Körner, 1995). However, in short rotation grass swards and nutrient-unlimited monospecific stands of deciduous tree seedlings the predicted increase of LAI has been found (Nijs *et al.*, 1989; Overdieck, 1993).

Plants grown under elevated CO_2 have almost always been found to produce tissues that contain more carbon and less nitrogen, even when subtracting starch accumulation (e.g. DeLucia *et al.*, 1985; Curtis *et al.*, 1989; Wong, 1990; Körner & Miglietta, 1994). It has been assumed that this CO_2-induced increase in carbon/nitrogen ratio (and possibly increased lignin content) will lead to reduced rates of decomposition, and thereby facilitate increased carbon sequestration to soils (Fig. 3.4). However, as illustrated by Fig. 3.4, there is another pathway by which CO_2 fertilization may influence the soil environment, namely through priming effects of increased rates of turnover in fine roots and exudation of low molecular weight organic compounds (sugars, amino acids) to the rhizosphere. These two avenues of carbon into the soil may have quite contrasting effects on the soil carbon pool, and currently it is still uncertain in which direction soil carbon pools actually will move. Leaf litter decomposition may not be affected at all (Norby *et al.*, 1986, 1995; Curtis *et al.*, 1989; Arnone &

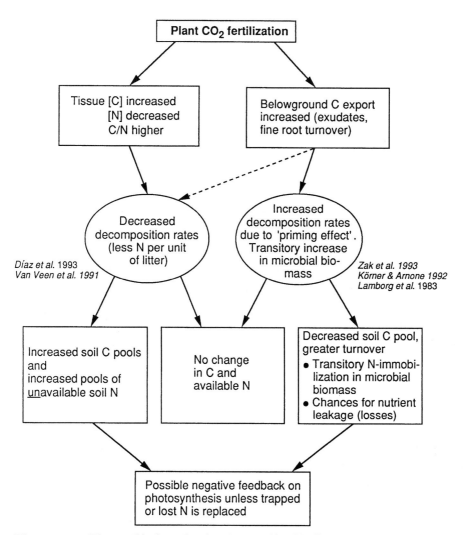

Figure 3.4 *The possible fate of carbon in CO$_2$ 'fertilized' ecosystems.*

Körner, 1995), but rhizosphere carbon enrichment may still be happening as may be concluded from the responses summarized in Table 3.3.

Finally, one of the 'classic' assumptions, perhaps indirectly related to the CO$_2$ problem, concerns the effect of possible concomitant climate warming on respiration. Increased atmospheric temperatures are assumed to induce increased respiratory rates by organisms (e.g. Houghton & Woodwell, 1989; Ryan, 1991; Agren *et al.*, 1991; Bonan & Sirois, 1992) and thereby could mitigate CO$_2$ effects on photosynthesis. The basis for this (unproven) assumption are temperature response curves measured in isolated tissues in the laboratory, neglecting the possibility of 'acclimation' (Precht *et al.*, 1973; Fig. 3.5). Stocker (1935) was possibly the first who noted with surprise that

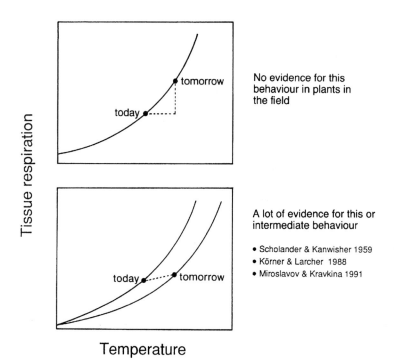

Figure 3.5 *Maintenance respiration and climate warming. The comparison of instantaneous laboratory response curves and* in situ *responses which account for acclimation.*

leaves of arctic shrubs in Greenland exhibit similar rates of respiration per unit leaf mass as leaves of tropical rainforest trees in Indonesia. Numerous later investigations clearly showed that *in situ* rates of respiration vary much less than might be expected from thermal site conditions and instantaneous response curve characteristics (e.g. Scholander & Kanwisher, 1959; Pearcy, 1977; Koike, 1987; Körner & Larcher, 1988; Criddle *et al.*, 1994; further literature in Friend & Woodward, 1990 and Lariganderie & Körner, 1995). Miroslavov and Kravkina (1991) showed that alterations in the number of mitochondria can account for this adjustment.

Little is known about the thermal acclimation of life processes in soils, but soils in areas that differ greatly in ambient temperatures reveal surprisingly similar rates of CO_2 evolution during the growing season. For instance, Raich & Nadelhoffer (1989) report a mean rate of soil CO_2 evolution for tropical forests of about 2 μmol CO_2 m^{-2} s^{-1}, and similar seasonal means (at lower natural soil temperatures) are reported for temperate and arctic soils (Lloyd & Taylor, 1994), although the duration of the season is much shorter, causing the annual sums of soil CO_2 evolution to be very different. Obviously an indirect effect of rising temperatures, namely the lengthening of the growing season, will increase the annual sum of respiratory losses in temperate and polar latitudes, just like it may increase annual photosynthetic yield (Dahl & Mork, 1959). The situation becomes even more difficult if one accounts for observations showing that CO_2 may directly influence tissue respiration (reduction: Bunce, 1990; Amthor et al., 1992; Wullschleger & Norby, 1992; Mousseau & Saugier, 1992; increase: Poorter et al., 1992; Thomas et al., 1993) but distinctions between growth and maintenance respiration are difficult, and data for ecosystem CO_2 evolution are scarce and still conflicting (reduction: Bunce & Caulfield, 1990; increase: Elliott et al., 1990; Navas et al., 1995; Hättenschwiler & Körner, 1996; see also Table 3.3).

From rates to pools

One of the most common assumptions in the discussion and prediction of long-term effects of elevated CO_2 and increased temperature is reflected in the synonymous use in the literature of the terms growth, primary production and biomass. Plant growth stimulation is almost always seen as a self-evident indicator of enhanced carbon sequestering, which is not true (Prentice, 1993). The carbon pool of an ecosystem is not correlated with its growth rate, as can easily be seen when old growth and early successional forests are compared. The mean residence time of carbon in the system is of greater importance than the speed of carbon cycling through the system. If plants were to grow faster, they might be expected to mature earlier and also die earlier Körner 1996. In forests, which form the only significant biomass carbon pool of the biosphere, the speed of closure and re-opening of gaps are to large extent determining the carbon pool per unit land area. There is no evidence that the carbon residence time in biomass is likely to increase in a CO_2-enriched world. Rather the reverse may be true, an idea which found recent support in the observation that tree turnover in tropical forests has been increased significantly in recent decades (Phillips & Gentry, 1994; Fig. 3.6). In some temperate zone mountain regions increasing tree ring width points in the same direction (White Mountains, California: Graybill & Idso, 1993; Vosges Mountains: Becker, 1991; Central Alps: Nicolussi et al., 1995). In contrast Kienast & Luxmoore (1988) and Graumlich (1991) found no clear evidence for a recent response in tree rings of temperate zone plants that could not be explained by conventional climatological data. The absence of such trends or even reversed trends are now also well documented for the whole boreal belt (Fennoscandia: Briffa et al., 1990; new data for Canada and Siberia: F. Schweingruber, personal communication).

Animals and microbes may strongly interfere with CO_2 effects on plant

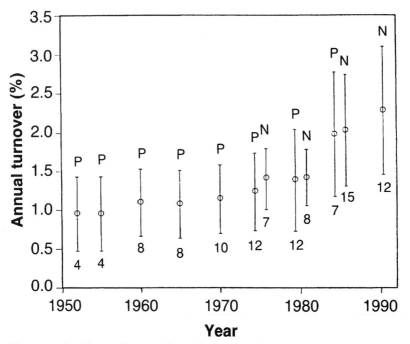

Figure 3.6 *First indirect evidence for enhanced carbon turnover in tropical forests as illustrated by recent increases of tree turnover (move of stems into larger diameter classes or death). After Phillips and Gentry (1994).*

growth, ecosystem dynamics and the ultimate carbon pool size per unit land area. With respect to herbivory we lack realistic experiments where herbivores are given a choice between various species. Compensatory feeding on CO_2-grown food (high C/N ratio) has been demonstrated (see review by Lincoln *et al.*, 1993), but in nature herbivores may exhibit 'alternative' feeding and select the remaining 'goodies' rather than feed more on the protein depleted food. The impact on ecosystems may be even more severe, but the mechanism is very different and will depend on the diversity of CO_2 responses in plant species. Whether mycorrhizae behave symbiontically or parasitically may also be influenced by CO_2 fertilization (Sanders, 1996), further emphasizing the need of experimental systems with a realistic soil microflora, if ecosystem responses to elevated CO_2 are to be understood.

Towards ecosystem physiology

A practical consequence of this list of fundamental shortcomings of simplified experiments is that we should apply our modern analytical technology to systems of higher complexity. We may still ask the traditional questions, but, in the sense of ecosystem physiology, answer them at levels of integration closer to the real world (Grace *et al.*, 1995). Achieving ecosystem responses by piecewise recombination of knowledge obtained from isolated partial systems is and will remain an illusion when natural ecosystems are considered. However, the synchronous study of processes in such partial systems in conjunction with the analysis of responses of the overall system may add a lot to our understanding of ecosystem functioning. We really need a top down approach in elevated CO$_2$ research, with the breakdown of complexity driven by clearly defined explanatory requirements at the uppermost level (just as in functional ecology in general; Odum, 1977). Even the most delicate molecular processes in the photosynthetic machinery are strongly dependent on processes at much larger scales, and thus cannot be considered to be 'fundamental (independent) drivers'.

The surprises that may be found in CO$_2$ enrichment experiments with complex systems in which competition for soil resources and light is permitted to occur can be illustrated by results obtained most recently with spruce model ecosystems in the Basel phytotron (Hättenschwiller & Körner, 1996). Young trees (*Picea abies*), forming dense stands, were allowed to grow for 3 years in natural soil and under three CO$_2$ levels: 280, 420 and 560 ppm. Two-thirds of the biomass found after 2 years was fixed under these experimental conditions while LAI more than doubled. Shoot photosynthesis under full light was consistently increased by 30% at 420 ppm compared with 280 ppm, with no further stimulation at 560 ppm. LAI showed a non-linear decrease with increasing CO$_2$ with the strongest decline between 280 and 420 ppm (quantified by both light extinction and litter production). Whole ecosystem net gas exchange measured in late summer of the second year was identical at all three CO$_2$ levels. Thus shoot gas exchange differences were almost perfectly balanced by allometric adjustments at the ecosystem level. It may be speculated that some of the worldwide observed needle loss in boreal and temperate zone conifer canopies may be due to such allometric responses to atmospheric CO$_2$ enrichment. The study further underlines that historical CO$_2$ changes possibly had greater effect than future ones will have – a consequence of the non-linearity of all CO$_2$ responses, in this case unravelled by studying the effect of more than two CO$_2$ levels (cf. Fig. 3.2).

Complex multispecies systems under elevated CO_2: first trends

In Table 3.3 an attempt is made to summarize some overall trends which have been observed in natural or semi-natural vegetation or model ecosystems exposed to elevated CO_2 for at least one season (data published by spring 1995). It turned out to be extremely difficult to compile such a list, since the source of information is very heterogeneous in almost any respect (e.g. methods of CO_2 enrichment, length of exposure, ecosystem level data provided or not, soil fertility). In total, there are only six studies in which CO_2 responses at ecosystem level have been considered in unfertilized, more or less natural soils, and in which the whole canopy was exposed to elevated CO_2 so that ecosystem processes could be related to the treatment. These are all studies with short stature vegetation (mostly graminoid systems). For the current purpose studies with isolated plant individuals, heavily fertilized plants, and rapidly expanding systems (which have not fully occupied the given area and soil space during the course of the experiment) were excluded. Since these criteria discriminate against all available data for forest plants (for obvious reasons no forest has ever been studied) I included available data for model ecosystems with young forest plants in which complete space occupancy was achieved, and for which canopy data per unit ground area were available. Hence, data presented here are for closed canopies and ecosystems in which competition for light and soil resources played an important role. Greenhouse or phytotron experiments have been included only if the considered biome would otherwise be missed out completely (as for boreal or tropical ecosystems). In order to make the table readable, a number of simplifications had to be made. For instance, only one 'overall' set of responses is presented if various facets of the same type of vegetation have been investigated (e.g. studies 6, 7 and 9). Also seasonal variations of trends could not be included (e.g. 2 and 9).

Before discussing the message of Table 3.3, it should be emphasized that one-third of the available positions are occupied by question marks, indicating a lack of information despite the availability of experimental facilities. Since so few experimental systems exist in the world which allow the study of ecosystem processes, it seems imperative to fill these gaps by the various research teams. There was insufficient information (less than three studies) for a number of key ecosystem processes (thus missing in Table 3.3) such as ecosystem nutrient cycling or losses, ecosystem level water relations, or the content and release of secondary plant compounds. Also phenological aspects are strongly under-represented despite the fact that they could easily have been detected.

The most obvious trends revealed by Table 3.3 can be summarized as follows. Almost all low fertility natural or model ecosystems studied so far under elevated CO_2

- show enhanced net CO_2 assimilation
- show no or small above-ground increase of biomass
- show no change in leaf area index
- show little or no stimulation of below-ground biomass
- show species specific biomass responses, with some species losing and other species gaining in abundance
- show no or insufficiently strong (i.e. not fully compensatory) stimulation of night-time CO_2 output (respiration).

As a result there is a missing link between the measured CO_2 balance and detectable presence of carbon in biomass.

This missing carbon may have been sequestered into the soil carbon pool or distributed among several pools or fluxes, each of which was too small by itself to be detected with sufficient accuracy. The evidence for soil CO_2 evolution is particularly weak but is of key significance to understand the carbon balance of an ecosystem. However, the current experimental data base for CO_2 responses of whole ecosystems indicates the likelihood of increased carbon sequestration to the soil. Sequestration into biomass appears to be small or negligible, except for fertile systems. In addition, Table 3.3 demonstrates reduced nitrogen concentrations in plant tissues growing under elevated CO_2. This trend is possibly not paralleled for phosphate, the demand for which has been shown to increase under elevated CO_2 (Conroy et al., 1992). Most but not all studies revealed increased concentrations of non-structural carbohydrates in the leaf canopy and in one case (11) increased contents of secondary plant compounds have been detected in the field. There is preliminary evidence (mostly unpublished; studies 3, 4, 6, 8, 9 and 12) that the total water consumption per unit land area is slightly reduced with twice current CO_2, causing water use efficiency of biomass production to increase. Such changes appear to be particularly important in water-limited ecosystems. For instance, in mediterranean grassland and short grass prarie, this appears to be the major effect of CO_2 enrichment (Jackson et al., 1994; Ch. Field, personal communication; Owensby et al., 1996).

It should be noted that five of the listed studies showed a positive above- and below-ground growth response: the semi-aquatic salt marsh, the fertile hayfield, the Kansas prairie, the closed canopies of beech seedlings growing under unlimited nutrient supply, and (though weakly expressed) the fertile tropical model ecosystem. However, none of these stimulations reached the

Table 3.3 *Community and ecosystem responses to elevated CO₂ in natural and semi-natural vegetation or model ecosystems*

References	Type of vegetation (geographic region)	No. of seasons	Soil fertility	Carbon fluxes			Structure/mass		Quality		Development			Community	
				Peak ecosystem photosynthesis	Night ecosystem respiration	Soil CO₂ evolution	Steady state LAI	Above-ground biomass	Below-ground biomass	Litter accumulation	Leaf N % d.w.	Leaf TNC % d.w.	Time of flowering or bud break	Leaf life-span	Differential species response
1	Arctic tundra (Alaska)	3	Low	•	•	•	•	•	•	?	?	?	?	?	+
2	Alpine grassland (Swiss Alps)[a]	3	Low	+	•	•	•	•	•	•	•	•	•	•	+
3	Montane conif. forest (model ecosystem)	2	Low	•	•	+	−	•	?	+	−	+	+	−	+
Humid temperate grassland															
4 a)	Infertile pasture (Switzerland)[a]	1	Low	+	•	•	(·)	•	•	•	−	+	•	•	+
5 b)	Fertile pasture (Quebec)	3	High	?	?	?	(+)	•	?	?	?	?	−	?	+
6 c)	Fertile hayfield (Switzerland)[a]	2	High	?	?	?	(+)	+	+	+	•	•	?	?	+
7	Saltmarsh (mixed C₃/C₄)[a]	6	High	+	−	?	+	+	+	+	•	+	?	+	+
8	Dry temperate grassland (Kansas)[a]	4	Medium	+	?	?	?	+	?	?	−	+	?	?	+
Mediterranean grassland															
9 a)	Annual grassland (California)[a]	2	Low	+	?	•	•	•	?	?	?	?	?	?	•

10 b) Annual grassland (S. France)	2 Low	+	+	+	•	+	+	•	?	−	+
Temperate deciduous forest											
11 a) Understorey shrubs (Maryland)	1 Low	?	?	(−)	(•)	?	−	•	?	−	?
12 b) Seedings (model ecosystems)	3 High	+	+	+	+	+	?	?	•	?	?
Humid tropical forest											
13 a) Model ecosystems (16 species, 100 days)	1 High	+	+	+	+	+	+	+	?	•	•
14 b) Model ecosystems (7 species, 500 days)	2 Low	?	•	•	•	+	•	•	?	+	+
Summary											
Low fertility systems		+	(•)	(•)	•	•	−	+	(•)	(•)	+
High fertility systems		+	(•)	(•)	(+)	+	−	+	(•)	(•)	+

+, increased; −, decreased; •, not changed or inconsistent; ?, not known; ( ), uncertain trend.

a Current GCTE core research project.

b Regular cutting

Reference: 1, Grulke *et al.* (1990) and earlier references therein; 2, Schöppi & Körner 1995; 3, Hättenschwiler & Körner 1996; 4, Leadley *et al.* (1996); 5, J. Stewart & C. Potvin (personal communication); 6, H. Blum (personal communication); 7, Curtis *et al.* (1989) and personal communication with B. Drake and P. Leadley; 8, Owensby *et al.* (1996) and references therein; 9, Jackson *et al.* (1994) and personal communication with Ch. Field; 10, Navas *et al.* (1995); 11, Cipollini *et al.* (1993); 12, Overdieck (1993): 13, Körner & Arnone (1992); 14, Arnone & Ch. Körner 1995.

evel reported for isolated plants (see first paragraph). Except for the grazed prairie, which may hold an intermediate position with respect to soil nutrients, or profit from improved water use efficiency, the other four ecosystems can be ranked as rather fertile, underlining the significance of the nutritional status when CO_2 effects are discussed. Of course, each individual study encounters specific environmental and biological conditions, and involves somewhat different methodological approaches, but the above may be the smallest common denominator with many other important changes still neglected at this point.

Conclusions

The basis of the common assumption that atmospheric CO_2 enrichment will influence the biosphere directly, and not only via a possible atmospheric warming effect, is the still unsaturated rate of leaf photosynthesis with respect to CO_2 found in most plant species. While this assumption is certainly true in the short term and on a leaf area basis, predictions of growth or biomass pools per unit land area are impossible unless responses of a multitude of other growth determinants (Table 3.1) are known. Most experimental approaches suffer from a CO_2-driven asynchrony of space filling and resource utilization (since expanding systems are commonly used), creating immediate differences between CO_2 treatments which would disappear once the available space is filled, or when easily available resources are exhausted. The photosynthetic response of a leaf or a leaf canopy, including that to elevated CO_2, is a somewhat insufficient basis for predicting growth, except if one assumes all other growth determinants to be irrelevant or constant. Each of the components of plant growth listed in Table 3.1 differs with species or even ecotype. Consequently, the breakdown of system complexity is unlikely to be practical, nor will it lead to more accurate predictions of overall system response (Odum, 1977).

The new challenge lies in studying the maximum complexity that we can handle experimentally (ecosystem physiology) and not in the breakdown to the 'cleanest' subsystem. Models which seek their mechanistic basis at the tissue level will remain unable to reliably predict ecosystem responses to elevated CO_2 and temperature, in particular, when age structured, multispecies systems are considered. Given the current uncertainties of model predictions, more evidence from experimental model ecosystems in the field and in the lab is urgently required. Within GCTE Focus 1 (ecosystem physiology), a network of core research projects studying the effect of elevated CO_2 on ecosystems is now established that hopefully will fill some of the gaps in our understanding of how the 'real world' might respond. Unfortunately, the major biomass carbon pools of the globe, i.e. forests, will remain under-represented in these efforts, simply due to the overwhelming technical difficulties of artificial CO_2 enrichment in forests. In addition to palaeo-ecological studies, there is a need for several large-scale, possibly multinational CO_2 enrichment projects on intact natural forests and the immediate utilization of forest ecosystems growing around natural CO_2 springs.

In conclusion, the statement by J. H. Brown, recently published in *Science* (1994), is very much to be agreed with: 'The ecology of the last several decades has been largely reductionist, striving to understand ecological systems by reducing their complexity. In my opinion, this approach is not working, or at least, is inadequate by itself. We must develop more holistic approaches that confront complexity directly.' This is the challenge for research within the GCTE agenda.

Acknowledgements

I am grateful to the authors who generously provided unpublished, new results which could be incorporated in Table 3.3. Two anonymous reviewers and Paul Leadley helped improving this text. The conclusions presented have developed under the influence of several CO_2 enrichment experiments funded by the Swiss National Science Foundation, namely the project 5001-035214 of the Swiss Priority Program of the Environment, and 31 30048.90, 4031-033431 and 4031-034220 (National Research Program 31).

References

Acock, B. & Allen, L. H. Jr (1985). Crop responses to elevated carbon dioxide concentrations. In *Direct Effects of Increasing Carbon Dioxide on Vegetation*, ed. B. R. Strain & J. D. Cure, pp. 53–97. Publ. ER-0238. Washington, DC: US Department of Energy.

Agren, G. I., McMurtrie, R. E., Parton, W. J. Pastor, J. & Shugart, H. H. (1991). State-of-the-art of models of production–decomposition linkages in conifer and grassland ecosystems. *Ecological Applications*, 1, 118–38.

Allen, L. H. (1979). Potentials for carbon dioxide enrichment. In *Modification of the Aerial Environment of Crops*, ed. B. J. Barfield & J. F. Gerber, pp. 500–19. ASAE Monograph 2. St Joseph, Michigan: ASAE.

Amthor, J. S., Koch, G. W. & Bloom, A. J. (1992). CO_2 inhibits respiration in leaves of *Rumex crispus* L. *Plant Physiology*, **98**, 757–60.

Amthor, J. S.)1995) Terrestrial higher-plant response to increasing atmispheric CO_2-concentration in relation to the global carbon cycle. *Global Change Biology*, **1**, 243–274.

Arnone, J. A. III & Körner, C. (1995). Soil and biomass carbon pools in model communities of tropical plants under elevated CO_2 *Oecologia*, **104**, 61–71.

Bazzaz, F. A. & Fajer, E. D. (1992). Plant life in a CO_2-rich world. *Scientific American*, **266**, 68–74.

Becker, M. (1991). Incidence des conditions climatiques, édaphiques et sylvicoles sur la croissance et la santé des forêts. In *Les Recherches en France sur le Deperissement des Forêts*, pp. 25–41. Programme DEFORPA, second report. Nancy: ENGREF.

Bonan, G. B. & Sirois, L. (1992). Air temperature, tree growth, and the northern and southern range limits to *Picea mariana*. *Journal of Vegetation Science*, **3**, 495–506.

Briffa, K. R., Bartholin, T. S., Eckstein, D., Jones, P. D., Karlen, W., Schweingruber, F. H. & Zetterberg, P. (1990). A 1400-year tree-ring record of summer temperatures in Fennoscandia. *Nature*, **346**, 434–9.

Brown, J. H. (1994). The ecology of coexistence (book review). *Science*, **263**, 995–6.

Bunce, J. A. (1990). Short- and long-term inhibition of respiratory carbon dioxide efflux by elevated carbon dioxide. *Annals of Botany*, **65**, 637–42.

Bunce, J. A. & Caulfield, F. (1990). Reduced respiration at elevated atmospheric carbon dioxide in three herbaceous perennials. *Bulletin of the Ecological Society of America*, (Supplement), **71**, 108.

Chen, J.-L., Reynolds, J. F., Harley, P. C. & Tenhunen, J. D. (1993). Coordination theory of leaf nitrogen distribution in a canopy. *Oecologia*, **93**, 63–9.

Cipollini, M. L., Drake, B. G. & Whigham, D. (1993). Effects of elevated CO_2 on growth and carbon/nutrient balance in the deciduous woody shrub *Lindera benzoin* (L.) Blume (Lauraceae). *Oecologia*, **96**, 339–46.

Coleman, J. S., McConnaughay, K. D. M. & Ackerly, D. D. (1994). Interpreting phenotypic variation in plants. *Tree*, **9**, 187–91.

Conroy, J. P., Milham, P. J. & Barlow, E. W. R. (1992). Effect of nitrogen and phosphorus availability on the growth response of *Eucalyptus grandis* to high CO_2. *Plant, Cell & Environment*, **15**, 843–7.

Criddle, R. S., Hopkin, M. S., McArthur, E. D. & Hansen, L. D. (1994). Plant distribution and the temperature coefficient of metabolisms. *Plant, Cell & Environment*, **17**, 233–43.

Cure, J. D. (1985). Carbon dioxide doubling responses: a crop survey. In *Direct Effects of Increasing Carbon Dioxide on Vegetation*, ed. B. R. Strain & J. D. Cure, pp. 99–116. Publ. ER-0238. Washington, DC: US Department of Energy.

Curtis, P. S., Drake, B. G., Leadley, P. W., Arp, W. J. & Whigham, D. F. (1989). Growth and senescence in plant communities exposed to elevated CO_2 concentrations on an estuarine marsh. *Oecologia*, **78**, 20–6.

Dahl, E. & Mork, E. (1959). Om sambandet mellom temperatur, anding og vekst hos gran (*Picea abies* (L.) Karst.). *Meddelelser fra Norske Skogforsoksvesen*, **53**, 83–93.

DeLucia, E. H., Sasek, T. W. & Strain, B. R. (1985). Photosynthetic inhibition after long-term exposure to elevated levels of atmospheric carbon dioxide. *Photosynthesis Research*, **7**, 175–84.

Diaz, S., Grime, J. P., Harris, J. & McPherson, E. (1993). Evidence of a feedback mechanism limiting plant response to elevated carbon dioxide. *Nature*, **364**, 616–17.

Eamus, D. & Jarvis, P. G. (1989). The direct effects of increase in the global atmospheric CO_2 concentration on natural and commercial temperate trees and forests. *Advances in Ecological Research*, **19**, 1–55.

Ehret, D. L. & Jolliffe, P. A. (1985). Photosynthetic carbon dioxide exchange of bean plants grown at elevated carbon dioxide concentrations. *Canadian Journal of Botany*, **63**, 2026–30.

Elliott, E. T., Hunt, H. W., Detling, J. K., Moore, J. C. & Reuss, D. E. (1990). The effects of elevated CO$_2$ and climate change on grasslands: potentially mineralizable soil carbon in intact sods. *Bulletin of the Ecological Society of America* (Supplement), **71**, 147.

Enoch, H. Z. (1978). The role of carbon dioxide in productivity of protected cultivation. *Acta Horticulturae*, **87**, 125–9.

Field, C. B., Jackson, R. B. & Mooney, H. A. (1995). Stomatal responses to increased CO$_2$: implications from the plant to the global scale. *Plant, Cell & Environment*, **18**, 1214–1225.

Friend, A. D. & Woodward, F. I. (1990). Evolutionary and ecophysiological responses of mountain plants to the growing season environment. *Advances in Ecological Research*, **20**, 59–124.

Gifford, R. M. (1992). Interaction of carbon dioxide with growth-limiting environmental factors in vegetation productivity: implications for the global carbon cycle. In *Advances in Bioclimatology* 1, pp. 24–58. Berlin, Heidelberg, New York: Springer.

Grace, J., Lloyd, J., McIntyre, J., Miranda, A. C., Meir, P., Miranda, H. S., Nobre, C., Moncrieff, J., Massheder, J., Malhi, Y., Wright, I. & Gash, J. (1995). Carbon dioxide uptake by an undisturbed tropical rain forest in southwest Amazonia, 1992 to 1993. *Science*, **270**, 778–780.

Graumlich, L. J. (1991). Subalpine tree growth, climate, and increasing CO$_2$: an assessment of recent growth trends. *Ecology*, **72**, 1–11.

Graybill, D. A. & Idso, S. B. (1993). Detecting the aerial fertilization effect of atmospheric CO$_2$ enrichment in tree ring chronologies. *Global Geochemistry Cycles*, **7**, 81–95.

Grulke, N. E., Riechers, G. H., Oechel, W. C., Hjelm, U. & Jaeger, C. (1990). Carbon balance in tussock tundra under

ambient and elevated atmospheric CO$_2$. *Oecologia*, **83**, 485–94.

Hättenschwiler, S. & Körner, Ch. (1996). System-level adjustments to elevated CO$_2$ in model spruce ecosystems. *Global Change Biology*, in press.

Hoffman, J. S. (1984). Carbon dioxide and future forests. *Journal of Forestry*, **82**, 164–7.

Hogan, K. P., Smith, A. P. & Ziska, L. H. (1991). Potential effects of elevated CO$_2$ and changes in temperature on tropical plants. *Plant, Cell & Environment*, **14**, 763–78.

Houghton, R. A. & Woodwell, G. M. (1989). Global climatic change. *Scientific American*, **260**, 18–26.

Kienast, F. & Luxmoore, R. J. (1988). Tree-ring analysis and conifer growth responses to increased atmospheric CO$_2$ levels. *Oecologia*, **76**, 487–95.

Kimball, B. A. & Idso, S. B. (1983). Increasing atmospheric CO$_2$: effects on crop yield, water use and climate. *Agriculture & Water Management*, **7**, 55–72.

Koch, G. W. & Mooney, H. A. (1996) *Carbon Dioxide and Terrestrial Ecosystems*. San Diego: Academic Press.

Kohlmaier, G. H., Bröhl, H., Siré, E. O., Plöchl, M. & Revelle, R. (1987). Modelling stimulation of plants and ecosystem response to present levels of excess atmospheric CO$_2$. *Tellus*, **39B**, 1–2.

Koike, T. (1987). The growth characteristics in Japanese mountain birch (*Betula ermanii*) and white birch (*Betula platyphylla* var. Japonica), and their distribution in the northern part of Japan. In *Human Impacts and Management of Mountain Forests*, ed. T. Fujimori & M. Kimura, pp. 189–200. Ibaraki, Japan: Forestry and Forest Products Research Institute.

Körner, Ch. (1993). CO$_2$ fertilization: the great uncertainty in future vegetation development. In *Vegetation Dynamics & Global Change*, ed. A. M. Solomon & H. H. Shugart, pp. 53–70. New York, London: Chapman & Hall.

Körner, Ch. (1994). Leaf diffusive conductances in the major vegetation types of the globe. In *Ecophysiology of Photosynthesis*, ed. E. D. Schulze & M. M. Caldwell, pp. 463–90. Springer Ecology Studies 100. Berlin, Heidelberg, New York: Springer.

Köner, Ch. (1995). Towards a better experimental basis for upscaling plant responses to elevated CO_2 and climate warming. *Plant, Cell & Environment*, **18**, 1101–1110.

Körner, Ch. (1996). Tropical forests in a CO_2-rich world. *Climate Change*, in press.

Körner, Ch. & Arnone, J. A. III (1992). Responses to elevated carbon dioxide in artificial tropical ecosystems. *Science*, **257**, 1672–5.

Körner, Ch. & Larcher, W. (1988). Plant life in cold climates. In *Plants and Temperature*, ed. S. F. Long & F. I. Woodward, pp. 25–57. Symposium of the Society of Experimental Biology, 42. Cambridge: The Company of Biologists Ltd.

Körner, Ch. & Miglietta, F. (1994). Long-term effects of naturally elevated CO_2 on mediterranean grassland and forest trees. *Oecologia*, **99**, 343–51.

Körner, Ch., Pelaez-Riedl, S. & van Bel, A. J. E. (1995). CO_2 responsiveness of plants: a possible link to phloem loading. *Plant, Cell & Environment*, **18**, 595–600.

Körner, Ch. & Bazzaz, F. A. (1996). *Carbon Dioxide, Populations and Communities*. San Diego: Academic Press.

Lamborg, M. R., Hardy, R. W. F. & Paul, E. A. (1983). Microbial effects. In *CO_2 and Plants: the Response of Plants to Rising Levels of Atmospheric Carbon Dioxide*, ed. E. R. Lemon, pp. 131–76. American Association of Advances in Science Symposium. Boulder, Colorado: West View Press.

Larigauderie, A. & Köener, Ch. (1995). Acclimation of leaf dark respiration to remperature in alpine and lowland plant species. *Annals of Botany*, **76**, 245–252.

Lawlor, D. W. & Mitchell, R. A. C. (1991). The effects of increasing CO_2 on crop photosynthesis and productivity: a review of field studies. *Plant, Cell & Environment*, **14**, 807–18.

Leadley, P. & Körner, Ch. (1996). Effects of elevated CO_2 on plant species dominance in a highly diverse chalk grassland. In *carbon dioxide, populations and communities*, ed. Ch. Körner & F. A. Bazzaz. Physiological Ecology Series, San Diego: Academic Press, in press.

Lieth, J. H., Reynolds, J. F. & Rogers, H. H. (1986). Estimation of leaf area of soybeans grown under elevated carbon dioxide levels. *Field Crops Research*, **13**, 193–203.

Lincoln, D. E., Fajer, E. D. & Johnson, R. H. (1993). Plant–insect herbivore interactions in elevated CO_2 environments. *TREE*, 8, 64–48.

Lloyd, J. & Taylor, J. A. (1994). On the temperature dependence of soil respiration. *Functional Ecology*, 8, 325–33.

Long, S. P. & Drake, B. G. (1991). Effect of the long-term elevation of CO_2 concentration in the field on the quantum yield of photosynthesis of the C_3 sedge, *Scirpus olneyi*. *Plant Physiology*, **96**, 221–6.

Long, S. P. & Drake, B. G. (1992). Photosynthetic CO_2 assimilation and rising atmospheric CO_2 concentrations. In *Crop Photosynthesis: Spatial and Temporal Determinants*, ed. N. R. Baker & H. Thomas, pp. 69–101. Amsterdam: Elsevier.

Melillo, J. M., McGuire, A. D., Kicklighter, D. W., Moore, B. III, Vorosmarty, C. J. & Schloss, A. L. (1993). Global climate change and terrestrial net primary production. *Nature*, **363**, 234–40.

Miroslavov, E. A. & Kravkina, I. M. (1991). Comparative analysis of chloroplasts and mitochondria in leaf chlorenchyma from mountain plants grown at different altitudes. *Annals of Botany*, **68**, 195–200.

Mooney, H. A. (1990). Address of the past president. Toward the study of the earth's metabolism. *Bulletin of the Ecological Society of America*, **71**, 221–8.

Mooney, H. A., Medina, E., Schindler, D. W., Schulze, E.-D. & Walker, B. H. (eds.) (1991). *Ecosystem Experiments.* Chichester, New York: John Wiley & Sons.

Mousseau, M. & Saugier, B. (1992). The direct effect of increased CO$_2$ on gas exchange and growth of forest tree species. *Journal of Experimental Botany*, **43**, 1121–30.

Navas, M.-L., Guillerm, J.-L., Fabreguettes, J. & Roy, J. (1995). The influence of elevated CO$_2$ on community structure, biomass and carbon balance of mediterranean old-field microcosms. *Global Change Biology*, **1**, 325–335.

Nicolussi, K., Bortenschlager, S. & Körner, Ch. (1995). A possibly CO$_2$-related increase in tree-ring width in subalpine *Pinus cembra* of the Central Alps. *Trees*, **9**, 181–9.

Nijs, I., Impens, I. & Behaeghe, T. (1989). Leaf and canopy responses of *Lolium perenne* to long-term elevated atmospheric carbon-dioxide concentration. *Planta*, **177**, 312–20.

Norby, R. J., Pastor, J. & Melillo, J. M. (1986). Carbon–nitrogen interactions in CO$_2$-enriched white oak: physiological and long-term perspectives. *Tree Physiology*, **2**, 233–41.

Norby, R. J., O'Neill, E. G. & Wullschleger, S. D. (1995). Belowground responses to atmospheric carbon dioxide in forests. In *Carbon Forms and Functions in Forest Soils*, ed. W. W. McFee & J. M. Kelly, pp. 397–418. Madison, Wisconsin: Soil Science Society of America.

Odum, E. P. (1977). The emergence of ecology as a new integrative discipline. *Science*, **195**, 1289–93.

Oechel, W. C. & Riechers, G. H. (1986). Impacts of increasing CO$_2$ on natural vegetation, particularly the tundra. In *Climate–Vegetation Interactions*, ed. C. Rosenzweig & R. Dickinson, pp. 36–42. Greenbelt, Maryland: Goddard Space Flight Center.

Overdieck, D. (1993). Erhöhte CO$_2$-Konzentration und Wachstum junger

Buchen (*Fagus sylvatica*). *Verhandlungen d. Gesellschaft f. Oekologie (Göttingen)*, **22**, 431–8.

Owensby, C. E., Ham, J. M., Knapp, A., Rise, C. W., Coyne, P. I. & Auen, L. M. (1996). Ecosystem-level responses of tallgrass prairie to elevated CO$_2$. In *Carbon Dioxide and Terrestrial Ecosystems*, ed. G. W. Koch & H. A. Mooney, pp. 147–162. San Diego: Academic Press.

Pearcy, R. W. (1977). Acclimation of photosynthetic and respiratory carbon dioxide exchange to growth temperature in *Atriplex lentiformis* (Torr.). *Plant Physiology*, **59**, 795–9.

Phillips, O. L. & Gentry, A. H. (1994). Increasing turnover through time in tropical forests. *Science*, **263**, 954–8.

Pierce, L. L., Running, S. W. & Walker, J. (1994). Regional-scale relationships of leaf area index to specific leaf area nitrogen content. *Ecological Applications*, **4**, 313–21.

Poorter, H., Gifford, R. M., Kriedemann, P. E. & Wong, S. C. (1992). A quantitative analysis of dark respiration and carbon content as factors in the growth response of plants to elevated CO$_2$. *Australian Journal of Botany*, **40**, 501–13.

Precht, H., Christopherson, J., Hensel, H. & Larcher, W. (1973). *Temperature and Life*. Berlin, Heidelberg, New York: Springer.

Prentice, I. C. (1993). Process and production. *Nature*, **363**, 209–10.

Raich, J. W. & Nadelhoffer, K. J. (1989). Belowground carbon allocation in forest ecosystems: global trends. *Ecology*, **70**, 1346–54.

Ryan, M. G. (1991). Effects of climate change on plant respiration. *Ecological Applications*, **1**, 157–67.

Sage, R. F., Sharkey, T. D. & Seemann, J. R. (1989). Acclimation of photosynthesis to elevated CO$_2$ in five C$_3$ species. *Plant Physiology*, **89**, 590–6.

Sanders, I. R. (1996). Plant–fungal interactions in a CO$_2$-rich world. In *carbon dioxide populations and communities*, ed. C. Körner & F. A. Bazzaz. San Diego: Academic Press, in press.

Schäppi, B. & Körner, Ch. (1995). Growth responses of an alpine grassland to elevated CO₂. *Oecologia* **105**, 43–52.

Scholander, S. I. & Kanwisher, J. T. (1959). Latitudinal effect on respiration in some northern plants. *Plant Physiology*, **34**, 574–6.

Schulze, E.-D., Kelliher, F. M., Körner, C., Lloyd, J. & Leuning, R. (1994). Relationships between maximum stomatal conductance, ecosystem surface conductance, carbon assimilation rate and plant nitrogen nutrition: a global ecology scaling exercise. *Annual Review of Systematics and Evolution*, **25**, 629–60.

Sionit, N. & Kramer, P. J. (1986). Woody plant reactions to CO₂ enrichment. In *Carbon Dioxide Enrichment of Greenhouse Crops*, ed. H. L. Enoch & B. Y. Kimball, pp. 69–85. Boca Raton: CRC Press.

Stocker, O. (1935). Assimilation und Atmung westjavanischer Tropenbäume. *Planta*, **24**, 402–45.

Strain, B. R. (1987). Direct effects of increasing atmospheric CO₂ on plants and ecosystems. *TREE*, **2**, 18–21.

Strain, B. R. & Cure, J. D. (ed.) (1985). *Direct Effects of Increasing Carbon Dioxide on Vegetation*. Publ. DOE/ER-0238. Washington, DC: US Department of Energy, Office of Energy Research.

Thomas, R. B., Reid, C. D., Ybema, R. & Strain, B. R. (1993). Growth and maintenance components of leaf respiration of cotton grown in elevated carbon dioxide partial pressure. *Plant, Cell & Environment*, **16**, 539–46.

Thornley, J. H. M., Fowler, D. & Cannell, M. G. R. (1991). Terrestrial carbon storage resulting from CO₂ and nitrogen fertilization in temperate grasslands. *Plant, Cell & Environment*, **14**, 1007–11.

Valle, R., Mishoe, J. W., Campbell, W. J., Jones, J. W. & Allen, L. H. Jr (1985). Photosynthetic responses of bragg soybean leaves adapted to different CO₂ environments. *Crop Science*, **25**, 333–9.

van Veen, J. A., Liljeroth, E., Lekkerkerk, L. J. A. & van de Geijn, S. C. (1991). Carbon fluxes in plant–soil systems at elevated atmospheric CO₂ levels. *Ecological Applications*, **12**, 175–81.

Walker, B. H. (1991). Whole-terrestrial ecosystem experiments in the context of global change: working group report. In *Ecosystem Experiments*, ed. H. A. Mooney, E. Medina, D. W. Schindler, E.-D. Schulze & B. H. Walker, pp. 237–43. SCOPE 45. Chichester, New York: John Wiley & Sons.

Werger, M. J. A. & Hirose, T. (1991). Leaf nitrogen distribution and whole canopy photosynthetic carbon gain in herbaceous stands. *Vegetatio*, **97**, 11–20.

Wittwer, S. H. (1984). Carbon dioxide levels in the biosphere: effects on plant productivity. *CRC Critical Review in Plant Science*, **2**, 171–98.

Wong, S. C. (1990). Elevated atmospherical partial pressure of CO₂ and plant growth. II. Non-structural carbohydrate content in cotton plants and its effect on growth parameters. *Photosynthesis Research*, **23**, 171–80.

Wong, S. C. & Dunin, F. X. (1987). Photosynthesis and transpiration of trees in a eucalypt forest stand: CO₂, light and humidity responses. *Australian Journal of Plant Physiology*, **14**, 619–32.

Woodward, F. I., Thompson, G. B. & McKee, I. F. (1991). The effect of elevated concentrations of carbon dioxide on individual plants, populations, communities and ecosystems. *Annals of Botany*, **67**, 23–38.

Wullschleger, S. D. & Norby, R. J. (1992). Respiratory cost of leaf growth and maintenance in white oak saplings exposed to atmospheric CO₂ enrichment. *Canadian Journal of Forest Research*, **22**, 1717–21.

Zak, D. R., Pregitzer, K. S., Curtis, P. S., Teeri, J. A., Fogel, R. & Randlett, D. L. (1993). Elevated atmospheric CO₂ and feedback between carbon and nitrogen cycles. *Plant & Soil*, **151**, 105–17.

Elevated CO_2 and terrestrial vegetation: Implications for and beyond the global carbon budget

F. A. Bazzaz, S. L. Bassow, G. M. Berntson and S. C. Thomas

Introduction

Approximately 6.0 Pg of carbon per year are emitted into the atmosphere due to fossil fuel combustion, cement manufacturing, and deforestation (Rotty & Marland, 1986; Houghton *et al.*, 1990; Watson *et al.*, 1990). As a result, the concentration of CO_2 in the atmosphere has increased from pre-industrial levels of 280 ppm to the current level approaching 360 ppm. In addition to playing an important role in modifying global climate as a greenhouse gas, this increase in atmospheric CO_2 will have, and probably already has had (Gifford, 1994), profound direct effects on plant growth and development in natural systems. Thus elevated CO_2 has a dual role in affecting global vegetation patterns. CO_2 and other greenhouse gases may alter climate sufficiently such that climatic zones may shift northward as well as higher in elevation, altering vegetation distributions considerably (e.g. Davis, 1989; Eamus & Jarvis, 1989; Graham *et al.*, 1990; Peters, 1991; Davis & Zabinski, 1992). CO_2 also has direct, species-specific effects on the growth and physiology of plants, and these direct effects have the potential to alter plant community structure and ecosystem function (e.g. Zangerl & Bazzaz, 1984; Williams *et al.*, 1986; Curtis *et al.*, 1989; Reekie & Bazzaz, 1989; Arp *et al.*, 1993; Fig. 4.1).

The nature and scope of the direct effects of elevated CO_2 on plants have been extensively and recently reviewed (e.g. Bazzaz, 1990; Woodward *et al.*, 1991; Rogers *et al.*, 1993). It is worth emphasizing that these effects span the entire hierarchical range of biological phenomena. At a biochemical level CO_2 is the principle substrate of the most abundant enzyme on the planet (Rubisco: e.g. Bowes, 1991). At an ecological level CO_2 is a widely diffusing plant resource (Bazzaz & McConnaughay, 1992). At a global geophysical level, CO_2 is a primary link between vegetation and climate (e.g. Tans *et al.*, 1990; Houghton, 1991). This wide range of scales illustrates both the importance and complexity of CO_2 interactions with vegetation.

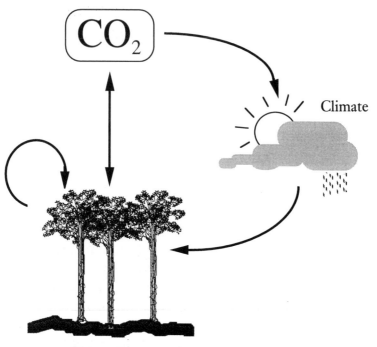

Figure 4.1 *A schematic illustration of the direct and indirect effects of elevated CO_2 on terrestrial vegetation. Arrows represent the following interactions: atmospheric CO_2 concentrations indirectly affect vegetation through the forcing of climate, leading to changes in temperature and precipitation as predicted by various GCMs. CO_2 also directly affects vegetation through photosynthesis, respiration, growth and reproduction. Vegetation takes up carbon, thereby acting as a sink for atmospheric CO_2, and an important feedback on the global carbon cycle. Finally, vegetation has important feedbacks on its own composition and function, and these also should be taken into account in long-term predictions of responses to rising CO_2.*

The present state of CO_2 research

Current national and international policy concerns have resulted in a strong research focus on potential feedbacks of elevated CO_2 with the global carbon cycle (e.g. Wisniewski & Lugo, 1992). This research has been given particular impetus by the fact that atmospheric CO_2 sources do not balance with estimated CO_2 sinks in terrestrial and marine systems. It is quite possible that this 'missing sink' in the global carbon budget is accounted for by CO_2 fertilization of terrestrial ecosystems (e.g. Gifford, 1994); however, the actual extent to which CO_2 fertilization occurs in any particular natural ecosystem is unclear. The importance of understanding the dynamics of carbon sinks in the biosphere has also been highlighted by recent analyses of the inter-annual fluctuations in the rate of CO$_2$ increase (e.g. Kohlmaier *et al.*, 1989; Kohlmaier, 1991; Sarmiento, 1993).

Elucidating the global carbon balance is undeniably the single dominant concern presently guiding research on the direct effects of CO$_2$ on natural vegetation. However, the carbon balance issue in and of itself seems unlikely to retain predominance as a research focus for an extended time. First, some progress has been made on closing the carbon budget. Part of the 'missing sink' may actually have been an overestimated source: revised estimates of tropical deforestation rates based on remote sensing indicate a smaller input from this source than had been previously assumed. Also, direct estimates for CO$_2$ flux from one temperate forest site (Wofsy *et al.*, 1993) and recent increases in production in European forests (Kauppi *et al.*, 1992) suggest that carbon uptake in mid-latitude forests may also account for some of the imbalance.

Emerging CO$_2$ issues

What research questions are likely to assume greatest importance in guiding CO$_2$ research in the near future? We contend that future CO$_2$ research should go beyond just quantifying putative direct links with the global carbon budget. In particular, we offer three general themes which we feel warrant future attention: (1) interactions of CO$_2$ and other anthropogenic factors; (2) scaling controlled experimentation to ecosystems; and (3) direct effects of elevated CO$_2$ on population biology and biodiversity. We acknowledge that this is not a complete list of important ecological implications of global change. Here we adopt a phytocentric perspective, and so do not address issues related to interactions of plants within higher trophic levels (for further discussions see Bazzaz & Fajer, 1992 and Fajer & Bazzaz, 1992).

In this paper we classify 'anthropogenic factors' in relation to CO$_2$ into two types: *interacting* and *overriding* factors. *Interacting* factors are those which in combination with CO$_2$ result in ecosystem changes that cannot be predicted from an understanding of the effects of elevated CO$_2$ or the other factor alone. This type of interaction often results when two (or more) factors both affect the processes that lead to whole system responses. We present two examples of potentially synergistic interactions between elevated CO$_2$ and other anthropogenically driven factors: temperature rise and nitrogen deposition. There are additional potential synergisms or antagonisms between elevated CO$_2$ and many other anthropogenic factors, including ozone (e.g. Mortensen, 1992), SO_2 (e.g. Carlson & Bazzaz, 1986), and UV radiation (e.g. Teramura *et al.*, 1990), though we will not elaborate on these issues here. *Overriding* factors, on the other hand, have such a great impact that the effects of elevated CO_2 may not be apparent. In this paper we discuss the implications of one general type of overriding factor, that of an altered landscape through deforestation, caused either directly or indirectly by human resource use (logging or local extinctions driven by climate change). We focus on the role elevated CO_2 may play in altering regeneration following such major disturbances. These sorts of interacting and overriding anthropogenic factors cannot be ignored in making projections of the future of the earth's terrestrial ecosystems.

Another fundamental challenge for quantifying the response of vegetation to elevated CO_2 is to link experimental results from controlled environments to long-lived, large-statured vegetation in complex natural assemblages. This scaling needs to be carefully considered. How similar are the responses of a small tree seedling (1, 2 or 3 years old) to those of a mature tree (as old as 200 years)? How similar are the responses of plants grown individually in controlled experiments to those growing in complex assemblages in natural environments? These questions are of critical importance to assessing our current understanding of future vegetation responses to elevated CO_2, and yet they still remain poorly developed within the scientific literature.

Finally, we briefly discuss direct effects of elevated CO_2 on the population biology and future diversity of vegetation. Over the long term, plant community composition and ecosystem function will be driven by the capacity of plants to reproduce and regenerate under altered environmental conditions. Although there is a relative paucity of studies on effects of elevated CO_2 on the reproductive biology of non-horticultural plants, there is some evidence that reproductive responses are not simply predictable in terms of enhanced photosynthesis or growth. Such variable reproductive responses are among the biological interactions that should be considered in a balanced appraisal of the long-term implications of rising CO_2 on plant diversity. These issues are of central importance in predicting future vegetation response to global change.

Elevated CO$_2$ and other global change factors

Temperature rise and elevated CO$_2$: How will plants cope?

Despite predictions that both CO$_2$ and air temperature will rise together, very few data are available to assess the potential for interactive effects of elevated CO$_2$ and temperature on most plant species. Limited data from studies with herbaceous species suggest that the combined effects of CO$_2$ and temperature are not necessarily additive and are therefore difficult to predict from a knowledge of their individual effects (e.g. Idso *et al.*, 1987; Idso & Kimball, 1989; see reviews by Long, 1991; Eamus, 1991; Farrar & Williams, 1991; Hogan *et al.*, 1991). Furthermore, many of the independent and interactive effects of CO$_2$ and temperature appear to be species specific (Bazzaz, 1990; Bazzaz & Miao, 1992; Bazzaz *et al.*, 1990; Coleman & Bazzaz, 1992; Ackerly *et al.*, 1992). As a result, simultaneous increases in CO$_2$ concentration and temperature may substantially alter species composition in natural communities (Morse & Bazzaz, 1994).

Small changes in temperature near a species' photosynthetic optima can lead to substantial changes in productivity (Larcher, 1983; Waring & Schlesinger, 1985; Woodward, 1987). For example, Grace (1988) concluded that an air temperature increase of only 1°C could result in a 10% increase in productivity in northern climates. However, as with the effects of CO$_2$, species differ in their growth sensitivity to elevated temperatures (Berry & Raison, 1981; Larcher, 1983; Grace, 1988), and the responses also vary with other factors such as water availability (Turner & Kramer, 1980; Berry & Raison, 1981). Thus to make any anticipatory statements regarding the vegetation response to climate warming, one must consider the species-specific effects of temperature in combination with elevated CO$_2$.

Mechanisms underlying plant responses to elevated CO$_2$ and temperature
Understanding the mechanisms of growth responses to elevated CO$_2$ and temperature is critical to making generalizations regarding responses of plants to climate change. Both CO$_2$ and temperature are known to affect plant carbon gain independently by altering processes at many levels of plant organization. Such processes include photosynthesis and respiration (e.g. Amthor, 1991; Bowes, 1991; Long, 1991; Stitt, 1991), leaf energy budgets and water use efficiencies (e.g. Eamus, 1991), carbon and nutrient allocation (e.g. Hogan *et al.*, 1991; Farrar & Williams, 1991; Coleman & Bazzaz, 1992), and leaf area production, phenology and architecture (e.g. Ford, 1982; Gaudillere & Mousseau, 1989; Reekie & Bazzaz, 1989; Ackerly *et al.*, 1992).

Changes in photosynthetic capacity under elevated CO_2 and temperature conditions may result from a number of interacting biochemical processes. Elevated CO_2 positively influences the carboxylation efficiency of C_3 plants by increasing the activation state of the primary carboxylation enzyme, Rubisco (Bowes, 1991; Stitt, 1991; Eamus, 1991; but see Sage *et al.*, 1988). The ratio of CO_2 to O_2 increases, reducing photorespiratory losses (Woodrow & Berry, 1988). Elevated air and leaf temperatures also influence the efficiency of carboxylation. The specificity of Rubisco for CO_2 in comparison to O_2 is lower at high temperatures, and the solubility of CO_2 relative to O_2 also decreases with temperature (Jordan & Ogren, 1984), both resulting in a reduction in carboxylation efficiency. Based on these contrasting effects of CO_2 and temperature on carboxylation, it is difficult to predict the interactive effects of elevated CO_2 and temperatures on plant carbon gain, although models may be valuable tools in this respect (e.g. Long, 1991).

Many plants grown at elevated CO_2 show a decline in photosynthetic enhancement through time. Several reasons for this decline have been proposed including: reductions in carboxylation efficiency caused by a decrease in the amount or activity of Rubisco, suppression of sucrose synthesis by an accumulation of starch, inhibition of the triose-phosphate carrier, and insufficient carbon sinks in the plant (Sasek *et al.*, 1985; Sage & Pearcy, 1987; reviewed in Arp, 1991; Bowes, 1991; Farrar & Williams, 1991; Stitt, 1991). Interactions between elevated CO_2 concentrations and temperatures may potentially influence the duration of photosynthetic enhancements. Higher temperatures lead to increased respiration rates, and increased sucrose synthesis, transport and utilization. These effects may in turn reduce the degree to which carbohydrates build up in sources and sinks and thus delay the onset and/or reduce the degree of feedback inhibition under elevated CO_2 (Farrar & Williams, 1991).

Whereas the effects of elevated CO_2 on photosynthetic processes are fairly well documented, the effects of elevated CO_2 concentrations on dark (non-photo) respiratory processes are still poorly understood (Ryan, 1991; Amthor, 1991). Many studies have reported reductions in plant respiration (e.g. Gifford *et al.*, 1985; Spencer & Bowes, 1986; Bunce, 1990; 1992); others have reported increases in respiration (e.g. Hrubec *et al.*, 1985; Poorter *et al.*, 1988; Nijs *et al.*, 1989); and yet others have reported no detectable changes in respiration caused by elevated CO_2 (Grulke *et al.*, 1990). Some of this difference may be accounted for by distinguishing the unit basis on which respiration is expressed: respiration expressed on a per gram of tissue basis may decline simply as a result of a higher leaf mass per unit leaf area (LMA) in elevated CO_2 atmospheres. In contrast, an increase in respiration rate in response to elevated temperatures is fairly well documented (Amthor, 1989; Ryan, 1991). It is thus difficult to predict, without more empirical data, how elevated CO$_2$ and temperature will interact to influence the ratio of photosynthesis to respiration, and how this ratio may vary among species.

Elevated CO$_2$ leads to higher leaf-level water use efficiency, via decreases in stomatal conductance (Lemon, 1983) and transpiration rate, and also via an increase in net photosynthesis (Norby & O'Neill, 1989; Eamus, 1991). In contrast, high temperatures may decrease water use efficiency due to greater vapour pressure deficits, and the increased need for transpirational cooling of warmer leaves (Jones, 1992). Therefore, it is possible that the magnitudes of CO$_2$ enhancement of water use efficiency may decrease with increasing temperature. It is critical, however, to consider the whole-plant perspective as well. Different patterns of net carbon gain as well as net water use efficiency may be seen when one considers the whole plant, not just leaf-level rates. For example, the increased growth resulting from elevated CO$_2$ may result in an increased total water use by an individual plant, even though the leaf-level water use efficiency has increased (e.g. Jones *et al.*, 1984; Eamus, 1991).

Species-specific implications of extreme temperatures

If the predicted mean global temperature rise is accompanied by an increase in the variability of temperatures, then the frequency of extremely hot periods as well as cold periods may increase (Mearns *et al.*, 1984; Rind *et al.*, 1989; Wigley, 1985; Mitchell *et al.*, 1990; Katz & Brown, 1992). Such extreme temperatures may be far more critical than a shift in the mean temperatures for determining species' future ranges. Extreme temperatures may damage plant tissue, and consequently limit the survival of certain plant species in a region. The implications of extreme temperature events are highly species-specific (Coleman *et al.*, 1991; Pastor, 1993; Bassow *et al.*, 1994). Elevated concentrations of CO$_2$ in the atmosphere alter plant allocation, physiology, and growth, which may accentuate or ameliorate the damage from extreme temperatures.

Transpirational cooling dissipates heat as latent heat of evaporation, keeping leaf tissue at or below air temperature in hot weather. However, growth in a high CO$_2$ atmosphere tends to lower foliar conductance rates which may cause higher leaf temperatures (Idso *et al.*, 1986, 1987), leading to more damage during an extremely hot period. However, elevated CO$_2$ concentrations may reduce the leaf area ratio (e.g. Bazzaz, 1990), thereby reducing the ratio of transpiring leaf area to water-acquiring roots, which may improve the plant's ability to cool under some conditions. If the

frequency of extremely high temperatures increases, the role that temperature extremes play in changing competitive interactions and thus affecting
community composition may increase in importance, as these temperatures
appear to severely alter plant survival and growth in some species.

The response of vegetation to rising CO_2 levels and temperatures may
not be simply predicted from the response of the plants to one or the other
factor. Rather, complex interactions make it necessary to consider these
combinations of factors simultaneously in order to make coherent statements about the effects of climate change on vegetation.

Nitrogen deposition in temperate forests: Feedbacks on the carbon cycle?

Along with increasing atmospheric CO_2 levels, industrial combustion of
fossil fuels has led to an increase in nitrogen and sulphur deposition
throughout North America and Europe (e.g. Galloway *et al*, 1984; Aber
et al., 1989). These anthropogenically derived nutrient inputs can lead to
enhanced forest productivity if addition rates are not excessive (Kauppi
et al., 1992) but may also result in forest decline if inputs become extreme
(Schulze, 1989). Net primary production in temperate deciduous forests of
northeastern United States is limited by the availability of nitrogen. This
has been demonstrated by the absence of measurable nitrification and low
net nitrogen mineralization (Aber *et al*., 1991, 1993) and by significant
increases in net primary productivity (NPP) with the acute addition of
nitrogen fertilizers (Lea *et al*., 1980). Because nitrogen limits the
productivity of these systems, nitrogen deposition from acid rain and
atmospheric aerosol and particle deposition may initially result in a 'fertilization' of these systems. It is theorized, however, that after an extended
period of time these forest ecosystems will become nitrogen saturated.

Increases in the supply of nutrients (especially nitrogen) can lead to
increased growth enhancements of trees with elevated CO_2 (e.g. Brown &
Higginbotham, 1986; Johnson *et al*., 1992; Brown, 1991; Wong *al*., 1992).
Thus, it is possible that increasing atmospheric CO_2 concentrations may
feedback positively with regionally enhanced nitrogen supply rates from
high levels of nitrogen deposition so that both the carbon and nitrogen sink
strength of forests is enhanced (see Fig. 4.2). It has been suggested that this
anthropogenic nitrogen fertilization may actually constitute a large portion
of the 'missing sink' for carbon (Hudson *et al*., 1994). Beyond its immediate
implications for enhanced NPP and altered nitrogen cycling, this positive
feedback has significant implications for community composition in
temperate deciduous forests (e.g. Vitousek, 1994). Preliminary evidence

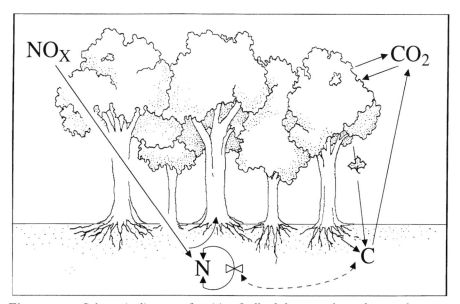

Figure 4.2 *Schematic diagram of positive feedback between elevated atmospheric CO₂ and nitrogen deposition on forest carbon and nitrogen dynamics. Simultaneously increasing the supply of both carbon and nitrogen to forests may increase their ability to sequester greater amounts of both carbon and nitrogen. (Figure drawn by E. J. Farnsworth.)*

demonstrates that tree seedlings of different species show differential patterns of growth enhancement under elevated CO_2 with different nutrient supply rates (Bazzaz & Miao, 1993). These differential modifications of growth enhancements under elevated CO_2 will probably alter species' competitive abilities during regeneration and thus affect future forest composition.

Total belowground inputs of carbon often increase with elevated atmospheric CO_2. This is of importance to global carbon budgets because the single largest pool of terrestrial carbon is in the soils. There is evidence that elevated CO_2 does not result in higher rates of root exudation per unit mass of root (Whipps, 1985; Norby *et al.*, 1987). Rather, total root mass increases under elevated CO_2 and thus substantially more carbon is transported into the soil (e.g. Norby *et al.*, 1992; Zak *et al.*, 1993). Increases in the total amount of carbon input into the soil may in turn increase the amount of carbon retained in the soil due to the preferential metabolism of readily decomposable material by microbes (Lekkerkerk *et al.*, 1990; van Veen *et al.*, 1991). Decreases in the decomposability of tissue due to increased C : N ratios may also play a role in increasing soil carbon retention (Strain & Bazzaz, 1983; Hobbie & Melillo, 1984.

Controlled environment studies of the effect of elevated atmospheric CO_2

on the dominant deciduous tree species in the Harvard Forest, where Aber and colleagues (1993) have done their work, indicated that many of these species show strong growth enhancements with elevated CO_2 (e.g. Bazzaz & Miao, 1993). These studies have also demonstrated that many of these species show enhanced responsiveness to elevated CO_2 with increasing nutrient supply. The increased input of carbon into these forests through enhanced photosynthesis and growth under elevated CO_2 may thus substantially increase the nitrogen sink capacity of these and other ecosystems exposed to high levels of nitrogen deposition. Thus, elevated CO_2 may ameliorate nitrogen saturation in forests exposed to high chronic nitrogen deposition and conversely, forests exposed to high levels of nitrogen deposition may be greater sinks for atmospheric CO_2.

Keeping things in perspective: The importance of land use and other factors in structuring future landscapes

The impact of elevated CO_2 and other anthropogenically derived components of global environmental change have recently been the object of scrutiny by many researchers interested in the ecological destiny of our planet. Often this research has been used to establish predictions about future carbon cycling and ecosystem composition (e.g. Norby *et al.*, 1992; Bazzaz & Miao, 1993; Vitousek, 1994). These extrapolations can be helpful in providing the information necessary to consider one potential outcome within the array of possible future worlds. However, the utility of predictions made on the basis of the observed physiological or ecological processes may be limited due to rapid changes in the landscape within which the studied processes are taking place. Knowledge of how mature phase forests alter patterns of carbon sequestration under elevated CO_2 is not relevant if the forest has been harvested for timber or cleared for agriculture. If temperature increases are so great that various plant species are living in temperatures far from their optima, their growth responses to elevated CO_2 may be trivial in comparison to the effects of temperature. An exhaustive list of direct and indirect anthropogenically derived factors that may 'obliterate' the effects of elevated CO_2 on plant growth and ecosystem structure and function is beyond the scope of this paper. We wish to point out simply that perspectives of global change that do not attempt to consider the direct as well as indirect effects of human presence on earth are of limited predictive value.

Elevated CO₂ in the context of a changing landscape: the importance of forest regeneration

With expanding use of land by humans and the potential of globally changing climatic conditions, it is possible that many of the forests we now study will be lost. One of the better studied cases of this process in the temperate zone are possible climate-driven shifts in species distribution. Changes in global temperature and precipitation patterns predicted from general circulation models may result in significant alterations in species distributions and community composition (Solomon, 1986; Davis and Zabinski, 1992; Davis, 1989; Davis *et al.*, 1986). If the climate changes very rapidly there may be large land areas with substantial tree mortality. In such a situation, regeneration will become increasingly important and large expanses of forested regions may become increasingly dominated by early successional vegetation.

Research in our laboratory on tree species from the Harvard Forest in central Massachusetts suggests that succession in temperate deciduous forests in an elevated CO_2 world may be substantially affected by the dominance of early successional tree species. The CO_2 responsiveness of early growth has been found to be very large in a number of early successional species in Eastern deciduous forest communities, including striped maple (*Acer pensylvanicum*) and grey birch (*Betula populifolia*) (Rochefort & Bazzaz, 1992; Bazzaz & Miao, 1993). These species do not attain large sizes, but often are important in suppressing the growth of seedlings and saplings of dominant canopy tree species (Burns & Honkala, 1990). In the evergreen forests of western North America, a similar differential enhancement in growth has been observed in the earlier successional red alder (*Alnus rubra*) (Arnone & Gordon, 1990) relative to the later successional Douglas Fir (*Pseudotsuga menziesii*) (Hollinger, 1987). Enhanced regeneration of early successional species under elevated CO_2 could thus constitute an important negative feedback on the regeneration of these forests to a more mature phase system (but see El Kohen *et al.*, 1993).

Preliminary analyses linking data from CO_2 responses of Harvard Forest species with a spatially explicit forest simulation model (SORTIE: Pacala *et al.*, in press) suggest that differential growth responses to elevated CO_2 may result in substantial changes in species composition (Bolker *et al.*, 1995). Moreover, such changes in species composition can have a relatively large impact on ecosystem-level properties such as carbon sink strength. This type of model should be applied to a greater variety of ecosystems to explore the potential role species' variations in CO_2 growth enhancement may play in altering biodiversity and ecosystem function.

The problem of scale: how do we make inferences about ecosystem processes from experiments on seedlings and individually grown plants?

Simulation models that have previously examined possible CO_2 fertilization effects (Shugart & Emanuel, 1985; Luxmoore *et al.*, 1990; Post *et al.*, 1992) have generally applied a single multiplicative growth enhancement to predict tree growth under elevated CO_2 (often termed β-factor). While these studies provide some valuable first approximations, this approach does not take into account several important empirical observations on tree responses to elevated CO_2. First, responses to elevated CO_2 differ substantially among tree species (Eamus & Jarvis, 1989; Bazzaz *et al.*, 1990; Rochefort & Bazzaz, 1992; Bazzaz & Miao, 1993; Poorter, 1993; Ceulemans & Mousseau, 1994). Second, the growth response values used in modelling efforts have generally been collected only for early growth responses of tree seedlings, which may greatly overestimate responses of older trees (Bazzaz *et al.*, 1993; but see Norby *et al.*, 1992). Third, growth responses to elevated CO_2 are greatly altered by the presence of neighbouring plants. Often, overall stand-level responses are lower than those obtained for trees grown without neighbours (Bazzaz & McConnaughay, 1992). A number of reviews have emphasized the need for field studies of forest responses to elevated CO_2 at the stand level (e.g. Eamus & Jarvis, 1989; Körner, 1993).

All studies aimed at predicting forest ecosystem responses to global atmospheric change face one fundamental challenge: the most direct experimental approach, that of fumigation experiments at the landscape level, is simply not feasible. Free-Air CO_2 Enrichment (FACE)-type experiments are severely limited by technological and cost constraints (Hendrey, 1993). Also, both FACE and branch-sized cuvette experiments (e.g. Dufrêne *et al.*, 1993; Barton *et al.*, 1993) necessarily sample a small portion of a tree's structure and life span, and may therefore greatly underestimate longer-term acclimation responses. In contrast, it is possible to investigate whole-plant responses of tree saplings in controlled CO_2 environments under field conditions. If our goal is to make predictions of future forest composition, carbon storage and flux in response to elevated CO_2, based on studies of sapling responses, then we must develop approaches which will allow us to scale these experiments to larger plants and longer time frames. These issues of scaling can be divided into two main levels: scaling in time and space, and scaling in ecological complexity. The former incorporates scaling from seedlings or saplings to mature trees, as well as scaling from small-scale, physiological processes to larger-scale, ecosystem-level processes. The second issue of scaling emphasizes the differences between individually grown plants' (saplings') responses and mixed community, stand-level responses.

The general issue of 'scaling up' from the leaf to the ecosystem level of integration has received considerable recent attention (e.g. Ehleringer & Field, 1993). A wide variety of modelling efforts initiated over the past decade are directed towards predicting ecosystem response to global change (e.g. Solomon, 1986; Pastor & Post, 1988; Reynolds *et al.*, 1992; Norman, 1993; Clark, 1993; Schimel, 1993; Pacala *et al.*, in press). While such models should in theory be based on physiological mechanisms, there are extreme challenges faced by such a reductionist or 'bottom-up' approach (cf. Jarvis, 1993).

An alternative approach is to examine empirical relationships that may exist between large-scale processes, such as net ecosystem carbon exchange, and physiological processes operating at the canopy, individual, or leaf level (e.g. Field, 1991; Field *et al.*, 1992; Wofsy *et al.*, 1993). In one such scaling, the diurnal and seasonal patterns of the canopy assimilation, estimated from leaf-level gas exchange, correspond closely with net ecosystem exchange measured by the eddy correlation technique (Fig. 4.3) (Wofsy *et al.*, 1993). The strong correspondence between the two approaches suggests that both methods provide reliable estimates of the net photosynthetic uptake by the temperate forest stand. However, if we are to extend this approach to scale gas exchange observations from controlled-environment experiments with tree seedlings and saplings to ecosystem-level gas exchange, we must address two distinct issues: (1) photosynthetic differences due to *developmental stage*, and (2) potential differences due to effects of the *experimental micro-environment* (Fig. 4.4) (S.L. Bassow & F.A. Bazzaz, unpublished data). Leaf-level physiological parameters have often been found to differ substantially between saplings and adult trees in forest tree species (e.g. Jurik *et al.*, 1988; Lei & Lechowicz, 1990; Hanson *et al.*, 1994). Such differences may in part reflect leaf-level acclimation to environmental conditions, particularly light (e.g. Boardman, 1981; Givnish, 1988), but fixed patterns of whole-plant ontogenetic change may also be important (e.g. Bazzaz, 1979). In the few studies that have explicitly addressed this issue, adult tree P_{max} values have been found to scale as a simple, linear function of sapling P_{max} (α; Koike, 1988; Thomas, 1993). Likewise, plants grown in controlled environment facilities have often been found to have reduced growth and lower photosynthetic capacity in comparison with field-grown plants (e.g. Bazzaz, 1973; Hickman, 1975; Gifford & Rawson, 1992). This effect can also be modelled as a simple linear function (β). By empirically determining these scaling factors, valuable information about ecosystem-level gas exchange under elevated CO$_2$ can be derived (Fig. 4.4).

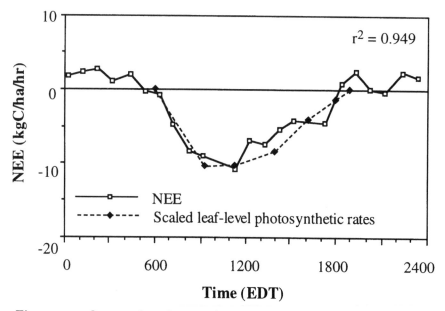

Figure 4.3 *Correspondence between the net ecosystem exchange (NEE) of CO_2 and scaled leaf-level photosynthetic rates. NEE is measured by the eddy-correlation technique. The scaled leaf-level photosynthesis assumes 62% red oak and 38% red maple, and LAI of three. One unit of LAI has photosynthetic rates of foliage measured at the top of the canopy, and the remaining two units of LAI have observed lower canopy photosynthetic rates. Respiration has been incorporated as a temperature-dependent function (Wofsy et al., 1993).*

The majority of studies addressing the direct CO_2 effects on natural vegetation have been conducted with individually grown plants. However, stand-level growth responses to elevated CO_2 may differ greatly from those of plants grown as isolated individuals (Bazzaz & McConnaughay, 1992; Morse & Bazzaz, 1994). Such differences may arise from the effects of neighbours on local resource availability, as well as inherent differences between individuals in responsiveness to elevated CO_2 (Woodward, *et al.*, 1991; Bazzaz & McConnaughay, 1992). A recent study in our laboratory with yellow birch seedlings found that biomass enhancement with elevated CO_2 was significantly reduced when the plants were grown together at a density commonly found for this species in the field (Fig. 4.5; Wayne & Bazzaz, 1995). System-level vegetation responses are more directly addressed by quantifying the responses of single or multi-species stands because plants rarely occur in isolation in their natural environments (cf. Bazzaz, 1990; Patterson & Flint, 1990; Woodward *et al.*, 1991; Körner, 1993). A number of studies have now examined net carbon exchange at the stand level in agricultural systems (Jones *et al.*, 1984, 1985*a,b*; Acock *et al.*,

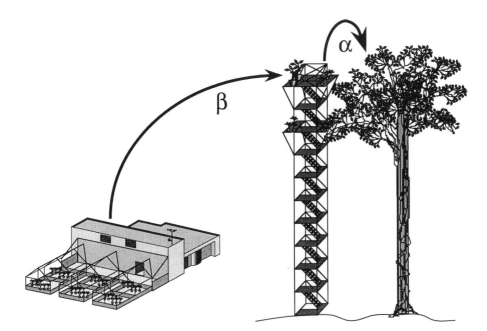

Figure 4.4 *Two components of scaling experimental results from controlled environment studies to potential mature tree response to rising CO_2 concentrations. As described in the text, α is an ontogenetic scaling factor between sapling and adult tree photosynthetic capacities, and β is an environmental scaling factor relating the photosynthetic capacities for saplings grown in elevated CO_2 in controlled environment to saplings in the field. By growing saplings on towers in the canopy, adjacent to mature tree foliage, differences in micro-environmental conditions surrounding leaves on saplings and leaves on the mature trees are minimized.*

1985; Allen *et al.*, 1989; Nijs *et al.*, 1989). However, parallel work on natural or semi-natural plant communities is limited to a few studies (Tissue & Oechel, 1987; Overdieck, 1990; Grulke *et al.*, 1990; Drake & Leadley, 1991; Oechel *et al.*, 1993; Körner & Arnone, 1992), most of which consider responsiveness of small-statured herbaceous systems.

Thus while it is impossible to do the quintessential experiment (short of simply waiting 100+ years), we may be able to use scaling approaches to improve our understanding of the potential effects of elevated CO_2 on large-statured, long-lived ecosystems. If a variety of approaches, including 'bottom-up' models and 'top-down' scaling relationships, yield similar results, we may have much greater confidence in predictions derived from any specific approach.

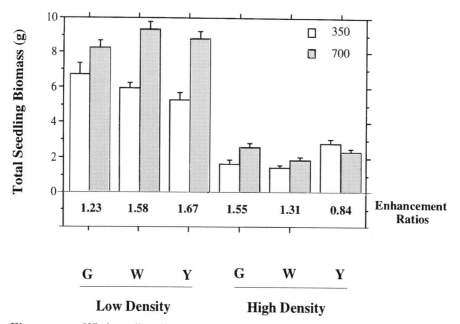

Figure 4.5 *Whole seedling biomass of three maternal families (G, W, Y) of yellow birch (*Betula alleghaniensis*) after 1 year of growth in ambient (350 μl l⁻¹) and elevated (700 μl l⁻¹) CO₂ in both low density (one plant per pot) and high density (144 plants m⁻²). The numbers below each pair of ambient/elevated bars represents the CO₂-induced growth enhancement (β, 700/350) for each of the three families in both density treatments (from Wayne & Bazzaz, 1995).*

Plant populations under elevated CO₂: Another critical dimension of long-term effects

In the context of globally increasing CO_2, we must be prepared to make predictions of effects on vegetation over a time scale of decades or longer. For most ecological systems such a time scale will necessarily encompass multiple generations of plants. This is clearly the case for most herbaceous systems; however, it should also be noted that while maximal life spans of forest trees are commonly 100+ years, generation times are often of the order of 10–60 years (e.g. Harper & White, 1974). One implication of multi-generational time scales is that CO_2 must be considered in the context of all phases of a plant's life cycle, including seed germination, seedling establishment, plant growth, reproductive onset, flowering, ovule fertilization, seed and fruit development, and seed dispersal. A second implication is that long-term responses to elevated CO_2 may be determined not only by individual or canopy-level physiology, but by population-level phenomena such as altered density-dependent reproduction (Bazzaz *et al.*, 1992) or accelerated plant senescence (St Omer & Horvath, 1983). Finally, a multi-generational time scale brings up the issue of potential evolutionary responses to increasing CO$_2$.

Elevated CO$_2$ and plant reproduction: Effects distinct from vegetative growth

From a population perspective, the future composition of plant communities and populations will be determined by the relative reproductive contributions of individual plants. In this regard, it has been clear for some time that elevated CO$_2$ atmospheres may alter plant reproduction in a fashion that is distinct from effects on vegetative growth and development. There is a fairly large body of work documenting effects of elevated CO$_2$ on the reproductive phenology. In some species elevated CO$_2$ accelerates floral initiation (Hovland & Dybing, 1973; Enoch *et al.*, 1976), while in many other species a significant delay has been documented (Hesketh & Hellmers, 1973; Marc & Gifford, 1984). In some cases either response may occur, depending on growth conditions (Reekie & Bazzaz, 1991). Other reported effects of CO$_2$ on reproductive development include altered sex expression (Imazu *et al.*, 1967) and diminished self-incompatibility (Palloix *et al.*, 1985).

Intriguingly, the varying effects of elevated CO$_2$ on plant phenology may be predictable from species' physiological characteristics. In a recent study of flowering phenology under inductive photoperiods for four short-day and four long-day species, Reekie and colleagues (1994) document a consistent pattern of delayed flowering in short-day plants and accelerated flowering in the long-day plants exposed to elevated CO$_2$ atmospheres. Although the physiological mechanisms accounting for this pattern are unclear, direct hormonal interactions (such as CO$_2$ inhibition of ethylene biosynthesis: cf. Horton, 1985) may be involved. It is worth noting that the magnitude of CO$_2$ effects on reproduction are, at best, loosely coupled with total biomass responses to elevated CO$_2$. Compiling the results of 10 years of research on old-field annual C$_3$ weeds from the Mid-west United States, Ackerly and Bazzaz (1995) demonstrated that there was a poor but positive relationship between the response of reproductive biomass and total biomass to elevated CO$_2$ levels (Fig. 4.6).

What are the consequences of altered reproductive timing on net reproductive output? From the point of view of optimal allocation theory, annual plants are expected to allocate all resources (e.g. photosynthate) to vegetative growth up to the point of reproductive onset, at which point most or all resources are invested in reproductive structures (Cohen, 1971; King &

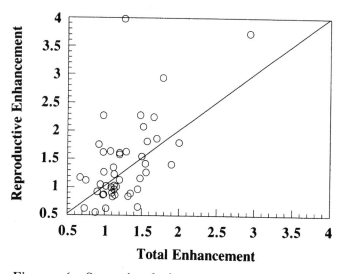

Figure 4.6 *Scatterplot of enhancement ratios for reproductive and total biomass for C_3 herbaceous plants of the Illinois old-field community. Enhancement is the ratio of biomass at elevated (typically 2× ambient) relative to that at ambient concentrations (≈ 350 µl l⁻¹). Each point represents the ratios calculated for one species under one set of environmental conditions (Ackerly & Bazzaz, 1995).*

Roughgarden, 1982; Chiariello & Roughgarden, 1984). If vegetative growth enhancements are small relative to phenological changes, an accelerated switch to reproductive growth could thus greatly reduce total reproductive output. In many temperate annual plants reproduction is terminated by an autumn frost. In this case delays in reproductive phenology might increase the probability of mortality prior to successful reproduction. Thus, either accelerated or delayed reproduction could negatively affect the long-term performance, particularly in annual plant species.

At present no study has specifically investigated the implications of altered phenology for net reproductive output under elevated CO_2. Although elevated CO_2 has generally been found to enhance reproductive output, there are several reports in the literature of decreased plant reproduction at elevated CO_2, including a number of studies in which annual plants have been grown to senescence. For example, the old-field annual *Abutilon theophrasti* has been found to have reduced reproduction at high CO_2 under a wide range of growth conditions (Garbutt & Bazzaz, 1984; Bazzaz *et al.*, 1992; S. Thomas, F. Jasienski & F. Bazzaz, unpublished data; E. Reekie, unpublished data). Other species exhibiting declines in reproduction at elevated CO_2 include *Bromus mollis* (Larigauderie *et al.*, 1988) and *Plantago lanceolata* (Wulff & Alexander, 1985). Although agricultural studies have almost always detected yield increases at elevated CO_2, there are some exceptions among crop species as well: cucumber (*Cucumis sativus* L.) is a notable example (Ito, 1978; Peet, 1986). In a study examining phylogenetic differences in reproductive responses to elevated CO$_2$, Farnsworth & Bazzaz (unpublished data) found that various measures of reproductive output were consistently negatively affected in three species of *Ipomoea*. These results suggest that phylogenetic affinity may be of considerable use in identifying the subset of species that may show reproductive declines at elevated CO$_2$.

From a life-historical perspective, there is very little data available on plant responses to elevated CO$_2$ beyond growth and reproductive output, although some attention has also been given to the issue of seed quality. Elevated CO$_2$ generally results in increases in the C : N ratio in vegetative tissues. We have documented such a nitrogen dilution effect in seeds of *Ambrosia artemisiifolia* (Garbutt *et al.*, 1990), although several other old-field annual species did not show pronounced differences. Seed nutrient content can have a substantial influence on the subsequent performance of seedlings, particularly under competitive conditions (Parrish & Bazzaz, 1985). We therefore speculate that species-specific changes in seed chemistry may result in substantial changes in community composition. Multiple-generation studies that examine seed germination, seedling viability, and long-term differences in species- and genotype-specific performance are needed to evaluate the long-term ramifications of such effects.

Plant population biology and microevolution under elevated CO$_2$

Some aspects of the effects of increasing CO$_2$ on plant populations may be predictable in terms of general theory regarding population-level responses to increasing resources (see also Bazzaz & McConnaughay, 1992). As well as accelerating plant growth, higher resource levels generally accelerate interactions among individuals. For example, density-dependent mortality generally occurs at a higher rate in higher fertility environments, a phenomenon sometimes called the 'Sukatshew effect' (Harper, 1977). Similarly, high resource levels have generally been found to accelerate the process of size hierarchy formation in plant populations, such that a smaller number of individuals come to proportionately dominate biomass and reproductive output within the population (Weiner, 1985; Weiner & Thomas, 1986). High resource levels may similarly have predictable effects on density–yield relationships; to the extent that intra-specific density dependence drives population dynamics, higher resource levels can operate to increase the amplitude of population fluctuations (Symonides *et al.*, 1986; Thrall *et al.*, 1989). A number of predictions regarding population-level responses to elevated CO_2 emerge from this logic. Elevated CO_2 may enhance the rate of size-hierarchy formation and self-thinning in plant monocultures. By increasing the variance in reproductive output among individual plants, elevated CO_2 may also act to reduce effective population size in genetic terms (cf. Heywood, 1986). Finally, population fluctuations may be exaggerated by elevated CO_2, increasing the likelihood of local population extinction.

A few recent studies have examined some of these predictions. Bazzaz and colleagues (1992) investigated the effects of elevated CO_2 on the density–yield function of *Abutilon theophrasti*, a species that has previously served as a model in several detailed analyses of plant population dynamics (e.g. Thrall *et al.*, 1989; Pacala & Silander, 1990). At high CO_2 levels there were pronounced declines in reproductive output at high densities in *Abutilon* monocultures. In a simple difference equation model of population dynamics, simulated *Abutilon* populations under elevated CO_2 exhibited greater oscillations and wider ranges of parameter values resulting in local extinction than did populations under ambient CO_2 conditions. In another recent study, Morse & Bazzaz (1994) investigated the interactive effects of temperature and CO_2 on population size structure in *Abutilon theophrasti* (C_3) and *Amaranthus retroflexus* (C_4). In both species there was evidence for a higher rate of size hierarchy formation and increased mortality with increasing CO_2 concentration and temperature.

Possible evolutionary effects of elevated CO_2 have also begun to be addressed. There is a growing body of evidence for genetic variation in CO_2 responses within natural plant populations (Wulff & Alexander, 1985; Curtis *et al.*, 1994). In some cases variability within species is similar in magnitude to the well-documented variability in CO_2 responses among species. As in Fig. 4.5, three different maternal lines of yellow birch seedlings had different enhancements of biomass due to elevated CO_2. The existence of such variation suggests that long-term responses of vegetation may be affected by evolutionary responses of plant populations; however, we are lacking any selection or population-age experiments to assess directly the potential for such evolutionary responses. In addition to genotype-specific effects, elevated CO_2 might also affect selection rates by altering fecundity distributions within plant populations (cf. Thomas & Bazzaz, 1993). If elevated CO_2 generally increases variability in reproductive output, this could operate to greatly accelerate the selection process. Further investigations are necessary in order to evaluate the potential significance of such evolutionary responses to future vegetation composition and carbon sink strength.

Biodiversity and conservation implications of direct CO$_2$ effects on vegetation

The impacts of potential global warming on biodiversity have received considerable attention from researchers and conservationists (e.g. Peters & Lovejoy, 1992). In contrast, very little emphasis has been placed on potential direct effects of elevated CO$_2$ on biodiversity issues until very recently. This general area of inquiry is still at a stage in which the ratio of speculation to data is very high. In the next few paragraphs we take a critical look at some recent speculations regarding the effects of rising CO$_2$ on biodiversity, and offer our perspective on this and related issues.

Many interactions may be of importance in determining future responses of biodiversity to rising CO$_2$. This point is well illustrated by the results briefly summarized above on species-specific reductions in reproduction. Even if such effects are relatively rare, a substantial reproductive impairment of even 1% of the world's flora must be considered a major threat. Moreover, if most species display enhanced growth and reproduction, species suffering adverse effects would be expected to be particularly at risk.

It has recently been argued that some aspects of the direct effects of CO$_2$ on vegetation (particularly increased water use efficiency) may work to *enhance* biodiversity in natural systems (Woodward & Rochefort, 1991; Rochefort & Woodward, 1992; Woodward, 1992). The logic behind this prediction is that elevated CO$_2$ may enhance water use efficiency, permitting the development of vegetation with a higher leaf area index (LAI). Across biomes, it is often the case that high LAI systems have a higher local diversity of plant species or higher taxonomic groups. Thus, for example, enhanced water use efficiency might favour rain forest over savanna vegetation, resulting in locally increased plant diversity. Further work is needed on two aspects of this scenario: it is not clear that elevated CO$_2$ enhances water use efficiency at the system level, nor is it clear that enhanced LAI would be expressed as a change in fundamental vegetation type over the time scales of interest. Woodward's scenario also illustrates the importance of taking spatial and temporal scales into account in any attempt to understand possible effects of rising CO$_2$ on global biodiversity. For example, at a local scale replacement of savanna by rain forest may increase plant diversity. However, at a regional scale existing rain forest species would displace existing savanna species, thus leading to a net loss in species diversity. In this scenario extinction rates may well be enhanced by rising CO$_2$, while speciation rates would be little affected due to the shortness of the time scale of interest (50–200 years).

Rising CO_2 seems likely to have a net negative impact on global biodiversity, although the relative magnitude of this effect is completely unknown. Can anything more specific be concluded? In any analysis of potential effects of rising CO_2 on biodiversity, it seems essential to note that CO_2 generally enhances the growth of individual plants, and that this growth enhancement is highly species-dependent. CO_2 fertilization of ecosystems will thus enhance the performance of responsive species in the system, probably at the expense of less responsive species, as previously noted. There is some evidence to suggest that growth responses to CO_2 are largest in species with high intrinsic growth rates early in ontogeny (cf. Poorter, 1993), a characteristic common among early successional species (Bazzaz, 1979; but see El Kohen *et al.*, 1993). Early successional species are also in many cases more tolerant of a broad range of environmental conditions (Bazzaz, 1987). In sum, CO_2 fertilization seems likely to accelerate plant competition for other resources, and in many systems relatively early successional species may be differentially favoured. Such a pattern could contribute to a loss of biodiversity in many systems.

Another aspect of CO_2 effects on biodiversity that should be taken into account is the fact that this change will take place gradually, rather than as a sudden step increase. Under these conditions the relative responses of species at intermediate levels of CO_2 are as important as responses to extremes (Ackerly & Bazzaz, 1995). The overall form of relationships between plant performance and CO_2 may be similar across the majority of species, being concave and monotonic in the range of 200–2000 ppm (e.g. Hunt *et al.*, 1991). However, species-specific differences in parameters of such functions can result in changes in species composition that are not predictable from growth at set reference levels (e.g. 350 and 700 ppm). In experiments examining differential performance of species in dense polycultures, it is not unusual to find pronounced non-linearities in species performance, such that a given species performs best (or worst) at some intermediate CO_2 level (cf. Bazzaz & Garbutt, 1988).

The foregoing points emphasize the importance and complexity of potential effects of increasing CO_2 on plant community structure and biodiversity. A few modelling efforts have taken some initial steps at handling this complexity. Initial efforts to include CO_2 fertilization in forest simulation models included only a single growth multiplier, or β-factor, in the model, and have focused on ecosystem-level processes (Shugart & Emanuel, 1985; Luxmoore *et al.*, 1990; Post *et al.*, 1992). Simulation studies have not to date examined in detail consequences for forest community composition. As mentioned in a previous section, spatially explicit simulation models may prove a valuable tool for exploring the impact of altered physiologies on species composition in a future elevated CO$_2$ landscape (e.g. Bolker *et al.*, 1995).

Conclusion

At the beginning of this chapter we posed the question 'What research questions are likely to assume greatest importance in guiding CO$_2$ research in the near future?' In answering this question, we offer four general themes of research, though indeed additional important research foci may emerge. To understand the overall impact of elevated CO$_2$ on terrestrial ecosystems we should address:

1. *Interactions with other anthropogenic effects.* The impact of elevated atmospheric CO$_2$ on individual plant growth and ecosystem function cannot be considered in isolation of other components of our changing planet.
   - Some components of global as well as regional change (e.g. temperature and chronic nitrogen deposition) may lead to non-additive and/or unpredictable effects
   - Some anthropogenically derived factors (e.g. land use patterns) may override effects of CO$_2$. In such cases, we need to consider the effects of CO$_2$ in the context of the changes brought about by these overriding factors.

2. *Scaling empirical observations of CO$_2$ effects.* Predictions of the future impact of increasing atmospheric CO$_2$ concentrations need to be made carefully. Much of our understanding of the impacts of elevated CO$_2$ on plant growth is derived from controlled experiments on small-statured individuals. We therefore must search for robust principles by which to scale our empirical observations to longer-lived, larger, and more complex and interactive systems.

3. *Population biology and evolution.* Elevated CO$_2$ will also probably affect plant population structure and dynamics, micro-evolutionary processes and plant community structure. Some species and genotypes will probably be favoured at the expense of others. These changes may in sum contribute to declining global biodiversity, although the magnitude of such an effect is highly uncertain at present.

4. *Species functional diversity and future ecosystem function.* Elevated CO$_2$ does not affect all plant species identically. A better appreciation of the role this functional variation plays on potential future ecosystem structure (e.g. species composition) and function (e.g. NPP) needs to be obtained through coordinated experiments and modelling exercises.

We have highlighted in this review several effects of CO_2 on terrestrial vegetation that have important policy implications. Our discussion is by no means comprehensive; equally profound ecological effects of elevated CO_2 may also result from shifts in other trophic levels, such as pollinators and herbivores (e.g. Bazzaz & Fajer, 1992; Lincoln *et al.*, 1993). At this point we wish to revisit the general theme of a previous paper that posed the question: 'Is carbon dioxide a "good" greenhouse gas?' (Fajer & Bazzaz, 1992). We strongly believe that the general argument that increasing CO_2 will lead to lusher natural ecosystems (the 'sanguine scenario' championed by some) has often been overstated. The issues raised here bolster the argument of Fajer & Bazzaz (1992) that caution is warranted in appraising a high-CO_2 world. It seems to us that the essence of 'anticipatory science' is to realistically consider potentially adverse as well as beneficial consequences of rising CO_2 and to evaluate critically possible consequences from a variety of perspectives. We believe that future CO_2 research must focus on issues beyond simply quantifying the global carbon budget – an arena which has recently greatly influenced research direction and funding. The global carbon cycle may be thought of as one currency that integrates the net effects of CO_2 on vegetation; however, we must not lose track of other currencies that are also of vital importance to the future of vegetation and life on earth.

Acknowledgement

We would like to acknowledge the contribution of our laboratory group to the research and ideas in this chapter. The research was in part supported by the United States Department of Energy. We also thank Bernard Saugier for comments on an earlier version.

References

Aber, J. D., Magill, A., Boone, R., Melillo, J. M., Steudler, P. & Bowden, R. (1993). Plant and soil responses to chronic nitrogen additions at the harvard forest, Massachusetts. *Ecological Applications*, **3**, 156–66.

Aber, J. D., Melillo, J. M., Nadelhoffer, K. J., Pastor, J. & Boone, R. D. (1991). Factors controlling nitrogen cycling and nitrogen saturation in northern temperate forest ecosystems. *Ecological Applications*, **1**, 303–15.

Aber, J. D., Nadelhoffer, K. J., Steudler, P. & Melillo, J. (1989). Nitrogen saturation in northern forest ecosystems. *Bioscience*, **39**, 378–86.

Ackerly, D. D. & Bazzaz, F. A. (1995). Plant growth and reproduction along CO_2 gradients: non-linear responses and implications for community change. *Global Change Biology*, **1**, 199–207.

Ackerly, D. D., Coleman, J. S., Morse, S. R. & Bazzaz, F. A. (1992). CO_2 and temperature effects on leaf area production in two annual plant species. *Ecology*, **73**, 1260–9.

Acock, B., Reddy, V. R., Hodges, H. F., Baker, D. N. & McKinnon, J. M. (1985). Photosynthetic response of soybean canopies to full-season carbon dioxide enrichment. *Agronomy Journal*, **77**, 942–7.

Allen, L. H. J., Boote, K. J., Jones, P. H., Jones, J. W., Rowland-Bamford, A. J., Bowes, G., Graetz, D. A. & Reddy, K. R. (1989). *Temperature and CO$_2$ effects on rice: 1988*. Washington, DC: US Department of Energy, Office of Energy Research, Carbon Dioxide Research Division.

Amthor, J. S. (1989). *Respiration and Crop Productivity*. New York: Springer-Verlag.

Amthor, J. S. (1991). Respiration in a future, higher-CO$_2$ world. *Plant, Cell and Environment*, **14**, 13–20.

Amthor, J. S. (1993). Effects of CO$_2$ enrichment on higher plant respiration. In *Design and Execution of Experiments on CO$_2$ Enrichment*, ed. E.-D. Schulze & H. A. Mooney. Brussels–Luxembourg: Commission of the European Communities.

Arnone, J. A. I. & Gordon, J. C. (1990). Effect of nodulation, nitrogen fixation and carbon dioxide enrichment on the physiology, growth and dry mass allocation of seedlings of *Alnus rubra* Bong. *New Phytologist*, **116**, 55–66.

Arnone, J. A. & Körner, C. (1993). Influence of elevated CO$_2$ on canopy development and red : far-red ratios in two-storied stands of *Ricinus communis*. *Oecologia*, **94**, 510–15.

Arp, W. J. (1991). Effects of source–sink relations on photosynthetic acclimation to elevated CO$_2$. *Plant, Cell and Environment*, **14**, 869–75.

Arp, W. J., Drake, B. G., Pockman, W. T., Curtis, P. S. & Whigham, D. F. (1993). Interactions between C$_3$ and C$_4$ salt marsh species during four years of exposure to elevated atmospheric CO$_2$. *Vegetatio*, **104/105**, 133–43.

Barton, C. V. M. & Jarvis, P. G. (1993). Description of branch bags and a system to control the CO$_2$ concentration within them. *Vegetatio*, **104/105**, 450–1.

Barton, C. V. M., Lee, H. S. J. & Jarvis, P. G. (1993). A branch bag and CO$_2$ control system for long-term CO$_2$ enrichment of mature Sitka spruce [*Picea sitchensis* (Bong.) Carr.]. *Plant, Cell and Environment*, **16**, 1139–48.

Bassow, S. L., McConnaughay, K. D. M. & Bazzaz, F. A. (1994). The response of temperate tree seedlings grown in elevated CO$_2$ atmospheres to extreme temperature events. *Ecological Applications*, **4**, 593–603.

Bazzaz, F. A. (1973). Photosynthesis of *Ambrosia artemisiifolia* L. Plants grown in greenhouse and in the field. *American Midland Naturalist*, **90**, 186–90.

Bazzaz, F. A. (1979). The physiological ecology of plant succession. *Annual Review of Ecology and Systematics*, **10**, 351–71.

Bazzaz, F. A. (1990). The response of natural ecosystems to the rising global CO$_2$ levels. *Annual Review of Ecology and Systematics*, **21**, 167–96.

Bazzaz, F. A. (1987). Experimental studies on the evolution of niche in successional plant populations: a synthesis. In *Colonization, Succession and Stability*, ed. A. J. Gray, M. J. Crawley & P. J. Edwards, British Ecological Society Symposium, vol. 26, pp. 245–72.

Bazzaz, F. A., Ackerly, D. D., Woodward, F. I. & Rochefort, L. (1992). CO$_2$ enrichment and dependence of reproduction on density in an annual plant and a simulation of its population dynamics. *Journal of Ecology*, **80**, 643–52.

Bazzaz, F. A., Coleman, J. S. & Morse, S. R. (1990). Growth responses of seven major co-occurring tree species of the northeastern United States to elevated CO$_2$. *Canadian Journal of Forest Research*, **20**, 1479–84.

Bazzaz, F. A. & Fajer, E. D. (1992). Plant life in a CO$_2$-rich world. *Scientific American*, **266**, 68–74.

Bazzaz, F. A. & Garbutt, K. (1988). The response of annuals in competitive neighborhoods: effects of elevated CO$_2$. *Ecology*, **69**, 937–46.

Bazzaz, F. A. & McConnaughay, K. D. M. (1992). Plant–plant interactions in elevated CO$_2$ environments. *Australian Journal of Botany*, **40**, 547–63.

Bazzaz, F. A. & Miao, S. L. (1993). Successional status, seed size, and response of tree seedlings to CO$_2$, light and nutrients. *Ecology*, **74**, 104–12.

Bazzaz, F. A., Miao, S. L. & Wayne, P. M. (1993). CO$_2$-induced growth enhancements of co-occurring tree species decline at different rates. *Oecologia*, **96**, 478–82.

Berry, J. A. & Raison, J. K. (1981). Response of macrophytes to temperature. (N. Lange Osmond and Ziegler) In *Physiological Plant Ecology. I. Responses to the Physical Environment*, ed. N. Lange *et al.*, pp. 278–338. Berlin: Springer Verlag.

Boardman, N. K. (1981). Comparative photosynthesis of sun and shade plants. *Annual Review of Ecology and Systematics*, **28**, 355–77.

Bolker, B. M., Pacala, S. W., Bazzaz, F. A., Canham, C. D. & Levin, S. A. (1995). Species diversity and ecosystem response to carbon dioxide fertilization: conclusions from a temperate forest model. *Global Change Biology*, **1**, 373–81.

Bowden, R. D., Melillo, J. M., Steudler, P. A. & Aber, J. D. (1991). Effects of nitrogen addition on annual nitrous oxide fluxes from temperate forest soils in the northeastern United States. *Journal of Geophysical Research*, **96**, 9321–8.

Bowes, G. (1991). Growth at elevated CO$_2$: photosynthetic responses mediated through Rubisco. *Plant, Cell and Environment*, **14**, 795–806.

Brown, K. R. (1991). Carbon dioxide enrichment accelerates the decline in nutrient status and relative growth rate of *Populus tremuloides* Michx. seedlings. *Tree Physiology*, **8**, 161–73.

Brown, K. & Higginbotham, K. O. (1986). Effects of carbon dioxide enrichment and nitrogen supply on growth of boreal tree seedlings. *Tree Physiology*, **2**, 223–32.

Bunce, J. A. (1990). Short- and long-term inhibition of respirator carbon dioxide efflux by elevated carbon dioxide. *Annals of Botany*, **65**, 637–42.

Bunce, J. A. (1992). Stomatal conductance, photosynthesis and respiration of temperate deciduous tree seedlings grown outdoors at an elevated concentration of carbon dioxide. *Plant, Cell and Environment*, **15**, 541–9.

Burns, R. M. & Honkala, B. H. (1990). *Silvics of North American Trees, vol. 2, Hardwoods*. Washington, DC: US Department of Agriculture.

Carlson, R. W. & Bazzaz, F. A. (1986). Plant response to SO$_2$ and CO$_2$. In *Sulfur Dioxide and Vegetation*, ed. W. E. Winner *et al.*, pp. 313–31. Stanford: Stanford University Press.

Ceulemans, R. & Mousseau M., (1994). Effects of elevated atmospheric CO$_2$ on woody plants. *New Phytology*, **127**, 425–46.

Chiariello, N. & Roughgarden, J. (1984). Storage allocation in seasonal races of an annual plant: optimal versus actual allocation. *Ecology*, **65**, 1290–301.

Clark, J. S. (1993). Scaling at the population level: effects of species composition and population structure. In *Scaling Physiological Processes. Leaf to Globe*, ed. J. R. Ehleringer & C. B. Field, pp. 255–85. San Diego: Academic Press.

Cohen, D. (1971). Maximizing final yield when growth is limited by time or by limiting resources. *Journal of Theoretical Biology*, **33**, 299–307.

Coleman, J. S. & Bazzaz, F. A. (1992). Effects of CO$_2$ and temperature on growth and resource use of co-occurring

C$_3$ and C$_4$ annuals. *Ecology*, **73**, 1244–59.

Coleman, J. S., Rochefort, L., Bazzaz, F. A. & Woodward, F. I. (1991). Atmospheric CO$_2$, plant nitrogen status and the susceptibility of plants to an acute increase in temperature. *Plant, Cell and Environment*, **14**, 667–74.

Curtis, P., Drake, B. G., Leadley, P. W., Arp, W. J. & Whigham, D. F. (1989). Growth and senescence in plant communities exposed to elevated CO$_2$ concentration on an estuarine marsh. *Oecologia*, **78**, 20–6.

Curtis, P. S., Drake, B. G. & Whigham, D. F. (1989). Nitrogen and carbon dynamics in C$_3$ and C$_4$ estuarine marsh plants grown under elevated CO$_2$ in situ. *Oecologia*, **78**, 297–301.

Curtis, P. S., Snow, A. A. & Miller, A. S. (1994). Genotype-specific effects of elevated CO$_2$ on fecundity in wild radish (*Raphanus raphanistrum*). *Oecologia*, **97**, 100–5.

Davis, M. B. (1989). Insights from paleoecology on global change. *Ecological Society of America Bulletin*, **70**, 222–8.

Davis, M. B., Woods, K. D., Webb, S. L. & Futyma, R. P. (1986). Dispersal versus climate: expansion of *Fagus* and *Tsuga* into the Upper Great Lakes region. *Vegetatio*, **67**, 93–103.

Davis, M. B. & Zabinski, C. (1992). *Changes in Geographical Range Resulting from Greenhouse Warming: Effects on Biodiversity in Forests*. New Haven: Yale University Press.

Drake, B. G. & Leadley, P. W. (1991). Canopy photosynthesis of crops and native plant communities exposed to long-term elevated CO$_2$. *Plant, Cell and Environment*, **14**, 853–60.

Dufrêne, E., Pontailler, J.-Y. & Saugier, B. (1993). A branch bag technique for simultaneous CO$_2$ enrichment and assimilation measurements on beech (*Fagus sylvatica* L.). *Plant, Cell and Environment*, **16**, 1131–8.

Eamus, D. (1991). The interaction of rising CO$_2$ and temperature with water use efficiency. *Plant, Cell and Environment*, **14**, 843–52.

Eamus, D. & Jarvis, P. G. (1989). The direct effects of increase in the global atmospheric CO$_2$ concentration on natural and commercial temperate trees and forests. *Advances in Ecological Research*, **19**, 1–55.

Ehleringer, J. R. & Field, C. B. (ed.) (1993). *Scaling Physiological Processes: Leaf to Globe*. San Diego: Academic Press.

El Kohen, A., Venet, L. & Mousseau, M. (1993). Growth and photosynthesis of two deciduous forest species at elevated carbon dioxide. *Functional Ecology*, **7**, 480–6.

Enoch, H. Z., Rylski, I. & Spigelman, M. (1976). CO$_2$ enrichment of strawberry and cucumber plants grown in unheated greenhouse in Israel. *Scientific Horticulture*, **5**, 33–41.

Fajer, E. D. & Bazzaz, F. A. (1992). Is carbon dioxide a 'good' greenhouse gas? Effects of increasing carbon dioxide on ecological systems. *Global Environmental Change*, **2**, 301–10.

Fajer, E. D., Bowers, M. D. & Bazzaz, F. A. (1992). The effect of nutrients and enriched CO$_2$ environments on production of carbon-based allelochemicals in *Plantago*: a test of the carbon/nutrient balance hypothesis. *American Naturalist*, **140**, 707–23.

Farrar, J. F. & Williams, M. L. (1991). The effects of increased atmospheric carbon dioxide and temperature on carbon partitioning, source–sink relations and respiration. *Plant, Cell and Environment*, **14**, 819–30.

Field, C. B. (1991). Ecological scaling of carbon gain to stress and resource availability. In *Response of Plants to Multiple Stresses*, ed. H. A. Mooney, W. E. Winner & E. J. Pell, pp. 35–65. San Diego: Academic Press.

Field, C. B., Chapin, F. S. III., Matson, P. A. & Mooney, H. A. (1992). Responses of terrestrial ecosystems to the changing atmosphere: a resource-based approach. *Annual Review of Ecology and Systematics*, **23**, 201–35.

Ford, H. (1982). Leaf demography and the plastochron index. *Biological Journal of the Linnean Society*, **17**, 361–73.

Galloway, J. N., Likens, G. E. & Hawley, M. E. (1984). Acid precipitation: natural versus anthropogenic components. *Science*, **226**, 829–31.

Garbutt, K. & Bazzaz, F. A. (1984). The effects of elevated CO_2 on plants. III. Flower, fruit and seed production and abortion. *New Phytology*, **98**, 433–46.

Garbutt, K., Williams, W. E. & Bazzaz, F. A. (1990). Analysis of the differential response of five annuals to elevated CO_2 during growth. *Ecology*, **71**, 1185–94.

Gaudillere, J. & Mousseau, M. (1989). Short term effect of CO_2 enrichment on leaf development and gas exchange of young populars (*Populus euroamericana* cv I214). *Acta Oecologica-Oecologia Plantarum*, **10**, 95–105.

Gifford, R. M. (1994). The global carbon cycle: a viewpoint on the missing sink. *Australian Journal of Plant Physiology*, **21**, 1–15.

Gifford, R. M., Lambers, H. & Morison, I. L. (1985). Respiration of crop species under CO_2 enrichment. *Physiologia Plantarum*, **63**, 351–6.

Gifford, R. M. & Rawson, H. M. (1992). *Investigation of wild and domesticated vegetation in CO_2 enriched greenhouses.* IGBP workshop on Design and execution of experiments on CO_2 enrichment, Weidenberg, Germany, 26–30 October 1992.

Givnish, T. J. (1988). Adaptation to sun and shade: a whole plant perspective. *Australian Journal of Plant Physiology*, **15**, 63–92.

Grace, J. (1988). Temperature as a determinant of plant productivity. (L. a. Woodward) *Plants and Temperature*.

Graham, R. L., Turner, M. G. & Dale, V. H. (1990). How increasing CO_2 and climate change affect forests. *BioScience*, **40**, 575–87.

Grulke, N. E., Riechkers, G. H., Oechel, W. C., Hjelm, U. & Jaeger, C. (1990). Carbon balance in tussock tundra under ambient and elevated atmospheric CO_2. *Oecologia*, **83**, 485–94.

Hanson, P. J., Samuelson, L. J., Wullschleger, S. D., Tabberer, T. A. & Edwards, G. S. (1994). Seasonal patterns of light-saturated photosynthesis and leaf conductance for mature and seedling *Quercus rubra* L. foliage: differential sensitivity to ozone exposure. *Tree Physiology*, **14**, 1351–66.

Harper, J. L. (1977). *The Population Biology of Plants*. London: Academic Press.

Harper, J. L. & White, J. (1974). The demography of plants. *Annual Review of Ecology and Systematics*, **5**, 419–63.

Hendrey, G. R., Lewin, K. F. & Nagy, J. (1993). Free air carbon dioxide enrichment: development, progress, results. *Vegetatio*, **104/105**, 17–31.

Hesketh, J. D. & Hellmers, H. (1973). Floral initiation in four plant species growing in CO_2 enriched air. *Environmental Control in Biology*, **11**, 51–3.

Heywood, J. S. (1986). The effect of plant size variation on genetic drift in populations of annuals. *American Naturalist*, **127**, 851–61.

Hickman, J. C. (1975). Environmental unpredictability and plastic energy allocation strategies in the annual *Polygonum cascadense* (Polygonaceae). *Journal of Ecology*, **63**, 689–701.

Hobbie, J. E. & Melillo, J. M. (1984). Comparative carbon and energy flow in ecosystems. Role of microbes in global carbon cycling. *Current Perspectives in Microbial Ecology*, **6**, 389–93.

Hogan, K. P., Smith, A. P. & Ziska, L. H. (1991). Potential effects of elevated CO_2 and changes in temperature on tropical plants. *Plant, Cell and Environment*, **14**, 763–78.

Hollinger, D. Y. (1987). Gas exchange and dry matter allocation responses to elevation of atmospheric CO_2 concentrations in seedlings of three tree species. *Tree Physiology*, **3**, 193–202.

Horton, R. F. (1985). Carbon dioxide flux and ethylene production in leaves. In *Ethylene and Plant Development*, ed. J. A. Roberts & G. A. Tucker, pp. 37–46. London: Butterworth.

Houghton, J. T. (1991). The role of forests in affecting the greenhouse gas composition of the atmosphere. In *Global Climate Change and Life on Earth*, ed. R. L. Wyman, pp. 43–56. New York: Chapman and Hall.

Houghton, J. T., Jenkins, G. J. & Ephraums, J. J. (ed.) (1990). *Climate Change: The IPCC Scientific Assessment*. Cambridge: Cambridge University Press.

Hovland, A. S. & Dybing, C. D. (1973). Cyclic flowering patterns in flax as influenced by environment and plant growth regulators. *Crop Science*, **13**, 380–4.

Hrubec, T. C., Robinson, J. M. & Donaldson, R. P. (1985). Effects of CO$_2$ enrichment and carbohydrate content on the dark respiration of soybeans. *Plant Physiology*, **79**, 684–9.

Hudson, R. J. M., Gherini, S. A. & Goldstein, R. A. (1994). Modeling the global carbon cycle: Nitrogen fertilization of the terrestrial biosphere and the 'missing' CO$_2$ sink. *Global Biogeochemical Cycles*, **8**, 307–33.

Hunt, R., Hand, D. W., Hannah, M. A. & Neal, A. M. (1991). Response to CO$_2$ enrichment in 27 herbaceous species. *Functional Ecology*, **5**, 410–21.

Idso, S. B. & Kimball, B. A. (1989). Growth response of carrot and radish to atmospheric CO$_2$ enrichment. *Environmental and Experimental Botany*, **29**, 135–9.

Idso, S. B. & Kimball, B. A. (1993). Tree growth in carbon dioxide enriched air and its implications for global carbon cycling and maximum levels of atmospheric carbon dioxide. *Global Biogeochemical Cycles*, **7**, 537–55.

Idso, S. B., Kimball, B. A., Anderson, M. G. & Mauney, J. R. (1986). Effects of atmospheric CO$_2$ enrichment on plant growth: the interactive role of air temperature. *Agriculture, Ecosystems and the Environment*, **20**, 1–10.

Idso, S. B., Kimball, B. A. & Mauney, J. R. (1987). Atmospheric carbon dioxide enrichment effects on cotton midday foliage temperature: implications for plant water use and crop yield. *Agronomy Journal*, **79**, 667–72.

Imazu, T., Yabuki, K. & Oda, Y. (1967). Studies on the influence of carbon dioxide concentration on the growth, flowering and fruit setting of eggplant (*Solanum melongena* L.). *Journal of the Japanese Society for Horticultural Science*, **36**, 222–7.

Jarvis, P. G. (1993). Prospects for bottom-up models. In *Scaling Physiological Processes. Leaf to Globe*, ed. J. R. Ehleringer and C. B. Field, pp. 115–26. San Diego: Academic Press.

Johnson, D. W., Ball, J. T. & Walker, R. F. (1992). Effects of CO$_2$, nitrogen, and phosphorus on *Ponderosa* pine seedlings. II. Nutrient uptake and soil interactions. *Bulletin of the Ecological Society of America*, **73**, 224.

Jones, H. G. (1992). *Plants and Microclimate – A Quantitative Approach to Environmental Plant Physiology*, 2nd edn. Cambridge: Cambridge University Press.

Jones, P., Allen, L. H. & Jones, J. W. (1985a). Responses of soybean canopy photosynthesis and transpiration to whole day temperature changes in different CO$_2$ environments. *Agronomy Journal*, **77**, 242–9.

Jones, P., Allen, L. H. Jr., Jones, J. W., Boote, K. J. & Campbell, W. J. (1984). Soybean canopy growth, photosynthesis, and transpiration responses to whole season carbon dioxide enrichment. *Agronomy Journal*, **76**, 633–7.

Jones, P., Allen, L. H., Jones, J. W. & Valle, R. (1985b). Photosynthesis and transpiration responses of soybean canopies to short- and long-term CO$_2$

treatments. *Agronomy Journal*, **77**, 119–26.

Jordan, D. B. & Ogren, W. L. (1984). The CO_2/O_2 specificity of ribulose-1,5-bisphosphate concentration, pH and temperature. *Planta*, **161**, 308–13.

Jurik, T. W. (1991). Population distributions of plant size and light environment of giant ragweed (*Ambrosia trifida* L.) at three densities. *Oecologia*, **87**, 539–50.

Jurik, T. W., Weber, J. A. & Gates, D. M. (1988). Effects of temperature and light on photosynthesis of dominant species of a northern hardwood forest. *Botanical Gazette*, **149**, 203–8.

Katz, R. W. & Brown, B. G. (1992). Extreme events in a changing climate: variability is more important than averages. *Climatic Change*, **21**, 289–302.

Kauppi, P. E., Mielikäinen, K. & Kuusela, K. (1992). Biomass and carbon budget of European forests, 1971 to 1990. *Science*, **256**, 70–4.

King, D. & Roughgarden, J. (1982). Multiple switches between vegetative and reproductive growth in annual plants. *Theoretical Population Biology*, **21**, 194–204.

Kohlmaier, G. H. (1991). Reply to Idso. *Tellus*, **43B**, 342–6.

Kohlmaier, G. H., Sire, E.-O. & Janacek, A. (1989). Modelling the seasonal contribution of a CO_2 fertilization effect of the terrestrial vegetation to the amplitude increase in atmospheric CO_2 at Mauna Loa Observatory. *Tellus*, **41B**, 487–510.

Koike, T. (1988). Leaf structure and photosynthetic performance as related to the forest succession of deciduous broadleaved trees. *Plant Species Biology*, **3**, 77–87.

Körner, C. (1993). CO_2 fertilization: the great uncertainty in future vegetation development. In *Vegetation Dynamics and Global Change*, ed. A. M. Solomon and H. H. Shugart. London: Chapman and Hall.

Körner, C. & Arnone III, J. A. (1992). Responses to elevated carbon dioxide in artificial tropical ecosystems. *Science*, **257**, 1672–5.

Larcher, W. (1983). *Physiological Plant Ecology*. Berlin: Springer-Verlag.

Larigauderie, A., Hilbert, D. W. & Oechel, W. C. (1988). Effect of CO_2 enrichment and nitrogen availability on resource acquisition and resource allocation in a grass, *Bromus mollis*. *Oecologia*, **77**, 544–9.

Lea, R., Tierson, W. C., Bickelhaupt, D. H. & Leaf, A. L. (1980). Differential foliar response of northern hardwoods to fertilization. *Plant and Soil*, **54**, 419–39.

Lee, H. S. J. & Barton, C. V. M. (1993). Comparative studies on elevated CO_2 using open-top chambers, tree chambers and branch bags. In *Design and Execution of Experiments on CO_2 Enrichment*, ed. E.-D. Schulze & H. A. Mooney. Brussels–Luxembourg: Commission of the European Communities.

Lei, T. T. & Lechowicz, M. J. (1990). Shade adaptation and shade tolerance in saplings of three *Acer* species from eastern North America. *Oecologia*, **84**, 224–8.

Lekkerkerk, L. J. A., Van de Geijn, S. C. & Van Veen, J. A. (1990). *Effects of Elevated Atmospheric CO_2 Levels on the Carbon Economy of a Soil Planted with Wheat*. New York: Wiley.

Lemon, E. R. (1983). *CO_2 and Plants*. Boulder, Colorado: National Center for Atmospheric Research.

Lincoln, D. E., Fajer, E. D. & Johnson, R. H. (1993). Plant–insect herbivore interactions in elevated CO_2 environments. *Trends in Ecology and Evolution*, **8**, 64–8.

Long, S. P. (1991). Modification of the response of photosynthetic productivity to rising temperature by atmospheric CO_2 concentrations: Has its importance been understood? *Plant, Cell and Environment*, **14**, 729–39.

Luxmoore, R. J., King, A. W. & Tharp, M. L. (1990). Approaches to scaling up physiologically based soil-plant models in space and time. *Tree Physiology*, **9**, 281–92.

Luxmoore, R. J., Tharp, M. L. & West, D. C. (1990). Simulating the physiological basis of tree-ring responses to environmental changes. In *Process Modeling of Forest Growth Responses to Environmental Stress*, ed. R. K. Dixon, R. S. Meldahl, G. A. Ruark & W. G. Warren. Portland, Oregon: Timber Press.

Marc, J. & Gifford, R. M. (1984). Floral initiation in wheat, sunflower, and sorghum under carbon dioxide enrichment. *Canadian Journal of Botany*, **62**, 9–14.

Mearns, L. O., Katz, R. W. & Schneider, S. H. (1984). Extreme high-temperature events: changes in their probabilities with changes in mean temperature. *Journal of Climate and Applied Meteorology*, **23**, 1601–13.

Mitchell, J. F. B., Mannabe, S., Tokioka, T. & Meleshko, V. (1990). Equilibrium climate change. In *Climate Change – The IPCC Scientific Assessment*, ed. J. T. Houghton, G. J. Jenkins & J. J. Ephraums. Cambridge: Cambridge University Press.

Morse, S. R. & Bazzaz, F. A. (1994). Elevated CO$_2$ and temperature alter recruitment and size hierarchies in C$_3$ and C$_4$ annuals. *Ecology*, **75**, 966–75.

Mortensen, L. M. (1992). Effects of ozone concentration on growth of tomato at various light, air humidity and carbon dioxide levels. *Scientia Horticuturae* (Amsterdam), **49**, 17–24.

Nijs, I., Impens, I. & Behaeghe, T. (1989). Leaf and canopy responses of *Lolium perenne* to long-term elevated atmospheric carbon-dioxide concentration. *Planta*, **177**, 312–20.

Norby, R. J., Gunderson, C. A., Wullschleger, S. D., O'Neill, E. G. & McCracken, M. K. (1992). Productivity and compensatory responses of yellow-poplar trees in elevated CO$_2$. *Nature*, **357**, 322–4.

Norby, R. J. & O'Neill, E. G. (1989). Growth dynamics and water use of seedlings of *Quercus alba* L. in CO$_2$-enriched atmospheres. *New Phytology*, **111**, 491–500.

Norby, R. J., O'Neill, E. G., Hood, W. G. & Luxmoore, R. J. (1987). Carbon allocation, root exudation and mycorrhizal colonization of *Pinus echinata* seedlings grown under elevated CO$_2$ enrichment. *Tree Physiology*, **3**, 203–10.

Norman, J. M. (1993). Scaling processes between leaf and canopy levels. In *Scaling Physiological Processes. Leaf to Globe*, ed. J. R. Ehleringer & C. B. Field, pp. 41–76. San Diego: Academic Press.

Oechel, W. C., Hastings, S. J., Vourlitis, G., Jenkins, M., Riechers, G. & Grulke, N. (1993). Recent change of arctic tundra ecosystems from a net carbon dioxide sink to a source. *Nature*, **361**, 520–3.

Overdieck, D. (1990). Direct effects of elevated CO$_2$ concentration levels on grass and clover in 'model ecosystems'. In *Expected Effects of Climatic Change on Marine Coastal Ecosystems*, ed. J. J. Beukema, W. J. Wolff & J. J. W. M. Brouns, pp. 41–7. Dordrecht: Kluwer Academic Publishers.

Pacala, S. W. & Silander, J. A. J. (1990). Field tests of neighborhood population dynamic models of two annual weed species. *Ecological Monographs*, **60**, 113–34.

Pacala, S. W., Canham, C. D. & Silander, J. A. (in press). Forest models defined by field measurements. I. The design of a northeastern forest simulator. *Canadian Journal of Forest Science*.

Palloix, A., Herve, Y., Knox, R. B. & Dumas, C. (1985). Effect of carbon dioxide and relative humidity on self-incompatibility in cauliflower, *Brassica oleracea*. *Theoretical and Applied Genetics*, **70**, 628–33.

Parrish, J. A. D. & Bazzaz, F. A. (1985).

Nutrient content of *Abutilon theophrasti* seeds and the competitive ability of the resulting plants. *Oecologia*, **65**, 247–51.

Pastor, J. (1993). Northward march of spruce. *Nature*, **361**, 208–9.

Pastor, J. & Post, W. M. (1988). Responses of northern forests to CO_2-induced climate change. *Nature*, **334**, 55–8.

Patterson, D. T. & Flint, E. P. (1990). Implications of increasing CO_2 and climate change for plant communities and competition in natural and managed ecosystems. In *Impact of CO_2, Trace Gases, and Climate Change on Global Agriculture*, ed. B. A. Kimball, N. J. Rosenberg and J. L. H. Allen, pp. 83–110. Madison, WI: American Society of Agronomists.

Payer, H.-D., Köfferlein, M., Seckmeyer, G., Seidlitz, H., Strude, D. & Thiel, S. (1993). Controlled environment chambers for experimental studies on plant responses to CO_2 and interactions with pollutants. In *Design and Execution of Experiments on CO_2 Enrichment*, ed. E.-D. Schulze & H. A. Mooney. Brussels–Luxembourg: Commission of the European Communities.

Peters, R. H. (1991). *A Critique for Ecology*. Cambridge: Cambridge University Press.

Peters, R. L. & Lovejoy, T. E. (1992). *Global Warming and Biological Diversity*. New Haven: Yale University Press.

Poorter, H. (1993). Interspecific variation in the growth response of plants to an elevated ambient CO_2 concentration. *Vegetatio*, **104/105**, 77–97.

Poorter, H., Pot, S. & Lambers, H. (1988). The effect of an elevated atmospheric CO_2 concentration on growth, photosynthesis and respiration of *Plantago major*. *Physiologia Plantarum*, **73**, 533–59.

Post, W. M., Pastor, J., King, A. W. & Emanuel, W. R. (1992). Aspects of the interaction between vegetation and soil under global change. *Water, Air, and Soil Pollution*, **64**, 345–63.

Reekie, E. G. & Bazzaz, F. A. (1989). Competition and patterns of resource use among seedlings of five tropical trees grown at ambient and elevated CO_2. *Oecologia*, **79**, 212–22.

Reekie, E. G. & Bazzaz, F. A. (1991). Phenology and growth in four annual species grown in ambient and elevated CO_2. *Canadian Journal of Botany*, **69**, 2475–81.

Reekie, J. Y. C., Hicklenton, R. P. & Reekie, E. G. (1994). Effects of elevated CO_2 on time of flowering in four short-day and four long-day species. *Canadian Journal of Botany*, **72**, 533–8.

Reynolds, J. F., Chen, J. L., Harley, P. C., Hilbert, D. W. & Tenhunen, J. D. (1992). Modeling the effects of elevated carbon dioxide on plants: extrapolating leaf response to a canopy. *Agricultural and Forest Meteorology*.

Rind, G., Goldberg, R. & Ruedy, R. (1989). Change in climate variability in the 21st century. *Climatic Change*, **14**, 5–37.

Rochefort, L. & Bazzaz, F. A. (1992). Growth response to elevated CO_2 in seedlings of four co-occurring birch species. *Canadian Journal of Forest Science*, **22**, 1583–7.

Rochefort, L. & Woodward, F. I. (1992). Effects of climate change and a doubling of CO_2 on vegetation diversity. *Journal of Experimental Botany*, **43**, 1169–80.

Rogers, H. H., Runion, G. B. & Krupa, S. V. (1993). Plant responses to atmospheric CO_2 enrichment with emphasis on roots and the rhizosphere. *Environmental Pollution*.

Rotty, R. M. & Marland, G. (1986). Production of CO_2 from fossil fuel burning by fuel type, 1860–1982. ORNL.

Ryan, M. G. (1991). Effects of climate change on plant respiration. *Ecological Applications*, **1**, 157–67.

Sage, R. F. & Pearcy, R. W. (1987). The nitrogen use efficiency of C_3 and C_4 plants. I. leaf nitrogen, growth and biomass partitioning in *Chemopodium album* L. and *Amaranthus retroflexus* L. *Plant Physiology*, **84**, 954–8.

Sage, R. F., Sharkey, T. D. & Seemann, J. R. (1988). The *in vivo* response of the ribulose-1,5-bisphosphate carboxylase activation state and the pool sizes of photosynthetic metabolites to elevated CO_2 in *Phaseolus vulgaris* L. *Plants*, **174**, 407–16.

Sarmiento, J. L. (1993). Atmospheric CO_2 stalled. *Science*, **365**, 697–8.

Sasek, T. W., Delucia, E. H. & Strain, B. R. (1985). Reversibility of photosynthetic inhibition of cotton after long-term exposure to elevated CO_2 concentrations. *Plant Physiology*, **78**, 619–22.

Schimel, D. S. (1993). Population and community process in the response of terrestrial ecosystems to global change. In *Biotic Interactions and Global Change*, ed. P. M. Karieva, J. G. Kingsolver & R. B. Huey. Sunderland, MA: Sinauer Associates.

Schulze, E.-D. (1989). Air pollution and forest decline in a spruce *Picea abies* forest. *Science*, **244**, 776–83.

Shugart, H. H. & Emanuel, W. R. (1985). Carbon dioxide increase: The implications at the ecosystem level. *Plant, Cell and Environment*, **8**, 381–6.

Solomon, A. M. (1986). Transient response of forests to CO_2-induced climate change: simulation modeling experiments in eastern North America. *Oecologia*, **68**, 567–79.

Spencer, W. & Bowes, G. (1986). Photosynthesis and growth of water hyacinth under CO_2 enrichment. *Plant Physiology*, **82**, 528–33.

St Omer, L. & Horvath, S. (1983). Elevated carbon dioxide concentrations and whole plant senescence. *Ecology*, **64**, 1311–14.

Stitt, M. (1991). Rising CO_2 levels and their potential significance for carbon flow in photosynthetic cells. *Plant, Cell and Environment*, **14**, 741–62.

Strain, B. R. & Bazzaz, F. A. (1983). Terrestrial plant communities. In *The*

Response of Plants to Rising Levels of Atmospheric Carbon Dioxide, ed. E. R. Lemon, pp. 177–222. Washington, DC: AAAS.

Symonides, E., Silvertown, J. & Andreasen, V. (1986). Population cycles caused by overcompensating density-dependence in an annual plant. *Oecologia*, **71**, 156–8.

Tans, P. P., Fung, I. Y. & Takahaski, T. (1990). Observational constraints on the global atmospheric CO_2 budget. *Science*, **247**, 1431–8.

Teramura, A. H., Sullivan, J. H. & Ziska, L. H. (1990). Interaction of elevated UV-B radiation and carbon dioxide on productivity, photosynthetic characteristics in wheat, rice and soybean. *Plant Physiology*, **94**, 470–5.

Thomas, S. C. (1993). Interspecific allometry in Malaysian rain forest trees. Ph.D. thesis, Harvard University, Cambridge, Massacusetts.

Thomas, S. C. & Bazzaz, F. A. (1993). The genetic component in plant size hierarchies: norms of reaction to density in a *Polygonum* species. *Ecological Monographs*, **63**, 231–50.

Thrall, P. H., Pacala, S. W. & Silander, J. A. (1989). Oscillatory dynamics in populations of an annual weed species. *Abutilon theophrasti. Journal of Ecology*, **77**, 1135–49.

Tissue, D. T. & Oechel, W. C. (1987). Response of *Eriophorum vaginatum* to elevated CO_2 and temperature in the Alaskan tussock tundra. *Ecology*, **68**, 401–10.

Turner, N. C. & Kramer, P. J. (1980). *Adaptations of Plants to Water and High Temperature Stress*. New York: Wiley.

van Veen, J. A., Liljeroth, E. & Lekkerkerk, L. J. A. (1991). Carbon fluxes in plant–soil systems at elevated atmospheric CO_2 levels. *Ecological Applications*, **1**, 175–81.

Vitousek, P. M. (1994). Beyond global warming: ecology and global change. *Ecology*, **75**, 1861–76.

Waring, R. H. & Schlesinger, W. H.

(1985). *Forest Ecosystems, Concepts and Management*. New York: Academic Press.

Watson, R. T., Rodhe, H., Oeschger, H. & Siegenthaler, U. (1990). Greenhouse gases and aerosols. In *Climate Change – The IPCC Scientific Assessment*, ed. J. Houghton *et al*. Cambridge: Cambridge University Press.

Wayne, P. M. & Bazzaz, F. A. (1995). Seedling density modifies growth responses of yellow birch maternal families to elevated carbon dioxide. *Global Change Biology*, **1**, 315–24.

Weiner, J. (1985). Size hierarchies in experimental populations of annual plants. *Ecology*, **66**, 743–52.

Weiner, J. & Thomas, S. C. (1986). Size variability and competition in plant monocultures. *Oikos*, **47**, 211–22.

Whipps, J. M. (1985). Effect of CO_2 concentration on growth, carbon distribution and loss of carbon from the roots of maize. *Journal of Experimental Botany*, **36**, 644–51.

Wigley, T. M. L. (1985). Impact of extreme events. *Nature*, **316**, 106–7.

Williams, W., Garbutt, K., Bazzaz, F. & Vitousek, P. M. (1986). The response of plants to elevated CO_2. IV. Two deciduous-forest tree communities. *Oecologia*, **69**, 454–9.

Wisniewski, J. & Lugo, A. E. (ed.) (1992). *Natural Sinks of CO_2*. Dordrecht: Kluwer Academic Publishers.

Wofsy, S. C., Goulden, M. L., Munger, J. W., Fan, S.-M., Bakwin, P. S., Daube, B. C., Bassow, S. L. & Bazzaz, F. A. (1993). Net exchange of CO_2 in a mid-latitude forest. *Science*, **260**, 1314–17.

Wong, S. C., Kriedemann, P. E. & Farquhar, G. D. (1992). $CO_2 \times$ nitrogen interaction on seedling growth of four species of eucalypt. *Australian Journal of Botany*, **40**, 457–72.

Woodrow, I. E. & Berry, J. A. (1988). Enzymatic regulation of photosynthetic CO_2 fixation in C_3 plants. *Annual Review of Plant Physiology and Plant Molecular Biology*, **39**, 533–94.

Woodward, F. I. (1987). *Climate and Plant Distribution*. Cambridge: Cambridge University Press.

Woodward, F. I. (1992). A review of the effects of climate on vegetation: ranges, competition, and composition. In *Global Warming and Biodiversity*, ed. R. L. Peters & T. E. Lovejoy. New Haven: Yale University Press.

Woodward, F. I. & Rochefort, L. (1991). Sensitivity analysis of vegetation diversity to environmental change. *Global Ecology and Biodiversity Letters*, **1**, 7–23.

Woodward, F. I., Thompson, G. B. & McKee, I. F. (1991). The effects of elevated concentrations of carbon dioxide on individual plants, populations, communities and ecosystems. *Annals of Botany*, **67** (Supplement 1), 23–38.

Woodwell, G. M. (1995). Biotic feedbacks from the warming of the Earth. In *Biotic Feedbacks in the Global Climatic System. Will the Warming Feed the Warming?* ed. G. M. Woodwell, pp. 3–21. New York: Oxford University Press.

Wulff, R. D. & Alexander, H. M. (1985). Intraspecific variation in the response to CO_2 enrichment in seeds and seedlings of *Plantago lanceolata* L. *Oecologia*, **66**, 458–60.

Zak, D. R., Pregitzer, K. S., Curtis, P. S., Teeri, J. A., Fogel, R. & Randlett, D. L. (1993). Elevated atmospheric CO_2 and feedback between carbon and nitrogen cycles. *Plant and Soil*, **151**, 105–17.

Zangerl, A. R. & Bazzaz, F. A. (1984). The response of plants to elevated CO_2. II. Competitive interactions between annual plants under varying light and nutrients. *Oecologia*, **62**, 412–17.

The role of vegetation in controlling carbon dioxide and water exchange between land surface and the atmosphere

E.-D. Schulze, F. M. Kelliher, Ch. Körner, J. Lloyd,
D. Y. Hollinger and N. N. Vygodskaya

Introduction

The atmosphere both controls and responds to the partitioning of solar energy into sensible and latent heat (evaporation) fluxes at vegetated surfaces, largely as a result of stomatal regulation. Stomatal control of water flow through plants from the soil to the atmosphere links regional and global climate systems to ecosystem water, nutrient and carbon cycles. Indeed, the role of vegetation in land surface–atmosphere exchange has gained increasing attention because of the potential importance of terrestrial ecosystems as a sink for atmospheric carbon (Jarvis & Dewar, 1993; Tans, 1993).

Recognizing the need for biophysical descriptions of land surface–atmosphere exchange processes especially in models, the operational plans of the International Geosphere Biosphere Programmes GCTE and BAHC recommended a quantification of maximum rates of water vapour and CO_2 exchange between terrestrial ecosystems and the atmosphere (Steffen et al., 1992; Bolle, 1993). Resultant models were intended to provide predictions of these maximum rates for different vegetation types at a global scale. In this chapter, we present a synopsis of data and models developed from the operational plans based principally on our own contributions (Kelliher et al., 1993, 1994a; Körner, 1994; Lloyd & Farquhar, 1994; Schulze et al., 1994a, 1995b).

One problem of global integration of terrestrial CO_2 and water fluxes is that even dominant vegetation types have not yet been comprehensively studied. For instance, despite the possibility of an unknown Northern Hemisphere terrestrial sink in the global carbon budget, an area of complete lack of information is the Siberian boreal forest. In the following we present (1) a description of processes involved in the exchanges of water and CO_2 from terrestrial surfaces and the atmosphere (Kelliher et al., 1993); (2) an analysis of the effects of certain parameters (LAI, nutrition) onto these processes (Kelliher et al., 1994a; Schulze et al., 1995b); (3) a global estimate of the maximum exchange rates (Schulze et al., 1994a; Lloyd & Farquhar, 1994;

Körner, 1994); and (4) a description of a case study on long-term carbon storage in boreal forest. The initial data of this case study originated from an expedition which was initiated by GCTE and which led to the deciduous coniferous forests in Eastern Siberia (Schulze *et al.*, 1995*a*).

The exchange of water and CO_2 between terrestrial surfaces and the atmosphere

Figure 5.1 summarizes the processes determining the exchange of water and CO_2 of terrestrial surfaces, which consist of a vegetation cover, characterized by plant height (*h*) and a leaf area index (*Δ*) which may be arranged in different canopy strata (*ξ*), and of ground surface, which receives solar radiation depending on leaf area index (R_g). We do not attempt to model plant height and leaf area index in this study because it depends on the species which form the plant cover and on climatic conditions, especially the plant water balance (Schulze, 1982). However, our analysis is based on the assumption that the partitioning of energy into latent heat flow (λE) is, in a complicated manner, determined by the nutrient status of the soil (*N*), which is a prime determinant of the photosynthetic capacity of leaves (Field & Mooney, 1986). For many ecosystems, nitrogen appears to be one of the most important factors constraining productivity, and only a certain fraction of this resource is allocated into the CO_2-fixing enzyme Rubisco. Nitrogen is also required for the construction of assimilatory organs and has physiological roles other than assimilation. Furthermore there is a close correlation between CO_2 assimilation and leaf conductance (Wong *et al.*, 1978), which is well established for many plant types and species (Körner *et al.*, 1979; Schulze & Hall, 1982). Although it is possible to calculate leaf evaporation (E_l) from leaf conductance, it is not possible to scale easily from leaves to canopies, because it is hard to define the exact climatic conditions within the canopy. Only recently have models been developed for this scaling exercise (Baldocchi & Harley, 1995).

Photosynthetic CO_2 assimilation rate (*A*) is related to the leaf stomatal conductance for CO_2 (g_{lc}) and the difference between the leaf internal and atmospheric CO_2 concentrations (c_i and c_a, respectively; Schulze *et al.*, 1994*a*; Fig. 5.1). As a model, we assume that maximum *A* (A_{max}) at light saturation, and with plentiful soil water, low air saturation deficit and moderate temperature, is a linear function of leaf nitrogen concentration *N* above a threshold value (N_t).

At the scale of a leaf, evaporation is controlled by the leaf stomatal conductance for water vapour (g_l) and air saturation deficit (*D*) at the

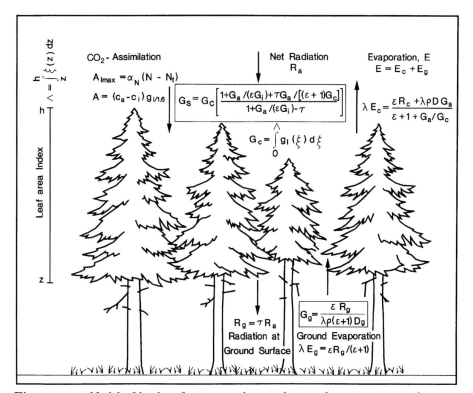

Figure 5.1 *Model of land surface–atmosphere exchange of water vapour and carbon dioxide for vegetation including a canopy of leaves (each of area ξ) of height h (base of canopy at height z) and leaf area index Δ, and a ground surface$_g$. Conceptually, leaf carbon assimilation rate (A) has a maximum rate (A_{max}) that is linearly related to the nitrogen concentration N above a threshold value N_t by slope αN. Leaf A is determined by the product of the stomatal conductance for CO_2 (g^{lc}) and the depletion of atmospheric CO_2 (c_a) in the leaf mesophyll (c_i). Maximum leaf stomatal conductance for water vapour (g_{lmax}) is integrated using ξ, Δ, and a radiation (R) interception submodel including τ, the fraction of R transmitted through the canopy, with radiation intercepted by the canopy (R_c) equal to radiation above the canopy (R_a) minus that at the ground surface (R_g), to estimate the maximum canopy conductance (G_{cmax}). Maximum canopy evaporation rate (E_{cmax}) is calculated using the Penman–Monteith equation with inputs of G_{cmax}, R_c, D and wind speed which determines the aerodynamic conductance (G_a). Air temperature determines the equation's physical coefficients λ = latent heat of vaporisation, ϵ = change of latent heat of saturated air with a change of sensible heat content, and ρ = air density. Maximum ground evaporation rate (E_{gmax}) is proportional to R_g. The maximum ground conductance for water vapour (G_{gmax}) is proportional to R_g/D_g where D_g is D at the ground surface.*

nominal evaporating surface. To scale up to a canopy of leaves, G_c can be
reasonably approximated by a parallel sum of g_l for the individual leaves (of
leaf area ξ) up to the canopy leaf area index (Δ; Raupach & Finnigan, 1988;
Raupach, 1994). Water vapour exchange or evaporation from a vegetated
surface (E including plant canopy and ground evaporation; E_c and E_g,
respectively) is governed by the available radiant energy supply (R_a), D,
atmospheric turbulence (commonly expressed theoretically through an aerodynamic conductance, G_a, though see Hollinger *et al.* (1994*b*) for direct
measurements) and stomatal control of the ability of the surface to transmit
water to the atmosphere (via conductances for the ground, G_g, plant
canopy, G_c, and bulk surface, G_s). This is expressed in the Penman–
Monteith equation which we use as a model of E_c (Fig. 5.1 and see Schulze
et al. (1995*b*) for a test of the E_c model using measurements made in a
Nothofagus forest tree canopy). Only a fraction of incoming radiation is
adsorbed by the canopy (R_c). This depends on leaf reflection, transmission
coefficients, and on total leaf area index (Δ). At least some radiation pene-
trates the canopy and reaches the ground (R_g) causing evaporation from
the soil. To model E_g, we assume an equilibrium rate of evaporation
proportional to the available energy at the ground surface (R_g; Kelliher
et al., 1992, 1994*a*). Following our earlier argument, we write an expression
for G_g from the ratio ($E_g{:}D_g$) where D_g is D at the ground surface (Schulze
et al., 1994*a*). To quantify how G_s differs from G_c because of the contribution of ground evaporation, we combined our equations for ground, plant
canopy and bulk surface evaporation rates to derive an expression for G_s in
terms of G_c and other plant canopy properties. The new term G_i,
proportional to the ratio ($R_a{:}D$), in this expression is the isothermal conductance (Monteith & Unsworth, 1990; Fig. 5.1).

It would be desirable to predict canopy assimilation and evaporation
from our knowledge of plant physiology. It is theoretically possible to
predict assimilation from nutrition, and, based on the close relation between
assimilation and conductance, to estimate canopy conductance. However,
published data are in many cases inadequate, because nutrition has been
measured only in few cases in combination with leaf gas exchange (Field &
Mooney, 1986). Further, plant ecologists have generally concentrated on
sun-leaves, and only rarely was the variation within the canopy studied
(Field, 1983). Likewise, micrometeorologists have previously not described
their system with respect to plant nutrition or leaf behaviour.

The relation between surface conductance and leaf area index

When relating measured data of maximum surface conductance to leaf area index no consistent pattern emerges at first glance (Fig. 5.2). It can be recognized, however, that at leaf area indices above 3 to 4, the upper and lower range is determined by the maximum stomatal conductance at the leaf level. Modelled canopy conductance (Kelliher *et al.*, 1994*a*) is expected to decrease as leaf area index declines below 4. However, the measured data of surface conductance do not follow this pattern because soil evapotranspiration contributes increasingly to the overall flux of water as leaf area index declines. Thus, maximum surface conductance is relatively conservative with respect to leaf area, and only a leaf area of 4 is required for maximal fluxes of water vapour in closed canopies.

Relations between plant nutrition and surface fluxes

To further explore the variation in surface conductance at high LAI, Fig. 5.3 investigates effects of plant nutrition on leaf and surface conductance, and the relation to surface assimilation. Only the relation between maximum surface assimilation and conductance is based on simultaneous measurements at specific sites (Fig. 5.3C). The other relations are based on aggregated data per vegetation type. In this case data on the x-axis may originate from a different set of measurements within the same vegetation type as the data on the y-axis (Schulze *et al.*, 1994*a*; Körner, 1994). In Plate 1 (opposite p. 204) stomatal conductance per leaf area was related to leaf nitrogen per dry weight. The underlying problem of effects of leaf specific mass and its photosynthetic component was discussed in detail by Schulze and colleagues (1994*a*).

It is important to recognize that the relation between assimilation and surface conductance does not show a separate behaviour for woody versus herbaceous, or cultivated versus natural vegetation types, but the variation within each of these groups is substantial. Data are well distributed with respect to A and G_s and indicate a close relationship with a slope of 1 μmol CO_2 m^{-2} s^{-1} per mm s^{-1} (Fig. 5.3C). Only some C_4 crops (e.g. maize) and very dense tropical canopies fall above the regression line. We would suggest that increases in leaf area above a LAI of 3 to 4 would lead to a greater increase in the flux of CO_2 than in the flux of water vapour.

The conservative relationship between A_{smax} and G_{max} in Fig. 5.3C indicates that only small variations would be expected to exist in the ratio of

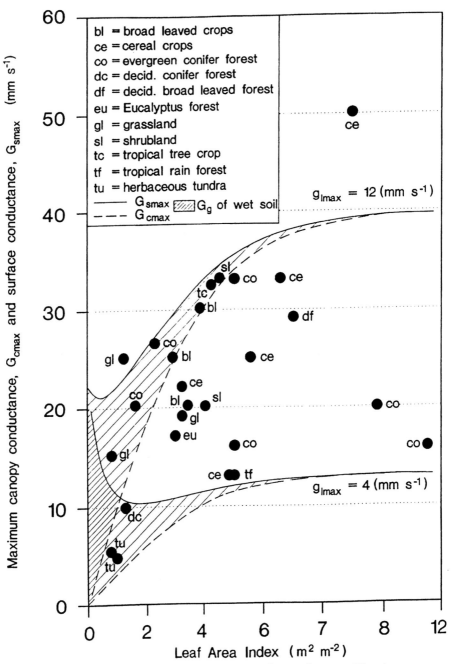

Figure 5.2 *The relationship between maximum surface conductance (G_{smax}) and one-sided leaf area index for 26 field studies with vegetation type codes described in the legend. Also shown are the relationships for modelled values of G_{smax} and maximum canopy conductance (G_{cmax}) using maximum leaf stomatal conductances (g_{lmax}) from the range of data in the global review of Körner (1994), and the maximum ground conductance (G_g) (After Kelliher et al., 1994a).*

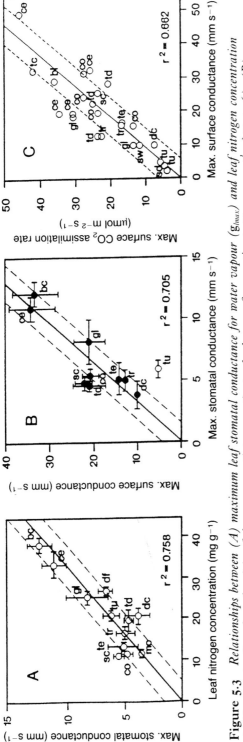

Figure 5.3 *Relationships between (A) maximum leaf stomatal conductance for water vapour (g_{lmax}) and leaf nitrogen concentration (solid regression line through origin of slope 0.3 mm s^{-1} per mg g^{-1}, standard error of g_{lmax} estimate = 1.4 mm s^{-1}, r^{2} = 0.76); (B) maximum surface conductance (G_{smax}) and g_{lmax} (regression slope = 3.0, standard error of G_{smax} estimate = 4.5 mm s^{-1}, r^{2} = 0.70); and (C) maximum surface carbon dioxide assimilation rate (A_{smax}) and G_{smax} (regression slope = 1.0 mmol CO_2 m^{-2} s^{-1} per mm s^{-1}, standard error of A_{smax} estimate = 6.8 mmol CO_2 m^{-2} s^{-1}, r^{2} = 0.62). Key: bc, broad-leaved crops; ce, cereal crops; co, evergreen conifer forest; dc, deciduous conifer forest; df, tropical deciduous forest; gl, grassland; mo, monsoonal forest; sc, sclerophyllous shrub; sd, savanna dry; sm, savanna met; tc, tropical tree crop (Hevea); td, temperate deciduous broad-leaved forest; te, temperate evergreen broad-leaved forest; tr, tropical rain forest; tu, herbaceous tundra. (After Schulze et al., 1994a.)*

Table 5.1 Global vegetation cover types

Type	No.	N	g_{smax}	G_{smax}	A_{smax}	c_c/c_a
Rice	4	27.8	8.4	25.1	26.3	0.57
Evergreen conifers	10, 11	11.0	5.5	20.6	21.6	0.53
Mixed forest	12, 13	15.8	4.6	13.8	14.5	0.63
Evergreen broadleaf tree	14, 19	13.4	4.0	12.1	12.7	0.65
Evergreen broadleaf crop	15	25.2	7.6	22.7	23.9	0.59
Evergreen broadleaf shrub	16	10.4	3.1	9.4	9.9	0.63
Deciduous conifers	17, 18	20.7	3.8	11.4	12.0	0.67
Deciduous broadleaf tree	20, 21	19.6	5.9	20.7	21.7	0.67
Deciduous tree crop	22	23.8	7.2	21.5	22.6	0.56
Drought deciduous forests	23, 24, 25, 26, 27, 28	27.1	8.2	24.5	25.7	0.63
Temperate grasslands	30, 31	25.2	7.7	23.0	24.2	0.57
Tropical savanna	32, 34, 35, 37, 39, 71, 73	18.9	5.7	17.1	17.9	0.50
Tropical pasture	33	17.1	5.2	15.4	16.2	0.41
Semi-arid rough grazing	36	10.7	3.2	9.7	10.1	0.59
Arable cropland	40, 41	36.0	10.8	32.5	34.1	0.59
Sugarcane	43	12.0	3.6	10.8	20.0	0.36
Maize	44	28.5	8.6	25.7	33.0	0.36
Gossypium	45	29.4	8.9	26.5	27.9	0.59
Irrigated cropland	47, 48	38.4	11.6	34.7	36.4	0.59
Tropical rainforest	50	16.5	5.0	14.9	15.6	0.65
Tropical tree crop	46, 49, 51	13.6	4.1	12.3	12.9	0.54
Tropical broadleaf trees	52	19.2	5.8	17.3	18.2	0.57
Tundra, bog, marsh	2, 61, 62	20.5	6.2	5.0	5.3	n.d.
Water, ice, desert, urban	0, 1, 3, 5, 70	n.d.	n.d.	n.d.	n.d.	n.d.

After Wilson and Henderson-Sellers (1985) with their affiliated numbers (No.) and associated values of leaf nitrogen concentration (N, mg g^{-1}) derived from the literature review of Schulze et al. (1994a). Also shown are maximum values of leaf stomatal conductance (g_{lmax}, mm s^{-1}), surface conductance (G_{smax}, mm s^{-1}), surface CO_2 assimilation rate (A_{smax}, mmol CO_2 m^{-2} s^{-1}) estimated from N according to the relations in Fig. 5.3. The ratio of leaf chloroplastic and atmospheric CO_2 concentrations (c_c and c_a, respectively) for C_3 plants (c_c/c_a) comes from the model of Lloyd & Farquhar (1994). n.d., no data.

chloroplastic to ambient partial pressures of CO_2 (c_c:c_a for C_3 plants if all plants operated near G_{smax} all the time (Farquhar & Sharkey, 1982). Although carbon isotope discrimination studies indicate some variability in C_3 plant c_c:c_a (Körner et al., 1991), on a global scale this ratio is indeed surprisingly constant. Plate 1 shows estimates of c_c : c_a for C_3 plants only, generated from the model of Lloyd and Farquhar (1994). This small variation in C_3 plant c_c:c_a on a global scale occurs despite large variations in key modulating variables such as leaf to air vapour pressure deficit. This low variability arises at least in part because plants growing in ecosystems with regular or periodic water supply are characterized by less conservative stomatal behaviour (in respect of the operation marginal cost of carbon gain) as evaporative demand increased (Lloyd & Farquhar, 1994).

Linear relations also exist in Fig. 5.3A and B between stomatal conductance and leaf nitrogen concentration (slope 0.3 mm s⁻¹ per mg N g⁻¹), as well as between maximum surface conductance and maximum stomatal conductance (slope 3 mm s⁻¹ in canopy conductance per mm s⁻¹ in leaf conductance). In this case the data points are not as evenly distributed, and the regressions are mainly driven by the differences between natural and anthropogenic vegetation types. It has even been suggested that natural vegetations have very similar (if not constant) stomatal conductance (Körner, 1994). Following the even distribution of data in Fig. 5.3C where each data point refers to simultaneous measurements of CO_2 and conductance, we think that the separation of natural and anthropogenic vegetations in Fig. 5.3A and B is in part due to the aggregation procedure (which we could not avoid because of lack of simultaneous data in the published literature) and also is a consequence of woody vegetation types being classified in greater detail than grasslands. Despite these difficulties, it is apparent that anthropogenic vegetation types operate at a higher nutritional level and at a higher rate of water loss.

Global estimates of vegetation nitrogen concentrations, and maximum conductances and maximum exchange rates of water vapour and CO_2

Global implications of the relations given in Fig. 5.3 emerge by combination with a world archive of vegetation cover data (Schulze et al., 1994a; Table 5.1). Leaf nitrogen concentrations are generally low for natural vegetation types, an exception being deciduous forests (Plate 2A, between pp. **204** and **205**). By contrast, for anthropogenic vegetation types of the Northern Hemisphere, especially in Europe, Eastern North America, India, and East

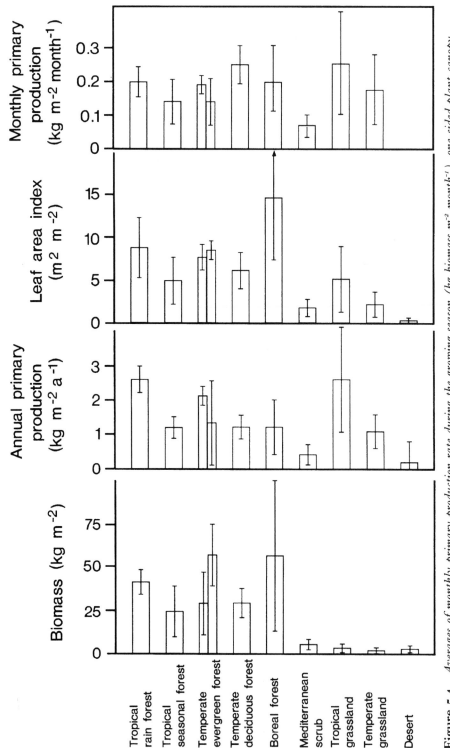

Figure 5·4 *Averages of monthly primary production rate during the growing season (kg biomass m⁻² month⁻¹), one-sided plant canopy leaf area index, annual primary production rate, and total (above- and below-ground) biomass (kg biomass m⁻²) for a range of natural vegetation types. Vertical lines are ranges. (After Schulze, 1982.)*

Asia, nitrogen is up to 2 to 4 times higher than that of natural vegetation because of land management (including fertilization) and anthropogenic nitrogen deposition (Galloway *et al.*, 1994). According to Plate 1, this difference between anthropogenic and natural vegetation continues for g_{lmax} (Plate 2B), G_{smax} (Plate 2C) and A_{smax} (Plate 2D). About 15–20% of the terrestrial Northern Hemisphere has high N and A_{smax} rarely obtained in the Southern Hemisphere during the growing season. We note that these high values of conductance and of water or CO_2 flux generally occur in regions of intense agricultural activity and/or atmospheric pollution, particularly nitrous oxide deposition (Galloway *et al.*, 1994).

In terms of feedback to the atmosphere, we wonder if widespread Northern Hemisphere changes from natural vegetation to highly fertilized seasonal crops which occurred this century may have caused and still be causing increasing seasonal variation in atmospheric CO_2 concentrations (Conway *et al.*, 1988; Schulze & Caldwell, 1994). This recognizes that global A_{smax} and the spatial variation in atmospheric CO_2 concentrations are a consequence of human activity. Consequently, we believe that arable land in Europe, North America and Asia is the principal site of maximum rates of surface–atmosphere carbon dioxide exchange in the Northern Hemisphere. Our nitrogen and A_{smax} predictions do not support the suggestion that the vast boreal forest has the 'potential' to be a major (missing) sink in the global carbon budget (Tans *et al.*, 1990; Jarvis, 1993; Jarvis & Dewar, 1993). However, A_{smax} does not necessarily correlate with net primary productivity (NPP) and carbon storage. The long-term carbon sequestration rates are determined by additional factors such as irradiance, temperature and soil water (rainfall), and air saturation deficit (e.g. Kelliher *et al.*, 1993). Forests and grasslands respond similarly to light and air saturation deficits, but large differences exist in their response to soil drought. Forests can maintain water fluxes at lower soil water contents in the upper soil layer than grasslands because of greater rooting depth. This may result in forests exhibiting less seasonality in productivity in climates characterized by periodic water supply.

Despite these differences in physiological and morphological characters, we note that net primary productivity during the growing season is surprisingly conservative for natural vegetation types (an exception being sclerophylls) and does not correlate with biomass per unit land area (Fig. 5.4; Schulze, 1982). Differences in annual net primary productivity amongst natural vegetation types generally appear to be within the bounds of statistical precision.

Nitrogen limitation of surface–atmosphere exchange rate: a case study of boreal forest

Recent direct measurements of ecosystem–atmosphere CO_2 exchange rates have demonstrated that even old-growth forest at pristine sites can achieve a net uptake of CO_2 over days or years (e.g. Hollinger *et al.*, 1994*a*). The possibility of such vegetation types accumulating carbon from the atmosphere contrasts with ecological convention, but it may also be attributed to 'fertilization' from increased atmospheric CO_2 concentrations. However, rather than simply accepting this conclusion, we suggest that nitrogen supply might constrain the utilization of additional atmospheric CO_2 even in an otherwise favourable ecosystem environment. We believe that such a limit may be of greater importance for some vegetation types in the long term. As a case study for one of the Northern Hemisphere and world's largest terrestrial ecosystems, we now explore measurements of carbon sequestration and nitrogen concentration in a chronosequence of pristine *Larix gmelinii* (deciduous conifer) stands representative of the Eastern Siberian boreal forest.

For *L. gmelinii* forest, G_{smax} and A_{smax} are 10 mm s^{-1} (Kelliher *et al.*, 1994*b*) and 4 mmol m^{-2} s^{-1} (Hollinger *et al.*, 1995*a*), respectively. These values are very low with respect to those of other global vegetation types, and comparable only with semi-arid grassland, evergreen broadleaf shrubs (mediterranean sclerophylls) and arctic tundra (Table 5.1). For example, in the middle of summer, the *L. gmelinii* forest had a net daily carbon uptake of 6 kg C ha^{-1} day^{-1} (0.05 mol m^{-2} day^{-1}; Hollinger *et al.*, 1995*b*). In spring and autumn, the seasonal cycle of atmospheric CO_2 concentration at boreal latitudes indicates that net CO_2 efflux (respiration exceeding photosynthesis) at night exceeds net CO_2 uptake (photosynthesis exceeding respiration) by day (Lloyd & Farquhar, 1994). Nevertheless, a crude estimate indicates that *L. gmelinii* may assimilate in the large natural range of 7.8×10^6 km^2 and in a 120-day growing season in the order of 0.6 Pg C (i.e. 0.6×10^{15} g C, Hollinger *et al.*, 1995*b*). This accounts for only about 20% of the estimated 'missing' global carbon budget sink that has been attributed by some to Northern Hemisphere terrestrial ecosystems (Tans *et al.*, 1990).

We now consider what limits carbon sequestration by the vast boreal forest of Eastern Siberia as well as the longevity of carbon storage in the ecosystem. For *L. gmelinii* forest, there was a remarkably constant quantity of nitrogen in the above-ground biomass for stands of age 50 to almost 400 years (Fig. 5.5) while N losses continue to decrease the total N pool. Trees were extremely slow-growing (average height and diameter in the

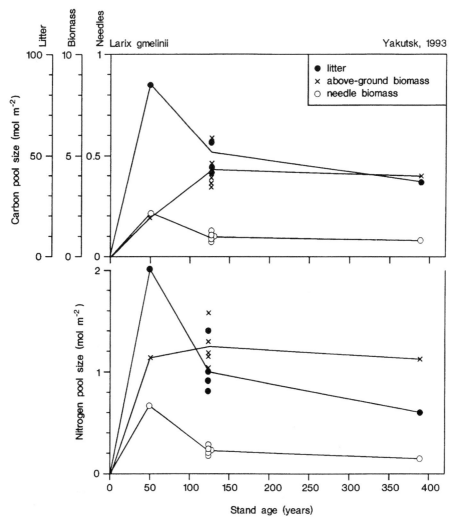

Figure 5.5 *Relationships between above-ground carbon and nitrogen pools and stand (tree) age in a chronosequence of* Larix gmelinii *stands in Eastern Siberia. Symbols in the legend define separate data for the forest floor litter, tree (needle) canopy, and total above-ground biomass. (After Schulze et al., 1995b.)*

380-year-old stand were only 11 m and 160 mm, respectively), and there was essentially no above-ground biomass accumulation between ages 130 and 380 years. This was attributed to a lack of available nitrogen, with about 80% of the above-ground nitrogen (and 98% of the above-ground carbon) immobilized in wood by age 130 years (Schulze *et al.*, 1995*a*). This conclusion was supported by graphical analysis indicating that the self-thinning process in these stands was not governed by the availability of radiation according to allometric theory (see Schulze *et al.*, 1995*a*). The immobilized nitrogen appears to be returned to a vegetation-available form at a significant rate only by recurrent fire. Thus fire is essential for the persistence of *L. gmelinii* forest or otherwise the vegetation would need to change towards a plant cover with lower N turnover. The frequency of fire in Eastern Siberia suggests that the *L. gmelinii* forest is unlikely to be a prolonged net carbon sink. We believe that carbon residence time in this boreal forest is only of an order of 100 years.

Conclusions

In scaling evaporation from a leaf to that of a surface (plant canopy and ground), maximum surface conductance is relatively conservative because of the compensating decrease in plant canopy evaporation and increase in ground evaporation from moist soil as leaf area index diminishes.

Under favourable meteorological and hydrological conditions, leaf nitrogen concentration sets the maxima of leaf stomatal conductance, surface conductance and surface carbon dioxide assimilation rate.

Large areas of high evaporation and carbon assimilation capacity exist in the Northern Hemisphere related to anthropogenic intervention through agriculture.

In terms of identifying a missing terrestrial ecosystem sink in the global carbon budget, carbon assimilation rate in the vast boreal forest of Eastern Siberia is extremely low because of limited available nitrogen. Carbon residence time in this forest is of the order of 100 years with recurrent return of carbon to the atmosphere by naturally caused fire.

Acknowledgements

E.-D. Schulze acknowledges the support by the German Federal Ministry for Food, Agriculture and Forestry and the Humboldt Foundation for making the Siberia Expedition possible. F. M. Kelliher thanks the New Zealand Foundation for Research, Science and Technology for their continued support of international atmospheric research.

References

Baldocchi, D. & Harley, P. (1994). Scaling carbon dioxide and water vapor exchange from leaf to canopy in a deciduous forest: model testing and application. *Plant Cell and Environment*.

Bolle, H. (1993). *Biospheric Aspects of the Hydrological Cycle* (BAHC): *the Operational Plan*. IGBP Report 27. Stockholm: International Geosphere-Biosphere Programme.

Conway, T. J., Tans, P., Waterman, L. S., Thoning, K. W., Masarie, K. A. & Gammon, R. H. (1988). Atmospheric carbon dioxide measurements in the remote global troposphere 1981–1984. *Tellus*, **40B**, 81–115.

Farquhar, G. D., von Caemmerer, S. & Berry, J. A. (1980). A biochemical model of photosynthetic CO_2 assimilation in leaves of C_3 species. *Planta*, **149**, 78–90.

Field, C. B. (1983). Allocating leaf nitrogen for the maximization of carbon gain: leaf age as a control on the allocation program. *Oecologia*, **56**, 341–7.

Field, C. B. & Mooney, H. A. (1986). The photosynthesis–nitrogen relationship in wild plants. In *On the Economy of Plant Form and Function*, ed. T. J. Givnish, pp. 25–56. Cambridge: Cambridge University Press.

Galloway, N. J., Levy, II H. & Kasibhatla, P. S. (1994). Year 2020: consequences of population growth and development on deposition of oxidized nitrogen. *Ambio*, **23**, 120–3.

Hollinger, D. Y., Kelliher, F. M., Byers, J. N., Hunt, J. E., McSeveny, T. M. & Weir, P. L. (1994*a*). Carbon dioxide exchange between an undisturbed old-growth forest and the atmosphere. *Ecology*, **75**, 134–50.

Hollinger, D. Y., Kelliher, F. M., Schulze, E.-D. & Köstner, B. M. M. (1994*b*). Coupling of tree transpiration to atmospheric turbulence. *Nature*, **371**, 60–2.

Hollinger, D. Y., Kelliher, F. M., Schulze, E.-D., Vygodskaya, N. N., Varlagin, A., Milukova, I., Byers, J. N., Sogachov, A., Hunt, J. E., McSeveny, T. M., Kobac, K. I., Bauer, G. & Arneth, A. (1995*a*). Initial assessment of multi-scale measures of CO_2 and H_2O flux in the Siberian taiga. *Global Change Biology*, (in press).

Hollinger, D. Y., Kelliher, F. M., Schulze, E.-D., Bauer, G., Vygodskaya, N. N., Byers, J. N., Hunt, J. E., McSeveny, T. M. & Arneth, A. (1995*b*). Forest––atmosphere carbon dioxide exchange in eastern Siberia. *Ecology* (in preparation).

Jarvis, P. G. (1993). Water losses of crowns, canopies and communities. *Water Deficits: Plant Responses from Cell to Community*, ed. J. A. C. Smith & H. Griffiths, pp. 285–316. Oxford: Bios Science Publishers.

Jarvis, P. G. & Dewar, R. C. (1993). Forests in the global carbon balance: from stand to region. In *Scaling Physiological Processes: Leaf to Globe*, ed. J. R. Ehleringer & C. B. Field, pp. 191–222. San Diego: Academic Press.

Kelliher, F. M., Hollinger, D. Y., Schulze, E.-D., Vygodskaya, N. N., Byers, J. N., Hunt, J. E. & McSeveny, T. M. (1994*b*). Evaporation from an eastern Siberian larch forest. Proc. Int. Symp. For. Hydrol. 1994, T. Ohta (ed.), Tokyo Japan, 24–28 October 1994 (in press).

Kelliher, F. M., Köstner, B. M. M., Hollinger, D. Y., Byers, J. N., Hunt, J. E., McSeveny, T. M., Meserth, R., Weir, P. L. & Schulze, E.-D. (1992). Evaporation, xylem sap flow, and tree transpiration in a New Zealand broad-leafed forest. *Agricultural and Forest Meteorology*, **62**, 53–73.

Kelliher, F. M., Leuning, R. & Schulze, E.-D. (1993). Evaporation and canopy characteristics of coniferous forests and grasslands. *Oecologia*, **95**, 153–63.

Kelliher, K. M., Leuning, R., Raupach, M. & Schulze, E.-D. (1994*a*). Maximum conductances for evaporation of global vegetation types. *Agricultural and Forest Meteorology*.

Körner, Ch. (1994). Leaf diffusive conductance in the major vegetation types of the globe. In *Ecophysiology of Photosynthesis*, ed. E.-D. Schulze & M. M. Caldwell, pp. 463–90. Berlin Heidelberg New York: Springer.

Körner, Ch., Farquhar, G. D. & Wong, S. C. (1991). A global survey of carbon isotope discrimination in plants from high altitudes. *Oecologia*, **74**, 623–32.

Körner, Ch., Scheel, J. A. & Bauer, H.

(1979). Maximum diffusive conductance in vascular plants. *Photosynthetica*, **13**, 45–82.

Lloyd, J. & Farquhar, G. D. (1994). ^{13}C Discrimination during assimilation by the terrestrial biosphere. *Oecologia* (in press).

Monteith, J. L. & Unsworth, M. H. (1990). *Principles of Environmental Physics*, 2nd edition. London: Edward Arnold.

Raupach, M. R. (1994). Vegetation–atmosphere interaction and surface conductance at leaf and regional scales. *Agricultural and Forest Meteorology*.

Raupach, M. R. & Finnigan, J. J. (1988). Single-layer models of evaporation for plant canopies are incorrect but useful, whereas, multilayer models are correct but useless. Discuss. *Australian Journal of Plant Physiology*, **15**, 705–16.

Schulze, E.-D. (1982). Plant life forms as related to plant carbon, water and nutrient relations. *Encyclopedia of Plant Physiology. Physiological Plant Ecology*, vol. 12B. *Water Relations and Photosynthetic Productivity*, ed. O. L. Lange, P. S. Nobel, C. B. Osmond & H. Ziegler, pp. 615–76. Berlin Heidelberg New York: Springer.

Schulze, E.-D. & Caldwell, M. M. (1994). Overview: Perspectives in ecophysiological research of photosynthesis. In *Ecophysiology of Photosynthesis*, ed. E.-D. Schulze & M. M. Caldwell, pp. 553–64. Berlin Heidelberg New York: Springer.

Schulze, E.-D. & Hall, A. E. (1982). Stomatal response, water loss and CO_2 assimilation rates of plants in contrasting environments. In *Encyclopedia of Plant Physiology. Physiological Plant Ecology*, vol. 12B, *Water Relations and Photosynthetic Productivity*, ed. O. L. Lange, P. S. Nobel, C. B. Osmond & H. Ziegler, pp. 181–230. Berlin Heidelberg New York: Springer.

Schulze, E.-D., Kelliher, F. M., Körner, Ch., Lloyd, J. & Leuning, R. (1994a).

Relationships between plant nitrogen nutrition, carbon assimilation rate, and maximum stomatal and ecosystem surface conductances for evaporation: a global ecology scaling exercise. *Annual Review of Ecology and Systematics*, **25**, 629–60.

Schulze, E.-D., Schulze, W., Kelliher, F. M., Vygodskaya, N. N., Ziegler, W., Kobak, K. I., Koch, H., Arneth, A., Kusnetsova, W. A., Sogachev, A., Issajev, A., Bauer, G. & Hollinger, D. Y. (1995a). Above-ground biomass and nitrogen nutrition in a chronosequence of pristine Dahurian *Larix* stands in Eastern Siberia. *Canadian Journal of Forestry Research*.

Schulze, E.-D., Leuning, R. & Kelliher, F. M. (1995b). Environmental regulation of vegetation surface conductance for evaporation. *Vegetatio*.

Steffen, W. L., Walker, B. H., Ingram, J. S. & Koch, G. W. (1992). *Global Change and Terrestrial Ecosystems: the Operational Plan*. IGBP Report 21. Stockholm: International Geosphere-Biosphere Programme.

Tans, P. P. (1993). Observational strategy for assessing the role of terrestrial ecosystems in the global carbon cycle: scaling down to regional levels. In *Scaling Physiological Processes: Leaf to Globe*, ed. J. R. Ehleringer & C. B. Field, pp. 179–90. San Diego: Academic Press.

Tans, P., Keeling, R. F. & Berry, J. A. (1993). Oceanic ^{13}C data. A new window on CO_2 uptake by the oceans. *Global Biogeochemistry Cycles*, **7**, 353–68.

Wilson, M. F. & Henderson-Sellers, A. (1985). A global archive of land cover and soils data for use in general circulation models. *Journal of Climatology*, **5**, 119–43.

Wong, S. C., Cowan, I. R. & Farquhar, G. D. (1978). Leaf conductance in relation to assimilation in *Eucalyptus pauciflora* Sieb. ex Spreng. Influence of irradiance and partial pressure of carbon dioxide. *Plant Physiology*, **62**, 670–4.

Integrated models of ecosystem function: A grassland case study

M. B. Coughenour and W. J. Parton

Introduction

A large number of ecosystem models have been used to evaluate the potential impact of global environmental change on natural and managed ecosystems. Ågren et al. (1991) present a state-of-the-art review of models of forests and grasslands used for climatic change research, while Parton et al. (1993) present a review of grassland models for impact assessment and climatic change research. The Ågren et al. review (1991) shows that there are a number of detailed plant physiological level models which describe plant production processes in great detail; however, the models are frequently not linked to nutrient cycling models which operate at longer time scales and thus limit the application of these models for assessing climatic change effects. They indicate that ecosystem level models include the interactive impact of plant on soil system; however, these models use simplified representations of the plant production processes and frequently do not include the direct impact of atmospheric CO_2 levels on important plant and soil processes. The Parton et al. review showed that grassland models have been used extensively for impact assessment and climatic change research, and noted that many of the models were limited by the simplistic way that they treated atmospheric CO_2 levels and generally did not deal with the impact of environmental changes on the animal production system. More recently Coughenour & Chen (1996) and D. X. Chen, H. W. Hunt & J. Morgan (personal communication) have developed physiologically based plant production models that are linked to nutrient cycling and soil organic matter models and thus reduce the limitations of many of the existing models for simulating the combined impact of climatic changes and increasing atmospheric CO_2 levels for long time periods (10–100 years).

Recently there has been much interest in comparing models used for climatic change assessment work. Ryan et al. (1995a,b) present a comparison of seven forest models. Each of the models used the same calibration data sets to test and parameterize their models for two coniferous biome sites and then simulated the potential impact of environmental change scenarios (combined weather and atmospheric CO_2 levels) on the two sites for a 120 year period. The results showed that the model predictions were quite different; however, there were many similarities between some of the models. One of the biggest uncertainties identified by the model comparisons is the question of how increased nutrient availability caused by increased decomposition rates may be offset by increased drought stress caused by higher temperatures, and how increased atmospheric CO_2 levels will modify these interactions. This work also indicated how difficult it was to compare models because of structural difference in the models and suggested that we need to develop a common model framework that will allow us to substitute different formulations for submodels and determine how differences in the formulations of the ecosystem process impact the response of the models.

Three ecosystem models (CENTURY, Parton *et al.*, 1993; BIOM-BGC, Running & Hunt, 1993; and TEM, Melillo *et al.*, 1993) are now being used to simulate the impact of climatic change for the continental United States using a one-half degree by half degree resolution (VEMAP project). The models are using the same climatic change scenarios, vegetation classification schemes, and soils data bases to drive the models and the results from the models are compared at the regional and national level. There are a number of other ongoing model comparison activities where wheat models (Ingram, 1994), litter decomposition models and soil organic matter models are being compared. This model intercomparison work will, it is hoped, lead to an improved understanding of how to formulate and structure ecosystem models.

There is substantial evidence to show that atmospheric CO_2 levels have been rising during the last 100 years (Keeling, 1986). The increase in atmospheric CO_2 has been correlated to the global increase in consumption of fossil fuels and land management practices such as clearing of forests and cultivation of arable land (Rotty & Marland, 1986; Houghton, 1988; Woodwell, 1988). Scientific evidence suggests that increasing atmospheric CO_2 levels have both direct and indirect effects on natural ecosystems (Field *et al.*, 1992; Bazzaz, 1990). Some of the direct effects of increasing atmospheric CO_2 on plants include: increasing the photosynthesis rate for C_3 plants (Pearcy & Bjorkman, 1983), decreasing the dark respiration rates (Amthor, 1989), and decreasing the stomatal conductance for both C_3 and C_4 plants. Indirectly, CO_2 alters the allocation of carbon to the roots (Larigauderie *et al.*, 1988; Luxmoore *et al.*, 1986; Sionit *et al.*, 1985) and decreases the N concentration in live plant leaves (Owensby *et al.*, 1993*b*). Stomatal responses to atmospheric CO_2 potentially affect many processes in grassland ecosystems. Stomata close in response to increased CO_2 concentrations (Ball *et al.*, 1987; Collatz *et al.*, 1992; Dougherty *et al.*, 1994), which decreases water loss through transpiration per unit leaf area. Elevated CO_2 increases or maintains the rate of CO_2 diffusion into the stomates, however. The net result of stomatal closure and increased CO_2 diffusion in both C_3 and C_4 species is an increase in water use efficiency (WUE), since less water is lost through transpiration per gram of carbon fixed (Coughenour & Chen, 1996). Increased WUE increases net primary production in water-limited environments, with many subsequent effects. Reduced transpiration rate can potentially increase soil moisture and reduce plant water stress. More favourable plant water status may decrease carbon allocation to roots, increase photosynthesis rate and reduce leaf senescence rate. Decomposition and soil N mineralization rates may be increased by greater soil moisture. The decrease in N concentration of aboveground leaves caused by increased atmospheric CO_2 should tend to decrease N mineralization from decomposing surface litter (e.g. Melillo *et al.*, 1982).

The response of plants to increasing atmospheric CO_2 levels is dependent upon light intensity, soil moisture, temperature and nutrient availability. The interactions among these different plant resources are quite complex and difficult to predict at the ecosystem level. The objective of this chapter is to evaluate the impact of increasing atmospheric CO_2 levels on grasslands using the linked GRASS-CSOM ecosystem model. This process-oriented ecosystem model considers the interactive impact of atmospheric CO_2 levels on photosynthesis, plant growth, nutrient cycling, allocation of plant carbon, leaf energy budget and soil water dynamics. The model was set up to simulate the impact of atmospheric CO_2 levels for shortgrass steppe in eastern Colorado and tallgrass prairie in eastern Kansas. The model has been tested extensively using observed field data from both sites. We will focus on evaluating the effect of changing the atmospheric CO_2 levels on plant production, soil water dynamics, water budgets and nitrogen mineralization.

Methods

The linked GRASS-CSOM ecosystem model was used to examine ecosystem responses to elevated CO_2. The first version of GRASS (Coughenour, 1984; Coughenour *et al.*, 1984) was revised (Coughenour *et al.*, 1993; Coughenour & Chen, 1996). The current photosynthesis submodel was derived from the C_3 model of Farquhar *et al.* (1980), with extensions to simulate C_4 photosynthesis (Chen & Coughenour, 1994), and with linkages to the stomatal conductance submodel of Ball *et al.* (1987). Transpiration is predicted as part of leaf energy balance. Labile carbon reserves and phenology, and growth and maintenance respiration are simulated. A greater fraction of fixed carbon is partitioned to roots under soil water stress and when plant N concentration decreases, as in Batchelet *et al.* (1989). Maintenance respiration is related to tissue nitrogen content (Ryan, 1991). Shoot maintenance respiration is decreased by 30% under doubled CO_2 (700 ppm) as in Coughenour & Chen (1996). Increasing CO_2 decreases plant N concentration (Owensby *et al.*, 1993*b*), so simulated plant N : C ratio is empirically decreased by 20% as CO_2 increases from 350 ppm to 700 ppm. Microclimatology is simulated as in Chen & Coughenour (1994). Soil moisture budgets are simulated by coupling leaf energy balance and transpiration routines from GRASS with a daily bare soil evaporation and soil water flux submodel (Parton, 1978). The GRASS plant growth and the daily abiotic submodels were linked with the decomposition and nutrient cycling submodels of CENTURY (CSOM, Parton *et al.*, 1993).

Model predictions compare favourably with biomass data (Fig. 6.1a) from a Colorado C_4 short-grassland (Coughenour & Chen, 1966). Colorado data were collected in 1970–1975 at the Central Plains Experimental Range near Fort Collins (Dodd & Lauenroth, 1979), which is dominated by *Bouteloua gracilis*. Simulation results also compare favourably with plant biomass dynamics data collected under the NSF-LTER research programme 1984–1990 at the Konza Prairie Research Natural Area (Fig. 6.1b) located near Manhattan, Kansas (Coughenour & Chen, 1996). This is a C_4-dominated tall-grassland dominated by *Andropogon gerardii*, with both unburned and annually burned treatments.

The GRASS-CSOM model, parameterized for the Konza grassland, was used to simulate CO_2-fertilization experiments conducted during 1989–1991 on an *A. gerardii*-dominated grassland near Manhattan, Kansas. In the field experiment (Owensby *et al.*, 1993*a*) and model runs presented here, standing crop was harvested annually in the dormant season. The field experiment employed open-top chambers to increase atmospheric CO_2 concentration to twice ambient level (700 ppm). A control chamber, supplied with air at the ambient CO_2 level (*c.* 350 ppm) was used to study chamber effects on microclimate and plant growth.

The open-top chambers altered the microclimate in several ways. Chambers reduced incident photosynthetically active radiation (PAR) by about 11% (Owensby *et al.*, 1993*a*), and model PAR was reduced accordingly. Dewpoint temperatures were elevated by about 1.5°C at mid-day. The corresponding increase in water vapour concentration was calculated in the model, and this was used to recalculate relative humidity. Windspeed in the

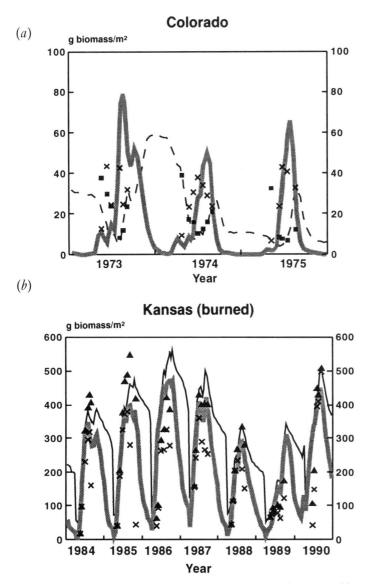

Figure 6.1 *Comparison of observed and simulated aboveground biomass dynamics at (a) the CPER Colorado shortgrass steppe site and (b) Kansas tallgrass prairie site. Thick line, ×s, simulated and observed live shoots; dashed line, filled squares, standing dead shoots; thin line, filled triangles, live plus dead (total) shoots. CPER data from IBP studies (Dodd & Lauenroth, 1979). Kansas data from Konza-LTER data bank (J. Briggs, personal communication).*

chamber was reduced to about 0.75 m/s. The low windspeed in the chambers decoupled the canopy air from the air above the canopy, so that canopy temperatures were about 2°C cooler, and above-canopy temperatures were about 5°C warmer than ambient (Ham *et al.*, 1995). A 2°C decrease in air temperature was employed in model runs. Since air was drawn from canopy level outside the chambers, and then pumped into the chambers at canopy level (Ham *et al.*, 1993, 1995), atmospheric resistances above the canopy outside the chambers were virtually applicable inside the chambers. Atmospheric resistance within chambers was empirically determined to be 15 s/m (Ham *et al.*, 1995). This resistance was used in place of the leaf boundary layer resistance normally calculated from canopy windspeed. Effects of reduced windspeed on heat fluxes were implicit in this empirical atmospheric resistance.

The GRASS model employs the empirical model of stomatal responses developed by Ball *et al.* (1987)

$$C_s = E_b + E_m \frac{A_N H}{C_a}$$

where A_N is assimilation rate, H is humidity and C_a is CO_2 concentration. The parameter E_m thus expresses sensitivity to CO_2. In previous model analyses (Chen *et al.*, 1993), stomates of Kansas tallgrasses grown in 700 ppm CO_2 had an E_m of 6.2 while plants grown at 350 ppm CO_2 had an E_m of 4.1. Model fits to the data of Knapp *et al.* (1993) yielded an E_m value of *c.* 6.2, assuming H of 50–55%. We report model results using both values of E_m. Recently, Dougherty *et al.* (1994) demonstrated limitations of the Ball *et al.* model. They found better fits ($r^2 = 0.84$–0.92 vs 0.71–0.75) by making C_s a non-linear function of atmospheric vapour pressure deficit. At low values of A_N the linearity between C_S and A_N is poor – stomata are more sensitive to C_a at low A_N. Thus, conclusions about the effects of CO_2 in the modelling reported here may be slightly conservative. Recently the Dougherty *et al.* model (RAPPS) has been linked with GRASS, which did improve accuracy (Sims *et al.*, 1995).

The model was tested against the 1989–91 field data from Kansas by simulating 1983–1988, and then invoking the chamber or CO_2 plus chamber treatments in 1989. Daily weather data 1983–1991 was used. Long runs of 35 years in Kansas were driven by daily weather data 1951–1986. Annual spring burning was assumed. Runs of 36 years at the Colorado site were driven by daily weather data 1946–1984.

Kansas CO_2 enrichment results

Aboveground and belowground net primary productivities (ANPP and BNPP) in Kansas were low in 1989 and high in 1990, according to the model (Figs. 6.2a, 6.3a). Similar among-year responses were observed in the field (Figs. 6.2b, 6.3b). ANPPs and BNPPs were both 8% lower with the higher stomatal sensitivity (E_m = 6.2) than with the lower sensitivity (E_m = 4.1). Chambers affected ANPP in two of the simulated years (Fig. 6.2a). Mean ANPP was 8% and 9% higher in chambers than in ambient conditions using low and high E_m values respectively, which was comparable to the average response observed in the field experiment. The model predicted that chambers stimulated BNPP by an average of 15% and 81% over the 3 years relative to ambient conditions using the low and high values for E_m.

Doubled CO_2 concentrations increased simulated ANPP in 2 years and increased BNPP in 2 years according to the model (Figs. 6.2a, 6.3a). Doubling CO_2 enhanced ANPP by 25% and 37%, with the larger response at higher E_m. Doubled CO_2 enhanced BNPP by 46% and 67% overall using low and high E_m values respectively. In comparison, peak aboveground standing crop in field plots was stimulated by 1–32% (Fig. 6.2b), and the average increase was 13%. BNPP was not measured in the field, but a two-fold increase in root ingrowth biomass was observed under doubled CO_2 in 1990 (Fig. 6.3b).

Chambers increased simulated mid-day photosynthesis (Ps) by 5–6% and daily Ps rates by 3–4% (Table 6.1). Modelled mid-day stomatal conductance (Cs) was increased by 14–17% and daily Cs was increased by 10–12%. Chambers decreased simulated mid-day transpiration rate per unit leaf area (Trl) by 1–2% and daily Trl by 8–9%. Simulated transpiration per unit ground area (Trg) was increased by 1% at mid-day and decreased by 5% daily. Doubling CO_2 inside chambers decreased predicted mid-day Ps rates by 1–4% and daily Ps rates by 1–5% (Table 6.1). Doubled CO_2 decreased mid-day stomatal conductances by 28–33% and daily Cs was decreased by 24–29%. Increased CO_2 caused transpiration rate per unit leaf area to decline by 23–28% at mid-day and by 22–26% daily. CO_2 decreased simulated transpiration per unit ground area by 1–5% at mid-day and on a daily basis Trg was decreased by 2% and increased by 3% with low and high E_m values respectively.

In comparison, actual measurements of leaf gas exchange showed 50% reductions in stomatal conductance under doubled CO_2 (Knapp *et al.*, 1993; Ham *et al.*, 1995), and a 22% reduction in evapotranspiration rate (Owensby *et al.*, 1993*a*; Ham *et al.*, 1995), as discussed below.

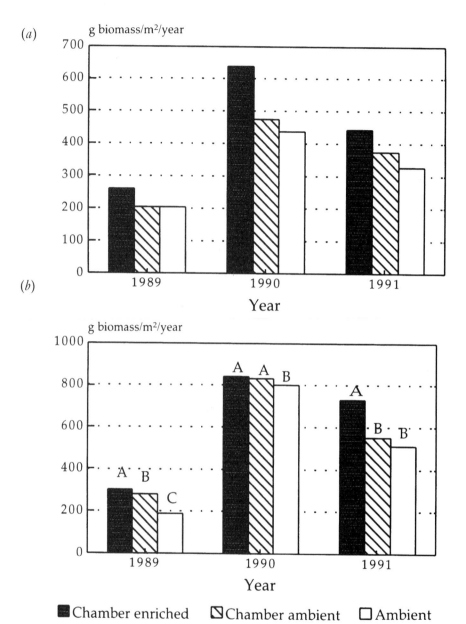

Figure 6.2   *Aboveground plant growth responses to chambers and to CO_2 in three years of experiments in Kansas using ambient conditions, open-top chambers supplied with ambient air, and open-top chambers supplied with 700 ppm CO_2. (a) Aboveground NPP simulated by the model compared with (b) observed peak standing crop in the field (Owensby et al., 1994). Bars with same letter were not significantly different.*

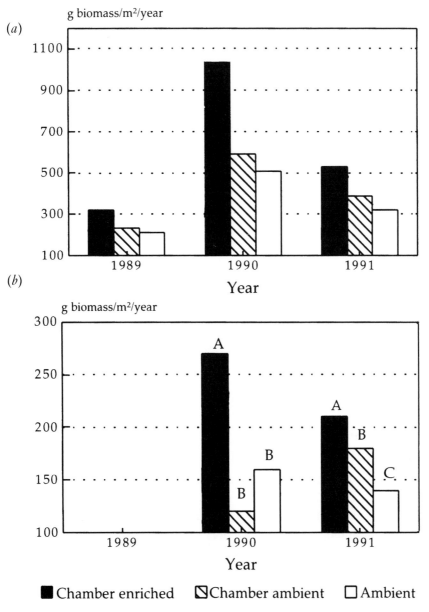

Figure 6.3 *Belowground plant responses in the same conditions as in Figure 6.1. (a) Belowground NPP simulated by the model compared with (b) observed biomass in root ingrowth bags (Owensby et al., 1994).*

Table 6.1 *Simulated effects of CO_2 and chambers on photosynthetic rates, stomatal conductance and transpiration and water use efficiency in Kansas tallgrass, April–September 1989–91. The top numbers were produced by using an E_m of 6.25 and the bottom numbers were results with an E_m of 4.1.*

Response variable	Mid-day			Daily		
	Ambient	Chamber		Ambient	Chamber	
	350ppm	350ppm	700ppm	350ppm	350ppm	700ppm
Ps (μmol m^{-2} s^{-1})	5.10	5.39	5.35	3.33	3.46	3.44
	5.02	5.27	5.08	3.30	3.39	3.23
Cs cm s^{-1}	0.180	0.211	0.141	0.153	0.172	0.104
	0.142	0.162	0.116	0.125	0.137	0.122
Trl: leaf area (mmol m^{-2} s^{-1})	2.44	2.95	1.74	1.44	1.32	0.98
	1.99	2.76	1.52	1.19	1.10	0.86
Trg: ground area (mmol m^{-2} s^{-1})	2.91	2.95	2.92	1.83	1.74	1.79
	2.73	2.76	2.61	1.74	1.66	1.63
WUE (μmol CO_2 mmol^{-1} H_2O)	2.09	2.25	3.07	2.31	2.62	3.51
	2.52	2.66	3.24	2.77	3.08	3.76

Ps, photosynthesis; Cs, stomatal conductance; Tr, transpiration rate; WUE, water use efficiency.

Long-term results

Doubling CO_2 led to 12% reductions in mid-day and 4% lower daily photosynthetic rates (Table 6.2) in 35-year simulations of the Kansas site using an E_m of 4.1. Corresponding stomatal conductances were reduced by 23% and 18%, transpiration rates on a leaf area basis were reduced by 20% and 19%, transpiration rates on a ground area basis were decreased by 6% and 4%, and WUE values were increased by 10% and 17%. Doubled CO_2 increased mid-day and daily photosynthetic rates by 21% and 19% and decreased corresponding stomatal conductances by 15% and 15% in simulations of the Colorado site (Table 6.2). Elevated CO_2 decreased transpiration rate on a leaf area basis by 18% at mid-day and by 7% daily. Transpiration rates per unit ground area were reduced by 6% and 6%. WUE values were increased by 48% at mid-day and by 42% daily.

Doubling CO_2 increased mean soil water content by 2% at the Kansas and by 5% at the Colorado site over the entire year, in 35–36 year runs (Table 6.2). Seasonal dynamics of soil water revealed that in Kansas, elevated CO_2 increased soil water most in June and July (Fig. 6.4a). By late

Table 6.2 *Simulated effects of doubling CO_2 concentration on photosynthetic rates, stomatal conductance and transpiration and water use efficiency in April–September in Kansas tallgrass, 1951–86 and Colorado shortgrass, 1949–84. E_m was 4.1 in the Kansas runs*

Response variable	Kansas				Colorado			
	Mid-day		Daily		Mid-day		Daily	
	350 ppm	700 ppm	350 ppm	700 ppm	350 ppm	700 ppm	350 ppm	700 ppm
Ps (μmol m^{-2} s^{-1})	6.00	5.29	3.64	3.49	9.66	11.71	6.69	7.98
Cs (cm s^{-1})	0.151	0.117	0.132	0.108	0.286	0.242	0.254	0.216
Trl: leaf (area mmol m^{-2} s^{-1})	2.15	1.72	1.36	1.11	6.13	5.01	3.26	2.72
Trg: ground area (mmol m^{-2} s^{-1})	3.66	3.45	2.40	2.31	2.31	2.16	0.96	0.63
WUE (μmol CO_2 mmol^{-1} H$_2$O)	2.79	3.08	2.68	3.14	1.58	2.24	2.06	2.94
Mean soil water content (cm)	–	–	38.20	37.44	–	–	15.00	15.82

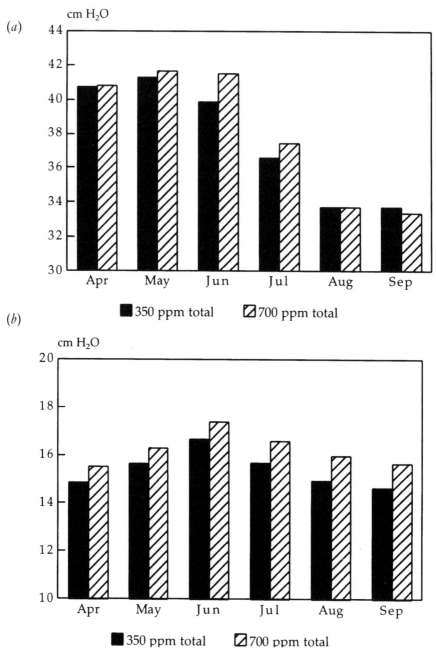

Figure 6.4 *Mean monthly soil water content over 35–36 year simulations for (a) Kansas and (b) Colorado. Soil depth was 185 cm in Kansas and 135 cm in Colorado.*

season and over winter, differences between CO_2 levels were small. In the
Colorado runs, CO_2 increased soil water most between July and September
(Fig. 6.4b). Differences existed all year.

Ecosystem water budgets revealed that elevated CO_2 reduced total transpiration losses by 3% over 35 years in Kansas and by 9% in Colorado
(Table 6.3). Bare soil evaporation loss was reduced by 36% under elevated
CO_2 in Kansas and by 14% in Colorado. Interception of rainfall was
increased by 20% in Kansas, and by 33% in Colorado, due to increased
leaf areas.

In 35–36 year simulations total net primary production (NPP) increased
asymptotically with CO_2 in Kansas and in Colorado (Fig. 6.5). Above 700
ppm, additional CO_2 increments had less effect than below 700 ppm. NPP
responses were closely paralleled by increases in WUE, the ratio of carbon
fixed to water lost through transpiration. Annual nitrogen mineralization
(Nmin) from microbial decomposition increased at both sites in response to
increasing CO_2 from 175 ppm to 350 ppm, then decreased slightly above
350 ppm. The initial increase, as discussed below, was due to increased
plant litter inputs. With the higher E_m value for Kansas, Nmin increased by
8% between 350 ppm and 700 ppm CO_2. Peak green leaf area indices
asymptotically increased with CO_2 at both sites. Transpiration rate per unit
leaf area declined asymptotically as CO_2 increased. Transpiration per unit
ground area decreased slightly with increasing CO_2 in Kansas and Colorado
above 350 ppm CO_2.

Discussion

Modelled aboveground responses in the Kansas CO_2 exposure study
compared favourably to field responses when averaged over all 3 years. Both
model and field data indicated that the largest CO_2 responses occurred in
1991, and the smallest responses occurred in 1989. There was a lack of
agreement in 2 years. Data indicated that CO_2 stimulated aboveground
growth very slightly in 1989, while the model predicted no stimulation in
1989. Data did not show a CO_2 response in 1990, but the model simulated
a positive response that year. CO_2 responses were small, however, and it
was difficult to distinguish model error from sampling error. Furthermore,
peak standing crop is not an exact measure of aboveground net primary
production.

The GRASS-CSOM model predicted that CO_2 would increase
belowground net primary production, but only weakly in the dry year of
1989. The response in the 1990 root ingrowth data was not matched by the

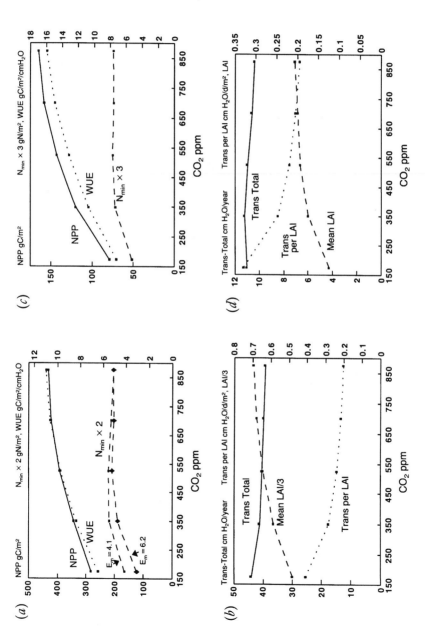

Figure 6.5 *Results of 35–36 year simulations at five different CO₂ concentrations. (a) Total annual net primary production (NPP), water use efficiency (WUE) and annual net nitrogen mineralization (Nmin) responses in Kansas tallgrass. (b) Transpiration per unit ground area (Trans-Total), mean transpiration per unit leaf area (Trans per LAI) and mean leaf area index (LAI) during the growing season in the Kansas simulations. (c) NPP, WUE and Nmin responses in Colorado shortgrass steppe. (d) LAI and transpiration responses in Colorado.*

Table 6.3. *Simulated water budgets. Mean annual flows for 1951–86 in Kansas tallgrass ($E_m = 4.1$) and 1949–84 in Colorado shortgrass (cm H_2O/year) at two CO_2 concentrations*

Site	CO_2 ppm	Precipitation (cm)	Interception (cm)	Drainage (cm)	Transpiration (cm)	Soil evaporation (cm)
Kansas	350	85.5	32.7	2.58	41.2	10.0
	700	85.5	39.1	2.33	39.9	6.4
Colorado	350	32.1	6.3	0.73	11.4	13.6
	700	32.1	8.4	1.6	10.4	11.7

model. The simulated response in 1991 was greater than the response indicated by root ingrowth data. Because root ingrowth data are indices and not actual measures of root growth, an exact fit between model and data may be unlikely.

Belowground primary production responded to elevated CO_2 more strongly than aboveground primary production in simulations as well as the field experiment (Owensby *et al.*, 1993*a*). In long-term simulations aboveground production was stimulated by 12% in Kansas while total NPP was increased by 24% between 350 ppm and 700 ppm CO_2. In Colorado aboveground net primary production was stimulated by 9% while total NPP was stimulated by 32%. A shift in allocation belowground is consistent with the functional balance hypothesis (e.g. Thornley, 1972), which states that as soil resources such as nitrogen and water become more limiting relative to aboveground resources such as light and carbon, allocation should shift belowground to increase soil resource uptake.

Reduced transpiration rate per unit leaf area, coupled with predictions of sustained rates of photosynthesis per unit leaf area under elevated CO_2, led to a 34% increase in daily WUE in Kansas with high E_m and 22% increase with low E_m (Tables 6.1, 6.2). In comparison, a 32% increase was calculated from concentrations and flow rates of CO_2 and H_2O in input and output streams of the open-top chambers (Ham *et al.*, 1995). Over long-term simulations WUE was increased by comparable amounts: 17% in Kansas and 42% in Colorado in April–September (Table 6.2) and by 27% and 39% year-long (Fig. 6.5).

Alterations in nitrogen use and soil nitrogen release can also contribute to positive NPP responses to CO_2. A decrease in senescing plant N : C ratios under elevated CO_2 means that less nitrogen is required to produce each gram of plant biomass. When N is limiting, more total biomass production would be expected. The N : C of senescing tissues is more important than that of live tissues because N is retranslocated during senescence. Increases in nitrogen mineralization (Nmin) under elevated CO_2 can also occur, or increases can offset negative effects of reduced N : C on decomposition. The initial increases from 175 ppm to 350 ppm CO_2, and from 175 ppm to 500 ppm in Kansas with low and high E_m, respectively (Fig. 6.5), were clearly due to increases in plant N inputs into the soil. Similar increases occurred in Colorado, Kansas and Kenya simulations as CO_2 was increased from 350 ppm and 700 ppm and as climates were made warmer and wetter than ambient (Coughenour & Chen, 1996). When plant N demands increase due to increased carbon fixation, despite reduced N : C ratios of dying plant tissues, then more N is returned to soil. Since perennial grasses have a high nitrogen use efficiency, plant uptake is only about 1.5–3.0 gN/year. A 10% higher uptake per year only represents 0.15–0.25 gN/year, for example, which is a very small fraction of the active cycling N (~0.15%). To increase annual plant uptake, and thus Nmin, by 10% only requires a very small increase in active cycling N.

Predictions of Nmin responses to CO_2 vary among models. Another model (GEM2; Chen *et al.*, 1995) predicted slight decreases in Nmin under a C_4 shortgrass but under a C_3 grass, elevated CO_2 increased Nmin by 3–15% under higher rainfall and warmer temperatures. The CENTURY model predicted lower Nmin under elevated CO_2 for Kansas but not Colorado (Hall *et al.*, 1995; D. S. Ojima, W. J. Parton and others, unpublished data). The exact reasons for the differences are unclear. Preliminary model experimentation suggests that the Nmin response is sensitive to the effect of CO_2 on plant N : C ratio, possibly due to the balance between negative effects of low N : C on decomposition rate and positive effects of low N : C on total litter inputs. The response varies among sites, and with precipitation and temperature (Coughenour & Chen, 1996). Thus, it is important to use standardized weather driving variables and management practices when attempting to compare different model results. For example, biomass was annually burned in Kansas as far as this chapter is concerned but not in CENTURY model runs.

Photosynthetic rates per square metre of leaf area were little affected or were decreased by CO_2, due to stomatal closure and reduced leaf N, in the Kansas simulations (Tables 6.1 and 6.2). Likewise, Knapp *et al.* (1993) found no effect of doubled CO_2 on *A. gerardii* photosynthetic rate. Therefore, increases in NPP under elevated CO_2 could not be attributed to photosynthetic responses. Photosynthetic rate was stimulated by CO_2 in the Colorado simulations (Table 6.2), mostly as an indirect result of improved water relations. Results at both sites demonstrated the potential for complex results of indirect effects of CO_2 on water budgets and primary production.

Ecosystem responses to CO_2 were strongly affected by stomatal responses. The model simulated 24–33% reductions in stomatal conductance in response to CO_2 doubling in the Kansas chambers (Table 6.1). Knapp *et al.* (1993), and Ham *et al.* (1995), observed 50% reductions in stomatal conductance under doubled CO_2. However, the observed percentage change in stomatal conductance was greater than the observed percentage change in total transpiration. Conductance data were probably biased upwards because they were collected on sunlit leaves near the top of the canopy (Ham *et al.*, 1995, pers. comm.). Conductances reported by Knapp *et al.* (1993) were measured in well-watered, illuminated laboratory conditions.

The model simulated a 5% reduction in daily transpiration rate in open-top chambers relative to ambient conditions, and no reduction was caused by elevated CO_2 (Table 6.1). In comparison, an energy balance model predicted that the chambers would reduce transpiration rate per ground area by 14%, and that 700 ppm CO_2 would reduce transpiration by another 15% (Owensby *et al.*, 1993*a*). The discrepancy in responses was mainly a result of different leaf area responses to CO_2 in the two models. While CO_2 did not increase green leaf area in the energy balance model, GRASS-CSOM predicted that leaf area would increase under elevated CO_2, thus offsetting the reduced transpiration rate per unit leaf area. In addition, GRASS-CSOM results are integrated over longer time periods. Similar problems can arise in comparing field data with model predictions. Gas flux measurements over a 34 day period indicated that doubled CO_2 reduced total ET rate by 22% (Ham *et al.*, 1995). If leaves remain green longer under elevated CO_2 as suggested by data from Kansas and elsewhere (Curtis *et al.*, 1989; Ham *et al.*, 1995), then mid-season reductions in transpiration should be offset by increases later in the season.

Transpiration per ground area was only reduced by 4% over a 35 year simulation of the Kansas site (Table 6.2). In contrast, the energy balance model indicated that 700 ppm CO_2 would reduce transpiration rates by 21% outside of chambers (Owensby *et al.*, 1993*a*). Transpiration rate per unit leaf area was reduced by 20% in GRASS-CSOM (Table 6.2). Therefore, had green leaf area index not increased in response to elevated CO_2, comparable reductions in transpiration per ground area would have been predicted.

Small responses of transpiration per unit ground area to CO_2, and small CO_2 effects on mean soil water contents, seemed to disagree with findings of lower mid-day leaf water potentials under elevated CO_2 (Knapp *et al.*, 1994). Mid-day leaf water deficits are likely to be more sensitive to CO_2 than soil moisture, however. Mid-day plant water potential is affected by internal plant water storage, and the relative rates of water uptake and loss during the day. A reduction in transpiration rate could reduce mid-day plant water deficit without any change in soil moisture due to more rapid depletion of plant than soil water. During the night, plant water content re-equilibrates with soil water, so pre-dawn water potentials are more indicative of soil water potential than mid-day values.

Increases in leaf area under elevated CO_2 closely compensated for decreases in transpiration rate per unit leaf area over a range of increasing CO_2 concentrations (Fig. 6.5). In the Colorado system, increased leaf area fully compensated for reduced transpiration per unit leaf area, so that transpiration per unit ground area was not affected by CO_2. At the Kansas site, the compensation was not complete, so transpiration per ground area was reduced slightly by elevated CO_2. Leaf area compensation should be expected if water loss per unit leaf area is reduced. In water-limited environments, plants that do not use available water will be outcompeted by plants that do. The model simulated compensation using photosynthetic rate and shoot : root allocation responses to soil water potential. These mechanisms are required for the model to respond realistically to moisture fluctuations. Increased root growth under elevated CO_2 increased belowground biomass, including rhizomes and crowns, which further contributed to shoot growth.

The mechanisms underlying positive shoot growth responses to CO_2 should weaken responses to CO_2 when water is not limiting. When moisture does not limit photosynthesis, reducing transpiration will not reduce the effects of water stress on photosynthesis and increased WUE will not increase primary production. Thus, in 1992 when growing season precipitation was above normal, temperatures were cool, and it was cloudy, CO_2 had no effect on mid-day leaf water potential (Knapp *et al.*, 1993) or peak shoot mass (Ham *et al.*, 1995). The model may have overestimated water stress and thus CO_2 effects in wet years because long-term mean cloud-cover data are used rather than daily data. This should be evaluated in future analyses.

Although conserved water may have little impact in a wet year, the effect may carry over into a subsequent dry year. Stored soil water, conserved from a wet year, may result in a total greater amount of water available for growth the next year. It is possible, moreover, that this could lead to greater total transpiration under higher CO_2, rather than lower total transpiration. Indeed, this carry-over phenomenon was observed in model simulations.

Effects of CO_2 on soil water contents over long simulations were relatively small but could be significant because soil water potential is non-linearly related to soil water content. CO_2 affected soil water most strongly during early to mid growing season in Kansas and mid to late season in Colorado (Fig. 6.4). Later in the season, the differences may narrow as leaves exposed to 700 ppm CO_2 continue to extract water, while leaves in 350 ppm CO_2 have already extracted as much water as they can. Although convergence of soil water between CO_2 levels takes longer in Colorado, it probably continues into November and begins again in March–May.

The small (4%) decreases in total transpiration over 35–36 years in ecosystem water budgets were compensated by increased losses to interception and deep soil drainage (Table 6.3). Interception losses were markedly increased by elevated CO_2, as a result of greater aboveground biomass. While drainage was small, the increase in Colorado was a significant fraction of the reduction in transpiration. The decreases in bare soil evaporation in both simulations under higher CO_2 were notable and were caused by increased shade from litter and shoot mass.

Summary

Interacting responses at the level of the whole plant and ecosystem were involved in net CO_2 responses. Stomatal responses alone would reduce water loss under elevated CO_2, but the ultimate consequences of reduced water loss for plant growth and soil processes are affected by a complex suite of interactions. Realistic models of plant growth in water-limited environments predict increased plant growth when water is more available. Higher amounts and longer durations of green leaves feed-back to enhance season-long water use. Nitrogen or other nutrient limitations could prevent increased growth and water use under elevated CO_2. However, plants require less nitrogen to produce an equivalent biomass at higher CO_2 levels. The negative effects of reduced N : C ratio on decomposition can be offset by an increase in N mineralization, which in turn can be supported by increased plant litter N inputs and a very slight increase in actively cycling N.

It would be useful to derive a simplified representation of primary production responses to CO_2 that could be used in aggregated models like CENTURY. Results here suggest that the main response to CO_2 in C_4 water-limited grasslands is an increase in water use efficiency. WUE can be increased by a direct stimulation of photosynthesis or by a decreased rate of water loss per unit leaf area. An increase in WUE can be achieved through reduced transpiration per unit of ground area as well, but this will be complicated by the extent of leaf area compensation. Actual responses may lie anywhere in between pure WUE and pure transpirational per unit leaf area response.

NPP increased despite little or no reduction in total transpiration per unit ground area over a season (Fig. 6.3, Table 6.2). Water saved through reduced transpiration rate per unit leaf area can be rapidly used to support new growth. In a discrete time step model, the water saved in the current time step may increase growth in the subsequent time step. This process may occur on a daily, hourly, monthly, or seasonal basis. Thus, NPP may increase despite a lack of reduction in transpiration per unit ground area over a month or season. If WUE is increased by decreasing transpiration rate alone, then there may be an unrealistic time lag in the compensatory NPP response in long time step models. Increased WUE and decreased transpiration rates will increase primary production under elevated CO_2 in vegetation that is not water-limited. The relative importance of increased NPP, decreased transpiration and leaf area compensation will probably vary among vegetation types. In slow-growing coniferous forests for example, leaf area compensation over short time frames is unlikely but seems possible over months to years. Grasslands represent the other extreme since they have rapid biomass turnover rates.

References

Ågren, G. I., McMurtrie, R. E., Parton, W. J., Pastor, J. & Shugart, H. H. (1991). State-of-the-art of models of production–decomposition linkages in conifer and grassland ecosystems. *Ecological Applications*, **1**, 118–38.

Amthor, J. S. (1989). Respiration and crop productivity. New York: Springer.

Ball, J. T., Woodrow, I. E. & Berry, J. A. (1987). A model predicting stomatal conductance: its contribution to the control of photosynthesis under different environmental conditions. In *Progress in Photosynthesis Research*, vol. IV, ed. I. Biggins, pp. 221–4. Dordrecht: Martinus Nijhof.

Batchelet, D., Hunt, H. W., Detling, J. K. (1989). A simulation model of intraseasonal carbon and nitrogen dynamics of blue grama swards as influenced by above and belowground grazing. *Ecological Modelling*, **44**, 231–52.

Bazzaz, F. E. (1990). The responses of natural ecosystems to the rising global CO$_2$ levels. *Annual Review of Ecology and Systematics*, **21**, 167–96.

Chen, D. & Coughenour, M. B. (1994). GEMTM: A general model for energy and mass transfer of land surfaces and its application at the FIFE sites. *Agricultural and Forest Meteorology*, **68**, 145–71.

Chen, D., Coughenour, M. B., Knapp, A. K. & Owensby, C. E. (1993). Mathematical simulation of C$_4$ grass photosynthesis in ambient and elevated CO$_2$. *Ecological Modelling*, **73**, 63–80.

Chen, D.-X., Hunt, H. W. & Morgan, J. A. (1995). Responses of a C$_3$ and C$_4$ perennial grass to CO$_2$ enrichment and climate change comparison between model predictions and experimental data. *Ecological Modeling* (in press).

Collatz, G. J., Ribas-Carbo, M. & Berry, J. A. (1992). Coupled photosynthesis–stomatal conductance model for leaves of C$_4$ plants. *Australian Journal of Plant Physiology*, **19**, 519–38.

Coughenour, M. B. (1984). A mechanistic

simulation of water use, leaf angles and grazing in East African graminoids. *Ecological Modelling*, **26**, 127–8.

Coughenour, M. B. & Chen, D. X. (1996). Assessing grassland responses to atmospheric change using linked ecophysiological and ecosystem process models. *Ecological Application* (in press).

Coughenour, M. B., Kittel, T. G. F., Pielke, R. A. & Eastman, J. (1993). Grassland/atmosphere response to changing climate: coupling regional and local scales. Final Report to US Department of Energy, DOE/ER/60932–3. Washington, DC: US Department of Energy.

Coughenour, M. B., McNaughton, S. J. & Wallace, L. L. (1984). Modeling primary production of perennial graminoids: Uniting physiological processes and morphometric traits. *Ecological Modelling*, **23**, 101–34.

Curtis, P. S., Drake, B. G., Leadly, P. W., Arp, W. J. & Whigham, D. F. (1989). Growth and senescence in plant communities exposed to elevated CO_2 concentrations on an estuarine marsh. *Oecologia*, **78**, 20–6.

Dodd, J. L. & Lauenroth, W. K. (1979). Analysis of the responses of a grassland ecosystem to stress. In *Perspectives in Grassland Ecology*, ed. N. R. French, pp. 43–58. New York: Springer.

Dougherty, R. L., Bradford, J. A., Coyne, P. I. & Sims, P. L. (1994). Applying an empirical model of stomatal conductance to three C_4 grasses. *Agricultural and Forest Meteorology*, **67**, 269–90.

Farquhar, G. D., Von Caemmerer, S. & Berry, J. A. (1980). A biochemical model of photosynthetic CO_2 assimilation in leaves of C_3 species. *Planta*, **149**, 78–90.

Field, C. B., Chapin III, F. S., Matson, P. A. & Mooney, H. A. (1992). Responses of terrestrial ecosystems to the changing atmosphere: A resource-based approach. *Annual Review of Ecology and Systematics*, **23**, 201–35.

Hall, D. O., Ojima, D. S., Parton, W. J. &

Scurlock, J. M. O. (1995). Response of temperate and tropical grasslands to CO_2 and climate change. *Global Ecology and Biogeography Letters* (in press).

Ham, J. M., Owensby, C. E. & Bremer, D. J. (1995). Fluxes of CO_2 and water vapour from a prairie ecosystem exposed to ambient and elevated atmospheric CO_2. *Agricultural and Forest Meteorology* (in press).

Ham, J. M., Owensby, C. E. & Coyne, P. I. (1993). Technique for measuring air flow and carbon dioxide flux in large open-top chambers. *Journal of Environmental Quality*, **22**, 759–66.

Houghton, R. A. (1988). The global carbon cycle (letter to the editor). *Science*, **241**, 1736.

Ingram, J. (1994). The wheat model intercomparison. IGBP NewsLetter No. 18, June.

Keeling, C. D. (1986). Atmospheric CO_2 concentrations. Mauna Loa Observatory, Hawaii 1958–1986. NDP-001/R1. Carbon Dioxide Information Analysis Center. Oak Ridge, Tennessee: Oak Ridge National Laboratory.

Knapp, A. K., Hamerlynk, E. P. & Owensby, C. E. (1993). Photosynthetic and water relations responses to elevated CO_2 in the C_4 grass *Andropogon gerardii*. *International Journal of Plant Sciences*, **154**, 459–66.

Larigauderie, A., Hilbert, D. W. & Oechel, W. C. (1988). Effect of CO_2 enrichment and nitrogen availability on resource acquisition and resource allocation in a grass, *Bromus mollis*. *Oecologia*, **77**, 544–9.

Luxmoore, R. J., O'Neill, E. G., Ellis, J. M. & Rogers, H. H. (1986). Nutrient-uptake and growth responses of Virginia pine to elevated atmospheric CO_2. *Journal of Environmental Quality*, **15**, 244–51.

Melillo, J. M., Aber, J. D. & Muratore, J. F. (1982). The influence of substrate quality of leaf litter decay in a northern hardwood forest. *Ecology*, **63**, 621–6.

Melillo, J. M., McGuire, A. D., Kicklighter, D. W., Moore, B.., Vorosmarty, C. J. & Schloss, A. L. (1993). Global climate change and terrestrial net primary production. *Nature*, **363**, 234–40.

Owensby, C. E., Coyne, P. I., Ham, J. M., Auen, L. M. & Knapp, A. K. (1993a). Biomass production in a tallgrass prairie ecosystem exposed to ambient and elevated levels of CO_2. *Ecological Applications*, **3**, 644–53.

Owensby, C. E., Coyne, P. E. & Auen, L. M. (1993b). Nitrogen and phosphorous dynamics of a tallgrass prairie ecosystem exposed to elevated carbon dioxide. *Plant, Cell and Environment*, **16**, 843–50.

Parton, W. J. (1978). Abiotic submodel of ELM grassland simulation model. In *Grassland Simulation Model*, ed. G. Innis. Berlin: Springer-Verlag.

Parton, W. J., Scurlock, J. M. O., Ojima, D. S., Gilmanov, T. G., Scholes, R. J., Schimel, D. S., Kirchner, T., Menaut, J.-C., Seastedt, T., Garcia-Moya, E., Kamnalrut, A. & Kinyamario, J. I. (1993). Observations and modeling of biomass and soil organic matter dynamics for the grassland biome worldwide. *Global Biogeochemical Cycles*, **7**, 785–809.

Pearcy, R. W. & Bjorkman, O. (1983). Physiological effects. In *CO$_2$ and Plants: The Response of Plants to Rising Levels of Atmospheric Carbon Dioxide*, ed. E. R. Lemon, pp. 65–106. Boulder, Colorado: Westview Press.

Rotty, R. M. & Marland, G. (1986). Fossil fuel consumption: recent amounts, patterns, and trends of CO_2. In *The Changing Carbon Cycle: A Global Analysis*, ed. J. R. Trabalka & D. E. Reichle. New York: Springer.

Running, S. W. & Hunt, E. R. (1993). Generalization of a forest ecosystem process model for other biomes, BIOME-BGC, and an application for global-scale models. In *Scaling Physio-*

logical Processes: Leaf to Globe, ed. J. R. Ehleringer & C. B. Field, pp. 141–58. New York: Academic Press.

Ryan, M. (1991). The effects of climate change on plant respiration. *Ecological Applications*, **1**, 157–67.

Ryan, M. G., Hunt, E. R. Jr., Ågren, G. I., Friend, A. D., Pulliam, W. M., Linder, S., McMurtrie, R. E., Aber, J. D., Rastetter, E. B. & Raison, R. J. (1995a). Comparing models of ecosystem function for temperate conifer forests. I. Model description and validation. In *Effects of Climate Change on Forests and Grasslands*, ed. J. Melillo & A. Breymeyer. New York: Wiley (in press).

Ryan, M. G., McMurtrie, R. E., Hunt, E. R. Jr., Friend, A. D., Pulliam, W. M., Agren, G. I., Aber, J. D. & Rastetter, E. B. (1995b). Comparing models of ecosystem function for temperate conifer forests. II. Simulations of the effect of climate change. In *Effects of Climate Change on Forests and Grasslands*, ed. J. Melillo & A. Breymeyer. New York: Wiley (in press).

Sims, P. L., Coughenour, M. B., Bailey, D. W. & Bradford, J. A. (1995). Evaluation of two gas exchange mathematical simulation models for C_4 grasses. In *Proceedings of the Fifth International Rangeland Congress*, ed. N. E. West. Logan, Utah.

Sionit, N., Strain, B. R., Hellmers, H., Riechers, G. H. & Jaeger, C. H. (1985). Long-term atmospheric CO_2 enrichment effects and the growth and development of *Liquidambar styraciflua* and *Pinus taeda* seedlings. *Canadian Journal of Forestry Research*, **15**, 468–71.

Thornley, J. H. M. (1972). A balanced quantitative model for root : shoot ratios in vegetative plants. *Annals of Botany*, **36**, 431–41.

Woodwell, G. M. (1988). The global carbon cycle (letter to the editor). *Science*, **241**, 1736–7.

Part three

Ecosystem structure

The importance of structure in understanding global change

H. H. Shugart

Introduction

Structure, in the context of ecosystem studies, can have a variety of meanings. An ecosystem (*sensu stricto*; Tansley, 1935) is an abstract 'system of definition' in the parlance of the systems scientist. Thus, the structure of an ecosystem is strongly related to the details of the definition of the particular ecosystem in question. To many, ecosystem structure betokens the look, the arrangement, or the physiognomy of an ecosystem. Thus, tropical rain forests, with their complex canopies and dominance of the tall tree life form, are of one structure. Monolayered, homogeneous, tall grasslands are a different structure. Cloud forests found in tropical montane situations and formed from short trees burgeoning with epiphytes are a third, still different, structure. However, to the community ecologist, the 'structure' of an ecosystem might connote the arrangement of interactions among the components of an ecosystem. The structure might be captured in terms of such features as the degree of branching in the nodes of food webs, the average number of competitors a population has, or the frequency of species and their relative rareness and commonness (Cornell & Lawton, 1992; DeAngelis, 1992).

Since the classic studies of A. S. Watt (1947) on 'pattern and process' in natural ecosystems, the idea that the spatial structure of ecological systems in two dimensions (in some cases, three dimensions) is derived from underlying ecological processes has been a central concept among vegetation scientists (Whittaker & Levin, 1977; White, 1979; Bormann & Likens, 1979; Whitmore, 1982). Perhaps because of the influence of the pattern and process view of the dynamics of vegetation, the 'structure' of an ecosystem is sometimes taken to suggest a static description of such features as the leaf layering, the biotic diversity or spatial pattern of ecosystems. While these 'structural' properties can be measured with a considerable level of quantification, they are seen as variables of description, not variables of dynamic change.

As ecology has moved from a largely descriptive science to a science with greater emphases on dynamics – the importance of understanding structural features of ecosystems has taken something of a lesser position to the importance of understanding the processes or the functions of ecosystems. It is sometimes easy to forget that 'structure and function' (as a classic biologist might use the term) or 'pattern and process' (a phraseology more frequently used by ecologists), are portrayals of two mutually causal agents.

Biology has a central tenet: the concept that the form or shape of entities is both modified by and a creator of function. Processes cause patterns to occur; patterns alter the magnitude and the direction of processes. Both are part of a *yang* and *yin* of a science that, for example and at different levels, associates the three-dimensional form of an enzymatic protein with its function in biochemical reactions, the morphology of the limb of an animal with the rate at which it can run, the length of a bird's first primary feather with the lift the wing will generate, the geometry of a flower with that of its potential pollinators, or the layering of a plant canopy with its ability to capture light.

Time and space scale and the response of ecosystems to change

An appreciation of space and time scale interactions in ecological systems has been identified as a necessary precursor to understanding how ecosystems will respond to large scale environmental change (O'Neill, 1988). Further, the experience in building interdisciplinary research teams indicates that an attention to space and time scales may not guarantee success, but not to do so seems to enhance the likelihood of failure (Shugart & Urban, 1988). This attention to scale has been highlighted in the development of hierarchy theory in ecology (Allen & Starr, 1982; O'Neill *et al.*, 1986) and may be its most important contribution.

Woodward (1987) considered the responses of different plant processes to various components of the characteristic cycles of variation in climate by aligning the plant processes that responded to different frequencies of input variation. Expansion and contraction of the ranges of different types of plants (late successional trees, herbaceous perennials, annuals) responded to annual or longer frequencies of climatic variation. The late successional trees were thought to require climatic cycles of multiple centuries to induce a contraction in their distribution. The responses of other types of plants were more rapid and the tendency was to expect range expansion to be driven by frequencies of variation that were at slightly higher frequencies than range contractions. These responses were not felt to be influenced by the monthly to daily variations in the climate even though this scale of variation accounts for a large proportion of the variance in climate. The daily to monthly periodic variations excite a different aspect of the plant response (flowering, germination) and the significant variation in the climate signal at minute-to-second periods are involved with another aspect of the plant response (stomatal opening, leaf gas exchange).

Time and space scale are of significant importance in understanding system dynamics. Several authors have attempted to categorize phenomena of interest in the time and space domain. One such example is Delcourt *et al.* (1983) who considered the time and space scale of different disturbance factors, the ecological mechanisms of these phenomena, and the patterns produced by the interactions between the disturbance and the ecological mechanisms. Delcourt *et al.* (1983) illustrated this in a three-part diagram in which the disturbances, biotic responses, and resultant patterns were indicated at the space and time intervals over which they were typically measured.

Different disturbances operate on the vegetation systems at different frequencies. Just as in the Woodward (1987) example above, inputs of different frequencies in time or space induce responses from different vegetation processes and can be thought of as producing different patterns. The time and space scales of plate tectonics in separating, moving and melding the continents over millions of years are an active disturbance on the evolution of the biota and on the diversity of the vegetation formations for continental-scale zones. Plate tectonics is sufficiently slow as to be a non-consideration in understanding the response of ecological succession to decade-to-century variations in wildfires and producing differences in local vegetation composition (types and subtypes).

In attempting to understand how the terrestrial surface might respond to large scale changes in the environment, ecologists are required to consider new environmental conditions operating on processes in unusual time and space scales. The potential scales of consideration are quite broad. Further, ecologists are realizing that the systems with which they deal have been subjected to a varied history of change at all time scales.

The scale-related consequences of structure

The geometry of plants can have a profound effect on the manner in which they function. Fig. 7.1 illustrates typical responses of net photosynthesis of plants to changes in the light level. Bazzaz (1979) illustrated the tendency for plants found in early succession to have high levels of carbon fixation at high light levels and lower levels of carbon fixation at low light intensities relative to late successional plants (Fig. 7.1a). This pattern, manifested at the leaf tissue level, has been implicated as a mechanism that gives plants different tolerances to shading and thus can be used to explain patterns in the successional patterns of ecological systems (e.g. Bazzaz, 1979; Bazzaz & Pickett, 1980; Kozlowski *et al.*, 1991). This is also one of the several basic responses of plant tissues to their environment that global changes in the environment (e.g. alterations of CO_2 levels in the atmosphere, temperature) might change.

Horn (1971) inspected the case of trees with equivalent tissue level responses to changes in light levels but with different geometries with respect to leaf layering and produced the whole plant net photosynthesis curves illustrated in Fig. 7.1b. These curves demonstrate the same pattern as that summarized by Bazzaz (Fig. 7.1a), but arise, not from differences in leaf tissue physiology and performance, but as a consequence of structure, alone. Plants can differ in their ability to utilize light due to their geometry (Horn, 1971) as well as their physiology (Bazzaz, 1979). It seems likely that combinations of physiology and geometry might amplify the effect of a seeming trade-off in the ability of a plant to use light at different light levels.

Körner (1993) reviewed over 1000 published papers to determine the response of plant systems at several different levels (single plant, cultivated plants, natural vegetation) to conditions of elevated CO_2. He found that at higher levels of organization one considered in judging the response (leaf photosynthesis, plant growth, ecosystem yield) and over longer periods of time of observation (from hours to years), the magnitude of the positive effects of elevated levels of CO_2 were attenuated. The causes of these responses are potentially many, ranging from the tendency for plants to outgrow their pots in longer-term greenhouse studies and thus slow their growth to a 'down regulation' of photosynthesis in high CO_2 conditions.

Science progresses by simplifying the systems under consideration. In experimental science, one attempts to simplify by controlling, to as great a degree as possible, all extraneous factors with respect to a given experimental objective. The reason that the experimentalist desires to control more complex interactions in the design of an experiment is because these interactions alter and confuse experimental outcomes. The responses that Körner (1993) documents are the sort expected in treating a fundamental response of an ecosystem component at progressively more complex levels of interactions. The structure and variation of structure in natural systems is one of the principal sources of single-effect-altering interactions.

One would expect that the structure of ecosystems should have the effect

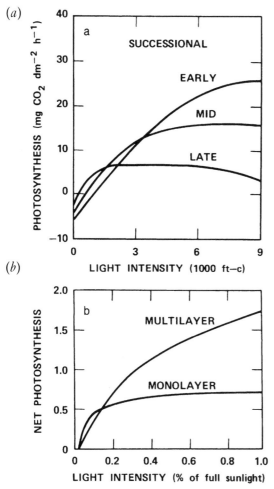

(a)

(b)

Figure 7.1 *Typical responses of net photosynthesis in plants as a function of light level. Note that at the tissue level and at the canopy level, there is a reversal of photosynthetic efficiency at high versus low light levels. (a) Idealized light saturation curves for early, mid, and late successional plants (from Bazzaz, 1979). (b) The effect of light on net photosynthesis in multilayered and monolayered trees (from Horn, 1971).*

of altering the manifestation of changes in rates or processes in terrestrial ecosystems. Indeed, there appears to be a hierarchical spectrum of structural effects that can alter the response that attends a change in a fundamental process (Table 7.1). These structurally controlled modifications of system response are non-trivial in many cases – in some of the example cases shown in Table 7.1 they can reduce the potential response of the system by 70% or more. Other changes in system structure can amplify potential responses in the same sorts of ways.

The examples in Table 7.1 are all examples in which the growth of

Table 7.1. *Structure of systems as a mediating factor in the system response to changes in fundamental processes. This table is intended to present examples of what are a wide range of structural responses that can either attenuate or amplify the process response*

Level of response	Structural change	Functional implication
Leaf tissue	Change in stomatal index of plants grown in different CO_2 environments. Effect can be seen in plant material collected before the industrial revolution and can be induced under laboratory conditions (Wood-ward, 1987)	Alteration of the stomatal conduction response of the plant. Implication that plant responses historically may be different from present responses (due to change in stomatal index induced by ambient atmospheric conditions)
Individual plant	Rates of leaf photosynthesis can increase of the order of 50% for C_3 plants in response to a doubling of CO_2 but rates of whole plant growth are often less than 20% of those for control conditions (Körner, 1993)	Structural considerations including photosynthate alloca-tion and internal interactions (e.g. with nutrients such as nitrogen) can moderate the carbon fixation at the leaf level
Plant stand	The increase in stand biomass is less by a factor of about 0.30 (Shugart & Emanuel, 1985) than the increase in growth of the individual plants comprising the stand	Stand interactions (compe-tition, shading etc.) reduce the stand level biomass increase (or yield) in response to increased growth rates of individual plants
Vegetated landscape	Mosaic properties of landscapes alter the stand biomass response to changed conditions (Bormann & Likens, 1979)	Landscapes can be thought of as mosaics in different stages of recovering from natural disturb-ances. Changes in plant and stand processes are mediated by the local state of disturb-ance recovery
Region	The terrestrial surface can alter-nate between being a source to a sink of carbon in the transient response to environmental change, even in cases in which the long term response to change is similar to the initial condition (Smith & Shugart, 1993)	Shifts in vegetation in response to change are delayed by large scale processes involving dispersal, recovery and other inertial effects.

plants and yield of natural ecosystems is reduced from what one might expect from a one-to-one response to increased photosynthetic performance. Of course, the effects of structure on ecosystem processes can also be positive as well as negative. One group of scientists who have very successfully manipulated structure of systems to amplify (rather than attenuate) growth responses in plants are agricultural scientists who have managed to use spatial patterns (optimal planting densities) and other interactions (phenology, timing of patterns of photosynthate allocation, etc.) to greatly increase the yields of important crops.

Ecosystem structure and global change

As will be demonstrated in the chapters that follow, the understanding of ecosystem processes involving productivity, biogeochemical cycling and physical interactions with the environment are all mediated by structural constraints at a wide range of organizational levels. Focus 2 in the GCTE programme has traditionally been strongly motivated to explore theoretical consequences of change through the use of tested ecological models designed to represent ecosystem responses at a hierarchy of spatial scales (patch, region or landscape, continental or global). The experimental and observational tests of the model-based theory being developed in Focus 2 (and the synthesis that these models represent) are an important aspect of the future directions in understanding the consequences of ecosystem structure.

One particularly vexing problem in predicting the response of the Earth's terrestrial vegetation to global change presents itself as a problem in developing models that can simultaneously produce predictions of the changes in both structure and function of ecosystems in response to global change.

For example, one approach to simulating the response of landscapes to change is to assume that the internal workings of the landscape are sufficiently well mixed to allow the landscape to be simulated in its entirety without resorting to consideration of the dynamics of the various component parts. Models based on an assumption that the important processes in an ecosystem can be approximated in aggregate without explicit consideration of spatial heterogeneity originate from initial attempts to formulate ecosystem models that have a rich history of development and application. Such models 'scale-up' the response of processes as they are understood at the smaller scale to the landscape or regional system response. For example, one might assume that the fluxes of heat, water and CO_2 associated with the functioning of a single leaf are duplicated by the sum of the responses of the billions of individual leaves comprising a vegetated landscape. In some cases, the underlying assumption is that the mathematical structure of particular biophysical and chemical reactions at the landscape level resemble those observed at a detailed level (perhaps with some differences in the model parameters). These models are usually the type of ecological models linked to other models of ocean or atmospheric dynamics to assess the feedbacks among these major Earth systems (Ojima, 1992).

There is a rich array of models of this class. For example, most models of element cycling in watersheds or other ecosystems tend to view the processes as being homogeneous within the system of definition. Indeed, traditional ecosystem models were often referred to as point models because they simulated the dynamics of ecosystems with no explicit references to spatial heterogeneity. Two types of these models that have been widely applied in the context of changing environmental conditions and the feedbacks between vegetation change and other global changes are those that simulate a regional plant canopy and those that are focused on the storage and transfer of material. Examples of each of these will be discussed in the remainder of this chapter.

In their application to larger space–scale problems, the different landscape models have strengths and weaknesses that are, to a degree, complementary. Many of the models that simulate the landscape elements for mosaic landscapes have demonstrated a capability to predict changes in the structure of vegetation associated with environmental change, but are limited in continental- and global-scale applications by a lack of the species-specific information that they require. Interactive landscape models have provided cautionary results that point to the potential importance of spatial interactions as factors that can alter the rates of processes to a significant degree, but these models tend to be data- and information-demanding to a degree that currently limits their applications to case studies. Models of ecosystem material flows scale up to larger scales relatively easily (given the appropriate base data sets), but have little in their internal mechanisms to change the ecosystem structure and the feedbacks of structure onto the underlying processes.

In part due to the increase in cheap computational power, chimeras that combine different models (sometimes with different underlying assumptions) in ways that compensate one another's weaknesses have been produced as solutions to these problems. The HYBRID model which uses a gap model to change the physical structure of the plant canopy (based on the allometry and the performance of individual plants) and a canopy model to capture the productivity of the canopy to determine the water use and carbon fixation is an example of such a development. Other mixed models, such as the DOLY model, are intended to tackle the difficult problem of mixing a material flow model with phenomena associated with canopy processes. The data to test such models, as well as the understanding of the numerical and analytical features of these models, are still a 'work in progress'.

Nevertheless, we are beginning to apply a variety of landscape models to the problem of understanding the ways that natural ecosystems might respond to large-scale environmental change. Some examples of these findings will be the topic of the next three chapters and are part of an ongoing research agenda.

In a practical sense, the two most important issues to be addressed in understanding the response of the terrestrial surface of our planet to environmental change are the understanding of the present and future role of humans in altering landscape pattern and function, and the effect of the altered concentration of CO_2 in the Earth's atmosphere on terrestrial processes – either indirectly through modification of the Earth's climate, or directly through alterations in fundamental plant processes related to photosynthesis.

References

Allen, T. F. H. & Starr, T. B. (1982). *Hierarchy: Perspectives for Ecological Complexity*. Chicago: University of Chicago Press.

Bazzaz, F. A. (1979). The physiological ecology of plant succession. *Annual Review of Ecology and Systematics*, **10**, 351–71.

Bazzaz, F. A. & Pickett, S. T. A. (1980). Physiological ecology of tropical succession: A comparative review. *Annual Review of Ecology and Systematics*, **11**, 287–310.

Bormann, F. H. & Likens, G. E. (1979). *Pattern and Process in a Forested Ecosystem*. New York: Springer.

Cornell, H. V. & Lawton, J. H. (1992). Species interactions, local and regional processes, and limits to the richness of ecological communities: A theoretical perspective. *Journal of Animal Ecology*, **61**, 1–12.

DeAngelis, D. L. (1992). *Dynamics of Nutrient Cycling and Food Webs*. London: Chapman and Hall.

Delcourt, H. R., Delcourt, P. A. & Webb III, T. (1983). Dynamic plant ecology: The spectrum of vegetation change in time and space. *Quaternary Science Reviews*, **1**, 153–75.

Horn, H. S. (1971). *The Adaptive Geometry of Trees*. Princeton, NJ: Princeton University Press.

Körner, C. (1993). CO_2 fertilization: The great uncertainty in future vegetation development. In *Vegetation Dynamics and Global Change*, ed. A. M. Solomon & H. H. Shugart, pp. 53–70. New York: Chapman and Hall.

Kozlowski, T. T., Kramer, P. J. & Pallardy, S. G. (1991). *The Physiological Ecology of Woody Plants*. San Diego: Academic Press.

O'Neill, R. V. (1988). Hierarchy theory

and global change. In *Scales and Global Change*. SCOPE 35, ed. T. Rosswall, R. G. Woodmansee & P. G. Risser, pp. 29–45. Chichester: John Wiley.

O'Neill, R. V., DeAngelis, D. L., Waide, J. B. & Allen, T. F. H. (1986). *A Hierarchical Concept of Ecosystems*. Princeton: Princeton University Press.

Ojima, D. (ed.) (1992). *Modeling the Earth System*. Boulder, Colorado: UCAR / Office for Interdisciplinary Earth Studies.

Shugart, H. H. & Emanuel, W. R. (1985). Carbon dioxide increase: The implications at the ecosystem level. *Plant, Cell and Environment*, 8, 381–6.

Shugart, H. H. & Urban, D. L. (1988). Scale, synthesis and ecosystem dynamics. In *Essays in Ecosystems Research: A Comparative Review*, ed. L. R. Pomeroy & J. J. Alberto, pp. 279–90. New York: Springer.

Smith, T. M. & Shugart, H. H. (1993). The transient response of terrestrial carbon storage to a perturbed climate. *Nature*, **361**, 523–6.

Tansley, A. G. (1935). The use and abuse of vegetational concepts and terms. *Ecology*, **16**, 284–307.

Watt, A. S. (1947). Pattern and process in the plant community. *Journal of Ecology*, **35**, 1–22.

White, P. S. (1979). Pattern, process and natural disturbance in vegetation. *Botanical Review*, **45**, 229–99.

Whitmore, T. C. (1982). On pattern and process in forests. In *The Plant Community as a Working Mechanism*. Special Publ. No. 1, British Ecological Society, ed. E. I. Newman, pp. 45–59. Oxford: Blackwell Scientific.

Whittaker, R. H. & Levin, S. A. (1977). The role of mosaic phenomena in natural communities. *Theoretical Population Biology*, **12**, 117–39.

Woodward, F. I. (1987). Stomatal numbers are sensitive to increases in CO_2 from pre-industrial levels. *Nature*, **327**, 617–18.

8 The application of patch models in global change research

T. M. Smith and H. H. Shugart

Introduction and background

The term 'patch model' as used in this chapter refers to models which examine vegetation dynamics at a spatial scale corresponding to the area occupied by a small number of mature individual plants. In many cases, this is approximately the size of a plot or quadrant used for vegetation sampling. There are several rather different modelling approaches that can be called patch models. For example, many of the models developed to summarize element cycles and carbon metabolism of ecosystems (notably the developments during the International Biological Programme of the 1960s and 1970s) were applied at relatively small spatial scales. Indeed, these models often were referred to by their originators as 'point models' to emphasize that they did not cover large spatial areas. These early ecosystem models were developed to duplicate experimental or observational results obtained from studies that were developed on relatively small tracts of land.

Patch models emphasize the dynamics of ecosystems at relatively small spatial scales. The initial reasons for this emphasis lay with a need to model at a spatial scale at which data are collected, and with the necessity to assume a degree of spatial homogeneity in the model formulation. Recently, a recognition of the importance of treating phenomena that do not 'scale up' easily to larger spatial scales has reinforced an interest in patch modelling.

An important class of early patch models were individual-based models of forest succession based on the growth of the individual trees. These models were developed by quantitatively oriented foresters and were focused toward practical issues in production forestry. Computer models that simulate the dynamics of a forest by following the fates of each individual tree in a forest stand also were developed initially in the mid 1960s. The earliest such model was developed by Newnham (1964) and this was followed by similar developments at several schools of forestry (Lee, 1967; Mitchell, 1969; Lin, 1970; Bella, 1970; Hatch, 1971; Arney, 1974). The models predicted change using a digital computer to dynamically alter a map of the sizes and positions of each tree in a forest (often a plantation). The essential approach was to represent the three-dimensional spatial inter-actions among individual trees. For this reason, these initial models are as complex and as detailed as their descendants. These models have become increasingly used as the computer power available to ecologists has increased (Munro, 1974; Shugart & West, 1980).

These dynamic mapping techniques were applied initially to even-aged plantations but applications in more complex forests soon followed. From these first modelling efforts, individual-based tree models tended to feature simplifications of these initial models. The early individual-tree-based simu-lators took what was known from yield tables and other data sets and developed a more flexible, quantitative methodology for prediction. Some of the earliest attempts to apply such models were very successful and produced results of surprising detail.

An important sub-category of individual-organism-based patch models that have been widely used in ecology (as opposed to traditional forestry applications) are the so-called gap models (Shugart & West, 1980). The first such model was the JABOWA model (Botkin *et al.*, 1972; Botkin, 1993) developed for forests in New England. Over the past 20 years, gap models have been developed for a wide variety of forest ecosystems, from boreal to tropical, and the general approach has been extended to non-forested ecosystems such as grasslands, shrublands and savannas.

Gap models feature relatively simple protocols for estimating the model parameters (Botkin *et al.*, 1972; Shugart, 1984). For many of the more common and particularly for the commercially important temperate and boreal forest trees, there is a considerable body of information on the performance of individual trees (growth rates, establishment requirements, height/diameter relations) that can be used directly in estimating the parameters of such models. Gap models have simple rules for interactions among individuals (e.g. shading, competition for limited resources) and equally simple rules for birth, growth and death of individuals. The simplicity of the functional relations in these models can be seen to have both positive and negative consequences. The positive aspects are largely involved in the ease of estimating model parameters for a large number of species. The negative aspects relate to the lack of physiological mechanism in the description of growth and environmental response. Much of the current research in the development of patch models is related to over-coming these limitations.

One of the research objectives defined in The Operational Plan (1992) of the Global Change and Terrestrial Ecosystems (GCTE) Core Project within the International Geosphere-Biosphere Programme (IGBP) is 'to develop patch models of ecosystem dynamics for global application, incorporating mechanistic information on the responses of plant processes to global change and the influences of these responses on ecosystem structure' (Task 2.1.3 of IGBP Report No. 21, Stockholm, 1992). As an initial step to meet this stated objective of GCTE, a workshop was held in Apeldoorn, Netherlands, the week of 28 March–1 April 1994. The objective of the workshop was to evaluate the application of individual-based patch models of vegetation dynamics to global change issues. The workshop brought together an international group of scientists involved in the development and application of patch models to discuss the current state of the science, examine the limitations of these models in addressing global change issues, and to make recommendations for future research and model development.

The results of this workshop are in press as a special issue of *Climatic Change*. In this chapter, we will provide a summary of the progress to date and recommendations for future work in the application and development of patch models for global change related issues. This chapter is divided into five sections relating to the workshop presentations and discussions: (1) Application of patch models to examine regional sensitivity to climate change; (2) Model comparison; (3) Current model limitations and associated research needs; (4) Recent developments; and (5) Future research.

Application of patch models to examining regional sensitivity to climate change

Patch models have been widely used to examine the sensitivity of ecosystems to climate change, both past and future. However, comparisons of these results have been hindered by differences among the studies in the scenarios used and the manner in which they have been applied. An important objective of the workshop was to conduct regional sensitivity analyses with existing patch models using a standard set of climate change scenarios. Participants were provided data for monthly precipitation and temperature predictions from four general circulation models and asked to run a transient change exercise that consisted of implementing an equilibrium $2 \times CO_2$ climate over a period of 100 years. The changes were to be realized at the rate of 1% per year. Models were run under current climate conditions for a time period sufficient to establish steady state conditions (usually hundreds of years), the climate change scenario was then introduced, and the model was run for an additional period to establish a new equilibrium (steady state). While the focus of the meeting was on the transient responses, it was considered important to evaluate the initial and following steady states as an aid to interpreting the transient results.

Table 8.1 *Authors, ecosystem types, and locations for which sensitivity analyses were conducted and presented at the IGBP/GCTE workshop, Apeldoorn, Netherlands, 28 March – 1 April 1994*

Authors and institution	Ecosystem type	Location
H. Bugmann Systems Ecology, Swiss Federal Institute of Technology Zurich, CH-8952 Schlieren, Switzerland	Deciduous/coniferous forest	European Alps
D. P. Coffin & W. K. Lauenroth Rangeland Ecosystem Science Department, Colorado State University, Fort Collins, CO 80523, USA	Grassland	Central USA
S. G. Cumming & P. J. Burton Department of Forest Science, University of British Columbia, Vancouver, B.C. V6T 1Z4, Canada	Coniferous forest	British Columbia
P. V. Desanker School of Forestry and Wood Products, Michigan Technological University, Houghton, MI 49931, USA	Miombo woodland	Malawi
K. Kramer & G. M. J. Mohren Institute of Forestry and Nature Research, IBN-DLO, 6700 AA Wageningen, The Netherlands	Decidous forest	The Netherlands
M. Linder & P. Lasch Potsdam-Institute for Climate Impact Research, Telegrafenberg, D-14412 Potsdam, Germany	Deciduous/coniferous forest	Northeast Germany
Philippe Martin Institute for Remote Sensing Applications, Commission of the European Communities, Joint Research Center, I-21020 Ispra (VA), Italy	Deciduous/coniferous forest	Minnesota, USA

Table 8.1 (cont.)

Authors and institution	Ecosystem type	Location
T. Oja & P. A. Arp Institute of Geography, University of Tartu, Estonia Department of Forest Resources, University of New Brunswick, Canada	Deciduous/coniferous forest	Ontario, Canada
Guofan Shao Department of Environmental Sciences, University of Virginia, Charlottesville, VA 22903, USA	Deciduous/coniferous forest	Northeastern China
A. K. Stevens, A. D. Friend & D. C. Mobbs Institute of Terrestrial Ecology, Bush Estate, Midlothian, EH26 0QB, UK	Coniferous forest	British Isles
M. Sykes Department of Plant Ecology, Lund University, S-223 61 Lund, Sweden	Coniferous forest	Southern Sweden
Mikhail Vidyushkin Center for Problems of Ecology and Productivity of Forests, RAS, Novocheremushkinskaya 69, Moscow 117418, Russia	Coniferous forest	Eurasia
Y. Xiaodong & Z. Shidong Institute of Applied Ecology, Academia Sinica, Shehyang 110015, China	Deciduous/coniferous forest	Northeastern China

Thirteen groups presented results at the meeting (Table 8.1) and 17 groups will contribute to the final workshop report. Areas represented were mostly forests. The actual sites simulated ranged from single locations to multiple locations throughout a region. Predicted ecosystem responses varied depending upon the specifics of the current climate, current vegetation structure, and the magnitude and seasonal distributions of precipitation and temperature changes. Three examples will be described to illustrate the kinds of results that were reported.

A mixed coniferous-deciduous forest at the Changbaishan Biosphere Reserve in the Peoples Republic of China had a dramatic response to changed climate with a major dieback of the existing forest followed by a pulse of regeneration leading to a new forest of only deciduous species (Fig. 8.1). A similar, but much less dramatic in terms of loss of biomass, response was predicted for a mixed coniferous-deciduous forest in the European Alps (Fig. 8.2). The transient sensitivity to climate change included a decrease of approximately 30% in stand biomass. Following the attainment of changed conditions (i.e. temperature and precipitation) biomass recovered, but species composition of the recovered forest was very different from the pre-climate change conditions.

Responses of C_3 and C_4 grasses in the central United States were predicted to show varying responses depending upon interactions between current and changed patterns of temperature and precipitation (Fig. 8.3). In the driest part of the region all sites, regardless of their temperature regime, were dominated by C_4 species before the onset of the change scenario. As a result of climate change the dominance of C_4 species increased on the two coolest sites and decreased on the warmest site (Fig. 8.3a). In the wettest

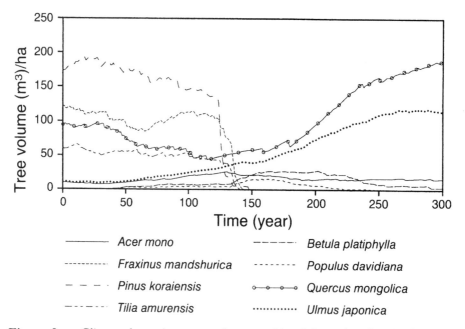

Figure 8.1 *Climate change impacts on the compositional dynamics of a mixed deciduous Korean pine forest at Changbaishan Biosphere Reserve in China. Changes in temperature and precipitation are based on the GFDL 2 × CO₂ equilibrium scenario and were applied using a linear interpolation between years 50 and 150 of the simulation. Climate was assumed to be constant following year 150.*

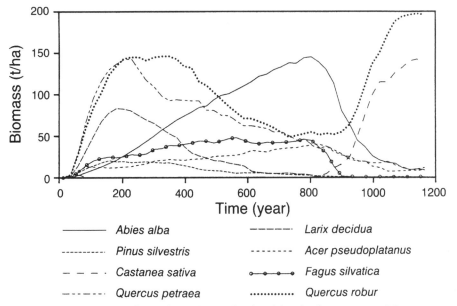

Figure 8.2 *Transient forest dynamics as simulated by the forest gap model FORCLIM (Bugmann, 1994) at the grid point 47.5°N 8.5°E. The simulation starts from bare ground, and the graph shows the average dynamics from 200 patches. The first 800 years are simulated under current climate. Between the years 800 and 900, a linear change of average temperature and precipitation takes place. The change scenario is based on the GFDL 2 × CO₂ equilibrium scenario. The climate after year 900 is assumed to be constant.*

part of the region, C$_4$ species dominated all three sites before and after the change scenario was implemented (Fig. 8.3c). Sites with an intermediate moisture regime showed the largest range of difference in dominance before the climate change was implemented and this range narrowed as a result of the new climatic conditions (Fig. 8.3b).

Two things were clear from the large collection of examples of the sensitivities of patch models to predictions of climate change. First, all the models predicted important changes in ecosystem structure as a result of climate change. The explanations for the predicted changes were often model-specific and traceable to elements of the formulations of the relationships. Second, all the groups had in-depth appreciation for the major sources of uncertainty associated with the sensitivity of their model. These ideas stimulated some of the most productive discussions of the workshop concerning the direction that both empirical research and model development need to take in order to produce the next generation of patch models.

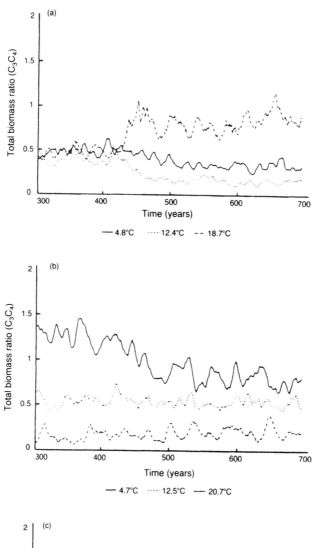

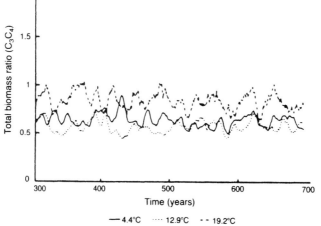

Model comparisons

One of the major points of discussion at the workshop was the development of a protocol for model comparison. In general, comparability of models has not been a high priority in the past where the primary applications have been site- or region-specific. The majority of patch models presented at the workshop share a common structure in the manner in which they simulate the demographics of plant populations. The models simulate the processes of establishment, growth and mortality of individual plants on a plot of ground which is scaled to the maximum size achievable by a single individual (i.e. size of plot or patch dependent on size of dominant species). Despite this similarity in structure, there is a great deal of variation among models in the functions used to describe plant growth and in the inclusion and description of environmental factors which influence the plant processes of growth and survival. Some of the differences among models are a result of the necessity to include important environmental processes influencing plant dynamics within a given region (e.g. permafrost formation in a boreal zone). However, much of the variation is due to different formulations of the same environmental processes (e.g. characterization of moisture availability and its effects on growth).

Many changes to the underlying structure of patch models have been to include additional aspects of species life history. In adapting the underlying patch model structure to different ecosystems or geographic regions, it has been necessary to include additional aspects of species life history as they relate to establishment, mortality and reproduction. Features such as chilling requirements, frost tolerance, herbivore defence, or specific requirements for germination (e.g. fire) may be necessary to simulate the dynamics of species composition for a given ecosystem. The necessity of including these ecosystem-specific species characteristics further complicates comparisons among models.

Variation in the methods for calculating species optimal growth functions can result in different growth parameters for the same species. There are a variety of approaches being used to define species-specific optimal growth

Figure 8.3 *Simulated total biomass (above- and belowground) ratios of C_3 and C_4 species at nine sites representing the existing ranges of mean annual temperature and mean annual precipitation in the central grassland region of the United States. Sites were selected to represent similar ranges of mean annual temperature under dry (a), intermediate (b), and wet (c) moisture conditions. Each of the three lines on each graph represents a site with a different mean annual temperature. Simulations were run for 700 years and the climate change scenario was implemented between years 400 and 500.*

curves (i.e. size as a function of time or size increment as a function of size). These methods include the original formulation based on simple silvicultural data (e.g. maximum observed size, age, height), empirically derived functions from remeasurement data or tree ring analyses, and the use of more physiologically detailed models of individual plant growth. This diversity of methods can (and does) result in different optimal growth functions being defined for the same species. The sensitivity of species composition, stand structure or biomass dynamics to these variations in parameterization are largely unknown.

Model results have been shown to be sensitive to the formulations used to describe environmental parameters. Results from a number of studies presented at the workshop show that the behaviour of patch models is quite sensitive to the exact mathematical representation of environmental conditions, especially those that are dependent on weather. For example, the representation of soil moisture is influenced by the model used to describe evapotranspiration and the manner in which soils are represented (e.g. one or n layers). In some cases this sensitivity becomes apparent only when the behaviour of the model variants was compared under a scenario of climate change. The divergence of predicted patterns using different formulations of the same processes has major implications on the utility of the models to assess climate change impacts.

As a continuing research priority, work needs to be undertaken to examine the consequences of differences in model structure on the patterns of plant dynamics predicted by the models. Since each patch model has been developed for the purposes of examining vegetation dynamics for a given site or region, the comparability among models has not been a major focus of concern. However, in the context of global change, it is essential that we understand the consequences of these variations among models on their response (sensitivity) to changing environmental conditions. This is particularly important if we are to use this network of models to make predictions on a regional or global basis.

Patch models have been developed for a wide range of ecosystems under a variety of environmental conditions, but little is known about the behaviour of several models under the same conditions. Such comparisons are not necessarily straightforward. Parameters required for models may vary. To overcome the current limitations of quantitative model comparisons across a wide range of ecosystems, we recommend that a systematic effort to parameterize and test the behaviour of patch models should be undertaken at a variety of sites. This systematic comparison of models should be performed along transects representing major environmental gradients, such as those transect studies established or being established under IGBP auspices.

There is also a clear need to support efforts towards the development of improved methods for documenting the nature and rationale for changes in model structure and any benefits in performance achieved. Many of the difficulties in comparing the predictions from various patch models stem from a lack of detailed information on the specific differences between models. Standards need to be developed for model documentation which includes specific comments on changes in model structure, tabulation of model parameters and compilation of validation data sets. This activity would be greatly aided by the Long-term Ecological Modelling Activity (LEMA) network proposed under Focus 2 of GCTE.

To seriously engage the problem of simulating the response of vegetation to large-scale environmental change, there is a need to compile and document data sets for model validation and comparison. Patch models make predictions concerning the long-term dynamics of composition and structure. Data for testing these dynamics are very limited. Efforts to gather paleoecological and paleoclimatic data to provide some long-term data against which the models can be tested should be a high priority.

If patch models are to contribute to the study of global change on larger scales, complexity even in the simpler models will constitute a severe problem. There is an obvious tradeoff in the complexity of model structure required to simulate the behaviour of vegetation at a given site (or region), and the ability to take a single model structure and apply it across sites or ecosystems. To overcome this inherent problem we suggest two distinct approaches could be followed: first, for regional or local applications of the model, the traditional approach of adding as much detail as required to reproduce the important features of the local vegetation cover should be followed. Second, for application on larger scales, where only the major patterns are of concern, simplified patch models should be developed, tested and applied.

Current model limitations and associated research needs

The scientific questions relating to the potential response of terrestrial vegetation to global environmental change pose new problems for the development and application of patch models of vegetation dynamics. The original development and application of patch models focused on examining the temporal (e.g. successional) dynamics of vegetation at a given site or along simple environmental gradients (e.g. elevation). Many of the simplifying assumptions that have been made for ease of characterizing basic plant processes and patterns may no longer hold under changing environmental conditions. The following points reflect the recognition of many of these limitations and the steps necessary to overcome them. Some of these activities involve further model development; however, many of these current limitations are a result of our basic lack of data and/or lack of understanding of key plant processes.

New research is essential to determine the effects of environmental factors such as temperature on whole plant growth and mortality. Although ecological theory suggests that the geographic distribution of a species is a function of both physiological tolerances (absolute limits or 'fundamental niche') and competition with other species (realized distribution or 'realized niche'), in the absence of empirical data on the relative importance of these two processes, the environmental response functions characterizing growth, establishment and survival were parameterized on distributional data. For example, the establishment and growth of a species was assumed to be zero for values of temperature (e.g. growing degree days) observed at the boundaries of its geographical distribution. Although questionable in the absence of empirical data on the response of various processes to temperature, the approach ensured that the predictions of the models were broadly consistent with observed latitudinal and altitudinal ranges. However, this simple temperature response function hides the direct response of many processes to temperature, such as photosynthesis and respiration, flowering, or response to extreme events such as drought or frost. Under a climate change, the balance of these phenomena may change, and thus better estimates of the thermal response of plants are needed.

It is relatively straightforward to measure the effects of temperature on physiological performance (e.g. photosynthesis and respiration), but extrapolation to longer-term responses at the level of the whole plant is currently difficult. The ability to incorporate species-specific environmental responses to establishment, growth, reproduction and survival is not an inherent problem of the patch models, rather it is a problem of available data. Recent regional studies which have included specific temperature responses relating to chilling requirements or frost tolerance within patch models have shown the sensitivity of model predictions under climate change scenarios. The characterization of species-specific environmental response functions to key plant processes is critical for advancing the ability of patch models to address global change issues related to climate change.

Beyond these fundamental problems arising from a lack of information about fundamental physiological responses of major species, experiments involving mature trees in natural stands are essential to settle issues of CO_2 fertilization. Information of the effects of CO_2 enrichment on plant species is virtually restricted to short-term studies of juvenile plants. Considerable uncertainty surrounds the extrapolation from juveniles to adults, including the tendency of some species to acclimate to enhanced CO_2. Moreover, there is a large degree of variation among species in the observed responses to CO_2. This diversity of response, both among species and with time, presents a problem with incorporating direct effects of CO_2 enrichment into patch models. Patch models simulate the dynamics of species composition and vegetation structure, explicitly considering the variation among species in their life history and environmental responses. This approach precludes the application of a simple scalar (such as the beta factor) which is applied equally to all species, since differential species responses could influence changes in relative abundance. The current limitation on including direct CO_2 effects into patch models is the lack of data on whole plant response at the species level. A number of studies based on the limited data for tree species which are available are currently under way. Preliminary results suggest that in the case where there is variation among species in CO_2 response, the resulting changes in net primary productivity differ from corresponding estimates where a single response function (such as a beta factor) is applied across species. It is encouraging to note the increasing emphasis in the GCTE Focus 1 Elevated CO_2 Research Programme on evolutionary, population and community responses to elevated CO_2.

The modelling of soil processes in patch models has improved but further work is needed. In general, patch models give little attention to soil processes. More recent models (e.g. LINKAGES: Pastor & Post, 1986) incorporate more complete models of the nutrient cycle to capture the complex nature of the feedback mechanisms between vegetation and soil dynamics. Decomposition rates and organic matter formation will both be affected by a changed climate, and the net result (C storage in soils) is therefore difficult to predict. Given the possibilities for feedbacks between changes in productivity/species composition and the rates of decomposition and nutrient cycling under changing environmental conditions, there is a need for explicit consideration of soil processes within the framework of patch models. Work on biogeochemical cycling within GCTE Focus 1 and soil organic matter dynamics in Focus 3 will provide useful data and further understanding which will aid future model development.

Patch models that comply with mass balance equations for carbon and major nutrients are essential for most applications related to global change. Few, if any, patch models include or are constrained by mass balance of carbon or other nutrients. For many applications, such as addressing carbon storage issues, models must incorporate nutrient cycling/mass balance of carbon and nitrogen. However, it is probably not necessary that all patch models have this capability. Initial studies which have linked patch models with ecosystem models of carbon and nitrogen dynamics (e.g. CENTURY: Parton *et al.*, 1988) which require mass balance have met with some difficulties related to the spatial scales at which the different processes are described (Lauenroth *et al.*, 1993). Given the spatial scale at which patch models operate, exogenous inputs and outflows from the patch will remain a problem in models of a single patch. Spatially extensive forms of the patch model (e.g. ZELIG: Smith & Urban, 1989; STEPPE: Coffin & Lauenroth, 1989, 1994; Coffin *et al.*, 1993) may be necessary to achieve reliable estimates of mass balance.

Insufficient attention has been given to recruitment and mortality processes in patch models. In earlier versions of patch models recruitment was represented very simplistically. All species from the area/region were assumed to be available for recruitment into the patch. Species which were inappropriate for the environmental conditions on the patch were filtered out via the environmental constraints on growth. Dispersal was largely ignored. Under conditions of environmental change, these assumptions (models) are inappropriate. Changes in species composition within an area or region will be related to rates of dispersal and species immigration. Current forms of the patch models which operate within a spatial framework (e.g. ZELIG: Smith & Urban, 1989; SORTIE: Pacala *et al.*, 1993) are explicitly modelling the process of dispersal and linking recruitment to canopy composition. Data are needed to develop more specific models of early recruitment processes in a realistic manner, and models of dispersal (including long-distance dispersal) need to be developed and tested.

Mortality has been shown to be a major controlling variable in patch models (openings or gaps are made free for occupancy by neighbours or for colonization). Few data are available, and in general are hard to obtain, linking mortality with tree condition and/or environmental conditions on the patch. Empirical estimates of mortality rates without an environmental context will be of limited use in model development.

A major challenge in patch model development is to integrate different life forms within the same model. The link between patch size and the size of the dominant life form being modelled presents a problem in modelling ecosystems in which multiple life forms are co-dominant. Patch models have been developed for both grasslands and shrublands; however, there remains much work to be done on integrating greatly differing life forms in the same model. In many situations involving mixed life forms the spatial distribution of the component species may be important, and thus the fundamental assumption made in the initial development of patch models (that there is an appropriate patch size in which every individual may be assumed to interact with every other individual) is invalid. Progress is being made in deriving spatially explicit versions of patch models.

Spatial and temporal coherences in disturbance regimes will need to be taken into account in exploring the impacts of global change. The approach of simulating the effects of fire on biomass and net primary productivity using an age-dependent fire return frequency applied to independent patches (the normal approach used in patch models) gives a markedly different result than a landscape-coupled model structure where the probability of fire initiation remains age-dependent but the spread of fire to adjacent patches is not. Such spatial interactions are not easy to incorporate into most patch models. This restriction may limit their validity under certain scenarios and future progress will be needed in developing spatially more extensive versions of patch models.

The role of animals and pathogens in vegetation dynamics needs to be incorporated more explicitly into the model framework. A limitation of most patch models is that they do not explicitly incorporate effects of animals/pathogens/symbionts as agents of mortality, pollination, dispersal, nutrient capture, etc. This may not be a problem when simulating existing ecosystems under current climate, as many of these factors may be implicitly incorporated into current parameterization of the models. However, their absence may become a major shortcoming for predicting effects of climate change where interactions and feedbacks may become more uncertain.

Patch models provide a promising means of assessing the impacts of biodiversity on ecosystem-level response to climate change. Patch models explicitly incorporate species level biodiversity and could be modified to include within-species genetic variation. These models can be used to address ecosystem consequences of biodiversity and complement experimental approaches. For example, consider the large interspecific variation in the response of tree species to CO_2. As atmospheric levels increase, species with the largest response to enhanced CO_2 should increase in abundance relative to others. Such changes would increase the mean capacity of 'forest trees' to sequester carbon.

Despite the opportunity and demand to add increasing complexity to patch models, we should do so only when demonstrably necessary. Computing resources are becoming less of a limit to model complexity. We have to be aware that this new facility in building models is not a sufficient reason to build complex models. A 'reductionist' approach enables one to understand mechanisms in their detail, but often it is not clear which mechanisms are truly responsible for ecosystem change. A complex model can be too complex to be understood in its details; imitating nature does not mean that we understand it. Predictions with simple (statistical) models are often as accurate or more accurate than with complex models. The problems of model validation are considerable without the addition of complexity for the sake of 'realism'.

Recent developments

Over the past two decades, the original formulation of the forest gap model (individual-based patch model) has been adapted to a wide variety of forested ecosystems. At the same time, the basic model structure has been modified to meet the needs of various research applications. The following examples represent some of the recent developments in research related to patch models which were presented at the workshop.

The original development of patch models focused on forest ecosystems. Over the past decade, patch models have been developed for a wide variety of forested ecosystems; however, only in recent years has the patch model framework been applied to grassland (Coffin & Lauenroth, 1989, 1990, 1994) and shrub and heathland (Prentice *et al.*, 1987) ecosystems. In addition to the smaller spatial scale at which these models operate (<1 m^2 as compared with 0.1–1.0 ha for forest patch models), the development of grassland patch models required a more detailed consideration of competition for below-ground resources.

Research on the development of spatially explicit patch models has progressed on two independent fronts: (1) explicit consideration of the position of each individual within a single patch; and (2) the linking of individual patches into a spatial network. Explicit consideration of the position of individuals within the patch (e.g. SORTIE: Pacala *et al.*, 1993; Busing, 1991) has allowed for studies which examine small scale patterns of species distribution within a forest stand resulting from feedbacks between canopy composition and patterns of recruitment. New model structures which place each patch being simulated into a spatial context (e.g. ZELIG: Smith & Urban, 1989; STEPPE: Coffin & Lauenroth, 1989, 1994) have allowed for patch models to be used to explore a range of questions relating to landscape processes and scale-related issues.

Traditionally, patch models are individual-based, simulating the establishment, growth and mortality of individual plants. Alternative cohort-based approaches have been developed which simulate age or size classes of individuals (e.g. Fulton, 1991). The major advantage of this approach is computational efficiency.

Incorporation of physiologically detailed models of individual plant growth.

Patch models have traditionally described plant growth using a species-specific optimal growth function. The optimal growth function defines expected annual growth increment as a function of plant size. This optimal annual growth increment is then modified by various functions which relate growth to the environmental conditions on the patch. In general, these functions (both optimal growth and environmental response) are empirically derived. Recent work has focused on replacing this simple approach of modelling annual growth with physiologically detailed models of individual plant growth which explicitly consider the processes of photosynthesis, respiration and carbon allocation. These models typically simulate net photosynthesis on an hourly or daily basis, with calculations of net primary production and carbon allocation on an annual basis. One such example is the HYBRID model (Friend *et al.*, 1992).

Although patch models have been developed for grasslands, shrublands and forests, only recently have patch models been developed which are capable of simulating the dynamics of ecosystems in which multiple life forms are co-dominant (e.g. tropical/subtropical savannas). Two basic approaches are being taken. The first approach involves using a spatially explicit structure where the position of each individual within a patch is considered and interactions among individuals are spatially dependent (e.g. Menaut *et al.*, 1990). The second approach involves the use of nested patches similar in concept to the approach of nested quadrats used in vegetation sampling. Smaller patches which simulate herbaceous and small woody plants are 'contained' within larger patches which are used to simulate larger woody plants. Larger patches define the environment for smaller patches while recruitment is handled as a statistical process from smaller to larger patches (e.g. VEGOMAT: Smith *et al.*, 1989; Lauenroth *et al.*, 1993).

In the original formulation of patch models, plant canopies are represented as a single layer. Newer formulations have included a more complex three-dimensional canopy geometry. The more complex canopy structures allows for the consideration of both self-shading and side-shading, a feature found to be important in simulating forests at high latitudes (i.e. boreal forest).

Traditionally, patch models have represented a forest (or grassland) as the statistical average of a number of independently simulated patches. By placing the simulated patches within a spatial grid, recent research has focused on examining landscape processes which involve spatially explicit interactions between patches such as dispersal, catchment hydrology, fire and insect outbreaks.

Linking patch models with ecosystem models of biogeochemical processes is an area of active research. Patch models have been linked with ecosystem models that simulate the dynamics of the carbon, nitrogen and hydrological cycles. The ecosystem models require information on features of the vegetation structure such as leaf area, biomass, litter input, and litter quality (e.g. $C:N$). These parameters are provided by the patch model. In return, the ecosystem model provides a description of certain environmental conditions on the patch, such as the availability of nitrogen, soil carbon and soil moisture.

Future research

The research agenda outlined below focuses on questions related to global change. Although the workshop participants believe that patch models have an important role to play in global change research, their utility as a research tool is not limited to global change issues. Patch models have a rich history in ecological research over the past two decades, in areas ranging from the development of ecological theory to applications in forest management and the conservation of biodiversity.

A major focus of research relating to global change is the attempt to understand how certain processes that operate at one spatial or temporal scale influence process and pattern at larger or longer spatial or temporal scales. For example, how do patterns of photosynthesis at the leaf level relate to patterns of growth at the level of the whole plant, or how do population and community dynamics of species composition and population structure relate to ecosystem-level processes of the cycling of water and nutrients at the ecosystem level? Individual-based patch models can help address many of these questions because they can simulate population, community, ecosystem and landscape dynamics as a function of the response of individual plants interacting with the environment of the patch.

Patch models have incorporated disturbances such as fire, hurricanes, herbivory and timber harvest into the model framework as agents of mortality. However, in the context of global change there is a need for a more mechanistic approach to modelling disturbance and an explicit consideration of human impacts related to land use. To address many of these features/processes using patch models requires that a spatial approach be adopted such as the one discussed above for examining landscape-level processes.

In addition to increasing atmospheric concentrations of CO_2 and the associated potential for climate change, there are additional potential stresses on plant function related to anthropogenic causes, such as increased levels of ozone, UV radiation and acid precipitation. Experimental studies have shown that there are significant interactions among these and other stress factors. These interactions may be a function of direct physiological effects on the plant or via indirect effects on ecosystem level processes such as decomposition and nutrient availability. The nature of these interactions must be understood and incorporated into the framework of models of plant processes if we are to be able to predict the response of ecosystems to changing environmental conditions relating to anthropogenic disturbance.

Patch models simulate the dynamics of species composition and stand structure using species-specific parameters describing life history and environmental response. Although this approach has proved feasible even for modelling species-rich ecosystems such as tropical/subtropical rain-forest, deriving parameters for even the dominant species at a continental to global scale appears unrealistic. To overcome this limitation, efforts are underway to develop a functional classification(s) of species based on similarities in characteristics (e.g. life history, environmental response) important to dominant ecosystem processes and environmental feedbacks. Patch models can, and should, play an important role in the theoretical and practical development of such classification systems. Patch models can be used to examine the consequences of aggregating species into groupings which are characterized by a single set of parameters, in contrast to predicted dynamics based on species-specific parameterizations.

Many of the problems involved in comparisons among patch models could be overcome with a more modular approach to model structure. Modular approaches which isolate important processes and functions related to plant growth (e.g. annual growth and species allometry), environmental feedbacks (e.g. canopy architecture and light extinction), and definition of environmental conditions on the patch (e.g. soil moisture and evapotranspiration) would aid in modification of the models to new sites and conditions as well as provide an easy protocol for model comparison and sensitivity analysis through substitution of components (e.g. different models of evapotranspiration).

Patch models have been developed for a wide range of ecosystems around the world; however, there are many ecosystems (even biomes) for which models have not been developed. One obvious area for future model development is in the southern hemisphere, specifically in the tropical regions of Africa, South America and Asia. The development of mixed life form models will provide for an expansion of the approach into tropical/subtropical savannas. This addition will allow the application of patch models to address potential long-term changes along the moisture gradient from grassland to forest. In addition, the recent development of patch models in herbaceous and shrub-dominated ecosystems should allow for the development of patch models for tundra ecosystems, providing the ability to address the important transition zone between tundra and boreal forests at the high northern latitudes.

One of the priority activities of IGBP GCTE Focus 2 is the development of dynamic models of global vegetation (DGVM). Two approaches for model development have been discussed, the bottom-up and the top-down (see Woodward, this volume). The bottom-up approach involves the scaling-up of patch models using some statistical sampling procedure to provide continental and global coverage. This approach would require a generalized patch model which is able to simulate the dynamics of all ecosystem types (e.g. boreal forest, tropical savanna, temperate grassland). This generalized patch model would use a functional classification of plants rather than address species composition. The development of such a modelling approach is dependent on future research and development of patch models relating to many of the activities outlined in this document.

The second approach, top-down, utilizes current 'equilibrium models' of global vegetation pattern. This class of models includes biogeographical models which relate the large scale patterns of climate and vegetation (e.g. Holdridge Life Zone Classification: Holdridge, 1967; BIOME: Prentice *et al.*, 1992), and global models of net primary productivity (e.g. TEM: Melillo *et al.*, 1993). Although capable of predicting changes in the state of vegetation following a climate change, these models do not address the time scale or processes associated with those changes; they do not represent the transient dynamics of vegetation involved in achieving these new equilibrium patterns. The top-down approach of DGVM development would modify these global vegetation models by defining functional types of vegetation which make up each of the ecosystems or biomes currently used to describe vegetation pattern/composition within global models. These functional plant types would then be assigned parameters relating to rates of growth, mortality, dispersal and other processes which influence the transient dynamics of vegetation in response to changing environmental conditions.

Both approaches to the development of DGVMs will rely heavily on patch models, either directly in the case of the bottom-up approach, or indirectly in the development of parameters for the functional plant types in the case of the top-down approach.

References

Arney, J. D. (1974). An individual tree model for stand simulation in Douglas fir. In *Growth Models for Tree and Stand Simulation*. Res. Notes 30, ed. J. Fries, pp. 38–46. Stockholm: Department of Forest Yield Research, Royal College of Forestry.

Bella, I. E. (1971). A new competition model for individual trees. *Forest Science*, **17**, 364–72.

Botkin, D. B., Janak, J. F. & Wallis, J. R. (1972). Some ecological consequences of a computer model of forest growth. *Journal of Ecology*, **60**, 849–73.

Bugmann, H. (1994). On the ecology of mountainous forests in a changing climate: A simulation study. PhD thesis no. 10 638, Swiss Federal Institute of Technology, Zurich, Switzerland.

Busing, R. T. (1991). A spatial model of forest dynamics. *Vegetatio*, **92**, 167–79.

Coffin, D. P. & Lauenroth, W. K. (1989). Disturbance size and gap dynamics in a semiarid grassland: A landscape-level approach. *Landscape Ecology*, **3**, 19–27.

Coffin, D. P. & Lauenroth, W. K. (1990). A gap dynamics simulation model of succession in the shortgrass steppe. *Ecological Modelling*, **49**, 229–66.

Coffin, D. P. & Lauenroth, W. K. (1994). Successional dynamics of a semiarid grassland: Effects of soil texture and disturbance size. *Vegetatio*, **110**, 67–82.

Coffin, D. P., Lauenroth, W. K. & Burke, I. C. (1993). Spatial dynamics in recovery of shortgrass steppe ecosystems. *Lectures on Mathematics in the Life Sciences*, **23**, 75–108.

Friend, A. D., Shugart, H. H. & Running, S. W. (1993). A physiology-based gap model of forest dynamics. *Ecology*, **74**, 792–7.

Fulton, M. (1991). A computationally efficient forest succession model: Design and initial tests. *Forest Ecology and Management*, **42**, 23–34.

Hatch, C. R. (1971). Simulation of an even-aged red pine stand in Northern Minnesota. PhD thesis, University of Minnesota.

Holdridge, L. R. (1967). *Life Zone Ecology*. San Jose, Costa Rica: Tropical Science Center.

Lauenroth, W. K., Urban, D. L., Coffin, D. P., Parton, W. J., Shugart, H. H., Kirchner, T. B. & Smith, T. M. (1993). Modeling vegetation structure–ecosystem process interactions across sites and ecosystems. *Ecological Modelling*, **67**, 49–80.

Lee, Y. (1967). Stand models for lodgepole pine and limits to their application. *Forest Chronicles*, **43**, 387–8.

Lin, J. Y. (1970). Growing space index and stand simulation of young Western hemlock in Oregon. PhD thesis, Duke University, Durham, NC.

Melillo, J. M., McGuire, A. D., Kicklighter, D. W., Moore, B., Vorosmarty, C. J. & Schloss, A. L. (1993). Global climate change and terrestrial net primary production. *Nature*, **363**, 234–40.

Mitchell, K. J. (1969). Simulation of growth of even-aged stands of white spruce. *Yale University School Forestry Bulletin*, **75**, 1–48.

Newnham, R. M. (1964). The development of a stand model for Douglas-fir. PhD thesis, University of British Columbia, Vancouver.

Pacala, S. W., Canham, C. D. & Silander, J. A. Jr. (1993). Forest models defined by field measurements: I. The design of a northeastern forest simulator. *Canadian Journal of Forestry Research*, **23**, 1980–8.

Parton, W. J., Stewart, J. W. B. & Cole, C. V. (1988). Dynamics of C, N, P and S in grassland soils: a model. *Biogeochemistry*, **5**, 109–31.

Pastor, J. & Post, W. M. (1986). Influence of climate, soil moisture and succession on forest carbon and nitrogen cycles. *Biogeochemistry*, **2**, 3–27.

Prentice, I. C., Cramer, W., Harrison,

S. P., Leemans, R., Monserud, R. A. & Solomon, A. M. (1992). A global biome model based on plant physiology and dominance, soil properties and climate. *Journal of Biogeography*, **19**, 117–34.

Prentice, I. C., van Tongeren, O. F. R. & de Smidt, J. T. (1987). Simulation of heathland vegetation dynamics. *Journal of Ecology*, **75**, 203–19.

Shugart, H. H. (1984). *A Theory of Forest Dynamics*. New York: Springer.

Shugart, H. H. & West, D. C. (1980). Forest succession models. *BioScience*, **30**, 308–13.

Smith, T. M., Shugart, H. H., Urban, D. L., Lauenroth, W. K., Coffin, D. P. & Kirchner, T. B. (1989). Modeling vegetation across biomes: Forest-grassland transition. In Forests of the World: Diversity and Dynamics (abstracts), ed. E. Sjogren. *Studies in Plant Ecology*, **18**, 240–1.

Smith, T. M. & Urban, D. L. (1989). Scale and resolution of forest structural pattern. *Vegetatio*, **74**, 143–50.

9 Climate change, disturbances and landscape dynamics

R. H. Gardner, W. W. Hargrove, M. G. Turner
and W. H. Romme

Introduction

The objectives of the Global Change and Terrestrial Ecosystems (GCTE) core project of the International Geosphere-Biosphere Programme (IGBP) are to predict the effects of changes in climate, atmospheric composition and land use on terrestrial ecosystems, and to determine how these effects lead to feedbacks to the atmosphere and physical climate system (Steffen *et al.*, 1992). These issues are challenging because they require studies of broad-scale patterns of ecological change which consider processes at many scales (Urban *et al.*, 1987; Levin, 1992). For instance, the linkage between climate, fire and terrestrial ecosystem dynamics requires an understanding of fine-grained details of fire ignition and spread, landscape scale patterns of vegetation and their response to fire, as well as changes in the long-term patterns of weather and climate.

The possibility of rapid changes in terrestrial landscapes as a result of drier climate conditions and altered fire regimes is recognized as a serious risk in many regions of the world (Bonan *et al.*, 1990; Overpeck *et al.*, 1991; Romme & Turner, 1991; Swetnam, 1993; Moreno & Oechel, 1994). The effect of climate change on fire regimes has been exacerbated by contemporary changes in land-use patterns and fire suppression efforts, causing fuels to accumulate and the risk of large fire to increase (Ales *et al.*, 1992; Davis & Burrows, 1993). The possible loss of large areas of forested land may result in significant changes in the pattern of vegetation regrowth and succession (Neilson, 1993) with many areas shifting to a new vegetation state (Overpeck *et al.*, 1990). It is the serious implications of these hypothesized effects of climate change that have motivated many IGBP/GCTE studies.

Our current understanding of the interaction between climatic conditions, fire regimes and forested landscapes is based on a variety of empirical and theoretical studies. Analysis of recent trends between fire and weather (e.g. Balling, Meyer & Wells, 1992; Renkin & Despain, 1992; Johnson & Wowchuk, 1993) allowed relationships between weather and fire to be evaluated over the duration of contemporary records (*c.* 100 years). Studies based on intensive sampling of stand histories and dendochronology (e.g. Romme, 1982; Romme & Despain, 1989; Swetnam & Betancourt, 1990) and stratigraphic charcoal data (Clark, 1990) have extended available records to nearly 800 years. Fires within stands of the giant sequoia (*Sequoiadendron giganteum*) produce the longest records of all, allowing climate variables and fire frequencies to be reconstructed for the past 2000 years (Swetnam, 1993).

As impressive and important as these studies are for establishing relationships between weather, fuels and fire, simulation models are necessary to evaluate the hypothesized effects of altered climate and fire regimes on ecosystems lacking long-term fire records. Baker *et al.* (1991) have used model simulations based on statistical relationships between stand age and risk of large fires to quantify changes in amount and spatial configuration of forested regions. Antonovski *et al.* (1992) developed a simple coarse-grained grid-based model that simulated the effects of climate change on fires on boreal systems over a 3000-year period. The results (which Antonovski *et al.* regarded as preliminary) showed unexpected changes in the area burned through time as a consequence of drier climate conditions. A more detailed model was developed by Davis and Burrows (1993) for simulating fire patterns in Mediterranean landscapes. Davis and Burrows adapted Rothermel's (1972) thermodynamic model of fire spread (that includes such factors as fuel type, fuel moisture, wind and topography) and simulated fire patterns in a fine-grained (0.9 ha per cell) grid of California chaparral over a 500 year period. Their tests of the sensitivities of this model to changes in spread rate and frequency of fire ignitions showed that the patch size and age distribution of fires were closely coupled to ignition frequency and the fragmentation of combustible vegetation.

Each of these models provides valuable insight into the long-term consequences of changes in landscape pattern as a result of the adjustment of climatically affected model parameters. However, none has simultaneously considered the effect that climate change has on the year-to-year variation in weather patterns, the subsequent effect that changes in the mean and variance of local weather conditions has on the severity and intensity of fire, and the mitigating or amplifying effects that landscape heterogeneity of fuels (including topography) can have on results. We have developed a model that links fine-grained details of vegetation pattern with the stochastic process of fire ignition and spread by wind and firebrands as a consequence of hypothesized changes in climate. We use this model here to (1) examine how changes in climate affect the frequency and extent of fires; and (2) project the long-term consequences of such changes on the spatial pattern of forested landscapes. It is hoped that these simulations will provide additional insight into the identification of subtle but significant effects of climate change on landscape patterns.

Simulating fires at landscape scales

The need to simulate the spatial effect of wildfires at landscape scales has resulted in the development of a variety of models which vary in their complexity and data requirements. For instance, a number of models have been based on concepts from percolation theory and cellular automata (e.g. Kessell, 1976; Green, 1983; Albinet et al., 1986; Ohtsuki & Keyes, 1986; von Niessen & Blumen, 1986; Turner et al., 1989; Green et al., 1990), while others have been developed from more detailed, thermodynamic description of fire spread (e.g. Rothermel, 1972; Kessell, 1976, 1979; Vasconcelos & Guertin, 1992; Davis & Burrows, 1993). Some models have relied on statistical relationships to predict the severity and extent of fire effects (e.g. Johnson & Van Wagner, 1985; Baker, 1992), while others are directly linked to GIS data bases (e.g. Kessell, 1979; Vasconcelos & Guertin, 1992; Clark et al., 1993). Our approach to the problem of simulating fire effects over broad spatial scales has been to develop an event-driven, probability-based model (W. W. Hargrove et al., unpublished observations) that efficiently simulates fire events over large areas and long periods of time. We recently developed algorithms that generate weather conditions during the fire season and linked the fire simulations to a vegetation recovery model, allowing us to test the sensitivity of landscape pattern to the long-term effects of climate induced changes in fire regimes.

EMBYR, the fire model

EMBYR is an event-driven, grid-based simulation model of fire ignition and spread in heterogeneous landscapes. Fire spread is simulated by examining each burning site (50 × 50 m) and determining spread to the eight adjacent sites as a function of fuel type, fuel moisture, wind speed and direction, and topography. In addition, burning sites distribute fire-brands to downwind sites, where the probability of ignition of new fires is a function of fuel type and moisture conditions. A qualitative index of fire severity was developed to identify cells where fire intensity was high enough to consume all the trees on the 50 × 50 m cell. Burn severity was calculated as a linear combination of fuel type, fuel moisture, wind speed and the rate the cell burned. The value of the severity index that results in a stand-replacing fire was calibrated by reconstructing the weather events of

1988, simulating the spread of fire, and comparing the simulation results with the proportion and distribution of sites that experienced a stand-replacing fire measured within a 1 × 1 km study site that lies approximately 2 km northeast of the Old Faithful weather station.

A set of model parameters was developed to represent weather conditions and distribution of fuels on the subalpine plateau of Yellowstone National Park. Simulation experiments revealed relationships between model parameters and landscape heterogeneity, and the results of repeated simulations were compared by evaluating risk (the cumulative frequency distribution of the area burned) as a function of the change in weather conditions.

The landscape

Yellowstone National Park (YNP) lies in the interior of the North American continent (mostly in the northwest corner of Wyoming, United States), straddling the Continental Divide. Elevations within the park range from 1600 m in the north where the Yellowstone River leaves the park and flows into Montana, to several peaks above 3000 m along the eastern and northern boundaries. Upper timberline is near 3000 m. Most of the park is between 2000 and 2700 m in the subalpine zone (Despain, 1990).

YNP is about 83% forested, with lodgepole pine (*Pinus contorta* Loudon var. *latifolia* Engelm.) the dominant species. The growth and development of lodgepole forests has been characterized by four stages (Romme & Despain, 1989; Despain, 1990). Young stands that regenerate after fire are classified as LP0 until the time of canopy closure (approximately 40 years). Small dead fuels within these stands are too sparse to carry fire readily and live fuels are usually too green to burn. Consequently, probability of fire spread in LP0 is low under all weather conditions. Young, even-aged lodgepole pine stands that are approximately 40 to 150 years post-fire are classified as LP1. Although tree densities are usually high within these stands, fuels on the forest floor usually are sparse, and crown fires will propagate only in extremely windy conditions. Even-aged closed canopies of lodgepole pine with a developing understorey of Engelmann spruce (*Picea engelmanni* Parry) and subalpine fir (*Abies lasiocarpa* (Hook.) Nutt.) are classified as LP2. This stage lasts from approximately 150 to 300 years after a fire. Fuels on the forest floor are sparse to moderate, and flammability begins to increase in the later portions of this stage. Finally, LP3 stands have pine, fir, and spruce of all ages and size classes. This stage persists until the next stand-replacing fire with young spruce, fir, and dead fuels accumulating in sufficient density to propagate crown fires under dry conditions even without wind.

The area chosen for the simulations (Fig. 9.1) was a 30 × 30 km (90 000 ha) section of land (UTM coordinates: north 3917975, south 4887975, east 550825, west 520825) that lies in the south central portion of the park (Fig. 9.1). The distribution of forest successional stages prior to the extensive fires in 1988 was available in GIS format based on interpretation of low-altitude aerial photographs taken in 1969 and 1972 by D. G. Despain (personal communication). The simulation area contained 78 723 ha of forested land (87.5%), with 56% of that area (43 899 ha) composed of mature LP3 stands. Only 7035 ha (9%) of the landscape was composed of non-flammable water, rock, etc.

Post-fire recovery of each cell is modelled as an independent Markov process. Parameters for the Markov model were developed from an extensive sampling of increment cores and fire scars within a 130 000 ha study area on

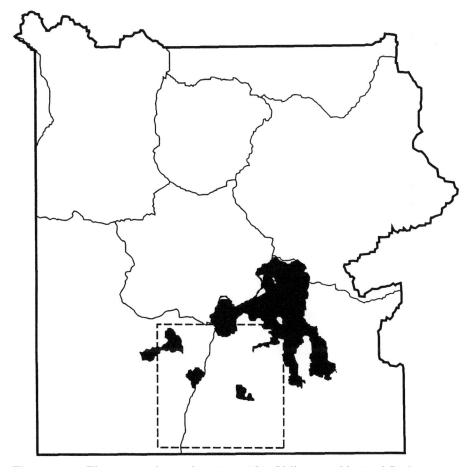

Figure 9.1 *The 30 × 30 km study region within Yellowstone National Park used for the simulations is indicated by the dotted line. Lakes are represented as solid black, and lines show the major roads within the park.*

the subalpine plateau of Yellowstone National Park that resulted in a map of stand age (time since fire) and successional stage in 1985 (Romme & Despain, 1989 and unpublished data). Transition probabilities were estimated by empirically fitting the transition of forest types by assuming that an LP0 stand forms if the burn severity index indicated that a stand-replacing fire occurred, and that LP0 sites can not transform into LP1 until they are at least 40 years of age. At the end of each fire season, each cell was examined and the Markov transition probabilities (0.94, 0.99 and 0.995 for LP0 to LP1, LP1 to LP2 and LP2 to LP3, respectively) used to determine the fuel class of each cell.

Weather generation

The weather in Yellowstone National Park is characterized by short cool summers and long, cold winters with precipitation highest in winter and spring (Dirks & Martner, 1982; Despain, 1987). The summers are usually dry with an average of about 32 mm of rain per month (Renkin & Despain, 1992), resulting in a fire season that lasts from mid June to late September, with large fires most probable during July through September (Overpeck *et al.*, 1990). Weather records used for the 1000 year simulations consist of 30 years of weather data recorded at the Old Faithful weather station and obtained from the WIMS/NIFMID (Weather Information Management System/National Interagency Fire Management Integrated Database) National Historical Fire Weather Database (US Forest Service, Missoula, MT, USA). Recorded variables include date, time, temperature, relative humidity, wind speed, wind direction, and estimates of 10 hour and 1000 hour time lag fuel moisture. The percentage fuel moisture estimates are calculations based on the National Fire Danger-Rating System (Deeming *et al.*, 1978) which measures long-term moisture conditions and serves as a drought severity indicator. The 1000 hour time lag fuel moisture (THFM) was a key variable describing fire dynamics within YNP. Observations of naturally burning fires from 1972 to 1987 revealed that fire spread was extremely unlikely when THFM >16%, but that severe crown fire activity and rapid spread were likely when THFM <13% (Renkin & Despain, 1992). Thus, we focused on these thresholds in fire behaviour relative to THFM (Fig. 9.2).

The simulation of fires over a 1000 year time period by EMBYR required a generation of a data set that realistically represented daily changes in fuel moisture, wind speed and direction, as well as the number and location of fire starts. It was also important for the generated weather to adequately represent seasonal patterns (for instance, relatively wetter spring and autumn

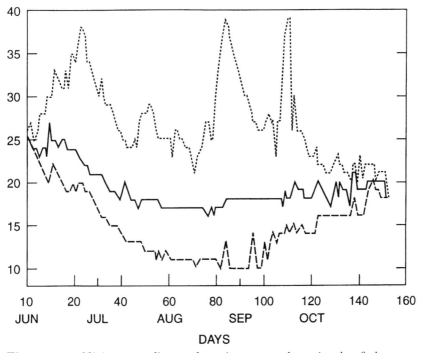

Figure 9.2 *Minimum, median, and maximum 1000-hour time lag fuel moisture (THFM) values summarized from 26 years of weather data taken at the Old Faithful weather station.*

weather, with drier periods in July and August). We designed a procedure to produce daily weather records for 1000 years by systematic sampling from the existing weather records. The objectives of the procedure were to: (1) maintain the appropriate relationships between the daily weather variables of wind speed and direction, fuel moisture and fire starts; (2) randomly select weather records with realistic autocorrelated changes in fuel moisture; and (3) produce weather sequences that provide an unbiased representation of extreme events. To achieve these objectives, an algorithm was developed that: (1) produces an autocorrelated Brownian motion with displacements at each step produced by Gaussian noise (Plotnick & Prestegard, 1993); (2) normalizes the trajectory to produce values that fall between randomly selected minimum and maximum percentiles of THFM (the average minimum and maximum percentiles for 26 years was 0.19 and 0.95, with standard deviations of 0.18 and 0.10 for the minimum and maximum values, respectively); (3) selects daily weather records by matching the randomly generated THFM percentile with the corresponding THFM record from the data (Fig. 9.3); and (4) writes the selected data record to a file for input to EMBYR.

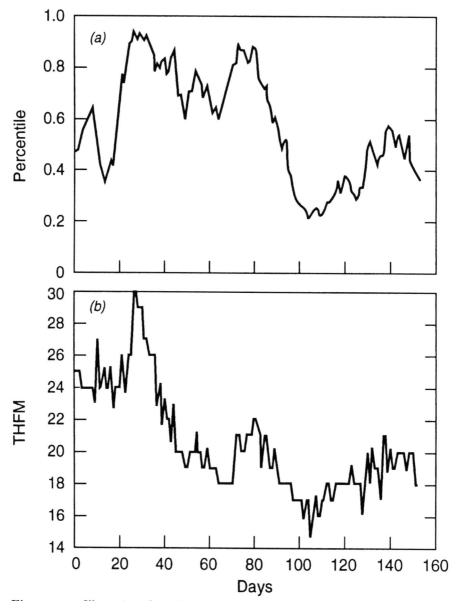

Figure 9.3   *Illustration of weather generation for a single fire season. (a) The percentiles for 1000-hour timelag fuel moisture (THFM) generated by a correlated random walk. (b) The corresponding THFM values selected from the weather records.*

The mid-point displacement method (Barnsley *et al.*, 1988) was used to generate the random trajectories of Brownian motion. This method produced a one-dimensional transect by randomly selecting the end points of the transect and then recursively dividing the line in half, resulting in a sequence of successively finer segments. The variance associated with the 'displacement' of each segment is reduced by $0.5^{(0.5H)}$ with each recursive division of the line, resulting in an average correlation between past and future increments that is equal to $2^{(2H-1)} - 1$ (Mandelbrot, 1983; Feder, 1988). The parameter H can range between 0.0 and 1.0, with values above 0.5 indicating persistence in trends (Feder, 1988). For our simulations, H was set to 0.7, producing an average correlation of 0.3. The use of the mid-point displacement method to generate percentiles of THFM seemed to capture the natural autocorrelation of THFM values over a 3 week period (Fig. 9.4). The matching of the generated percentile with percentiles of THFM from observed records preserves within-season patterns of fuel moisture change (Figs. 9.2 and 9.3). The selection of the entire record based on this single variable resulted in consistent relationships between fuel moisture, wind speed and direction, and fire ignitions.

The weather sampling procedure was used to create three scenarios by adjusting the mean value from which yearly minimum and maximum percentiles of THFM were generated (Table 9.1). For the 'nominal' scenario, the mean minimum and maximum percentiles were estimated from the historical data to be 0.20 and 0.75, respectively. The 'wetter' scenario was created by increasing these percentiles by 0.05, while the 'drier' scenario was created by decreasing the mean minimum and maximum percentiles by 0.05.

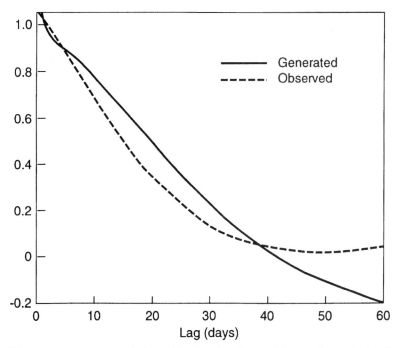

Figure 9.4 *Autocorrelation of simulated and recorded 1000-hour timelag fuel moisture (THFM) values.*

Table 9.1. *Mean minimum and maximum THFM[a] percentiles for generating autocorrelated weather sequences for the three weather scenarios*

	Weather scenario		
	Wetter	Nominal	Drier
Minimum[b]	0.25	0.2	0.15
Maximum[b]	0.80	0.75	0.70

[a]THFM is the percentage moisture content of large ($\geq$ 7.6 cm diameter) fuels that require approximately 1000 hours to equilibrate with atmospheric humidity (Deeming *et al.*, 1978).
[b]The standard deviations for the minimum and maximum were 0.20 and 0.35, respectively, for all weather scenarios.

Fire starts

The location of each fire start was randomly selected with probabilities estimated from the historical records (dating back to 1972) of natural fires within the study region. Because records were not extensive for this region of the park (the total number of records equals 262), probabilities for the entire area were estimated by aggregating the 50 × 50 m cells into larger units of 2750 m on a side (1406.25 ha). A two-dimensional spline with tension (Mitas & Mitasova, 1988; Mitasova & Mitas, 1993) was then used to interpolate a smooth surface through these points at a resolution of 50 × 50 m, with the constraint that all locations except lakes and rivers had a small but finite (0.0002) possibility of ignition. The ignition probabilities were greater at higher elevations, but the relationship between fire starts and elevation was complex due, in part, to the pattern of storms within YNP and the irregular distribution of non-flammable sites (<8% of the total area, or 7035 ha).

A separate FORTRAN program preselected locations for fire starts from the spatial probability distribution and wrote this information to a data file (starts.dat). During each simulation, when the weather record indicated that a start occurred, EMBYR reads the next record from starts.dat to obtain the coordinates for the fire start.

Model simulations were tested by comparison against historical records of fires that occurred during the 1979 and 1981 seasons. We know that the rate and extent of fire spread are sensitive to the variation of weather conditions, especially wind speed and direction (W. W. Hargrove *et al.*, unpublished observations). Because detailed weather records for historical fires are not available, it does not seem reasonable to compare each fire with a single model simulation. However, Monte Carlo simulations with EMBYR can be used to characterize the frequency distribution of stand-replacing fires (W. W. Hargrove *et al.*, unpublished observations). When the actual fire events of 1979 and 1981 were compared with the frequency distribution of Monte Carlo simulations produced by EMBYR, the actual fires could not be distinguished from the simulations. Further details and an illustration of this comparison with EMBYR simulations can be viewed on the World Wide Web at URL http:// www.esd.ornl.gov/ern/embyr/embyr.html.

Simulation of alternative climates

EMBYR was used to conduct 1000 year simulations for each of three weather scenarios: nominal, reflecting the weather of the past 20 years, drier than present, and wetter than present. Monte Carlo simulations with 10 replicates were conducted for each scenario. Data for each simulation included a gridded map of the distribution and successional stage of forested sites, the topography of the region, the probability surface for selecting locations for fire starts, and the randomly generated weather data for each weather scenario. During the simulation, EMBYR examined the weather records to determine the date for the next ignition, a random ignition site was selected from the fire starts probability layer, and the spread of the fire was simulated until the fire was extinguished. The severity of the fire (the effect of the fire on the vegetation) was recorded and, at the end of the year, the vegetation map was updated via the Markov recovery model. The number of ignitions, the size of the fires, the number of burned sites and changes in the amount of LP0, LP1, LP2 and LP3 were recorded each year during the simulation.

Spatial patterns of the forest vegetation at the end of each 1000 year simulation were quantified to estimate the cumulative effect of fires on the amount and arrangement of LP3 stands. The statistics of interest were the size of the largest cluster of LP3 sites, the total number of LP3 clusters, and the area-weighted average cluster size (Saw). Individual sites were labelled as members of the same cluster if any one of the eight neighbouring sites (that is, adjacent or diagonal neighbours) were of the same successional stage. The more common method of cluster identification excludes the four diagonal neighbours and is inconsistent with EMBYR's rule for fire spread to the eight adjacent and diagonal neighbours.

Results

Variation in minimum and maximum THFM percentiles from the wetter to drier weather scenarios (Table 9.2) resulted in a 4.5% decrease in the frequency of days within the wettest THFM class (>16%). The frequency of

Table 9.2 *Results of the weather generation procedure. Percentage of days in three THFM^a classes and muber of ignitions during the fire season (June 1 to October 15) for a single iteration of 1000 years of weather*

	Weather scenario			THFM class
	Wetter	Nominal	Drier	
Percentage of days				
	95.1	93.4	90.6	>16%
	4.2	5.5	7.6	12–16%
	0.7	1.1	1.8	<12%
Number of ignitions				
Total ignitions	7588	7838	8315	
Mean daily ignitions	0.05	0.051	0.054	

days within the moderate class (12–16%) increased by 3.4% and the frequency of days within the driest class (<12% THFM) increased from 0.7% to 1.8%. Although the total number of ignitions increased by 8.7% from wettest to driest, the average number of ignitions per day was large enough in all three scenarios to ensure that ignitions did occur (Table 9.2).

Among the three scenarios, fewer fires occurred in the wetter scenario (Table 9.3). The cumulative frequency distribution of fire sizes (Table 9.3) shows little difference for percentiles ≥75, but the distribution is skewed above the 75th percentile, with different patterns for the extreme events. For instance, the 95th and 99th percentiles for the wetter scenarios are nearly 1.5 times larger than the drier scenario (Table 9.3). Occurrence of less frequent but more extreme events in the wetter scenario was confirmed by examining the return interval of fires in three different size classes (Table 9.4). The average number of years between all fire events anywhere within the simulated landscape was 7.2, 5.7 and 4.4 for the wetter, 'nominal' and drier scenario, respectively. However, fires larger than 20 000 ha occurred once every 1000 years in the wetter scenario, but only once every 1626 years in the drier scenario (Table 9.4). The time required to burn an area equal to the total forested region of the landscape used in the simulation (Fig. 9.1, total forested area = 78 723 ha) was longest in the wetter scenario (2043 years) and shortest in the drier scenario (1626 years, Table 9.4).

The average area burned per year can be estimated by regressing the total area burned from the beginning of the simulation against time (years). Results of these regressions (Table 9.5) showed that rates of burn ranged from 145.6 ha/year for the wetter scenario to 176.3 ha/year for the drier scenario, with the 'nominal' scenario having an intermediate rate of 158.6 ha/year. The R^2 for all three regressions were high and the standard errors of the slopes were

Table 9.3 *Selected percentiles of the area burned per year (ha) and the total number of fires resulting from 10 sets of 1000 year simulations of EMBYR for three weather scenarios*

	Weather scenario		
Percentile	Wetter	Nominal	Drier
1	0.25	0.25	0.25
5	0.25	0.25	0.25
25	1.5	1.0	1.0
50	8.5	6.8	7.8
75	294.8	265.8	357.8
95	7141.0	6403.8	4727.8
99	18384.0	16279.0	13125.0
Total Number	1391	1744	2256

*a*THFM is the percentage moisture content of large (≥ 7.6 cm diameter) fuels that require approximately 1000 hours to equilibrate with atmospheric humidity (Deeming *et al.*, 1978). The classes represent wet conditions (>16%), moderately dry conditions (12–16%), and very dry conditions (<12%).

low (Table 9.5), resulting in differences between slopes being highly significant ($P<0.0001$).

The total area of LP3, the oldest age class of lodgepole pine stands, at the end of each 1000 years indicated that the amount of LP3 was highest in the wetter scenario and lowest in the drier scenario (Table 9.6). The reduction in amount of LP3 from wetter to drier conditions results in a corresponding increase in younger lodgepole stands when the climate was drier. As a consequence, the coefficient of variation for LP0 stands was highest with a general decline in relative variability as the forest stands increased in age. However, the sudden transformation of LP3 stands to LP0 resulting from a stand-replacing fire caused the coefficient of variation of LP3 stands to be somewhat greater than for LP2 stands (Table 9.6).

The spatial arrangement of LP3 sites was analysed at the end of each 1000 year simulation (the computational expense associated with this spatial analysis made it impractical to perform the analysis more frequently). Total number of LP3 clusters increased by 40% from the wetter to the drier scenarios (Table 9.7). The coefficient of variation of the total number of clusters (that is, the per cent variability between model simulations) decreased by over 50% from the wetter to drier scenario. The increase in fragmentation produced by drier conditions was indicated by the increased number of

Table 9.4 *The rotation period and return interval of fires of three different size classes resulting from 10 sets of 1000 year simulations of EMBYR for three weather scenarios*

	Weather scenario		
	Wetter	Nominal	Drier
Fire return interval (years)[a]			
All fires	7.2	5.7	4.4
>2500 ha	71.4	52.1	45.5
>20000 ha	1000	1250	1667
Rotation period (years)[b]	2043	1818	1626

[a]Fire return interval is the mean number of years between successive fires of a given size class occurring anywhere within the study area.

[b]The rotation period is the estimated time required to burn an area equal to the total forested region of the landscape used in the simulation (Fig. 9.1).

Table 9.5 *Regression analysis of the cumulative area burned from 10 sets of 1000 year simulations of EMBYR for three weather scenarios*

	Weather scenario		
Regression[a]	Wetter	Nominal	Drier
β(hay/year)	145.6	158.6	176.3
SE	1.06	0.68	0.50
R^2	0.93	0.97	0.98

[a]The statistics were estimated by regressing the cumulative area burned (ha) on the year the burn occurred. The units for β, the slope of the regression, are hectares per year. SE is the standard error of the slope and R^2 is the ratio of regression sum of squares to total sum of squares. The statistics were estimated by SAS's GLM procedure (SAS Institute, 1990). Differences among the slopes are significant at $P<0.0001$.

clusters and the decrease in the size of the largest cluster (from 15 370 ha to less than 2000 ha from wetter to drier conditions). The area-weighted cluster size (Saw, Table 9.7) further illustrates the effect of fire on forest fragmentation, showing a 20-fold decline in Saw from the wetter to drier conditions.

Table 9.6 *Forest cover at the end of 10 sets of 1000 year simulations of EMBYR for each of three weather scenarios*

Forest cover (ha)	Weather scenario		
	Wetter	Nominal	Drier
LP0 mean/year	8738	9486	10 440
CV[a]	87.9%	82.1%	73.6%
LP1 mean/year	14 568	16 066	17 304
CV[a]	35.5%	33.1%	29.5%
LP2 mean/year	24 832	26 715	27 872
CV[a]	16.6%	16.5%	16.5%
LP3 mean/year	30 251	26 073	22 642
CV[a]	22.5%	23.9%	23.4%

[a]CV is coefficient of variation defined as the standard deviation/mean × 100

Discussion and conclusions

Small changes in climate, and the subsequent climatically induced changes in fire regime (Clark, 1989), can result in rapid shifts in pattern and process at landscape scales (Clark, 1990; Baker, Egbert & Frazier, 1991). Although many terrestrial systems may be fire-adapted, the complex interaction between weather, fuel characteristics, history of fire suppression, and the landscape pattern of fuels make landscape-scale effects of altered regimes difficult to predict. The simulations reported here explored the long-term effects of climate-altered fire regimes on the pattern of forested areas within Yellowstone National Park. The results showed that small changes in weather parameters can produce significant shifts in the age structure and spatial arrangement of the forest. The 'drier climate' scenario had more frequent fires (Table 9.3), burned a larger area in a shorter time (Table 9.4), but had fewer extreme fire events (Table 9.3). The result was a reduction in the amount and an increase in the fragmentation of mature forest (Table 9.7). The 'wetter climate' scenario had fewer fires with a longer period required to burn the area being studied (Table 9.4). However, because wetter conditions allowed larger areas of contiguous patches of mature forest to form (Table 9.7), the risk of larger fires increased under wetter climates. These results are remarkably similar to those of Swetnam (1993) whose 2000 year fire records in giant sequoia stands showed that frequent small fires occurred during warm periods but more widespread fires occurred during cooler climate conditions.

Climate projections from global climate models do not clearly indicate if

Table 9.7. *Spatial pattern of forest cover of LP3 stands at the end of 10 sets of 1000 year simulations of EMBYR for each of three weather scenarios*

	Weather scenario		
Spatial pattern[a]	Wetter	Nominal	Drier
Total number	9560	9783	13 329
CV[b]	38.9%	33.5%	15.9%
Largest cluster (ha)	15 370	12 245	1997
CV[b]	84.3%	90.6%	52.0%
Saw (ha)	10 204	7552	490
CV[b]	115%	139.7%	79.9%

[a]'Total number' refers to the total number of clusters of LP3 sites. 'Largest cluster' is the single largest number of connected sites for each simulation. Saw, the area-weighted average cluster size of all fires, is calculated as: Saw $= \Sigma S_i^2 / \Sigma S_i$, where S_i is the cluster size of burned sites. Clusters were identified as LP3 sites connected by the next-nearest neighbour rule.
[b]CV is coefficient of variation defined as the standard deviation/mean × 100.

increasing levels of atmospheric CO_2 will create wetter or drier conditions in the Yellowstone region. However, it does appear that warm conditions in Yellowstone are more likely, with the probability of summer and autumn drought increasing, making weather conditions for large fires more probable (Overpeck *et al.*, 1990; Flannigan & Van Wagner, 1991). A recent analysis of 1895–1989 weather records for Yellowstone (Balling, Meyer & Wells, 1992) revealed such a trend: summer temperatures generally have increased, spring precipitation has declined, and variations in burn area within YNP have been significantly related to these changes. However, projecting the long-term biotic response to these changes is directly dependent on the adequacy of the methods used to generate daily weather records of sufficient length and detail to simulate the long-term response of vegetation. The adequacy of the weather generator for fire modelling is especially important because 'weather, as it controls fuel moisture, is more important than differences in elevation and vegetation composition within the subalpine forest in determining the rate of spread, intensity and fuel consumption' (Fryer & Johnson, 1988). Care must also be taken to observe subtle relationships among weather variables: for instance, temporal and spatial patterns of rainfall can be more important in explaining burned area than cumulative precipitation (Flannigan & Harrington, 1988). The advantages of the methods we have used are (1) temporal correlations in weather are maintained (Fig. 9.4); (2) the instrumental record preserves important relationships among weather variables; and (3) parameters controlling the weather generation process (Table 9.1) can be empirically estimated. The disadvantages of this method are those imposed by the limitations of the available instrumental records and the possible importance of inter-annual trends and cycles in weather (Yeakley *et al.*, 1994). No matter what method is used to produce long sequences of weather records, many uncertainties remain. For instance, climate induced changes in wind will be particularly important, but it is not currently known if winds will increase or decrease as a result of climate change (Flannigan & Van Wagner, 1991). The inclusion of methods for accurately simulating wind fields (e.g. Zack & Minnich, 1991) will be an important improvement.

Fire-dominated landscapes are never truly in an equilibrium state (Pickett & White, 1985; Picket *et al.*, 1989; Baker *et al.*, 1991; Turner *et al.*, 1993). However, the spatial patterns of stand-replacing fires showed a recurring pattern (Fig. 9.5) that is affected by climate and which, in turn, is likely to affect the process of plant reestablishment, landscape-scale changes in species abundance and diversity, and ecosystem fluxes of matter and energy (Davis & Burrows, 1993; Turner & Romme, 1994). The non-equilibrium conditions produced by the continual build-up and combustion of fuels illustrates the self-organized criticality of the Yellowstone landscape that is similar to that predicted from consideration of very simple fire models (Bak, Tang & Wisenfeld, 1988; Bak, Chen & Tang, 1990). These self-organized systems are unpredictable over short time intervals, but long-term dynamics are driven by the rate of fuel accumulation and the frequency of weather conditions that allow fires to occur. Therefore, the effect of climate change on fire regimes (Table 9.5; Turner & Romme, 1994), combined with the build-up of fuels during long periods without fire (Romme & Despain, 1989), may result in rapid loss of large areas of mature forest (Flannigan & Van Wagner, 1991). If drier weather conditions prevail, then more frequent but predictable fire events should occur after the initial loss of large areas of mature forest. The effect on species diversity of rapid shifts in the age structure and spatial arrangement of forests have not yet been fully explored (Suffling, Lihou & Morand, 1988; Romme & Turner, 1991). However, present knowledge suggests that species associated with older forest successional stages (LP3) may be strongly influenced by the size and spatial arrangement of LP3 patches. Our results showed that these patches are likely to become substantially smaller under drier climatic conditions, and suggest that these species may be vulnerable to reduced population sizes, genetic changes, and local extinctions.

Results of a number of studies indicated that climatically induced changes in the fire regime may lead to substantial changes in the frequency and extent of disturbance (Suffling *et al.*, 1988; Clark, 1990; Graham,

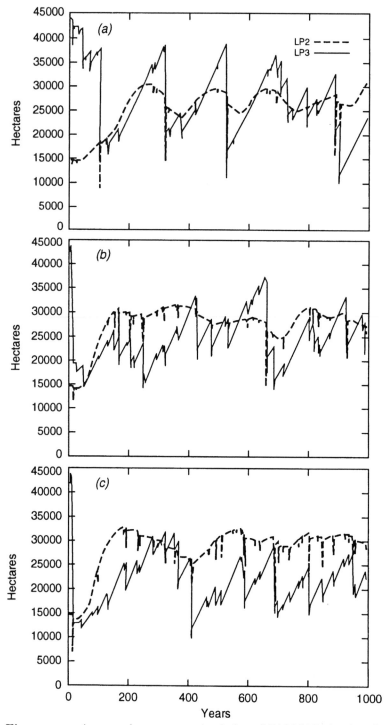

Figure 9.5   *An example 1000-year simulation of EMBYR showing the changes in LP2 and LP3 for (a) the wetter, (b) nominal, and (c) drier weather scenarios.*

Turner & Dale, 1990; Johnson *et al.*, 1990; Romme & Turner, 1991; Balling *et al.*, 1992; Davis & Burrows, 1993). These changes may affect landscapes in two ways. First, as suggested by our long-term simulations, the characteristic mosaic of successional stages across the landscape may be altered from that which has characterized the landscape during the past few centuries. Change in the abundance and distribution of old and young forest stands has implications for the rest of the biota, including changes in herbaceous species composition, an increased probability of invasion of new species, and shifts in the distribution and abundance of fauna. Field data we collected following the 1988 fires indicate widespread establishment of seedlings of aspen (*Populus tremuloides*) across burned areas of the Yellowstone Plateau, far from the previous range of aspen (W. H. Romme *et al.*, unpublished data). Second, other ecological processes, such as nutrient losses and changes in nutrient cycling (Kauffman, Cummings & Ward, 1994; Trabaud, 1994), soil erosion and hydrological dynamics (Meyer *et al.*, 1992) may be dramatically changed as a consequence of more frequent fires. Our simulations have not accounted for these changes, but studies of other ecosystems suggest that mature plants might survive short-term climate fluctuations (Malanson *et al.*, 1992) with species that are better adapted to a changed climate becoming established following fire (Dunwiddie, 1986). The linkage of a model of post-fire forest re-establishment that incorporates the climatic requirements for germination and growth of potential dominant tree species with EMBYR would permit potential community shifts to be explored.

Many issues must be resolved before the effects of climate change and disturbance regimes can be effectively linked. Changes in weather patterns due to increased levels of atmospheric CO_2 may have a positive feedback effect on climate change (Auclair & Carter, 1993), and increases in erosion are likely to affect soil moisture and fertility levels (Meyer *et al.*, 1992). If large areas of forests continue to burn (as occurred again in 1994 in the west and northwest United States) then the effect of altered albedo will be an important factor in local and regional weather patterns (Pielke & Avisser, 1990). These changes may, in turn, alter the pattern of species that can quickly repopulate disturbed areas. Although species composition after large fires in YNP largely reflects the species present before the fires occurred (Anderson & Romme, 1991), the long-term effect of large fires on local edaphic factors suggests that succession in some areas may be dominated by the ability of plants to migrate to suitable germination sites (Prentice, 1986).

The YNP landscape has undergone major changes in climate and vegetative composition in prehistoric times. For example, during the most recent glacial period (*c.* 20 000–16 000 BP) the upper timberline in the Yellowstone region was 600–1200 m lower than it is today (Barnosky *et al.*, 1987). Yellowstone was considerably warmer than present during the early Holocene (*c.* 10 000–14 000 BP), and Douglas fir (*Pseudotsuga menziesii*) apparently grew at higher elevations (Barnosky *et al.*, 1987). Future climate changes are uncertain, but tools such as EMBYR permit scientists and land managers to explore the implications of alternative climate scenarios and rates of climate change.

It is the goal of Focus 2 of GCTE to model the complex suite of impacts and responses of vegetation that result from climate change (Steffen *et al.*, 1992). The challenge is to develop and identify appropriate models that can handle the suite of ecosystem effects of repeated impacts (Smith, Leemans & Shugart, 1994), and link vegetation changes with animal population dynamics and socio-economic development. New methods and extensive data are rapidly being developed to address a variety of global change issues. The linkage of models and data in such a fashion that they address processes at many scales will be necessary before the implications of climate change on disturbance regimes can be understood.

Acknowledgements

Research for this chapter was funded by grants from the National Science Foundation (BSR-9018381) and the Ecological Research Division, Office of Health and Environmental Research, US Department of Energy, under contract no. DE-AC05-84OR24400 with Martin Marietta Energy Systems, Inc.

Special thanks are due to Don Despain and George McKay for making the GIS data for Yellowstone available. We also appreciate the many helpful discussions concerning the history and behaviour of fire in Yellowstone and the northern Rockies with Don Despain, Don Perkins, Roy Renkin, Richard Rothermel, Patricia Andrews and James K. Brown.

References

Albinet, G., Serby, G. & Stauffer, D. (1986). Fire propagation in a 2-D random medium. *Journal de Physique*, **47**, 1–7.

Ales, R. F., Martin, A., Ortega, F. & Ales, E. E. (1992). Recent changes in landscape structure and function in a mediterranean region of SW Spain

(1950–1984). *Landscape Ecology*, **7**, 3–18.

Anderson, J. E. & Romme, W. H. (1991). Initial floristics of lodgepole pine (*Pinus contorta*) forests following the 1988 Yellowstone fires. *International Journal of Wildland Fire*, **1**, 119–24.

Antonovski, M. Ya., Ter-Mikaelian, M. T.

& Furyaev, V. V. (1992). A spatial model of long-term forest fire dynamics and its application to forests in Western Siberia. *A Systems Analysis of the Global Boreal Forest*, ed. H. H. Shugart, R. Leemans & G. B. Bonan. Cambridge: Cambridge University Press.

Auclair, A. N. D. & Carter, T. B. (1993). Forest wildfires as a recent source of CO_2 at northern latitudes. *Canadian Journal of Forest Research*, **23**, 1528–36.

Bak, P., Chen, K. & Tang, C. (1990). A forest-fire model and some thoughts on turbulence. *Physics Letters A*, **147**, 297–300.

Bak, P., Tang, C. & Wiesenfeld, K. (1988). Self-organized criticality. *Physical Review A*, **38**, 364–74.

Baker, W. L. (1992). The landscape ecology of large disturbances in the design and management of nature reserves. *Landscape Ecology*, **7**, 181–94.

Baker, W. L. (1993). Spatially heterogeneous multi-scale response of landscapes to fire suppression. *Oikos*, **66**, 66–71.

Baker, W. L., Egbert, S. L. & Frazier, G. F. (1991). A spatial model for studying the effects of climatic change on the structure of landscapes subject to large disturbances. *Ecological Modelling*, **56**, 109–25.

Balling, R. C., Meyer, G. A. & Wells, S. G. (1992). Climate change in Yellowstone National Park: is the drought-related risk of wildfires increasing? *Climatic Change*, **22**, 35–45.

Barnosky, C. W., Anderson, P. M. & Bartlein, P. J. (1987). The Northwestern US during deglaciation: vegetational history and paleoclimatic implications. In *North America and Adjacent Oceans during the Last Glaciation. The Geology of North America*, vol. K-3, ed. W. F. Ruddiman & H. E. Wright Jr, pp. 289–321. Boulder, Colorado: Geological Society of America.

Barnsley, M. F., Devaney, R. L., Mandelbrot, B. B., Peitgen, H.-O.,

Saupe, D. & Voss, R. F. (1988). *The Science of Fractal Images*. New York: Springer.

Bonan, G. B., Shugart, H. H. & Urban, D. L. (1990). The sensitivity of some high-latitude boreal forests to climatic parameters. *Climatic Change*, **16**, 9–29.

Clark, J. S. (1989). Ecological disturbance as a renewal process: theory and application to fire history. *Oikos*, **56**, 17–30.

Clark, J. S. (1990). Fire and climate change during the last 750 years in northwestern Minnesota. *Ecological Monographs*, **60**, 135–59.

Clarke, K. C., Olsen, G. & Brass, J. A. (1993). Refining a cellular automaton model of wildfire propagation and extinction. In Proceedings, Second International Conference on the Integration of Geographic Information Systems and Environmental Modeling, Breckenridge, CO, USA.

Davis, F. W. & Burrows, D. A. (1993). Spatial simulation of fire regime in mediterranean-climate landscapes. In *The Role of Fire in Mediterranean-type Ecosystems*, ed. M. C. Talens, W. C. Oechel & J. M. Moreno. New York: Springer.

Deeming, J. E., Burgan, R. E. & Cohen, J. D. (1978). *The National Fire-Danger Rating System*. USDA Forest Service General Technical Report INT-39.

Despain, D. G. (1987). The two climates of Yellowstone National Park. *Proceedings of the Montana Academy of Science*, **47**, 11–19.

Despain, D. G. (1990). *Yellowstone's Vegetation: The Consequences of History and Environment in a Natural Setting*. New York: Roberts Rinehart, Inc.

Dirks, R. A. & Martner, B. E. (1982). *The Climate of Yellowstone and Grand Teton National Parks*. US Department of the Interior, National Park Service Occasional Paper no. 6, pp. 1–26.

Dunwiddie, P. W. (1986). A 6000-year record of forest history on Mount Rainier, Washington. *Ecology*, **67**, 58–68.

Feder, J. (1988). *Fractals.* New York: Plenum Press.

Flannigan, M. D. & Harrington, J. B. (1988). A study of the relation of meteorology variables to monthly provincial area burned by wildfire in Canada 1953–80. *Journal of Applied Meteorology,* **27**, 441–52.

Flannigan, M. D. & Van Wagner, C. E. (1991). Climate change and wildfire in Canada. *Canadian Journal of Forest Research,* **21**, 66–72.

Fryer, G. I. & Johnson, E. A. (1988). Reconstructing fire behaviour and effects in a subalpine forest. *Journal of Applied Ecology,* **25**, 1063–72.

Graham, R. L., Turner, M. G. & Dale, V. H. (1990). How increasing atmospheric CO_2 and climate change affects forests. *BioScience,* **40**, 575–87.

Green, D. G. (1983). Shapes of simulated fires in discrete fuels. *Ecological Modelling,* **20**, 21–32.

Green, D. G., Tridgell, A. & Gill, A. M. (1990). Interactive simulation of bushfires in heterogeneous fuels. *Mathematical and Computer Modelling,* **12**(12), 57–66.

Hargrove, W. W., Gardner, R. H., Turner, M. G., Romme, W. H. & Despain, D. G. (manuscript). Simulating fire patterns in heterogeneous landscapes.

Johnson, E. A. & Van Wagner, C. E. (1985). The theory and use of two fire history models. *Canadian Journal of Forest Research,* **15**, 214–20.

Johnson, E. A. & Wowchuk, D. R. (1993). Wildfires in the southern Canadian Rocky Mountains and their relationship to mid-tropospheric anomalies. *Canadian Journal of Forest Research,* **23** 1213–22.

Kauffman, J. B., Cummings, D. L. & Ward, D. E. (1994). Relationships of fire, biomass and nutrient dynamics along a vegetation gradient in the Brazilian cerrado. *Journal of Ecology,* **82**, 519–31.

Kessell, S. R. (1976). Gradient modeling: a new approach to fire modeling and wilderness resource management. *Environmental Management,* **1**, 39–48.

Kessell, S. R. (1979). *Gradient Modelling: Resource and Fire Management.* New York: Springer.

Levin, S. A. (1992). The problem of pattern and scale in ecology. *Ecology,* **73**, 1943–67.

Lewis, W. M. Jr. (1986). Nitrogen and phosphorus runoff losses from a nutrient-poor tropical moist forest. *Ecology,* **67**, 1275–82.

Malanson, G. P., Westman, W. E. & Yan, Y.-L. (1992). Realized versus fundamental niche functions in a model of chaparral response to climatic change. *Ecological Modelling,* **64**, 261–77.

Mandelbrot, B. B. (1983). *The Fractal Geometry of Nature.* San Francisco: W. H. Freeman.

Meyer, G. A., Wells, S. G., Balling, R. C. Jr. & Tull, A. J. T. (1992). Response of alluvial systems to fire and climate change in Yellowstone National Park. *Nature,* **357**, 147–50.

Mitas, L. & Mitasova, H. (1988). General variational approach to the interpolation problem. *Computers and Mathematics with Applications,* **16**, 983–92.

Mitasova, H. & Mitas, L. (1993). Interpolation by regularized spline with tension. I. Theory and implementation. *Mathematical Geology,* **25**, 641–55.

Moreno, J. M. & Oechel, W. C. (eds.). (1994). *The Role of Fire in Mediterranean-type Ecosystems.* New York: Springer.

Neilson, R. P. (1993). Transient ecotone response to climatic change: some conceptual and modelling approaches. *Ecological Applications,* **2**, 385–95.

Ohtsuki, T. & Keyes, T. (1986). Biased percolation: forest fires with wind. *Journal of Physics A. Mathematical and General,* **19**, L281–7.

Overpeck, J. T., Bartlein, P. J. & Webb, III, T. (1991). Potential magnitude of future vegetation change in Eastern North America: Comparisons with the past. *Science,* **254**, 692–5.

Overpeck, J. T., Rind, D. & Goldberg, R. (1990). Climate-induced changes in forest disturbance and vegetation. *Nature*, **343**, 51–3.

Pickett, S. T. A., Kolasa, J., Armesto, J. J. & Collins, S. L. (1989). The concept of disturbance and its expression at various hierarchical levels. *Oikos*, **54**, 129–36.

Pickett, S. T. A. & White, P. S. (eds.) (1985). *The Ecology of Natural Disturbance and Patch Dynamics*. New York: Academic Press.

Pielke, R. A. & Avissar, R. (1990). Influence of landscape structure on local and regional climate. *Landscape Ecology*, **4**, 133–55.

Plotnick, R. E. & Prestegaard, K. (1993). Fractal analysis of geologic time series. In *Fractals in Geography*, ed. N. Lam & L. DeCola, pp. 207–22. Englewood Cliffs, NJ: Prentice-Hall.

Prentice, I. C. (1986). Vegetation responses to past climatic variation. *Vegetatio*, **67**, 131–41.

Renkin, R. A. & Despain, D. G. (1992). Fuel moisture, forest type, and lightning-caused fire in Yellowstone Naional Park. *Canadian Journal of Forest Research*, **22**, 37–45.

Romme, W. H. (1982). Fire and landscape diversity in subalpine forests of Yellowstone National Park. *Ecological Monographs*, **52**, 199–221.

Romme, W. H. & Despain, D. G. (1989). Historical perspectives on the Yellowstone fires of 1988. *BioScience*, **39**, 695–9.

Romme, W. H. & Turner, M. G. (1991). Implications of global climate change for biogeographic patterns in the greater Yellowstone ecosystem. *Conservation Biology*, **5**, 373–86.

Rothermel, R. C. (1972). *A Mathematical Model for Predicting Fire Spread in Wildland Fuels*. USDA Forest Service, Intermountain Forest and Range Experiment Station Research Paper, INT-115, Ogden, Utah, USA.

SAS Institute (1990). *SAS User's Guide: Statistics*. Cary, NC: SAS Institute Inc.

Smith, T. M., Leemans, R. & Shugart, H. H. (1994). *The Application of Patch Models of Vegetation Dynamics to Global Change Issues*. Proceedings of a workshop, Apeldoorn, Netherlands 28 March–1 April, 1994. Amsterdam: Kluwer.

Steffen, W. L., Walker, B. H., Ingram, J. S. I. & Koch, G. W. (1992). *Global Change and Terrestrial Ecosystems: The Operational Plan*. IGBP Report 21. Stockholm: The International Geosphere-Biosphere Programme, International Council of Scientific Unions.

Suffling, R., Lihou, C. & Morand, Y. (1988). Control of landscape diversity by catastrophic disturbance: a theory and a case study of fire in a Canadian boreal forest. *Environmental Management*, **12**, 73–8.

Swetnam, T. W. (1993). Fire history and climate change in giant sequoia groves. *Science*, **262**, 885–9.

Swetnam, T. W. & Betancourt, J. L. (1990). Fire–Southern Oscillation relations in southwestern United States. *Science*, **249**, 1017–20.

Trabaud, L. (1994). The effect of fire on nutrient losses and cycling in a *Quercus coccifera* garrigue (southern France). *Oecologia*, **99**, 379–86.

Turner, M. G., Gardner, R. H., Dale, V. H. & O'Neill, R. V. (1989). Predicting the spread of disturbance across heterogeneous landscapes. *Oikos*, **55**, 121–9.

Turner, M. G. & Romme, W. H. (1994). Landscape dynamics in crown fire ecosystems. *Landscape Ecology*, **9**, 59–77.

Turner, M. G., Romme, W. H., Gardner, R. H., O'Neill, R. V. & Kratz, T. K. (1993). A revised concept of landscape equilibrium: Disturbance and stability on scaled landscapes. *Landscape Ecology*, **8**, 213–27.

Urban, D. L., O'Neill, R. V., & Shugart,

H. H. (1987). Landscape ecology. *Bio-Science*, **37**, 119–27.

Vasconcelos, M. J. & Guertin, D. P. (1992). FIREMAP – Simulation of fire growth with a geographic information system. *International Journal of Wildland Fire*, **2**, 87–96.

von Niessen, W. & Blumen, A. (1988). Dynamic simulation of forest fires. *Canadian Journal of Forest Research*, **18**, 805–12.

Yeakley, A. A., Moen, R. A., Breshears, D. D. & Nungesser, K. (1994). Response of North American ecosystem models to multi-annual periodicities in temperature and precipitation. *Landscape Ecology*, **9**, 249–60.

Zack, J. A. & Minnich, R. A. (1991). Integration of geographic information systems with a diagnostic wind field model for fire management. *Forest Science*, **37**, 560–73.

10 Linking the human dimension to landscape dynamics

I. R. Noble

Introduction

Much of human endeavour is devoted to planning and managing landscapes. We live out our daily lives in landscapes; agriculture, pastoralism, urban planning and conservation management are all forms of landscape management.

While large-scale and abrupt land-use changes have occurred in the past, changes in land-use practice were generally slow and cumulative. The pace and the global magnitude of these changes has increased since the onset of the industrial revolution and particularly in this century. Larger populations, more powerful technologies and a greater variety of land uses have increased our rate of modification of the landscape enormously.

Most societies have recognized that *laissez-faire* is not an appropriate method of landscape management and that some form of broader land-use planning is needed. But to achieve this planning, and to achieve it in an ever more rapidly changing social system and landscape, we need to have certain resources available. It can be suggested that there are four major resources or skill required.

1. We need information on current land use. Human societies have always been able to find methods to gather this information, be it the Domesday Book or the current spate of geographic information systems (GIS) and remote sensing systems that abound in replicate in most government agencies. Much of our effort in landscape planning and landscape ecology is spent dealing with the technology, the challenges and the aesthetics of remote sensing and GIS. There is no denying their utility in mapping and planning land allocation but currently they are not the limiting step in achieving better landscape management and utilization.

2. We need information about alternative land uses and how individuals and societies make decisions as to which uses will be given priority.

3. We need to improve our ability to predict the consequences of

particular allocations. Whereas in the past, changes in land use were usually introduced slowly after a series of small scale trial and error experiments, we can now make major changes in land use very rapidly. These changes interact with the changes in climate and society and can thus trigger a cascade of consequential changes. The last-mentioned two issues will be the main theme of this chapter.

4. We need a methodology for implementing preferred land uses through our administrative systems; this is beyond the scope of this chapter.

Most scientists involved in biophysically based research ('natural scientists') have little contact with the large literature dealing with issues (2) and (3) above. This is not the place to attempt to summarize the various theories and methodological approaches. Instead this chapter will raise, from the point of view of a natural scientist, some of the major issues that natural scientists should be mindful of in seeking a closer collaboration with their social science colleagues.

This chapter does not refer very often to specific impacts of the scenarios of global change over the next few decades. Obviously global change will accentuate the importance of achieving sound land-use planning, but, as with so many other aspects of global change the first step in dealing with the impacts is to learn how better to deal with current conditions.

The dynamics of landscapes

Within the current paradigm of IGBP, GCTE and the Land Use and Cover Change (LUCC) Core Project, landscape changes of greatest relevance are those that either:

- significantly change the patterning of vegetation across large areas of the earth and thus affect the feedbacks between land surfaces and the atmosphere (e.g. large areas of clearing in forests);
- significantly change the propagation of disturbances across land surfaces and thus provide a feedback accentuating further changes (e.g. landscape patterns that enhance or diminish the spread of fires); or that
- significantly affect the life styles, movement and consumption patterns of the societies that live in them.

To make progress with these issues we need to develop a better understanding of the biophysical responses of landscapes to forcing functions arising from global change; of the impacts of human actions on landscapes, and the response of human societies to the changes they are experiencing or expecting.

Our current understanding of the biophysical aspects of the dynamics of landscapes has been reviewed by Naveh and Lieberman (1984), Forman and Godron (1986) and Turner and Gardner (1991) among others. We are making some progress in understanding the driving forces and sensitivities in these processes, but many problems remain relating to the complexity of interactions between the behaviour and life histories of organisms that live in the landscape, their dispersal and the impact of disturbances in a spatial mosaic. Even when we begin to understand some of these processes at one scale there are problems in carrying information and models across scales and the transmutation of outputs that occur.

However, whatever the progress we make with the application of the physical, biological and mathematical sciences to landscapes, it will amount to little unless we can incorporate the impacts of human activity. In cases where the human actions can be well described, we can usually draw upon our biophysical understanding of landscape processes to incorporate these actions into our predictions of landscape change. The vastly more difficult problem is ascertaining just what the responses of human societies to global change will be.

Is it feasible to predict how humans will respond to global change?

There has been a sense of frustration in the biophysically based scientific community ('natural scientists') involved in IGBP as to why more social science colleagues have not joined forces in tackling the issues of global change. It is possible that the 'predictive modelling' paradigm (Bella *et al.*, 1994) adopted by the vast majority of natural scientists within IGBP is part of this problem. Whereas the natural scientists tend immediately to approach a problem by attempting to develop a predictive model, many social scientists argue that within their areas of expertise precise prediction is impossible and not a priority. This point is succinctly made by Torry (1983) where he argues that feedbacks between environment and society are too complex to make predictions, although we can make some generalizations, and we can take precautionary steps to avoid the worst impacts; but these steps must take into account the multiple factors affecting society. The problem is made more difficult by the reflexive nature of human societies. That is, quality predictions of social trends will be acted upon by those societies and may well have the effect of modifying those very trends and thus invalidating the prediction. This makes the whole discussion of driving forces, causation and explanation complex (Turner, 1990). In many areas of social science the whole issue of 'prediction' is problematical because of the connotations of positivist science. Some argue that to reduce human behaviour to a series of natural or physical forces is to ignore the human free agency (see Sack, 1990 for a discussion of these issues).

Thus, natural scientists in the GCTE community must appreciate the caution with which many social scientists will approach the 'predictive modelling' viewpoint that is the dominant paradigm of the IGBP. The challenge to progress in predictive modelling of landscape changes is likely to remain largely with a small subset of natural scientists and social scientists. Natural scientists may apply their own modes of thought to the problem but, while this may prove effective, they cannot neglect the considerable relevant experience, and cautions, provided by the social sciences. This chapter will now explore some of these issues.

The most common approach by natural scientists to predicting land-use change is to classify possible reactions of human societies to changed circumstances and to assume some decision model of behaviour. For example, in dealing with the options for a rural society when faced with a changed climate, the possible responses include: change in agricultural practices, abandonment of agriculture, migration, doing nothing and acceptance of the consequences.

The current models used in regional and global change projections (e.g. Vloedbeld & Leemans, 1993; Dale *et al.*, 1994) reflect this approach and then go on to assume that informed and rational decisions will be made. Their models are based on algorithms that seek the most suitable land nearby for expansion, or reallocate land use to the most suitable class. These assumptions are logical in that they meet the oft-stated goals of various societies and Occam's razor. But are human decisions informed and rational?

These issues have been confronted many times by the social sciences and there are many well-established theories or hypotheses about the way that land will be allocated and how settlements will form. Among the oldest and best known are von Thunen models which first appeared early in the nineteenth century (see Haggett, 1965; Morrill, 1974). Most are based on an optimization approach, i.e. humans are rational beings and their settlement patterns will reflect this. Many human settlements of widely different sizes fit particular models; many more do not for reasons that can be readily found *post hoc*. There is an enormous literature modifying and elaborating on these approaches (e.g. Turner *et al.*, 1990) and natural scientists should be careful of wasting effort reinventing these and other theories. In researching this chapter, an interesting example was found from a geography textbook (Haggett, 1965) which described cellular automata models of diffusion published by Yuill in 1965 and seems to have anticipated many of the methods and findings published more recently in the landscape ecology literature. Are natural scientists simply reinventing the wheel with a different number of corners?

What can we learn from the past: Are human societies responsive to and affected by climate?

The social sciences have drawn heavily on the interpretation of case studies to seek some insights into the main factors modifying human societies. There have been many studies of the response of human societies to climate changes in the past. Most case studies refer to the less developed agricultural societies of the past which may bear some resemblance to less developed countries of the present. But there must be awareness that significant changes have occurred such as cash cropping, international trade, famine relief, externally subsidized warfare and so on, all of which change the options and responses of modern societies.

The book *Climate and History: Studies of Past Climates and their Impact on Man* (Wigley *et al.*, 1981) is based on a conference held in 1979. It contains an illustrative collection of case studies. The cases vary from hill farming in Scotland, Vikings in Greenland, medieval Castile, Roman Africa, agriculture in Maine to the fall of the Ancien Regime in France. The responses described and the conclusions drawn about them vary greatly and many are restrictive in their analysis in that they do not fully integrate both social and ecological aspects. However, they provide an insight into the difficulties in constructing a predictive model of the response of human societies to global change and a few are presented here as examples of the difficulties in making generalizations about human actions.

The failure of the Norse settlements in Greenland

McGovern (1981) documented the failure of the managing hierarchy of the Norse settlements of Greenland (AD 985–1500) to cope with climactic stress. As conditions in the settlement deteriorated an ecclesiastical elite exercised increasing control over the community. The managers persisted in raising stock along the fjords even during periods of extremely cold years. They failed to redirect activities to more appropriately based marine activities (fishing and whaling) as did the Inuit. They also failed to adopt skin boats or even skin clothing and the demand for wood and wool placed unsustainable demands on their resources. McGovern argues that there is a general principle here; namely, as conditions become more marginal there is a tendency to invest more resources in management, until a collapse occurs.

Droughts of Brittany

Sutherland (1981) provides a direct contrast to the example above. A series of droughts and cold weather in northern France during the 1780s have been well documented and their role in the eventual fall of the Ancien Regime much debated. Sutherland argues that the Brittany peasantry of 1780–9 felt few long-lasting effects of a series of bad harvests because they had reserves mostly from church-controlled tithes. It also appears that the landlords were adaptable in that they were more lenient with rents during the bad periods. There was, nevertheless, considerable debt and hunger, but the young and aged bore most of the increased mortality, and the population was able to rebound quickly when better seasons occurred. Sutherland concludes that the peasantry were much more independent and the ruling elite more flexible in mitigating the effects of bad seasons than is usually assumed. Thus we have contrasting views about the response of human societies to changing climates or extreme weather patterns.

Jewish pogroms of Castile

Mackay (1981) challenges the link that is often drawn between climate and human action. He describes the unrest against Jews in Castile in the fifteenth century that some have attributed to a series of bad harvests at about that time. Mackay questions whether any simplistic stimulus–response hypothesis, and especially those relating to climate, is adequate. He doubts the logic of the response of attacking Jews for events in which they took little part and argues that the peoples' attitude to the Jews and their response was affected by many factors other than climate.

Climate change and European history

Anderson (1981) in his review article takes this conclusion one step further and concludes 'if there had been no climatic change, then the history of Europe would not have been different in any general sense'. He goes on to argue that 'Marginal areas by definition would have been affected, perhaps profoundly, but, except for the Scandinavians, peoples at the margins of cultivation in Europe were of little political or economic significance to the mainstream of European history.' But he also concludes that 'By contrast, local, and even regional, histories could well have been different in the absence of changes in climate.' In small communities the shocks of changing weather patterns are not diffused by the operation of the market and state agencies.

The 'lessening' and 'catastrophe' hypotheses

Anderson's (1981) conclusion must be re-examined in terms of Bowden et al.'s (1981) 'lessening' and 'catastrophe' hypotheses. The lessening hypothesis states that societies are able, through technology and social organization, to lessen the impacts of minor climatic stresses that have return periods of less than 100 years. Bowden and colleagues find support for the lessening hypothesis in several drought-prone societies (Tigris–Euphrates, Great Plains and the Sahel). But the insulating effects of this lessening process may increase vulnerability to a major climate stress leading to a collapse of societies that had up until then seemed to be coping. This would suggest that we must be cautious in interpreting the apparent resilience of societies to climate change, even over long periods, as an invulnerability to it.

Bowden and colleagues found it difficult to find sufficient examples and sequences of climatic and social information to test their hypotheses thoroughly. This is frequently a problem and its solution calls for more effective collaboration to achieve a synergism between the natural and social sciences.

Hill farming in Scotland and retrodiction

Parry (1981) also questioned the reliability of inductive conclusions based on spatial and temporal coincidences of climatic and social changes. He set out to test the links more rigorously in a case where both climatic and social data were available. He asked whether climate change is a 'contingently necessary condition' for economic change and suggests that retrodiction – i.e. rational reconstructions of history based on thorough understanding of the processes linking climate, agriculture and settlement – is an effective approach to answering such questions. There are two important steps in retrodictive studies. First, changes consequent on a change in climate, such as changes in agricultural output and the resultant changes in the capacity of an area to support the population, are predicted based on our best understanding of the relationship between climate and crop yields, etc. These predictions are then tested against independent historical data which may show evidence of famine, increased migration or altered trading patterns. Retrodiction is not often used in historical studies because it is rare to have two independent sources of information; however, these are more common in climate-history studies. Parry (1981) demonstrates that retrodiction does lead to good predictions of the permanent abandonment of farmland in hill farming areas of Scotland from AD 1300 to 1600.

There are many difficulties that remain in retrodictive studies. Climatic reconstructions are often highly uncertain and agricultural practices and crop varieties have changed. The circumstances of individual farmers were different from the present and thus their decision making (e.g. to change crops, abandon the land) may be quite different from those that would be made today. Nevertheless retrodiction seems to be a fruitful field of cooperation between natural scientists and social scientists.

It is an area rich in examples: for example Australia is a continent recently invaded by competing tribes of bureaucrats in the separate colonies. The early settlement of the Australian colonies was extensively documented and archived both in terms of agricultural occupancy, yields, and the bureaucratic and public debate about settling or abandoning new lands (e.g. Meinig, 1962). Settlement by Europeans was recent enough that there are reasonably reliable records of meteorological conditions. Comparative studies could be extended to other continents, for example between Australia and Argentina which have similar environments, similar isolation, but different cultural backgrounds. In Australia there was stubborn resistance by the aborigines to European occupation, but it was quickly suppressed because of the vastly different military capacities. What opportunities are there for comparisons with South Africa where cultures (European and Zulu) of widely different agricultural technologies but not so vastly different military capacities clashed?

What is it that GCTE needs?

One of the tasks of GCTE is to predict the effects of changes in climate, atmospheric conditions and land use on terrestrial ecosystems. Land-use changes are recognized as being important in driving changes in terrestrial ecosystems (and possibly in atmospheric systems) but the GCTE Operational Plan (Steffen *et al.*, 1992) largely assumes that changes in land use will be derived or predicted externally to the GCTE programme. As the GCTE programme has progressed, it has become clear that landscape change is an important factor in the dynamics of terrestrial ecosystems and that without good predictive models of these changes the entire programme will be weakened. Thus, the need and the urgency for progress in this area has increased.

There is currently a joint core project between IGBP-IHDP (Human Dimensions Programme) on Land Use and Cover Change (LUCC) which has as a major objective the task of better understanding human-driven changes to ecosystem composition and structure. The approach includes reconstruction of past land-cover and land-use changes, case studies and modelling. The modelling foci will require lengthy work to reach the level of specificity that GCTE desires. In the short run, carefully designed case studies are more likely to lead to insights into regional outcomes and models suited to some of GCTE's needs.

Conclusions

Predicting the impacts of changes in biophysical driving forces on land-scapes and land cover is a challenging task in itself. However, no matter how successful we are in this task, it will only be a partial success unless we are able to take into account human-related driving forces. Those involved in GCTE should not assume that predictive models taking human actions into account can be delivered to them by LUCC, IHDP or any other agencies. The reflexive nature of human societies and the issues raised by positivist approaches are likely to mean that there will always be considerable differences in the way social scientists approach predictive modelling.

There are some achievable steps that will aid progress. A thorough reconstruction of the land-use changes over the past few decades will help all workers in global change in understanding driving forces and responses. A statement of the human responses that seem certain, those that are likely, those that are possible, and those about which we know too little even to predict a direction of change (cf. the IPCC report, Houghton et al., 1990) would greatly benefit all parties. Similarly, more precise information from natural scientists about the net value of particular landscape changes will help in the process of land-use planning and help ensure that appropriate responses can be made by societies to the challenges of global change (Moreno, 1994).

We may have to accept that the best that can be done is to identify those parts of the human-related system that are most likely to show a significant response and to suggest the directions of those responses. Indeed, it may be possible to arrive at a classificatory scheme for human responses somewhat akin to ecologists' attempts to derive functional classifications of organisms. This will mean that scenarios of our future are much more open-ended than some would wish, but that is the nature of human societies.

Acknowledgements

I thank the referees and my social science colleagues for their comments, suggestions and tolerance of an ecologist making a foray into the ecotone between our disciplines.

References

Anderson, J. L. (1981). History and climate: some economic models. In *Climate and History*, ed. T. M. L. Wigley, M. J. Ingram & G. Farmer, pp. 337–55. Cambridge: Cambridge University Press.

Bella, D. A., Jacobs, R. & Li, H. (1994). Ecological indicators of global climate change: a research framework. *Environmental Management*, **18**, 489–500.

Dale, V. H., O'Neill, R. V., Southworth, F. & Pedlowski, M. (1994). Modelling effects of land management in the Brazilian Amazonian settlement of Rondônia. *Conservation Biology*, **8**, 196–206.

Forman, R. T. T. & Godron, M. (1986). *Landscape Ecology*. New York: John Wiley & Sons.

Haggett, P. (1965). *Locational Analysis in Human Geography*. London: Edward Arnold.

Houghton, J. T., Jenkins, G. J. & Ephraums, J. J. (1990). *Climate Change: The IPCC Scientific Assessment*. Cambridge: Cambridge University Press.

Mackay, A. (1981). Climate and popular unrest in late medieval Castile. In *Climate and History*, ed. T. M. L. Wigley, M. J. Ingram & G. Farmer, pp. 356–76. Cambridge: Cambridge University Press.

McGovern, T. H. (1981). The economics of extinction in Norse Greenland. In *Climate and History*, ed. T. M. L. Wigley, M. J. Ingram & G. Farmer, pp. 404–33. Cambridge: Cambridge University Press.

Meinig, D. W. (1962). *On the Margins of the Good Earth: The South Australian Wheat Frontier, 1869–1884*. Chicago: Rand McNally.

Moreno, J. M. (1994). *Global Change and Landscape Dynamics in Mediterranean Systems*. Madrid: IGBP Committee of Spain.

Morrill, R. L. (1974). *The Spatial Organization of Society*, 2nd edn, p. 267. Massachusetts: Duxbury Press.

Naveh, Z. & Lieberman, A. S. (1984). *Landscape Ecology: Theory and Application*. New York: Springer.

Parry, M. L. (1981). Climate change and the agricultural frontier: a research strategy. In *Climate and History*, ed. T. M. L. Wigley, M. J. Ingram & G. Farmer, pp. 319–36. Cambridge: Cambridge University Press.

Sack, R. D. (1990). The realm of meaning: The inadequacy of human-nature theory and the view of mass consumption. In *The Earth Transformed by Human Action*, ed. B. L. Turner, W. C. Clark, R. W. Kates, J. F. Richards, J. T. Mathews & W. B. Meyer, pp. 659–71. Cambridge: Cambridge University Press.

Steffen, W. L., Walker, B. H., Ingram, J. S. & Koch, G. W. (eds.) (1992). *Global Change in Terrestrial Ecosystems: The Operational Plan*. Stockholm: The International Geosphere Biosphere Programme and the International Council of Scientific Unions.

Sutherland, D. (1981). Weather and the peasantry of Upper Brittany, 1780–1789. In *Climate and History*, ed. T. M. L. Wigley, M. J. Ingram & G. Farmer, pp. 434–49. Cambridge: Cambridge University Press.

Torry, W. A. (1983). Anthropological perspectives on climate change. In *Social Science Research and Climate Change*, ed. R. S. Chen, E. Boulding & S. H. Schneider, pp. 208–27. Dordrecht: Kluwer.

Turner, B. L. (1990). Editorial Introduction. In *The Earth Transformed by Human Action*, ed. B. L. Turner, W. C. Clark, R. W. Kates, J. F. Richards, J. T. Mathews & W. B. Meyer, pp. 655–7. Cambridge: Cambridge University Press.

Turner, B. L., Clark, W. C., Kates, R. W., Richards, J. F., Mathews, J. T. &

Meyer, W. B. (eds.) (1990). *The Earth Transformed by Human Action.* Cambridge: Cambridge University Press.

Turner, M. G. & Gardner, R. H. (1991). *Quantitative Methods in Landscape Ecology.* New York: Springer.

Vloedbeld, M. & Leemans, R. (1993). Quantifying feedback processes in the response of terrestrial carbon cycle to global change: the modelling approach of IMAGE-2. *Water, Air and Soil Pollution,* **70**, 615–28.

Wigley, T. M. L., Ingram, M. J. & Farmer, G. (eds.) (1981). *Climate and History.* Cambridge: Cambridge University Press.

Yuill, R. S. (1965). *A Simulation Study of Barrier Effects in Spatial Diffusion Problems.* Discussion papers, vol. 5. Michigan: Inter-University Community of Mathematical Geographers.

11 Landscape diversity and vegetation response to long-term climate change in the eastern Olympic Peninsula, Pacific Northwest, USA

L. B. Brubaker and J. S. McLachlan

Introduction

Many of the processes controlling vegetation responses to climate change are modified by physical characteristics of the landscape. For instance, major mountain ranges, large bodies of water, and the large-scale patterns of river drainages may affect potential migration routes of species during periods of climate change (Davis *et al.*, 1986). Smaller-scale topographic and edaphic variations may result in environmental diversity that mitigates or accentuates the effect of climate on the survival/extinction of local populations (Perry, 1992; Murphy & Weiss, 1992). As a result, predicting how future climate change will affect vegetation calls for an understanding of how geomorphologic diversity across different spatial scales affects processes controlling the survival and spread of species.

This chapter discusses fossil pollen evidence of past vegetation responses to climate change in the Pacific Northwest, USA. Fossil pollen records have been a rich source of information about large-scale patterns of vegetation change in many regions of the earth. Although the bulk of this research emphasizes the role of climate in causing broad changes in the geographic distribution of plant biomes (e.g. Wright *et al.*, 1993), the effects of edaphic factors and migrational barriers on such responses have been documented in some areas by studies designed to investigate smaller-scale controls of vegetation (e.g. Brubaker, 1975; Pennington-Tutin, 1986; Davis *et al.*, 1994; Spear *et al.*, 1994). Previous investigations of fossil pollen in the Pacific Northwest have concentrated on the response of low-elevation forests to regional climate changes since deglaciation *c.* 13 500 BP (e.g. Barnosky, 1981; Tsukada *et al.*, 1981; Leopold *et al.*, 1982; Barnosky *et al.*, 1987; Cwynar, 1987; Whitlock, 1993). Virtually nothing is known of variations in forest history at different geomorphologic settings at low elevation or of the general history of forests in montane areas prior to 6000 BP (Dunwiddie, 1986). Thus, the role that landscape diversity has played in vegetation change is largely unknown in this geomorphologically diverse region.

This topic has gained attention recently as investigations into the consequences of the increasing homogeneity and fragmentation of regional forests combined with the potential for future warmer climates have pointed to the importance of landscape-scale processes (Franklin *et al.*, 1991; Murphy & Weiss, 1992; Peters, 1992). The effect of warmer climates on high-elevation forests is of particular concern in the Pacific Northwest because the magnitude of mean annual temperature increases predicted by most scenarios of global climate change could cause uphill migrational shifts in forest zones that would exceed the altitude of local mountains and eliminate subalpine forests (Franklin *et al.*, 1991). At lower elevations, landscape management practices, which simplify forest patterns and composition and reduce the connectivity of the landscape as a whole, are increasingly viewed as threats to the ability of species to survive the added stress of climatic warming (Franklin *et al.*, 1991).

The fossil pollen records discussed in this chapter reflect vegetation responses to post-glacial climate change at stand to landscape spatial scales (*c.* 10 to 10^4 ha) in the Olympic Peninsula region of western Washington, USA. This is an area of heterogeneous topography and strong interactions between land and marine influences. The area is small enough that it has experienced the same regional climate changes and that seed dispersal limitations to species movement should not have been important controls of vegetation responses to such changes. Differences in vegetation history within this area, therefore, reflect the effects of landscape factors on vegetation responses to large-scale climate variations. The vegetation history of low-elevation forests is described first and compared to records from three contrasting landscape locations, which represent a small portion of the regional topographic diversity. Overall, these records corroborate the general conclusions of other Quaternary paleorecords, that species responses to climate change are individualistic and involve large population fluctuations (e.g. Davis, 1986; Webb, 1986; Prentice, 1992). They also provide evidence that elevational, physiographic and geomorphic features of the landscape can mitigate effects of climate change, allowing species to survive long periods at favourable landscape locations when they are eliminated from regional vegetation types.

Physiography, climate and vegetation of the study area

The study area is the northeastern portion of the Olympic Penninsula and a nearby island, Orcas Island, in Washington State, USA (Fig. 11.1). The Olympic Peninsula is a broad, mountainous landmass (*c.* 150 km across) separated from the mainland by a complex waterway, Puget Sound

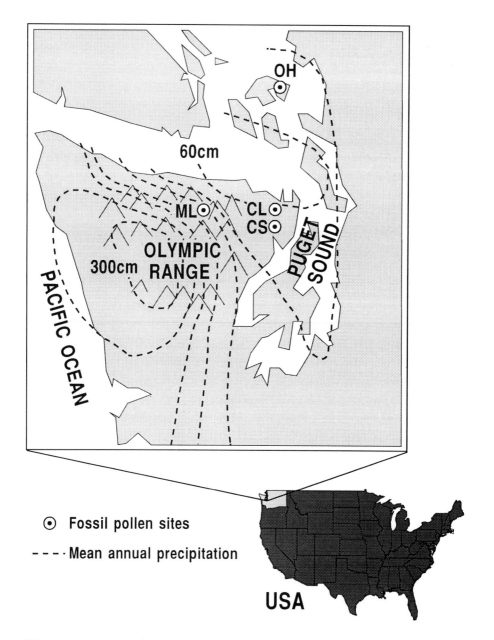

Figure 11.1   *Map of study area showing location of fossil pollen sites and gradients in mean annual precipitation in the Olympic Peninsula region of western Washington, USA. CL, Crocker Lake; CS, Cedar Swamp; OH, Orcas Hollow; ML, Moose Lake.*

(Fig. 11.1). The area is dominated by the Olympic Mountains, a glacially carved mountain range reaching nearly 3000 m, surrounded on three sides by a narrow fringe of coastal lowlands. An extensive lowland, the Puget Lowlands, extends from the eastern shore of the Puget Sound to the Cascade Mountains, a prominent north–south mountain range that forms the eastern boundary of the region defined here as western Washington.

Western Washington is characterized by mild summer and winter climates (Phillips, 1972). Summers are dry and cool because the Subtropical High Pressure System located over the North Pacific Ocean blocks the path of storm tracks and brings relatively cool airmasses across the region. In the winter, southwesterly airflows associated with the South Pacific Low Pressure System result in wet and warm conditions. Within western Washington, the effects of these large-scale climatic controls are strongly modified by interactions with mountain ranges (Fig. 11.1). The Olympic Mountains deflect marine air masses upward, causing an increase in precipitation and decrease in temperature as elevation increases on windward mountain slopes. Air masses warm as they descend leeward slopes of the mountain range, causing a pronounced rain shadow on eastern portions of the Peninsula. As a result of prevailing wind directions, southwest–northwest precipitation and temperature gradients are particularly pronounced across the Olympic Mountains (Henderson *et al.*, 1989).

These precipitation and temperature gradients have resulted in a prominent forest zonation on the Olympic Peninsula (Fig. 11.2; Henderson *et al.*, 1989). The *Picea sitchensis* zone is a temperate rain forest along coastal areas and in low mountain valleys of the western Olympics. *Tsuga heterophylla* and *Thuja plicata* are common associates with *Picea sitchensis*; neither of these species is adapted to fire. The *Tsuga heterophylla* zone, which forms a narrow forest band above the *Picea sitchensis* zone in the western Olympics, is the most extensive forest zone in northern, eastern and southern portions of the Peninsula, extending from sea level up to subalpine forest zones. Although *Tsuga heterophylla* and *Thuja plicata* are the most common species in late-successional forests, *Pseudotsuga menziesii*, a long-lived early successional species, typically comprises a large component of the biomass of old stands. The predominant disturbance to this forest type is stand-replacing fires with a long return interval (200–400 years). The *Abies amabilis* and *Tsuga mertensiana* zones occur above the *Tsuga heterophylla* zone except in the northeastern Olympics. These zones are characterized by heavy snows and short, cool growing seasons. Fires and other large-scale disturbances are relatively rare. The *Abies lasiocarpa* zone is found above the *Tsuga heterophylla* zone in the most extreme portion of the rain shadow in the northeastern Olympics. Snow depths are less than in western high-

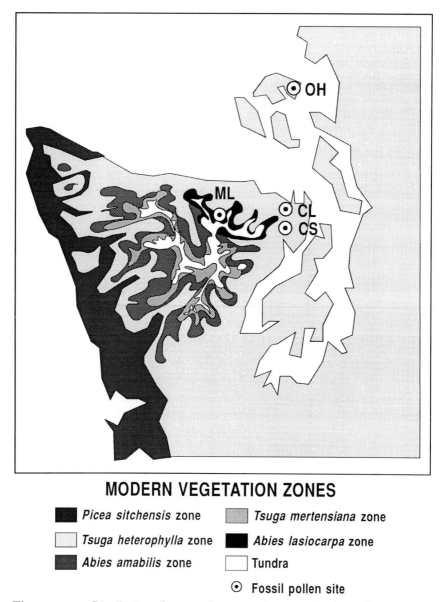

MODERN VEGETATION ZONES

■ *Picea sitchensis* zone   ▨ *Tsuga mertensiana* zone

□ *Tsuga heterophylla* zone   ■ *Abies lasiocarpa* zone

▨ *Abies amabilis* zone   □ Tundra

⊙ Fossil pollen site

Figure 11.2   *Distribution of present-day vegetation zones in the study area.*

elevation forests and summer droughts can be severe. Stand-replacing fires
are also relatively common. A narrow band of tundra is present on mountain peaks above the *Tsuga mertensiana* and *Abies lasiocarpa* zones. The
upper limit of tree growth is generally controlled by the depth of the winter
snowpack, which is quite variable due to the irregular topography. As a
result, the location of the treeline follows major topographic features, and in
many areas trees extends well above forested slopes along windswept ridges.

Conversely, non-forested patches occur on flat topographic locations within *Tsuga mertensiana* and *Abies lasiocarpa* zone forests where deep snow accumulations linger into the growing season and prevent tree seedling establishment.

Pollen site locations

We discuss pollen records from several topographic and physiographic settings. The source area of vegetation contributing pollen to each site depends on several factors, including the surface area of the site. The description below indicates the general spatial scale of our pollen records, based on our understanding of pollen rain on the Olympic Peninsula (unpublished data) and published pollen diagrams from the Puget Sound region (Barnosky, 1981; Tsukada *et al.*, 1981; Leopold *et al.*, 1982; Barnosky *et al.*, 1987; Cwynar, 1987).

Regional lowland vegetation

Crocker Lake

Crocker Lake is a *c.* 26 ha kettle lake, 60 m above sea level, in an area of small hills on the northeastern corner of the Olympic Peninsula (Fig. 11.1). It is located within the *Tsuga heterophylla* zone and currently surrounded by extensive stands of *Pseudotsuga menziesii* and *Alnus rubra*, which established on hillsides after cutting of the original forest. Pollen data from this site (Fig. 11.3) represent a regional-scale vegetation record that is similar to other low-elevation records in the Puget Lowland (McLachlan, 1994).

Diverse landscape locations

Cedar Swamp

Cedar Swamp is a forested wetland (*c.* 10 ha) about 3 km south of Crocker Lake (Fig. 11.1). This site is currently occupied by a second growth *Alnus rubra* and *Thuja plicata* stand. Cut stumps on the site indicate that the original stand was dominated by large *Thuja plicata* and *Tsuga heterophylla*. This site was vegetated between 13 000 and 10 000 BP and 6000–0 BP, resulting in a pollen record of local plant communities during those periods. Between 10 000 and 6000 BP, local water depths increased, eliminating terrestrial vegetation and resulting in a regional pollen source area similar to that of Crocker Lake. In general, this record (Fig. 11.4) reveals the composition of vegetation on geomorphologically controlled wet sites located within a landscape matrix of upland hill slopes.

CROCKER LAKE

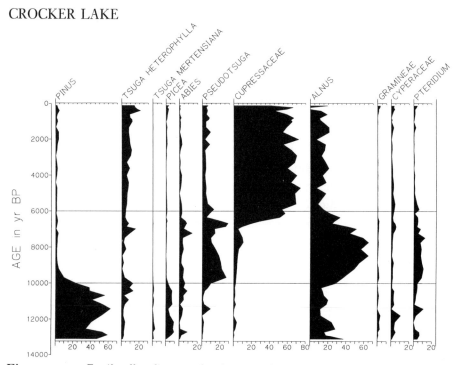

Figure 11.3 *Fossil pollen diagram showing post-glacial variations in major pollen taxa at Crocker Lake. Taxonomic designation (species, genus, family) represents finest level of pollen identification.*

Orcas Hollow

Orcas Hollow is a small depression (*c.* 15 m in diameter) on a bedrock plateau, *c.* 200 m above sea level, near the top of Mountain Constitution, Orcas Island. Soils are thin and rocky. The site is presently surrounded by a stand of *Pinus contorta* with *Pseudotsuga menziesii* and *Tsuga heterophylla*. Its small-scale pollen record (Fig. 11.5) reflects local forest composition in a particularly dry landscape setting resulting from combined effects of thin, coarse soils and the location of Orcas Island in the Olympic rain shadow. These conditions were present throughout the post-glacial period recorded by fossil pollen.

Moose Lake

Moose Lake is a small, glacially carved lake (*c.* 5 ha) in the bottom of a narrow, steep-sided valley (*c.* 1500 m above sea level) in the northeastern Olympic Mountains (Fig. 11.1). It is located within the *Abies lasiocarpa* zone and local forests are dominated by *Abies lasiocarpa* with minor components of *Tsuga mertensiana* and *Chamaecyparis nootkatensis*. The fossil pollen record (Fig. 11.6) at this site reflects the general composition of high-elevation forests in the rain shadow forests of the Olympic Peninsula.

CEDAR SWAMP

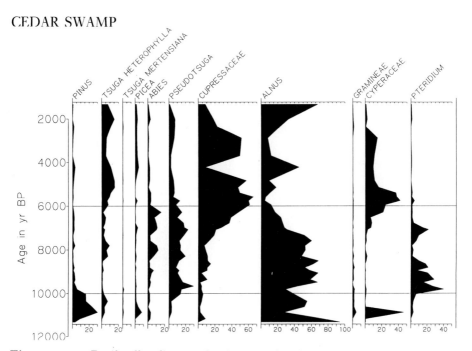

Figure 11.4 *Fossil pollen diagram showing post-glacial variations in major pollen taxa at Cedar Swamp. Taxonomic designation (species, genus, family) represents finest level of pollen identification.*

Regional vegetation and climate history

The post-glacial vegetation history of the Pacific Northwest is best known in the Puget Lowland, where the largest number of pollen diagrams is available (e.g. Barnosky, 1981; Tsukada *et al.*, 1981; Leopold *et al.*, 1982; Tsukada & Sugita, 1982; Barnosky *et al.*, 1987; Cwynar, 1987; Whitlock, 1993). Since the primary emphasis of previous work has been to document the effects of regional climatic changes, study sites have been chosen to record large-scale characteristics of the vegetation. The vegetation history at Crocker Lake is very similar to that of sites elsewhere in the Puget Lowland, indicating a close similarity in the vegetation of the northeastern Olympic Peninsula and the Puget Lowland throughout the post-glacial period (McLachlan, 1994). The understanding of the long-term climate changes causing major shifts in vegetation of this region has been substantially improved in recent years through the use of general circulation models, which simulate major features of the earth's climate system under different configurations of global boundary conditions (e.g. latitudinal and

ORCAS HOLLOW

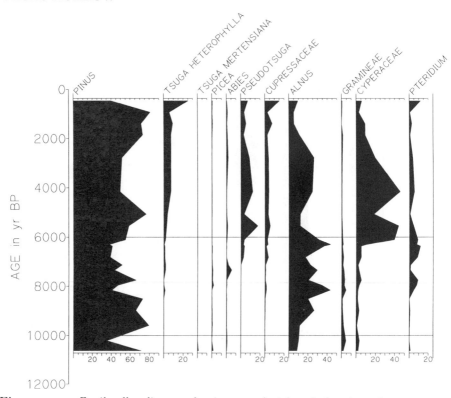

Figure 11.5   *Fossil pollen diagram showing post-glacial variations in major pollen taxa at Orcas Hollow. Taxonomic designation (species, genus, family) represents finest level of pollen identification.*

seasonal solar radiation, sea surface temperature, extent of ice sheets). This section summarizes post-glacial climate changes in the Pacific Northwest inferred from simulations by the National Center for Atmospheric Research Community Circulation Model, as recently reviewed by Thompson *et al.* (1993), and discusses major features of low-elevation forest changes in the eastern Olympic Peninsula and Puget Lowlands, as illustrated by the Crocker Lake pollen record (McLachlan, 1994).

13 000 – 10 000 BP

This period corresponds to the transition from glacial to interglacial climates in western North America. The influence of the North American ice sheets on airflow over the Pacific Northwest decreased as ice sheets diminished in size and the location of the jet stream moved northward. These changes in large-scale controls resulted in relatively cold, moist climates in the Pacific Northwest. Although most of the species present in

MOOSE LAKE

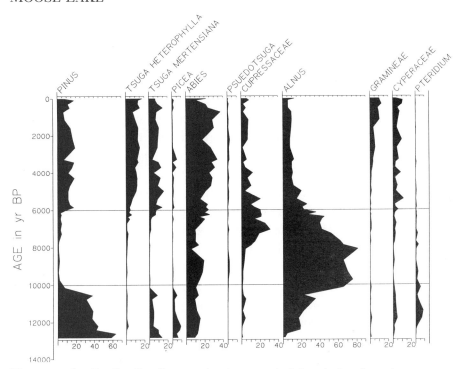

Figure 11.6   *Fossil pollen diagram showing post-glacial variations in major pollen taxa at Moose Lake. Taxonomic designation (species, genus, family) represents finest level of pollen identification.*

modern lowland forests were present on the regional landscape, a mixed forest–tundra biome apparently dominated lowlands for several millennia following deglaciation. Macrofossil evidence (leaves, seeds), which document the local occurrence of a species, indicate that conifers with differing present-day ecological requirements and disjunct modern ranges co-occurred within relatively small areas. For example, *Pinus contorta* (wide climatic and edaphic tolerance), *Picea sitchensis* (rich soils and cool, maritime climate) and *Abies lasiocarpa* (nutrient-poor soils and cold, continental climates) apparently grew in the same forest stands during this period. This unusual assemblage of species may reflect individual species responses to combinations of environmental factors that are not common today. For example, the coarse, unstable soils of recently deglaciated landscapes may have favoured *Pinus contorta* and *Abies lasiocarpa*, whereas the transitional climate with both marine and continental influences may have allowed *Picea sitchensis* and *Abies lasiocarpa* to grow in close proximity. As today at high elevations, small-scale topographic variations were probably important controls over the distribution of plant communities. Forests probably occupied the warmest microsites with early snow melt, while tundra communities were restricted to cold, low-lying sites where snow packs persisted into the growing season.

10 000–6000 BP

According to paleoclimatic simulations for this period, global-scale boundary conditions resulted in warm, dry summers and cold winters in the Pacific Northwest. Summer droughts were probably more severe than present because an increase in summer radiation at high-northern latitudes intensified the Subtropical High Pressure System in the North Pacific Ocean and more effectively blocked storm tracks over the region. Even though summer temperatures increased during this period, a decrease in winter insolation caused winter temperatures to remain colder than present. The regional lowland vegetation changed abruptly at the beginning of this period, when species adapted to warm climates and frequent disturbances became common and cold tolerant species generally disappeared.
Pseudotsuga menziesii and *Alnus rubra* became particularly abundant, but the characteristics of these forests are not well understood (Tsukada *et al.*, 1981; Barnosky, 1981; Cwynar, 1987). Though the possibility that both species inhabited upland cannot be excluded, it is more likely that they occupied different landscape locations with different disturbance regimes. *Pseudotsuga menziesii*, a relatively drought-tolerant conifer with thick bark, probably dominated uplands where fires were frequent. An increase in the abundance of sediment charcoal and spores of *Pteridium*, a common fern after fire, suggests that fires were common on a regional scale. In contrast, *Alnus rubra*, a riparian species sensitive to drought stress on upland sites, probably occupied frequently disturbed river and stream banks. Although riparian flooding events are not recorded in lake sediments, the possibility of increased flooding is suggested by the paleoclimatic simulations, which indicate a greater contrast between winter and summer temperatures than at present. Such conditions might have resulted in rapid melting of snowpack and extensive flooding of riparian zones. Overall, the regional lowland forests of this period differed substantially from modern *Tsuga heterophylla* zone forests. Mesic, late-successional species such as *Thuja plicata* and *Tsuga heterophylla* were apparently rare due to the combined effects of frequent fires and dry regional climate. Fluctuations in the pollen percentages of several species at Crocker Lake and several Puget Lowland sites suggest changes in regional forest composition near the end of this period, but the spatial extent and causes of these changes are not understood at present.

6000–0 BP

The cool, moist climate that characterizes modern Pacific Northwestern environments was established by *c.* 6000 BP, when global climatic controls reached near-present configurations. *Thuja plicata* (macrofossil evidence of Cupressaceae pollen) and *Tsuga heterophylla* increased in abundance and drought-tolerant, disturbance-adapted species declined at this time, resulting in a regional forest composition typical of pre-EuroAmerican landscapes. Fires undoubtedly remained the most important stand-replacing disturbances on upland sites, but the return interval of these events was probably longer than during the previous period. Less frequent fires would have allowed the development of old-growth stand structures characterized by large trees and abundant woody debris (Brubaker, 1991). Riparian disturbances probably became less common than previously, contributing to the regional decline in *Alnus rubra*.

Vegetation history on diverse landscape locations

This section summarizes major differences in the pollen records between Cedar Swamp, Orcas Hollow and Moose Lake and that of Crocker Lake. Comparisons between vegetation histories at different elevational, geomorphologic, and edaphic settings represented by these sites provides evidence of the moderating effects of landscape diversity in vegetation responses to climate change. Because these sites represent a small sample of the landscape variability within the study area, their pollen records should be considered a first glimpse at the effect of landscape heterogeneity on processes of vegetation change.

13 000–10 000 BP

The vegetation in the study area showed greatest similarity across landscape types during this period. The abundance of *Pinus contorta* pollen in the Crocker, Moose and Orcas pollen diagrams indicates that this species was common at a wide variety of soils and local climatic environments in keeping with its broad modern ecological tolerances. It apparently was associated with different species at each location, however. It was virtually the only conifer at the Orcas Hollow site, but shared dominance with *Abies lasiocarpa* at high elevation near Moose Lake, and occurred with *Picea sitchensis* and a variety of other conifers at lowland sites. Differences between the Crocker Lake and Cedar Swamp records, however, suggest that *Pinus contorta* was less common on wetland than upland sites at low elevations in the northeastern Olympic Peninsula. In contrast to the wide, regional distribution of *Pinus contorta*, other species seem to have been restricted to specific elevations or topographic settings. For example, *Tsuga heterophylla* and *Picea sitchensis* were apparently only common at mesic low elevation sites, whereas *Tsuga mertensiana* was predominantly at high elevations.

10 000–6000 BP

The climate changes that brought nearly complete species replacement to low-elevation forests at the beginning of this period affected other landscape locations differently. For example, *Pinus contorta* populations apparently disappeared at all sites except on Orcas Island, where the Orcas Hollow diagram indicates the persistence of local populations. Similarly, *Abies lasiocarpa* was dramatically restricted on the general landscape, but remained a dominant component of high-elevation forests near Moose Lake. Different species of *Alnus* were favoured by climate and disturbance regimes at a variety of locations. *Alnus rubra* was favoured by a dry climatic environment on Orcas Island as well as on the northeastern Olympic Peninsula and Puget Lowlands. *Alnus sinuata* was the predominant *Alnus* species at high elevation. This shrub is common in riparian sites and avalanche paths today and may have been favoured by an increase in flooding and avalanches caused by rapid warming and snow melt associated with the more extreme temperature fluctuations between winter and summer during this period. In contrast to the increase in *Alnus* at a variety of landscape locations, *Pseudotsuga menziesii* did not significantly expand its range during this period. It did not spread upslope as might be expected during a period of regional drought, for example, even though it was favoured by drought at low elevations. Another prominent difference among sites is represented by the major increase in *Chamaecyparis nootkatensis* (identification based on leaf macrofossils) near the end of this period at Moose Lake. Although minor fluctuations in tree species populations are also suggested at some low-elevation sites, these changes appear to have been less extensive than those at Moose Lake.

6000–0 BP

As in previous periods, the effects of regional climate change varied at different landscape locations. However, in general, the fossil pollen records at all sites indicate vegetation responses to increased moisture and/or decreased temperature. Although pollen percentages at low-elevation pollen diagrams such as Crocker Lake indicate that the modern regional forest composition was established throughout the Puget Lowland at the beginning of this period, pollen percentages at Moose Lake and at high-elevation sites in the central Cascades (Dunwiddie, 1986) do not reach modern values until *c.* 3000 BP. Another contrast to the regional lowland record is the surprising stability in several forest species compared with previous periods. For example, *Abies lasiocarpa* persisted as the dominant tree species at high elevations; *Pinus contorta* remained important at Orcas Island; and *Alnus rubra* continued to be abundant as part of the local vegetation at Cedar Swamp.

Discussion

Even the small number of fossil pollen records reported here reveal general features of vegetation responses to long-term climate variations and the moderating effects landscape diversity can have on such large-scale responses. Some of these features are discussed below.

Individualistic species response

Post-glacial pollen records from the eastern Olympic Peninsula region clearly indicate that vegetation responds to climate variations through changes in the abundance and distribution of individual species rather than by displacements of intact vegetation zones. Even though most species of modern forests were present in the Pacific Northwest at the time of glacier recession *c.* 13 500 BP (Whitlock, 1993), modern forest zones did not become established at any of the study sites until *c.* 6000 years ago. Furthermore, there is no evidence that modern forest zones existed prior to this time at different locations within the Pacific Northwest. For example, *Tsuga heterophylla* zone forests, which have dominated lowlands since *c.* 6000 BP, were not present at higher elevations in the Olympic Mountains or at more northern latitudes (Matthewes, 1973) during the warmer-than-present climates 10 000–6000 BP. Similarly, subalpine forest zones in the northeastern Olympics and central Cascades (Dunwiddie, 1986) indicate these zones are recent species associations. On the other hand, forest types that were generally stable for several millennia in the past (e.g. lowland *Pseudotsuga menziesii* and *Alnus rubra* forests 10 000–6000 BP) do not have counterparts on modern landscapes. The underlying ecological interactions resulting in individualistic species behaviour and the reorganization of forests are often not well understood from observations on modern landscapes. For instance, reasons for the simultaneous regional increase of *Pseudotsuga menziesii* and *Alnus rubra* 10 000–6000 BP are difficult to understand because these species have different drought tolerances and are adapted to different types of disturbance. Likewise, the disparate behaviour of *Chamaecyparis nootkatensis* and *Tsuga mertensiana* at high elevation (Moose Lake) is not readily explained by the close similarity in their modern distributions, in high snowpack areas of *Tsuga mertensiana* zone forests.

Thus, despite the clarity with which modern forest zones can be identified and related to contemporary environmental gradients in the study area (Henderson *et al.*, 1989), these biological associations have not been maintained in the face of past climatic changes. The lack of long-term stability in plant associations is a common theme of fossil pollen records worldwide (Davis, 1986; Webb, 1987; Prentice, 1992; Wright *et al.*, 1993) and is thought to reflect the unique trajectory of global climatic changes. Different configurations of large-scale controls of the earth's climate have resulted in different combinations of climatic conditions over time (e.g. COHMAP project members 1988; Wright *et al.*, 1993). Apparently the individualistic responses of species to different aspects of climate have, in turn, resulted in unique vegetation associations (e.g. Graumlich & Brubaker, 1994). These inferences from the fossil record also imply that vegetation responses to future climate change will be complex and individualistic. Furthermore, paleorecords from the Pacific Northwest and elsewhere suggest that just as it is sometimes difficult to understand the causes of past vegetation changes based on the current understanding of species requirements, it may be difficult to correctly predict species responses to future change from observations of present-day landscapes. Ultimately, accurate predictions of vegetation responses to future climate scenarios will require a broader knowledge of species–environment interactions than can be obtained from studies under present-day climates. It may, therefore, be necessary to obtain such information by investigations of species behaviours in experimentally altered environments.

Population fluctuations and landscape heterogeneity

Fossil pollen records from the study area contain numerous examples of dramatic fluctuations in tree population sizes over the past 13 000 years. *Pinus contorta* was the dominant tree species across a wide range of landscape types in the Pacific Northwest 13 000–10 000 BP, but is now present only in relatively small populations at widely scattered locations. Modern *Pinus contorta* forest on Orcas Island is one of these populations. Similarly, modern stands of *Abies lasiocarpa* in the northeastern Olympic Mountains near Moose Lake represent remnants of a wide-ranging late-glacial population. *Alnus rubra*, which until recent disturbance by logging was largely restricted to small riparian or wetland populations such as at Cedar Swamp, was very abundant 10 000–6000 BP. Conversely, *Thuja plicata* and *Tsuga mertensiana* are presently common in low- and high-elevation forest zones, respectively, but prior to *c*. 6000 BP were barely detectable in the pollen record.

Since the Olympic Mountains and Orcas Island are separated from the mainland by broad lowlands and extensive waterways, it is unlikely that these population fluctuations represent large-scale population migrations into and out of the study area. Instead, they probably resulted from the expansion and contraction of local populations. The nearly 13 000-year pollen record of *Abies lasiocarpa* at Moose Lake and *Pinus contorta* at Orcas is evidence that species can persist as relatively small populations at particularly favourable locations on the landscape despite major changes in the earth's climate system. Such populations represent seed sources for rapid population expansions if climate change results in favourable conditions on a wider variety of landscape sites. Local seed sources, for example, may have allowed the rapid expansion of *Thuja plicata* into lowland forests *c*. 6000 BP. Such inferences from the fossil pollen record also emphasize the need for global-change research to address a variety of questions regarding small species populations. For example: What are the consequences of population bottlenecks to genetic structures of local populations, genetic variations within species, and the ability of a species to respond to future change? What are the minimum population sizes and numbers necessary for long-term survival? Do disturbances make the survival of species in restricted populations more or less likely over the long term?

Landscape diversity and sensitivity to climate change

The fine-scale geomorphologic heterogeneity of the Pacific Northwest seems to buffer the effects of long-term climate change on species' ability to survive in the region. Differences in slope, elevation, aspect, proximity to the ocean, etc. form a complex physical template that modifies regional changes in climate and climate-related disturbance into a mosaic of environments, altering the sensitivity of landscapes to long-term climate change. The regional high-elevation record from Moose Lake, for instance, indicates a distinct vegetation period characterized by abundant *Chamaecyparis nootkatensis c*. 8000–6000 BP. This is consistent with the findings of Dunwiddie (1986), who observed that high-elevation communities in the Cascade Range appear to be more sensitive to climate change than those at low elevation. The regional, low-elevation record at Crocker Lake also suggests that lowland vegetation is less sensitive to climate change. The *Chamaecyparis* period of Moose Lake, for instance, is only observable in the Crocker Lake diagram only as a series of minor fluctuations in species abundances rather than a distinct and prolonged reorganization of the community.

Within the general lowland context, the fine-scale pollen records from Cedar Swamp and Orcas Hollow reveal the extent to which sensitivity of vegetation to regional climatic changes can vary with hydrologic and edaphic conditions. The pollen record from Cedar Swamp suggests wetland vegetation has been more sensitive than upland vegetation to hydrologic variability (flooding and changing hydroperiod) caused by changing seasonality and magnitude of precipitation and temperature. By contrast, the Orcas Hollow site records very little vegetation response to climatic change of any magnitude since deglaciation. In this case, edaphic conditions seem to have limited the effects of large climatic changes on forest composition by favouring one species over all others irrespective of macroclimate.

Variability in the sensitivity of landscape types to the effects of climate change reflects the uniqueness of individual landscape components. Treeline sites are sensitive to changes in temperature and seasonal precipitation (Rochefort *et al.*, 1994), wetlands are sensitive to changing hydrologic factors (Walker, 1970) and infertile soils may mute the effects of climate change on vegetation (Brubaker, 1975). The proximity of such diverse resource patches in a small area such as the Olympic region is responsible for the heterogeneous and dynamic landscapes that characterize mountain areas in general (e.g. Spear *et al.*, 1994). Under extreme climates, however, local controls on vegetation composition imposed by the natural variability of complex landscapes such as these can be overwhelmed by larger-scale environmental conditions. The similarity between forest communities following deglaciation (13 000–10 000 BP) illustrates the limits of the abilities of landform to buffer the effects of climate.

Conclusion

The forest histories recorded at the four study sites illustrate some of the ways that landscape heterogeneity can modify the influence of climate change on vegetation. Future paleoecological studies in this and other regions should sample a wide variety of landscape types to better document the ways specific environments respond to changing climate. Such information will be invaluable for managers and policy-makers attempting to mitigate the effects of future climate change on natural systems. For now, even the results from a small number of sites in the eastern Olympic Peninsula indicate some general parameters for understanding the effects of climate change on a complex landscape. Of particular importance to landscape planning, vegetation zones appear to break up in unpredictable ways in response to interactions between the idiosyncrasies of climate change and the physical template. Thus, it seems prudent to maximize the ability of communities to dissolve and reorganize. Geomorphologically complex biological reserves should be able to buffer the effects of climate change on biological diversity more efficiently than simple ones (Murphy & Weiss, 1992), but the effectiveness of reserves ultimately depends on the connectivity of the landscape as a whole. Various methods of improving the biological viability of simplified cultural landscapes have been proposed (Franklin *et al.*, 1991; Peters, 1992) and will have to be implemented over wide areas. The paleorecords from the Olympic Peninsula region indicate that such efforts to restore landscape complexity should help buffer the effects of climate change on species extinction.

Acknowledgements

This research was supported by a grant from the National Park Service Global Change Project for the Olympic National Park, WA.

References

Barnosky, C. (1981). A record of late Quaternary vegetation from Davis Lake, southern Puget Lowland, Washington. *Quaternary Research*, 16, 221–39.

Barnosky, C. W., Anderson, P. M. & Bartlein, P. (1987). The Northwestern US during deglaciation: vegetational history and paleoclimatic implications. In *North America and Adjacent Oceans during the Last Deglaciation*, ed. W. F. Ruddiman & H. E. Wright, Jr, pp. 289–321. Boulder: Geological Society of America.

Brubaker, L. B. (1975). Postglacial forest patterns associated with till and outwash in northcentral Upper Michigan. *Quaternary Research*, 5, 499–527.

Brubaker, L. B. (1991). Climate change and the origin of old-growth Douglas-fir forests in the Puget South Lowland. In *Wildlife and Vegetation of Unmanaged Douglas-fir Forests*, General

Technical Report PNW-GTRR-285, ed. L. F. Ruggiero, K. B. Aubry, A. B. Carey & M. H. Huff, pp. 17–24. Portland: USDA Forest Service.

COHMAP project members (1988). Climatic changes of the last 18,000 years: Observations and model simulations. *Science*, 241, 1043–52.

Cwynar, L. C. (1987). Fire and forest history of the North Cascade Range. *Ecology*, 68, 791–802.

Davis, M. B. (1986). Climatic instability, time lags, and community disequilibrium. In *Community Ecology*, ed. J. Diamond & T. J. Case, pp. 269–84. New York: Harper and Row.

Davis, M. B., Sugita, S., Calcote, R. R., Ferrari, J. B. & Frelich, L. E. (1994). Historical development of alternate communities in a hemlock-hardwood forest in northern Michigan, USA. In *Large-scale Ecology and Conservation Biology*, ed. R. May, N. Webb & P.

Edwards. Oxford: Blackwell Scientific.

Davis, M. B., Woods, D. D., Webb, S. L. & Futyma, R. P. (1986). Dispersal versus climate: Expansion of *Fagus* and *Tsuga* into the Upper Great Lakes region. *Vegetatio*, **67**, 93–103.

Dunwiddie, P. W. (1986). A 6000-year record of forest history on Mount Rainier, Washington, *Ecology*, **67**, 58–68.

Franklin, J. F., Swanson, F. J., Harmon, M. E., Perry, D. A., Spies, T. A., Dale, V. H., McKee, A., Ferrell, W. K., Gregory, S. V., Lattin, J. D., Schowalter, T. D., Larsen, D. & Means, J. E. (1991). Effects of global climate change on forests in north-western North America. *Northwest Environmental Journal*, **7**, 233–54.

Graumlich, L. J. & Brubaker, L. B. (1994). Long-term records of growth and distribution of conifers: integration of paleoecology and physiological ecology. In *Ecophysiology of Coniferous Forests*, ed. W. Smith & T. Hinckley. London and New York: Academic Press.

Henderson, J. A., Peter, D. H., Lesher, R. D. & Shaw, D. C. (1989). *Forested Plant Associations of the Olympic National Forest*. PNW R6 ECOL Technical Paper 001-88. Portland: USDA Forest Service.

Leopold, E. B., Nickman, R., Hedges, J. I. & Ertel, R. Jr. (1982). Pollen and lignin records of Late-Quaternary vegetation, Lake Washington. *Science*, **218**, 1305–7.

Mathewes, R. W. (1973). A palynological study of postglacial vegetation changes in the University Research, south-western British Columbia. *Canadian Journal of Botany*, **51**, 2085–103.

McLachlan, J. S. (1994). Local and Regional Vegetation Change on the Northeastern Olympic Peninsula during the Holocene. Masters Thesis, University of Washington, Seattle.

Murphy, D. D. & Weiss, S. B. (1992). Effects of climate change on biological diversity in western North America:

Species losses and mechanisms. In *Global Warming and Biological Diversity*, ed. R. L. Peters & T. E. Lovejoy, pp. 355–68. New Haven: Yale University Press.

Pennington-Tutin, W. (1986). Lags in the adjustment of vegetation to climate caused by the pace of soil development: Evidence from Britain. *Vegetatio*, **67**, 104–18.

Perry, D. A. (1992). Key processes at the stand to landscape scale. In *Implications of Climate Change for Pacific Northwest Forest Management*, ed. G. Wall, pp. 51–8. Occasional Paper No. 15. Waterloo: Department of Geography Publication Series, University of Waterloo.

Peters, R. L. (1992). Introduction. In *Global Warming and Biological Diversity*, ed. R. L. Peters & T. E. Lovejoy, pp. 3–14. New Haven: Yale University Press.

Phillips, E. L. (1972). The climates of Washington. In *Climates of the States*, vol. II. Washington, DC: National Oceanographic and Atmospheric Administration. Water Information Center.

Prentice, I. C. (1992). Climate change and long-term vegetation dynamics. In *Plant Succession: Theory and Practice*, ed. D. C. Glenn-Lewin. London: Chapman and Hall.

Rochefort, R. M., Little, R. L., Woodward, A. & Peterson, D. L. (1994). Changes in the distribution of subalpine conifers in western North America: a review of climate and other factors. *The Holocene*, **4**, 89–100.

Spear, R. W., David, M. B. & Shane, L. C. K. (1994). Late Quaternary history of low- and mid-elevation vegetation in the White Mountains of New Hampshire. *Ecological Monographs*, **64**, 85–109.

Swanson, F. J., Franklin, J. F. & Sedell, J. R. (1990). Landscape patterns, disturbance, and management in the Pacific Northwest, USA. In *Changing Landscapes: An Ecological Perspective*, ed. I. S. Zonnewald & R. S. Forman. New York: Springer.

Thompson, R. S., Whitlock, C., Bartlein, P. J., Harrison, S. P. & Spaulding, W. G. (1993). Climatic changes in the western United States since 18 000 BP. In *Global Climates since the Last Ice Age*, ed. H. E. Wright, J. E. Kutzbach, T. Webb III, W. F. Ruddiman, F. A. Street-Perrott & P. J. Bartlein, pp. 468–513. Minneapolis: University of Minnesota Press.

Thorson, R. M. (1980). Ice-sheet glaciation of the Puget Lowland, Washington, during the Vashon state (late Pleistocene). *Quarternary Research*, 13, 303–21.

Tsukada, M., Sugita, S. & Hibbert, D. M. (1981). Paleoecology in the Pacific Northwest. I. Late Quaternary vegetation and climate. *Internationale Vereinegung für theorretische und angewandte Limnologie*, 21, 703–37.

Tsukada, M. & Sugita, S. (1982). Late Quaternary dynamics of pollen influx at Mineral Lake, Washington. *Botanical Magazine*, 95, 401–18.

Walker, D. (1970). Direction and rate in some British Post-glacial hydroseres. In *Studies in the Vegetation History of the British Isles*, ed. D. Walker & R. G. West, pp. 117–39. Cambridge: Cambridge University Press.

Webb, T. III (1986). The appearance and disappearance of major vegetational assemblages: long-term vegetational dynamics in eastern north America. *Vegetatio*, 69, 177–87.

Whitlock, C. (1993). Vegetational and climatic history of the Pacific Northwest during the last 20 000 years: Implications for understanding present-day biodiversity. *The Northwest Environmental Journal*, 8, 5–28.

Wright, H. E. Jr., Kutzbach, J. E., Webb, T. III, Ruddiman, W. F., Street-Perrott, F. A. & Bartlein, P. J. (eds.) (1993). *Global Climates since the Last Glacial Maximum*. Minneapolis: University of Minnesota Press.

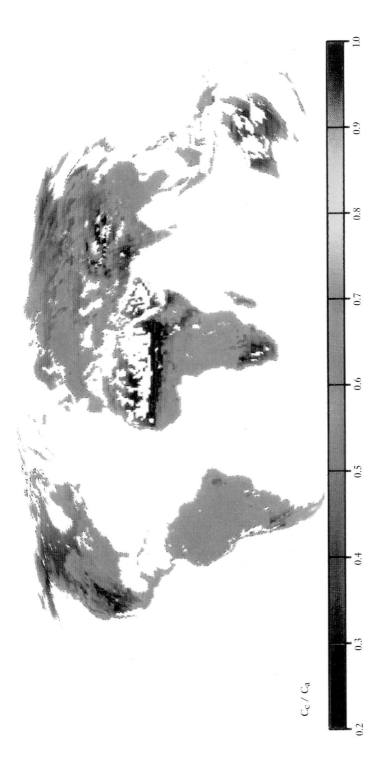

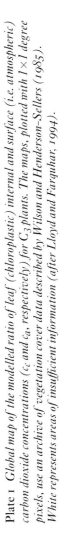

C_C / C_a

Plate 1 *Global map of the modelled ratio of leaf (chloroplastic) internal and surface (i.e. atmospheric) carbon dioxide concentrations (c_c and c_a, respectively) for C_3 plants. The maps, plotted with 1×1 degree pixels, use an archive of vegetation cover data described by Wilson and Henderson–Sellers (1985). White represents areas of insufficient information (after Lloyd and Farquhar, 1994).*

a

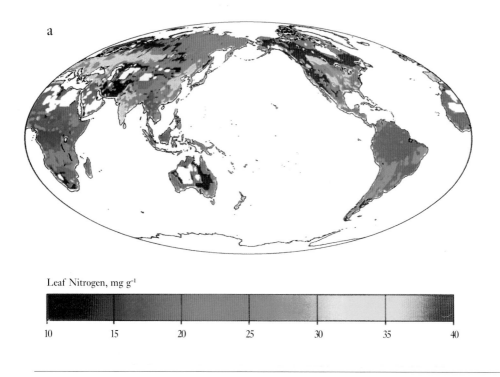

Leaf Nitrogen, mg g^{-1}

10	15	20	25	30	35	40

b

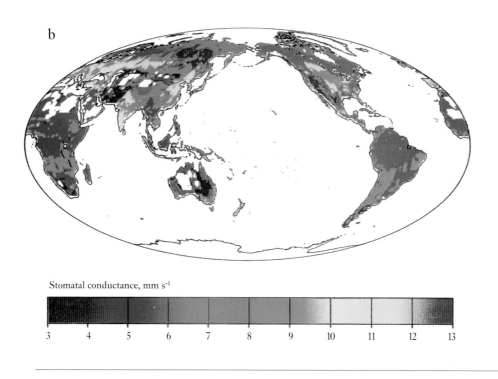

Stomatal conductance, mm s^{-1}

3	4	5	6	7	8	9	10	11	12	13

Plate 2 *Global maps of modelled* (a) *leaf nitrogen concentration* (N, mg g^{-1}); (b) *maximum leaf stomatal conductance for water vapour* (g_{max}, mm s^{-1})

c

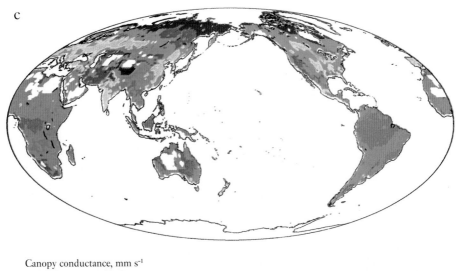

Canopy conductance, mm s⁻¹

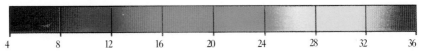

| 4 | 8 | 12 | 16 | 20 | 24 | 28 | 32 | 36 |

d

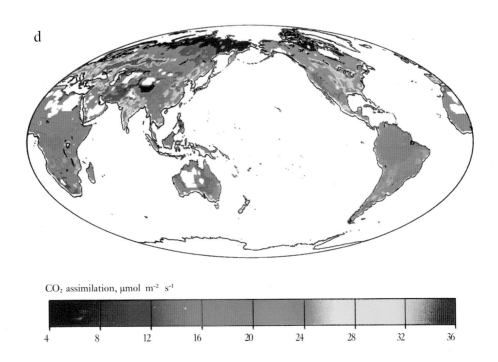

CO₂ assimilation, μmol m⁻² s⁻¹

| 4 | 8 | 12 | 16 | 20 | 24 | 28 | 32 | 36 |

Plate 2 *(continued)*
(c) *maximum surface (plant canopy and ground) conductance* $(G_{s\ max},\ mm\ s^{-1})$ *and*
(d) *maximum surface carbon dioxide assimilation rate* $(A_{s\ max},\ \mu mol\ CO_2\ m^{-2}\ s^{-1})$.
The maps, plotted with 1×1 degree pixels, use an archive of vegetation cover data described
by Wilson and Henderson-Sellers (1985). White represents areas of insufficient informa-
tion (after Schulze et al., 1994a).

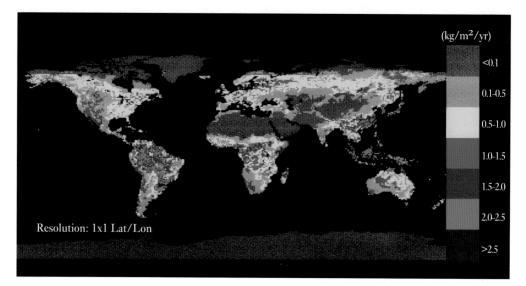

Plate 3 *Net primary production simulated by the EF class model, Biome-BGC, using a daily climate file for the year 1987. Leaf area index (LAI) was based on AVHRR/NDVI.*

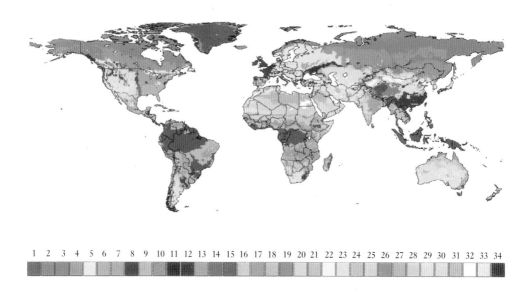

Plate 4 *Simulated distribution of major biomes of the world (MAPSS). The vegetation classes are listed in Table 23.1. Reprinted from the* Journal of Vegetation Science, *Neilson and Marks, 1994.*

a

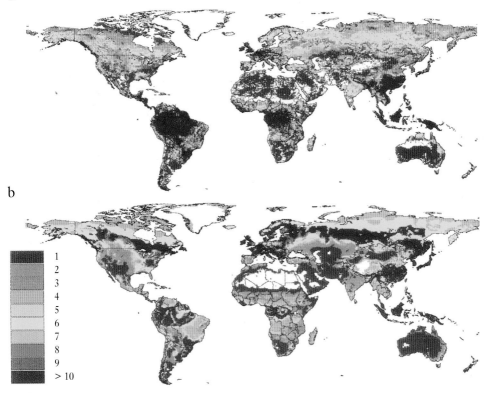

b

1
2
3
4
5
6
7
8
9
> 10

Plate 5 *Simulated maximum potential, all-sided leaf area index (LAI) from* (a) *Biome–BGC and* (b) *MAPSS.*

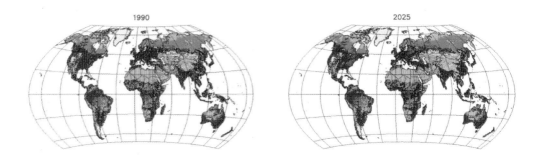

Plate 6 *Changes in land cover under 'conventional wisdom' assumptions for population, technology and economic development.*

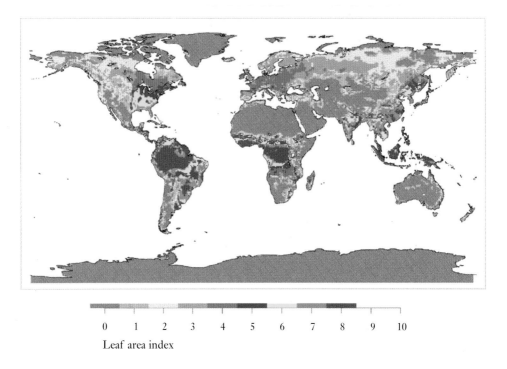

Leaf area index

Plate 7 *Predictions of current-day leaf area index (from Woodward et al., 1994).*

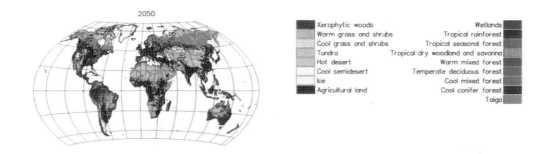

2050

Xerophytic woods	Wetlands
Warm grass and shrubs	Tropical rainforest
Cool grass and shrubs	Tropical seasonal forest
Tundra	Tropical dry woodland and savanna
Hot desert	Warm mixed forest
Cool semidesert	Temperate deciduous forest
Ice	Cool mixed forest
Agricultural land	Cool conifer forest
	Taiga

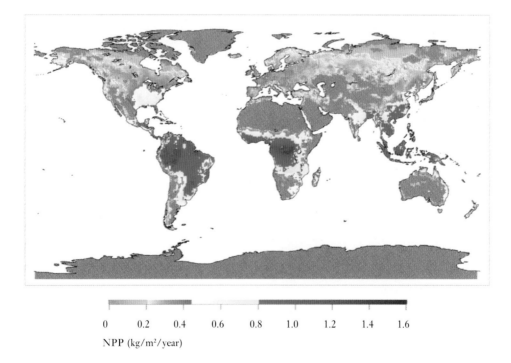

NPP (kg/m²/year)

Plate 8 *Predictions of current-day net primary productivity (from Woodward* et al., *1994).*

Net primary productivity 1990

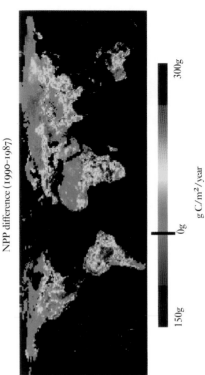

kg C/m²/year

0   2

NPP difference (1990–1987)

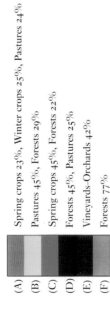

g C/m²/year

150g   0g   300g

Plate 10 TOP: *Diagnostic model for NPP estimation. fPAR is derived from the GVIs of 1990.* BOTTOM: *NPP differences between 1990 and 1987. Orange and red colours indicate a lower estimation of NPP for 1987, and magenta indicates a higher one.*

Land Cover Classification – Southwestern France

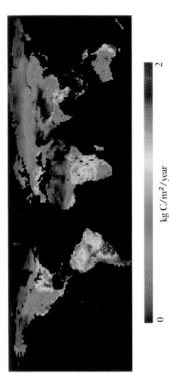

Atlantic Ocean
Garonne valley
Landes forest
Massif Central (old mountains)
Toulouse
Mediterranean Sea
Adour basin
Bouconne forest
Lannemezan plate
Pyrenees (young mountains)

45N
44N
43N
1W   0   1E   2E   3E

(A)
(B)
(C)
(D)
(E)
(F)

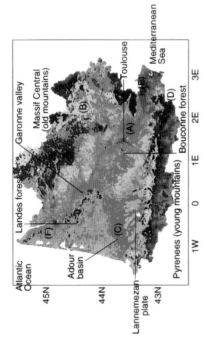

(A) Spring crops 23%, Winter crops 25%, Pastures 24%
(B) Pastures 45%, Forests 29%
(C) Spring crops 45%, Forests 22%
(D) Forests 45%, Pastures 25%
(E) Vineyards–Orchards 42%
(F) Forests 77%

Plate 9 *Automatic land cover classification (24 classes) of the southwestern part of France, using temporal NDVI profiles during 1987–9. The colour scale also refers to land use classes used in Figure 28.1.*

Part four

Agriculture, forestry and soils

The work of Focus 3

P. B. Tinker and J. S. I. Ingram

The rationale for Focus 3

Introduction

GCTE Focus 3 has three main areas of interest: agriculture, forestry and soils. Agriculture and forestry occupy a special position both in the world and in IGBP. They are the first two industries, and initially they gave work to and supported almost the whole of the population. As development continues, this proportion shrinks; in developed parts of Western Europe, only 2% of the population may be currently thus engaged, but in developing countries it may still be over 90%. This wide range corresponds to an equal range in intensification and technology, with high-input arable culture at one end, and simple shifting cultivation at the other. This great range is a problem for Focus 3, as this diverse and fragmented industry will face a bewildering array of problems as global change impacts upon it. Forestry also has a rather wide range, from single extraction of wood and other products from near-virgin forest up to regularly established, managed and exploited plantations.

Whatever the level of complexity or productivity, the agriculture and forestry industries are not optional – they are essential. Food and fibre needs are basic, and this need puts Focus 3 in a rather different position from the rest of GCTE or indeed from IGBP. It is therefore worth looking at the basic global food situation at the end of the twentieth century, to determine the context within which global change will operate.

It is reasonable to draw a distinction between developed and developing world agriculture, though there is an almost continuous range of conditions between them. These are facing very different situations. Developed world agriculture has overproduction and surpluses, and many policy initiatives have been introduced to reduce production. The environmental problems are troublesome rather than critical, with pollution from agrochemicals and damage to landscape values being the most common (Tinker, 1987). Given that developed agriculture has access to capital and to scientific advice and is not under pressure from an increasing population, it is comparatively well placed to face the potential problems of global change. Temperature changes are predicted to be highest in the high latitudes, so it is likely that the appropriate crops in any given region will change, new lands will be taken into use and new techniques will have to be developed and mastered. If the GCM predictions are right, the geographical changes will be massive (Parry *et al.*, this volume), and it will be difficult to predict, control and coordinate the responses that are needed. Large impacts within a major existing production area – e.g. the American Midwest – may cause limited food shortages, but the technical competence and scientific background to developed-world agriculture should allow it to cope with these changes.

It is much less easy to be optimistic about developing-world agriculture. The current GCM scenarios for the tropics do not predict large changes in temperature, but differences in amount, distribution and variability of rainfall may well be disruptive. The lack of detailed scientific knowledge and of capital, together with the very variable state of the advisory services and governmental support generally, make prediction of how impacts will be dealt with very difficult. One form of global change – land-use change into agricultural land – has been in progress in the tropics for many years. The confusion and conflict produced by this suggest that more fundamental climate changes will cause great stress.

It is very important to see the possible impacts of global change in the context of food security over the next 25 or 30 years. The increasing demand for food by a growing population in the developing world has been a major concern ever since World War II. On average *per capita* food production has increased due to extended irrigation and the techniques of the Green Revolution, i.e. better varieties, fertilizers and plant protection. There are, however, marked regional differences. Despite this, major famines in recent times have only occurred where there is war or social chaos; otherwise the world community has coped with shortages. However, population is still increasing and will continue to do so at roughly 1 billion per decade for the next three decades (Arizpe *et al.*, 1992). There is some hope that this may be the peak of the growth rate, but that is not yet certain. Virtually all this increase will occur in the developing world, which will also demand a better diet and life style. The net result is that the cereal needs of the developing world are expected to grow from around 1 billion tonnes per year now to about 2.5 billion tonnes by 2025 (Crosson & Anderson, 1992; Yudelman, 1993), while Cao *et al.* (1995) calculate that for China, even if full potential production is achieved, meeting the country's future food needs will remain a major concern. It is not possible to predict whether this projected demand will be met, and we do not discuss this

here. The important point is that it will be very difficult to meet this food
need in the best case, and the added problems and perturbations of global
change could easily make it impossible. It is the addition and interaction of
these various pressures that is threatening.

The developing world also faces environmental problems of a much
greater magnitude than those of developed agriculture. These mainly
concern soil damage, such as loss of nutrients and soil organic matter,
serious erosion, salinization, and desertification (see Table 12.1); and land
use pressures, which may damage forestry, and may cause marginal land to
be used for agriculture. These dangers are both to the environment and to
agricultural productivity, and global change may intensify all of them.

Impact mechanisms of global change

The existing and expected further increase in atmospheric CO_2 by $c.$ 0.5%
per annum is of enormous interest to agriculture, particularly the possibility
that it may increase yields per hectare. This topic has been extensively
discussed by Gifford $et\ al.$ (this volume), although at present it is too early
to come to any conclusion, as the yield increases appear to depend greatly
on the plant species and the conditions of growth.

Certainly a great deal of research on elevated CO_2 has already been done
in growth chambers, tunnels, open-topped chambers or other enclosed
spaces (see Table 12.2; and Kimball, 1983). From this the general
consensus is that yields may increase by up to 25–40% with twice current
ambient CO_2 concentrations, and that plants growing under high CO_2
produce more dry matter per unit amount of water respired, because of
effects on the stomata. However, there are questions of 'acclimation' of
plants over long exposures to high CO_2, and results from work in artificial
conditions are notoriously difficult to extrapolate to crops in the field. In
particular, the effects of pests and diseases may be very different. Also,
these figures for responses were usually obtained under conditions of full
nutrition and water supply, which may not apply in practical agriculture.
Better information on the effects on yield are now becoming available, in
particular from the important Free Air CO_2 Enrichment (FACE) ring
experiment in Arizona. With the workshops that are being organized on
this topic by GCTE Focus 1, we hope that this issue will be much clearer
within the next few years, as it is crucial to determine the size of potential
yield increases. However, it must be borne in mind that these values are for
'2 × present' CO_2, a concentration that will probably not be reached until
the end of the next century (IPCC, 1994). Also, even a full increase of 30–
40% in yield will not alone meet the need for possibly 150% more food by
2025 mentioned above.

Table 12.1 *Areas of different developing country regions moderately, severely or extremely affected by various types of human-induced land degradation (in million ha)*

Type	Land area		
	Africa	Asia	South and Central America
Water erosion	170	315	77
Wind erosion	98	90	16
Nutrient loss	25	104	3
Salinization	10	26	–

From Oldeman *et al.* (1990)

Table 12.2 *Per cent change +95% confidence limits in four key variables of 10 major crops due to a doubling of ambient CO_2 concentration*

Crop	Transpiration	Photosynthesis	Biomass	Yield
Corn	−26+6	+ 4+13	+ 9+5	+ 29+64
Wheat	−17+17	+27+20	+31+16	+ 35+14
Soybean	−23+5	+42+10	+39+5	+ 29+8
Sorghum	−27+16	+ 6+16	+ 9+29	–
Barley	−19+6	+14+[a]	+30+17	+ 70+[a]
Cotton	−18+17	+13+19	+84+126	+209+[a]
Rice	−16+9	+46+[a]	+27+7	+ 15+3
White potato	−51+24	–	−15+[a]	+ 51+111
Sweet potato	–	–	+59+18	+ 83+12
Alfalfa	–	–	+57+277	–

From Rogers & Dahlman, (1993).
[a]Data points too few to calculate.

The global change with the greatest capacity to affect agriculture is undoubtedly climate change. Temperature increase will be greatest at high latitudes, and this should greatly increase the productivity of land there, as well as bringing much new land potentially into cultivation (Parry *et al.*, this volume). The practical problems of extending spring wheat cultivation onto the soils of a degenerating boreal forest should not, however, be under-estimated. Even apparently benign changes such as this will introduce massive perturbations, with movements of people, demands for huge amounts of capital and the development of appropriate techniques and tech-nology. Like most other productive industries, agriculture and forestry operate best within a stable environment, and massive change itself can be detrimental in the short term, even if ultimately beneficial. The more worrying aspect of climate change is, however, change in rainfall and hydrology. Over vast areas a relatively small decrease in rainfall would have serious effects, with lowered or less stable yields, which could lead to a change in farming systems or even total abandonment of the land. In other areas a change in distribution of rainfall, or even an increase, could severely disturb the existing systems; heavy rain at the time of crop harvesting can be almost as damaging as drought, and the effect on soil workability could be important in many areas (Rounsevell & Brignall, 1994).

This discussion has so far concerned the effects of global change on production systems. We cannot, however, ignore the feedback interactions of agriculture and forestry on global change. The carbon budget of the world is still uncertain, but there is a very strong possibility that the forests are acting as a net sink for CO_2. Even if agricultural crops photosynthesize more rapidly because of high CO_2, they have little ability to store the extra carbon, and so, in themselves, cannot act as a useful long-term carbon sink. It does, however, seem possible that this enhanced photosynthesis by crops or grasses could increase soil organic matter through increased below-ground allocation and root exudation, although this is by no means certain (Fisher *et al.*, 1994). There is also a real possibility of storing some fossil carbon in forests, although the areas required to store all carbon emissions would be enormous.

Agriculture affects the production of trace greenhouse gases in two ways. Firstly, release from soils can be increased, e.g. methane from anaerobic soils such as in rice paddies, and nitrous oxide as a consequence of using more nitrogen fertilizer (Bouwman, 1990). Secondly, release from the clearing of land for agricultural purposes (including both the burning of vegetation and the subsequent tillage) accounts for a variety of inputs (both gaseous and particulate) to the atmosphere (Bouwman, 1990; Tinker *et al.*, 1994).

The adaptation to climate change

Because agriculture and forestry are industries, any changes they suffer will impact upon the socioeconomic structure of a country. Indeed, so strong is this perception that in the 1995 IPCC review process, agriculture and forestry are seen in a largely socioeconomic context. In fact impact adaptation will be both a technical and socioeconomic problem, involving changes in farming systems, crops, cultivars and inputs. This need is very much in the forefront of the GCTE Focus 3 priorities, discussed below.

The greatest problem concerning impact assessment is the uncertainty of climate change at the local level. The farmers work on their fields, and even national average changes are of little use to them or their agricultural advisors. Regional climate models, with a grid size of 50 km can now be nested inside GCMs. While they may be able to predict climate at this scale with improved accuracy, the exact details of weather patterns are so important for crop growth that a strategy to meet the problems of global change will still have to be based upon scenarios, probability and risk, rather than detailed prediction. All farmers are familiar with the risks of weather, to a degree depending upon their location. Most are not familiar with the idea of a long-term trend in climate superimposed on the normal irregular annual or decadal variation. A whole set of adaptation strategies must therefore be developed for them to apply as the real changes become more apparent.

However, Focus 3 cannot carry out all possible research on the impact of global change on agriculture, or even a reasonable fraction of it. A vast amount of largely site-specific research will be done by national or international agencies, as the outlines of global change become clearer, and its impact on agriculture and forestry become more likely. The role of Focus 3 is to do strategic and underpinning research, so that the applied and adaptive work done later will be more effective and economical. For this reason, in planning GCTE, great stress has been laid on coordinating research networks and on modelling. In addition to simulating crop growth, such models must be developed to incorporate the effects of pests, diseases and weeds, and soil and management variables, i.e. to simulate the cropping system, and ultimately the farming system. In this work, GCTE has been fortunate to form close links with other programmes working in the tropics, for example the Tropical Soil Biology and Fertility Programme of IUBS and UNESCO-MAB, and the 'Alternatives to Slash and Burn' programme led by a consortium of CGIAR Institutes.

There are three important factors which will be needed in both developed and developing countries to help cope with the effects of climate change. The first is a good database on the various cultivars, and alternative crops, that can be used in the existing and potential farming systems. The change of cultivars, and corresponding changes in appropriate agronomic treatments, is the first line of defence. There is already a great deal of information about crop cultivar behaviour in national and international organizations, even if it is not organized into a single unified database. The 'mandate' system under which the various CGIAR Centres have responsibility for one or more crops will help in this regard, as will the creation of the International Plant Genetic Resources Institute. Focus 3 has not been designed to undertake work of this nature, although it will of course draw upon all available information.

However, it is not possible to have total information about how a cultivar, species or mixture will grow under all possible conditions. The second, complementary approach therefore is to have crop, rangeland and forest models that are robust over a wide range of climatic and atmospheric conditions and preferably incorporating disease and pest effects. With these, the behaviour of crops, trees and grasses can be examined for a number of environmental scenarios, and the effect of changed cultivars, species or mixtures can be tested. The development of such models, so that response of single crops to global change can be more accurately predicted, is the main current thrust of Focus 3. This emphasis on the dependability of single crop models is seen as laying the best foundations for more elaborate models that will deal with farming systems, food security and socio-economics.

The third need is to have established, at least approximately, the climatic limits beyond which a given farming system (as distinct from a particular crop or variety) is not viable; this of course will depend upon the level of government economic support. Much work on farming systems by FAO and other organizations is very relevant to this problem, and Focus 3 aims to establish collaboration as and where appropriate.

Forward planning, largely based upon modelling, seems to be crucial to a rational approach to global change. There are instances of long-term trends in climate which in the historic past have always reversed themselves over periods of decades. Typical cases were the Midwest American drought of the 1930s, or the Sahelian drought of the 1970s. It is sobering to read of the disorganized and baffled attempts to cope with these initially. We are now better warned that some change is likely, but we are little better off in terms of predicting exactly what will happen in a particular locality, and how individual farmers should respond to this. Economic failure can, however, be greatly delayed by policy interventions, such as subsidized prices, income support, etc. The information that Focus 3 is collecting on the behaviour of crops, pastures and forests is therefore only one aspect of the overall approach, but it is intended to be an important input to the formation of policy.

Using a much wider approach, the impact of global change on world food supply and trade can be examined. The chapter by Parry *et al.* (this volume) makes an excellent start at putting together all the many variables involved, to predict the final effects on prices and (assuming a rational world) on land allocation. The model also includes population growth and increase in wealth, so it addresses the wider issues discussed in this chapter. Further, it acts as a bridge between the biophysical approach of IGBP, and the socioeconomic work that will be done by the International Human

Dimensions Programme (IHDP). It gives a framework into which improvements in data, models and theories can be fitted as they emerge.

Modelling the effects of global change is not the same as predicting it with certainty. There are uncertainties because the various GCMs do not agree in detail, and the differences can be important for agriculture and natural vegetation. The detail in the GCM outputs is also not sufficient to determine impacts accurately. It is therefore necessary to decide on a set of probable scenarios consistent with generally predicted changes, and work to these. At least the UKMO model now produces output on a 50 km grid, which reduces the latter problem. Beyond this, there are the uncertainties associated with the impact of rising CO_2 levels. These are likely to be crop-specific and require more research in the field. Finally, there is the vexed question of adaptation – the ability of farmers to change their procedures, or even their farming systems, so as to reduce the impacts on yield and profit. The estimation of these adaptations, for farming communities of vastly different skill and resource level, is little more than guesswork at present.

Forward 'prediction' (in the sense used above) depends upon two techniques. The first uses crop models to estimate effects on yield, and then these may be linked to economic and food models to determine impacts on nutrition, prices and gross national products (Parry & Rosenzweig, 1994; Strzepak & Smith, 1996). The second route uses broad-scale economic proxies for change in crop productivity. For example, Mendelsohn, Nordhaus & Shaw (1994) use capital value of land and its relationship to climatic parameters to predict effects on the crop value. Darwin *et al.* (1995) use a rather more complex procedure based on crop growth duration in different land classes. Both these models are interesting, but seem to need more biophysical input, especially as the procedures are extended from the USA to the rest of the world. Focus 3 needs to become more deeply involved in this area of research.

The structure of Focus 3

When designing Focus 3, much thought went into preparing a comprehensive programme of research that addresses the major issues identified above. To this end, the Focus covers four major production systems: monocrop agriculture (including a representative selection of seven food crops normally grown in monoculture), pasture and grazing systems, multi-species cropping (including agroforestry and rotational systems), and forestry. These have been grouped into three *Activities*.

Table 12.3 *The structure of GCTE Focus 3*

Activity 3.1	*Effects of global change on key agricultural systems*
Task 3.1.1	Global change experiments on key crops
Task 3.1.2	Modelling growth of key crops under global change
Task 3.1.3	Predicting golbal change impacts on pastures and rangelands, and the resulting effects on livestock production
Activity 3.2	*Changes in pests, diseases and weeds*
Task 3.2.1	Pest distribution, dynamics and abundance under global change
Task 3.2.2	Disease distribution, dynamics and severity under global change
Task 3.2.3	Weed distribution, dynamics and abundance under global change
Activity 3.3	*Effects of global change on soils*
Task 3.3.1	Global change impact on soil organic matter
Task 3.3.2	Soil degradation under global change
Task 3.3.3	Global change and soil biology
Activity 3.4	*Effects of global change on multi-species agroecosystems*
Task 3.4.1	Agroecosystem complexity, productivity and sustainability
Task 3.4.2	Long-term agroecological experiments and on-farm monitoring of global change research
Task 3.4.3	Modelling complex agricultural systems
Activity 3.5	*Effects of global change on managed forests*
Task 3.5.1	Experimental and observational studies of managed forests
Task 3.5.2	Modelling global change impact on function, structure and productive capacity of managed forests

As GCTE planning has progressed, new Activities have been assigned the next sequential number. While this avoids the possibility of confusion brought about by re-numbering Activities and Tasks, it must be stressed that the number does not imply importance.

Two further Activities crosscut these three major areas: one addresses the effect of global change on pests, diseases and weeds and the other addresses the effect of global change on soils. The full list of the Focus 3 Activities and their associated Tasks is given in Table 12.3.

The science in Focus 3

Key agricultural systems

Arable

If food security is an issue, then the emphasis must be on the world's main crops. There are only some 20 main food species, and Focus 3 has initially selected seven of these for concentrated attention: wheat, rice, cassava, potato, sorghum, groundnuts and maize. The list thus contains the major cereals, root crops and a legume. Initially, the main attention was on wheat, because of the wide geographical range over which this crop is grown, and the existence of a large and active research community on its agronomy, physiology and modelling. It therefore became the subject for the first Crop Network, in which we have tried out and developed ideas and systems. The underlying idea is always to develop our understanding to the point where the crop models are so reliable across a wide range of conditions that they can be used for confident extrapolation under global change conditions. This work covers monocrop or simple time-rotational systems.

The Wheat Network is now well developed, with 15 models (or sub-models), 43 pertinent data sets and a number of major research groups from a wide range of countries. The most interesting result has been from a joint run of all the models on two common datasets, from Europe and the USA respectively. The variation was surprisingly large, suggesting that many models have been adjusted and validated for a rather narrow range of climatic conditions (Goudriaan, this volume). The aim is to analyze this variation so that the most successful components of different models can be identified. Ultimately we need models that are dependable under a wide range of conditions, and that can be amended or tuned for different cultivars for use in models of the type used by Parry *et al.* (this volume) and others. On the other hand, a similar test with models of the rice crop gave much smaller spread. As yet it is not clear whether the rice models are inherently more similar, whether the basically very uniform environment in which wetland rice grows makes agreement easier to reach, or whether there is some other reason. This raises questions over how many models in global change studies have been exposed to this type of joint testing.

The other main aim for the principal crops is to determine the yield gain that may be obtained from the increase in atmospheric CO_2. If the crop yield increases discussed above do come about, they will only happen gradually, as CO_2 levels increase. It is now essential that the CO_2 response should be tested in conditions more closely approximating to the field, by using techniques such as FACE. The only major installation of this type working with arable crops is in Arizona which, after a series of experiments on cotton, is now carrying wheat. The most recent results (B. A. Kimball, personal communication) show that wheat yields were increased by 8% or 21%, depending upon water supply. These values are rather small compared with published controlled-environment results, but they are in agreement with data of Weigel *et al.* (1994), who concluded that wheat is particularly well adapted to the present CO_2 level in the atmosphere and that it therefore shows a particularly small response to elevated CO_2. It is of the highest importance that this type of experimentation should continue, and be undertaken elsewhere. FACE installations studying various problems of grassland are already operating: one in Zurich for the project 'CO_2-related processes in grassland ecosystems', and one for the project 'CO_2 and climate change effects on New Zealand pasture'; both are GCTE Core Research.

Pastures and rangelands

This topic covers a wide spectrum from the high-intensity grasslands of Northwest Europe to the extensive but low-yielding rangelands of Australia, Africa or Central Asia. Despite these differences, the topic is run as one Task. The main differences lie in the productivity, the level of inputs and the number of native species present.

Where a species mixture is present, the response to global change is more complex and interesting, and more difficult to predict than in monoculture. There will be species shifts, in addition to direct effects on growth rates, and both may affect use by animals and productivity. In some senses such systems relate to the complex systems in Activity 3.4 (Table 12.3) described below, and to the problems of Focus 4. As in all other GCTE Tasks, there have to be priorities; here they are temperate high-intensity pastures with low species number, and temperate/subtropical mixed pastures, especially with both C_3 and C_4 species, because these react differently to elevated CO_2 concentrations.

Four GCTE Core Research Projects (CRP) have been planned for the Task: forage and animal production (paddock/landscape scale); land use, commercial production systems (enterprise scale); land use, subsistence systems; and greenhouse gas fluxes. Each CRP will consist of (i) a set of contributing projects that will form a research network, (ii) a coordination committee that will guide each network, and (iii) a schedule of activities (e.g. sensitivity studies, model inter-comparisons, synthesis workshops, data collection) that will form the specific workplan for each CRP. The four CRPs will be launched sequentially with a phased approach, the first was launched in mid 1995.

Pastures and rangelands are particularly important in providing animal products directly rather than by feeding grain to animals. They often make use of land that could not reasonably be used in any other way. However, the low economic yield per hectare in low-intensity types mean that practical options to adapt to global change are few.

Effects of global change on multi-species agroecosystems

The work of the Crop Networks (as discussed above) relates to the main crops of the world, and mainly in regard to their use in continuous monoculture (as with rice) or simple rotations of annual monocultures. However, many of the world's farmers grow crops in complex spatial or time mixtures, and often with animals as an essential component of their farming systems. Basically this Activity is intended to address the low-intensity farming systems of developing countries. The social and economic situation of such farmers is also normally quite different to that of high-intensity farmers, and avoidance of risk is high priority. For all these reasons, research into these complex systems will be difficult. However, it is quite certain that in many places such systems are the only ones that are both productive and sustainable, at least over a moderate time period.

The concept of 'sustainable development' has led to an increased interest in these types of system as a means of stabilizing food production and conserving natural resources and environments, particularly in the tropics. Research within this Activity is concerned with exploring whether the use of numerous crop species in a field agroecosystem conveys any advantage in the face of disturbance such as global change. An underlying hypothesis is that the relationship between the number of plant species in an agroecosystem and the efficiency and stability of ecosystem functions (such as production and nutrient cycling) is hyperbolic; an increase from a monocrop to an intercrop of three, four or five species may be expected to significantly affect functional integrity; beyond this relatively small number additional species have only marginal effects (Swift & Anderson, 1993).

One of the most common practices in tropical ecosystems is intercropping, the growing of two or more crops in the same field at the same time. Intercropping often, although not invariably, produces higher yields than sole crops (Francis, 1986; Vandermeer, 1989). Intercrops are also thought to reduce farmer's risk (Rao & Willey, 1980), yet this function too has been questioned (Vandermeer & Schultz, 1990). Depending on the particular intercrops and the site involved, intercropping may promote enhanced nutrient utilization, pest control, weed control, and other agricultural functions, although it is not possible to generalize that any such functions are universally a consequence of intercropping. Most experiments have been limited to analysing interactions between two species in relation to the trade-offs between yield and competition for resources. Within farms, however, these interactions occur within a much more heterogeneous environment and a broader context of agricultural goals on the part of the farmer.

Traditional farmers often maintain many more species than two in their fields, occupying complex patterns in space and time. They may also delib-erately maintain diversity among cultivars of a major crop species. Such practices can serve a variety of purposes: for diversity of product, to spread risk through the cropping season, or to suit different microenvironments. For instance the Apatani tribe in northeastern India, who traditionally are involved with wet rice cultivation, have selected different rice genotypes to suit sites of varied nutrient status within a landscape. Waste recycling from the village is a key element in maintenance of soil fertility, so soil nutrient status is higher closer to the village. In flooded plots closer to the village, a long duration rice cultivar is grown in combination with the maintenance of fish to capitalize upon the high nutrient levels. Farther away, a more nutrient use-efficient, but shorter-duration, rice cultivar is grown, but without pisciculture (Ramakrishnan, 1994).

GCTE research on multi-species systems (Activity 3.4 in Table 12.3) has three major components, developed as separate but interdependent Tasks. First, it needs experimental studies to explore the types of relationships discussed above: between plant species number and the efficiency, stability and resilience in ecosystem function. Second, long-term experiments make it possible to assess patterns of response of agricultural systems to changing circumstances, such as global change. Data from such experiments can also be used to validate the predictions of simulation models. A diversity of such experiments have been initiated at various times during the last century, many in developing countries. Some are still ongoing, while others are aban-doned or at a very low level of management; the overall number is uncer-tain, but it may reach several hundreds. Furthermore the full data sets from many have never been published in the open literature. In view of the importance of such experiments, it is crucial to maintain those appropriate for global change research, and to initiate new ones, where necessary in conjunction with other interested parties.

The third component is simulation modelling of multi-species ecosystems. This is potentially the most realistic way of extrapolating predic-tions of change beyond the short term, provided models can be constructed with enough rigour. Whilst there is now a comparatively effective (but by no means adequate) suite of such models for individual crops such as wheat, maize or rice, the modelling of more complex systems such as inter-crops and agroforestry (with the inclusion of competitive and/or synergistic effects), is less well advanced, although considerable progress has been made using the same principles and approaches as for the sole crop models. More rapid advancement will come from an interactive effort between modellers tackling the issues of simulating complex cropping systems than if they operate in isolation. Further strength will be gained from linkages with both the monocrop and ecological modelling activities also being promoted in GCTE. A further dimension to the modelling activities comes from the need to assess the economic and social implications of global effects on agroecosystem performance, including modifications in agricultural practice. This requires the development of economic–ecological models of land-use and agroecosystems change. This is a relatively new but fast-growing area of research which requires collaboration with social scientists, and this will be achieved through linkages with the International Human Dimensions Programme and scientists in the CGIAR system.

Effects on managed forests

Forests cover a vast area of the world, and most are exploited and managed in various ways. The impact of global change is already large – occasionally devastating – through land-use change, while subsequent CO_2 effects and climate change can further impact the systems. The long time-scale over which forests mature makes both experimental research programmes and forest management difficult. Whereas annual crops can in principle be adapted from year to year, a forest may take a century or more to reach maturity, and several times as long to reach an equilibrium climax. The prediction of impacts is thus particularly important, but the lack of detailed local climate projections makes this work especially difficult.

Forests fulfil many needs in addition to that of providing timber, though the latter is a huge industry. Leisure uses are increasing, as is the importance of forests as water catchments or bio-reserves for conservation purposes. Forests also have great ecological interest as the dominant vegetation over much of the Earth. Work on forests therefore is spread throughout GCTE, and is also important in BAHC and GAIM.

The underlying strategy for the Focus 3 research programme is to establish experiments in a representative range of forest types, growing under different conditions. Where possible, similar experiments will test a series of hypotheses about the effects of global change on forests. There is particular interest in how global change will affect water relations, which determines where forests can grow, and carbon allocation, which determines speed of growth and nutrition. Experimental work should be aimed not only at providing empirical information, but also at understanding the physiological control mechanisms and key processes underlying forest response to environmental factors (Linder *et al.*, this volume). Thus experiments will be designed also to produce datasets to help develop and validate models that will be used for predicting how forests will respond to various scenarios. Forest modelling has already made considerable progress, and there is a good basis to build upon. However, the long-term nature of forests means that purpose-designed experiments and trials are not always possible. The transect approach (as in Foci 1 and 2) will be used heavily in developing the field programme, where a range of environmental conditions are used as surrogates for manipulative experiments.

Pests, diseases and weeds

Global change of all types will have consequences for the biological competitors of crops, i.e. the pests, diseases and weeds (Cannell & Knight, 1992; Tinker 1993). Different responses to CO_2 increase may markedly alter the crop–weed balance, and increase the impact of species that were formerly not regarded as weeds. CO_2 change may also alter the chemical composition of crops resulting in a different protection to attack. Climate change can alter insect–pest and fungal disease attack even more sharply. The adaptation to global change is therefore not simply to change crops or cultivars to accept a new climate, but also to resist a whole new suite of pests and diseases.

There is a great deal of uncertainty about how far we can operate by 'climatic analogies'. This means that if a crop or farming system is moved into a new area because it has an appropriate climate following climate change, it will be attacked by the same pests, diseases and weeds as in its previous home. This is by no means always clear, and indeed introductions of crops into new areas often develop a different suite of pests and diseases, even if the climate is fairly similar. There may be differences in day length or in soil type from those in the area of origin that affects pathogen attack, so that this is not identical with earlier situations. The impact of global change on pathogens and weeds, in a situation where farming systems and crops are changing rapidly, could be extremely complex, and calls for very rapid response in research. As much as possible of general underpinning research must therefore be done now.

Given the importance and complexity of this subject, it is managed in GCTE as an independent Activity. It nevertheless occupies a cross-cutting role, spanning GCTE research in major food crops, forestry and multi-species cropping systems. Initially it had less priority than the basic Crop Networks, but it is now developing rapidly. The interaction of the Crop

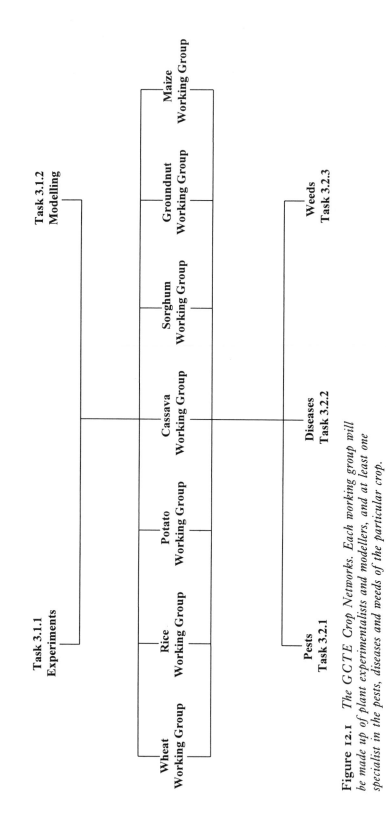

Figure 12.1 *The GCTE Crop Networks. Each working group will be made up of plant experimentalists and modellers, and at least one specialist in the pests, diseases and weeds of the particular crop.*

Networks and the pathogen/weed Activity has been given a great deal of thought because they are so closely linked in farming systems. A structure has been developed (see Fig. 12.1) to allow flexible but close relationships between them. In particular, it is very important to include the pathogen and weed models within crop models, in addition to the work on population biology models of the pests and diseases. The implication is that first attention will be given to the pests, diseases and weeds of our seven priority crops. What is said about pathogens automatically applies also to weeds, which may well be a more pervasive and difficult problem than pathogens.

With global change, there may be unprecedented movement of organisms. Already there is extensive transfer and invasion, due largely to human mobility (Williamson et al., 1986). With changing climates the agroecological zones themselves will shift geographically. There may be more displacement and turbulence in cultivated species and their pests than in any period since agriculture began.

Teng et al. (this volume) outline the rapid and exciting development of combined pest–crop, disease–crop and weed–crop models in the last few years. This involves knowing the population biology of the pest and disease organism or weed to estimate the level of infestation. The most critical part is then to determine a mechanistic and quantitative link between the pest and crop growth, by identifying the point of damage, be it leaf damage, withdrawal of phloem sap or weed competition. Given defined weather parameters both pest population and activity and crop growth under this attack need to be modelled. This work has largely been carried out on wetland rice, grown under good agronomic conditions; this Activity will build upon it to address the more complex situation where both water and nutrient stress and pest attack occur simultaneously on the crop.

Soils

Global change can affect soils directly, but the largest impact is via changes in the vegetation cover (Tinker & Ineson, 1990). While changes in climate and CO_2 concentration are expected to alter the cover over time, changes in land use have an immediate impact, which can be very dramatic, and which may greatly alter the susceptibility to water or wind erosion. One must also consider the impact of global change on the soil biota and on their function, especially their control of the element cycles. There is thus first the biodiversity question: will the soil biota migrate at a sufficient rate to avoid extinction and to meet the needs of crops established in new areas? Second, how will their functioning be affected? Many of the biologically mediated element cycles occur via changes in soil organic matter, which only slowly moves into equilibrium with a new state of the soil.

It was therefore concluded that the most important issues were soil organic matter dynamics, soil degradation processes and soil biology. These are all fundamental to soil behaviour and fertility, and they are interconnected. A soil that is protected against degradative processes, has an adequate biomass and faunal population, and has an adequate level of soil organic matter is in fairly good condition in terms of its use for forest or crop growth or to support natural vegetation. A sufficiency of soil organic matter, of adequate C/N ratio, is also an important safeguard against both structural failures and erosion and nutrient shortage. Organic matter, especially in tropical soils, is probably the best general-purpose indicator of soil health.

However, there are also wider implications. Soil is a major store of carbon, with roughly twice the amount of carbon in the world's vegetation, and changes in organic matter are an important part of the global budget. Treating soil organic matter as a 'black box' is also inadequate, though the basic reasons for soil organic matter being so stable, or the processes that produce it, are still not understood. This is part of the reason for the interest in soil biology, which is the third Task in this Activity (see Heal *et al.*, this volume). A detailed study of how soil biota, at all levels, influences the metabolism of soil organic matter is important in modelling and predicting the overall process. This Task also deals with the formation and consumption of greenhouse gases as a normal part of the soil biological activity. For example, recent work has shown that any injection of nitrogen into forest soil, from precipitation, fertilizers or tree felling, sharply increased N_2O emission rates (P. Ineson, personal communication).

The purpose of the soil organic matter programme is to do experimental work to produce datasets on soil organic matter changes, and to use these and existing sets to test, validate and improve soil organic matter models. The identification and collation of suitable datasets for these purposes is now being undertaken within GCTE's worldwide network 'SOMNET'. The aim is, again, to produce models that we can rely upon to predict change in a variety of scenarios. This will include research into the effect of changed inputs of organic materials – litter, roots, exudates – into soil under global change conditions, and also the expected change in chemical composition, e.g. the lignin/nitrogen ratio. The coupling between vegetation cover (land use), net primary productivity and soil organic matter needs to be defined for changes in CO_2, temperature and rainfall. Of these, the effect of changes in CO_2 is by far the least well known, and the combined effect of all three has received almost no attention yet. It is now a major focus of elevated CO_2 research. The different effects of injecting the same amount of organic material as litter, exudates or root material also need to be better defined. There is some evidence (GCTE Science Conference, see Chapter 1) that the additional carbon fixed and translocated into the soil may sometimes be almost wholly converted back into CO_2 in a very short time, but this conflicts with much earlier experience, according to which some soil organic matter is always produced when microbial activity is enhanced by organic matter addition.

The increasing temperature expected in high latitudes should have an effect on the large stores of soil organic matter in these soils (Tinker & Ineson, 1990) and a decline in soil carbon is expected. Very recent work on upland cool-temperate soils in England has shown that increasing temperatures rapidly increase the level of dissolved organic carbon, but decrease the level of nitrate, in the soil solution (P. Ineson, personal communication). Thus GCTE Core Research shows that the consequences of soil warming are complex.

Soil degradation, especially soil erosion, is the most serious consequence of land-use change at present (Norse *et al.*, 1992). For all the alarming visual impact of a felled forest, in almost all cases this will regrow if the soil is not damaged, and shifting cultivation has proceeded on this basis for millennia. The really irreversible damage occurs when the topsoil is stripped away, compacted or impoverished to a degree that regrowth cannot occur rapidly enough to provide cover. A vicious cycle is then initiated, leading to progressively worse degradation. The loss of useful land from these processes is serious (see Table 12.1), but needs better quantification. The GCTE Core Research in this Task (Valentin, this volume) has a particular focus in West Africa, because of the wide range of soils, topography and rainfall there. There are also good records of recurrent droughts. The mechanisms of the various forms of erosion are now reasonably clear, but a great deal more needs to be done to model erosion processes. Land-use changes already have major impacts on erosion hazards; they exemplify the potential interactions between increasing population pressure, diminishing food security and change in climate outlined above. One of the most surprising results of current work is that percentage runoff (i.e. water lost to the vegetation) increases as rainfall declines, due to soil crusting processes. The soil degradation Task is intended to deal with both water and wind erosion in the areas where they are most severe. The main thrust will not be to repeat erosion experiments in the field, but to monitor field situations and use existing data to refine current erosion models and make them more reliable at different scales.

Because land use is so strongly involved in this subject, close links are being established with the developing International Human Dimensions Programme (IHDP), and in particular with the joint IGBP-IHDP Core

Project 'Land-Use and Cover-Change' (LUCC) (see below). The aim of this collaboration in the soils work is to understand the imperatives that drive people to put their soil at risk, and hence to alter them. The technical understanding of erosion processes can suggest ameliorative measures, but if farmers persist in unwise activities, erosion must follow.

Finally, the GCTE Task dealing with soil biology has just been restructured. Originally, this focused almost wholly on the production of the greenhouse gases CO_2, CH_4 and N_2O. However, it was decided that this overlapped too closely with the interests of IGAC, and the new title was adapted to widen the remit. There are four areas where global change impact may occur: the direct interactions of soil biota and higher plant growth, the decomposition reactions and formation of soil organic matter, the impact on population biology of soil biota, and the production of trace greenhouse gases. All these are related, and the detailed programme of the Task is currently being developed. This will include a joint Focus 3–Focus 4 structure to coordinate studies in the functional role of soil biota, and how global change will impact upon it; the inclusion, where possible, of soil biological studies in elevated CO_2 experiments; and the joint IGAC-GCTE development of a nitrogen transformation model for predicting both biogenic trace gas production and nutrient release.

Conclusions

Focus 3 is unique in GCTE in being linked directly with two major world industries. It is absolutely clear that global change will affect many industries, from insurance to transport, but no other ones have been included in IGBP, for obvious reasons; the environment, however, is of overwhelming importance for agriculture and forestry. This unique involvement with economic and social matters demands that we must link more closely with IHDP, and in particular with LUCC. The impacts on crops are often discussed in a biophysical sense, but these crops are the livelihood of farmers and foresters, and the basis for industries in which millions of other people find employment, and on which literally all of us depend. The reaction of agricultural and forest communities to global change will depend on the socioeconomic, policy and biophysical contexts.

Focus 3 has a remit which is breathtaking in its size. The research programme is purposely planned to help in predicting the impacts of global change and in ameliorating its effects – it therefore aims to be practically useful in due course.

References

Arizpe, L., Costanza, R. & Lutz, W. (1992). Population and natural resource use. In *An Agenda of Science for Environment and Development into the 21st Century*, ed. J. C. I. Dooge, G. T. Goodman, J. W. M. la Rivière, J. Marton-Lefèvre, T. O. O'Riordan & M. Praderie, pp. 61–78. Cambridge: Cambridge University Press.

Bouwman, A. F. (1990). Exchange of greenhouse gases between terrestrial ecosystems and the atmosphere. In *Soils and the Greenhouse Effect*, ed. A. F. Bouwman, pp. 61–126. Chichester: John Wiley & Sons.

Cannell, M. E. & Knight, J. D. (1992). Effects of climatic change on the population dynamics of crop pests. *Advances in Ecological Research*, **22**, 117–55.

Cao, M., Ma, S. & Han, C. (1995). Potential productivity and human carrying capacity of an agro-ecosystem: an analysis of food production potential of China. *Agricultural Systems*, **47**, 387–414.

Crosson, P. R. & Anderson, J. R. (1992). *Resources and Global Food Prospects.* World Bank Technical Paper 184. Washington, DC: World Bank.

Darwin, R., Tsigas, M., Lewandrowski, J. & Raneses, A. (1995). World agriculture and climate change: economic adaptations. Agriculture economic report no. 703. Washington, DC: USDA Economic Research Service.

FAO (1993). *The State of Food and Agriculture 1993.* FAO Agriculture Series, 26. Rome: Food and Agriculture Organization of the United Nations.

Fisher, M. J., Rao, I. M., Ayarza, M. A., Lascano, C. E., Sanz, J. I., Thomas, R. J. & Vera, R. R. (1994). Carbon storage by introduced deep-rooted grasses in the South American savannas. *Nature*, **371**, 236–8.

Francis, C. A. (1986). *Multiple Cropping Systems.* New York: Macmillan.

IPCC (1992). *Climate Change: The Supplementary Report to the IPCC Scientific Assessment*, ed. J. T. Houghton, B. A. Callender & S. K. Varney. Cambridge: Cambridge University Press.

IPCC (1994). *Radiative Forcing of Climate Change: The 1994 Report of the Scientific Assessment Working Group of IPCC*, ed. B. Bolin, J. T. Houghton, & L. G. M. Filho. London: UK Meteorological Office.

Kimball, B. A. (1983). Carbon dioxide and agricultural yield: An assemblage and analysis of 430 prior observations. *Agronomy Journal*, **75**, 799–88.

Mendelsohn, R., Nordhaus, W. D. & Shaw, D. (1994). The impact of global warming on agriculture: a Ricardian analysis. *American Economic Review*, **84**, 753–71.

Norse, D., James, C., Skinner, B. J. & Zhao, Q. (1992). Agriculture, land use and degradation. In *An Agenda of Science for Environment and Development into the 21st Century*, ed. J. C. I. Dooge, G. T. Goodman, J. W. M. la Rivière, J. Marton-Lefèvre, T. O. O'Riordan & M. Praderie, pp. 79–90. Cambridge: Cambridge University Press.

Oldeman, L. R., Hakkeling, R. T. A. & Sombroek, W. G. (1990). *Global Assessment of Soil Degradation.* Wageningen: ISRIC.

Parry, M. L. & Rosenzweig, C. (1994). Potential impact of climate change on the world's food supply. *Nature*, **367**, 133–8.

Ramakrishnan, P. S. (1994). The Jhum agroecosystem in north-eastern India: A case study of the biological management of soils in a shifting agricultural system. In *The Biological Management of Tropical Soil Fertility*, ed. P. L. Woomer & M. J. Swift, pp. 189–207. Chichester: John Wiley.

Rao, M. R. & Willey, R. W. (1980). Evaluation of yield stability in intercropping: studies with sorghum/pigeon pea. *Experimental Agriculture*, **16**, 105–16.

Rogers, H. H. & Dahlman, R. C. (1993). Crop responses to CO_2 enrichment. *Vegetatio*, **104/105**, 117–31.

Rounsevell, M. D. A. & Brignall, A. P. (1994). The potential effects of climate change on autumn soil tillage opportunities in England and Wales. *Soil and Tillage Research*, **32**, 275–89.

Strzepek, M. L. & Smith, J. B. (1996). *As Climate Changes: International Impacts and Implications*. Cambridge: Cambridge University Press for the US Environmental Protection Agency.

Swift, M. J. & Anderson, J. M. (1993). Biodiversity and ecosystem function in agricultural systems. In *Biodiversity and Ecosystem Function*, ed. E.-D. Schultz & H. Mooney, pp. 17–38, Berlin: Springer.

Tinker, P. B. (1987). Efficiency of the agricultural industry in relation to the environment. In *Environmental Management in Agriculture*, ed. J. R. Park, pp. 7–20. London: Belhaven Press.

Tinker, P. B. (1993). Climate change and its implications. In *Global Climate Change: Its Implications for Crop Protection*, ed. D. Atkinson, pp. 3–12. Farnham: BCPC.

Tinker, P. B. & Ineson, P. (1990). Soil organic matter and biology in relation to climate change. In *Soils on a Warmer Earth*, ed. H. Scharpenseel, M. Schomaker & A. Ayoub, pp. 71–88. Amsterdam: Elsevier.

Tinker, P. B., Ingram, J. S. I. & Struwe, S. (1994). Effects of slash-and-burn agriculture and deforestation on climate change. In *Symposium ID-6: Alternatives to Slash-and-Burn Agriculture*, ed. P. A. Sánchez & H. van Houten, pp. 15–25. Transactions of the ISSS 15th World Congress, Acapulco.

Vandermeer, J. H. (1989). *The Ecology of Intercropping*. Cambridge: Cambridge University Press.

Vandermeer, J. H. & Schultz, B. (1990). Variability, stability and risk in intercropping: some theoretical considerations. In *Agroecology: Researching the Ecological Basis for Sustainable Agriculture*, ed. S. R. Gliessman, pp. 206–32. New York: Springer.

Weigel, H. J., Manderscheid, R., Jäger, H.-J. & Mejer, G. J. (1994). Effects of season-long CO_2 enrichment on cereals. I. Growth performance and yield. *Agriculture, Ecosystems and Environment*, **48**, 231–40.

Williamson, M. H., Cornburg, H., Holdgate, M. W. & Gray, A. J. (eds.) (1986). Biological invasions. *Philosophical Transactions of the Royal Society, Series B*, **314**, 503–4.

Yudelman, M. (1993). *Demand and Supply of Foodstuffs up to 2050 with Special Reference to Irrigation*. Colombo: International Irrigation Management Institute.

Agriculture and global change: Scaling direct carbon dioxide impacts and feedbacks through time

R. M. Gifford, D. J. Barrett, J. L. Lutze and A. B. Samarakoon

Introduction

If 'global change' refers to actual or prospective anthropogenic changes in land use, atmospheric composition, nutrient deposition, climate and UV-B radiation levels, then clearly accurate prediction of aggregated impacts of global change at any location on Earth is a massive task well beyond the field's current modelling capabilities. Impacts will change through time. The interactive effects among the global change variables can be complex and are largely uninvestigated. Furthermore future land-use change is particularly unpredictable as it is both localized and derives from myriad specific decisions by individual land-users. Given this situation, we suggest that the best analytical approach is to investigate the impacts of those attributes that are the most certain, the most immediate, the most uniformly widespread, and that have most substantial effects first. After these are established, the repercussions of more tenuous and location-specific aspects can be explored in the context of the more certain changes, as experimental and observational findings become available.

For agricultural impacts the increase in atmospheric CO_2 concentration is the global change that fits all the above priority criteria most strongly. The other changes are either less certain, less general or uniform, less predictable, of lesser potential impact, or will exert their impacts later. Accordingly, this paper focuses mostly on the CO_2 effects.

Scaling in space and time

Scaling up from experimental observations to the real world is a recurring theme which has two aspects: scaling in *space* and scaling in *time*. Although there are inevitable linkages between the two, most recent discussion of scaling has in practice centred on space, particularly in the contexts of hydrology, satellite imagery interpretation, and global steady state modelling (Ehrleringer & Field, 1993). Scaling in time from seconds to aeons is an equally important aspect which tends to become lost in steady state models considering the effects of a notional CO_2 doubling. Conceptualizing the continuous change in atmospheric composition as a series of small discrete annual step changes, temporal scaling has four aspects. One aspect is the time-marching sequence of repercussions, internal to the ecosystem (or agro-ecosystem) of concern, of a small step CO_2 increase; the second is the time-marching impacts of other changes in the environment which are also due to a step CO_2 increase (such as temperature, humidity, rainfall); the third is the continuous series of step-changes in CO_2 concentration together with other interacting global change attributes that are not caused by the CO_2 increase (e.g. UV-B, N deposition); and the fourth aspect is the repercussions for the ecosystem of the feedbacks onto the environment of the impacts deriving from the first three aspects. In this chapter the first of these aspects is emphasized with some reference to interactions and feedbacks.

A single step increase in atmospheric CO_2 concentration has a sequence of repercussions for ecosystems because each ecosystem process has a different characteristic relaxation time. The outputs of fast processes become the inputs to slower processes. And slower processes may diminish the expression of faster processes in the medium term but not necessarily in the longer term. At each successive longer time-scale there are more processes interacting as inputs to yet slower processes. Table 13.1 depicts some of these; some are proven effects, others are possibilities yet to be adequately investigated.

The further down Table 13.1, the longer the time-scale and the less certain is the nature of the processes and the sign of their effects. Although the term 'acclimation' in this context is often associated with leaf-level photosynthetic down-regulation, one could describe all the processes from level 4 onwards as aspects of ecosystem acclimation to elevated CO_2 concentration. It is the imponderables of various types of ecosystem acclimation, involving offsetting effects of both negative and positive feedbacks, that make prediction of the effect of elevated CO_2 so difficult.

Ecosystem adjustments to high CO_2 on different time-scales

A small selection of the processes listed in Table 13.1 is discussed below using examples from crops, pastures and rangelands where possible.

Table 13.1 *A sequence of some possible phenomena occurring over a range of time scales which may follow from a step increase of atmospheric CO_2 concentration*

Level 1 (seconds—minutes)

+	Increased leaf photosynthesis rate (different for different species)
+	Decreased stomatal conductance (different for different species)

Level 2 (hours—days)

+	Faster absolute and relative growth rate
+	Higher carbohydrate content of plant tissues
+	Less transpiration

Level 3 (days—weeks)

+	Higher plant C : N ratio and faster growth
+	Reduced soil water use
−	Down-regulation of photosynthetic capacity
?	Decreased specific leaf area
+	Variation in root morphology

Level 4 (weeks—months)

−	Diminution of the relative growth rate increase
±	Increased leaf area index and increased soil water use
+	Larger root system and greater soil nutrient extraction
?	Change in C-partitioning between leaf, root, stem
+	Increased litter fall
+	Increased C : N ratio of litter
?	Changed lignin content of live and dead matter
?	Changed secondary metabolites in live and dead matter
?	Depletion of soil-available nutrient
?	Changed soil water status
?	Increased specific exudation rate of organic acids and sugars into soil
?	Changed incidence of pests and diseases

Level 5 (months—years)

+	Increased standing biomass
+	Increased soil organic matter
?	Changed rate of soil organic matter decomposition
±	Increased microbial biomass and nutrient tie-up by microbial biomass
?	Changed competitive relations between plants and with other organisms

Table 13.1 (cont.)

Level 6 (years–decades)	
?	Changed species composition
?	Changed fire frequency in fire-prone ecosystems
+	Increased symbiotic and non-symbiotic N fixation
+	Increased mobilization of 'unavailable' soil nutrients
?	Increased accumulation of minerals in soil organic matter and biomass
Level 7 (decades–centuries)	
+	Stabilization of species composition
+	Increased accumulation of organic forms of C, N and other minerals in the whole ecosystem

The symbols indicate changes which would be expected, at the timescale indicated, to increase productivity and C storage if they were to occur alone all else equal (+), or decrease productivity and/or C storage if they occurred alone all else equal (−), or involve opposing processes the balance of which could be positive or negative (±).

Down-regulation of leaf photosynthetic capacity

Sometimes leaf photosynthetic capacity has been observed to decline relative to control plants when plants are grown in elevated CO_2 concentrations. Expression of such down-regulation is clearly complex and does not appear to be universally explained by any single proposed cause such as inadequate pot size (Arp, 1991; Thomas & Strain, 1991), build-up of starch (Cave *et al.*, 1981), deactivation of Rubisco (Sage *et al.*, 1989; Rowland-Bamford, *et al.*, 1991), down-regulation of the Rubisco gene expression mediated by increased soluble carbohydrate levels (Webber *et al.*, 1994), and associated redistribution of N from photosynthetic apparatus to other sinks such as the roots (Stitt, 1993). Contradictory evidence exists for each proposed mechanism. Photosynthetic down-regulation can be only weakly, or not at all, related to starch or sugar levels in leaves (Yelle *et al.*, 1989; Xu *et al.*, 1994; Barrett and Gifford, 1995a). It can occur without loss of Rubisco activity per unit leaf area, or can even occur despite an *increased* percentage of activation of Rubisco (Xu *et al.*, 1994). Photosynthetic down-regulation can be present at some stages of plant or leaf growth but not others. For example, in pea and soybean grown with abundant water and nutrient, high CO_2 concentration caused down-regulation in old soybean leaves but not young, and in young pea leaves but not old (Xu *et al.*, 1994). Nutrient limitations and sink growth restrictions may enhance the tendency for down-regulation, but not necessarily so (Bunce, 1992). For example, down-regulation of photosynthetic capacity occurred in cotton leaves under elevated CO_2 concentration but was unrelated either to pot size or to phosphate supply (Barrett & Gifford, 1995a). Xu et al. (1994) suggested that down-regulation of photosynthesis in elevated CO_2 concentration might be an expression of interaction between processes leading to the re-optimization of the deployment of within-plant resources at three levels of competition operating on different timescales: namely (1) competition within the individual leaf between require-ments to build the carboxylation machinery and the light harvesting machinery; (2) between the photosynthetic machinery as a whole and the plant's carbohydrate utilization machinery (i.e. source–sink competition); and (3) between the plant's carbon acquisition and storage system and its mineral acquisition system (i.e. shoot function–root function competition).

While it might be said that when photosynthetic down-regulation occurs it is an expression of the plant trying to resist in some way the increased growth potential offered by elevated CO_2 concentration, it seems more likely from the evidence that it is an expression of the plant re-organizing itself to *maximize* the utilization of all its scarce resources including CO_2 (cf. Bloom et al., 1985). When CO_2 is made less scarce, a plant can deploy resources to acquire more of the other resources which are rendered relatively more scarce. However, the 'carbon/nutrient balance' concept has been chal-lenged too by data indicating that reduced tissue N levels at high CO_2 are just an expression of the downward drift of plant N concentration with plant size (Coleman et al., 1993). Nevertheless, our own evidence is that reduced tissue N levels are persistent over years in grasses and that it is expressed even in naturally senesced litter (Lutze & Gifford, 1995, see below).

In summary, it is far from clear under what circumstances photosyn-thetic down-regulation acts to diminish the short-term response to elevated CO_2 concentration or indeed whether in the long term it reduces the growth response to elevated CO_2 concentration or not.

Decline of relative growth rate in high CO_2: What does it imply?

It has been observed many times that the relative growth rate (RGR) of plants is increased under elevated CO_2 concentration, relative to low CO_2 controls, for only a short time after the start of growth in elevated CO_2 concentration. Thereafter RGR can be similar to, or even lower than, that of control CO_2 plants (e.g. Hurd, 1968, for tomato; Neales & Nicholls, 1978, for wheat; Mauney et al., 1978, for several crop species; Sionit et al., 1982, for soybean and radish; Wong, 1990, for cotton; Norby & O'Neill, 1991, for yellow poplar; Watson & Graves, 1993, for several temperate grasses; Baxter et al., 1994, for several montane grasses). A *decrease* in

Table 13.2 *Average relative growth rate (RGR), and total plant dry weights at the end of the growth period, for* Danthonia richardsonii *(Wallaby grass) between 41 and 98 days from sowing at four growth-limiting nitrogen supply rates (J. L. Lutze & R. M. Gifford, unpublished data) and* Gossypium hirsutum *(cotton) between 26 and 31 days from sowing at three growth-limiting phosphorus supply rates (D. J. Barrett & R. M. Gifford, unpublished data)*

N or P supply rate (mg plant^{-1} day^{-1})	RGR (g g^{-1} day^{-1})			Final dry weight (g)		
	360 ppm CO$_2$	720 ppm CO$_2$	Ratio (720/360)	360 ppm CO$_2$	720 ppm CO$_2$	Ratio (720/360)
D. richardsonii						
0.4 (N)	0.059	0.053	0.90	2.24	2.74	1.22
0.8 (N)	0.061	0.056	0.91	5.16	6.46	1.25
1.2 (N)	0.063	0.058	0.92	7.19	9.08	1.26
1.6 (N)	0.063	0.061	0.97	7.58	11.74	1.56
G. hirsutum						
0.02 (P)	0.051	0.026	0.51	0.76	0.87	1.15
0.06 (P)	0.088	0.059	0.67	1.22	1.31	1.07
0.19 (P)	0.089	0.071	0.80	1.59	1.82	1.15

In each case plants were grown with and without CO$_2$ enrichment as single spaced plants with all other nutrients supplied in non-growth limiting quantities. RGR increased with nitrogen ($P<0.01$) and phosphorus ($P<0.001$) supply but CO$_2$ enrichment decreased RGR at all N ($P<0.001$) and P ($P<0.05$) supplies. Final dry weight increased at elevated CO$_2$ at all rates of N ($P<0.001$) and P supply ($P<0.05$).

relative growth rate under elevated CO$_2$ concentration is illustrated for cotton and the Australian native grass, Wallaby grass, in Table 13.2. The data show that the plants had grown larger under high CO$_2$ concentration even though the RGR had declined to a smaller value than the controls. They also show that the effect was independent of N and P nutrition. It has long been thought that '*the absence of a sustained, long-term effect of CO$_2$ enrichment on [relative] growth rate in these experiments is somewhat paradoxical when the well-established short-term, stimulatory effects of CO$_2$ enrichment on individual leaf photosynthesis are considered*' (Neales & Nicholls, 1978).

The disappearance of an advantage in RGR under elevated CO$_2$ concentration has often been taken to imply that any continuing response of dry weight growth per plant to elevated CO$_2$ concentration derives only from the change in initial capital, due to a brief initial growth response to high

CO_2, and that the plant compensates to oppose the effect of high CO_2 on photosynthesis. If so, then we would expect community productivity or biomass to be unresponsive to CO_2.

Countering that view, it is recognized that RGR has poor resolution for analysing differences in plant growth (Norby & O'Neill, 1991). Owing to the power of compound interest it takes only a small, perhaps statistically non-significant, CO_2-induced difference in RGR to have a large effect on plant dry weight during sustained exponential growth (Gifford & Morison, 1993). So the small difference in RGR required to have a big effect on absolute growth may be hidden by experimental or biological variability. Nevertheless, during truly exponential growth, a loss of initial treatment difference in RGR does imply some sort of metabolic or partitioning adjustment in response to high CO_2 (Masle et al., 1993). It has been suggested that the proposed compensation might be partly to do with photosynthetic down-regulation (e.g. Wong, 1990), or in part because the plant acts in some other way to minimize the impact of improved photosynthetic performance of leaves, such as by reducing the partitioning of dry matter into leaf weight or area (Badger, 1992).

However, seedlings usually grow truly exponentially for only a few days. Normally, RGR declines with increasing plant size (Evans, 1993). So if a CO_2 treatment enhances growth it would be expected to cause more rapid decline in RGR when growth is departing from exponentiality. Thus the loss of an initial CO_2 effect on RGR does not necessarily imply a metabolic or partitioning feedback effect has occurred. It may arise solely from the mathematics of growth that is under transition from the exponential to the non-exponential for plants having a constant photosynthetic stimulation and unchanged partitioning (Goudriaan, 1994). One reason for departure from exponential growth is self-shading of leaves, be it for individual plants or plants in a community. The following simple heuristic model illustrates the effect of this on growth attributes including RGR.

In the numeric simulation model (see Table 13.3 for its parameters, constants, variables and equations) a plant is growing from a seedling in a population of 100 seedlings m^{-2} giving a ground area, $A = 100$ cm^2 per plant. The model runs with daily time-steps starting with initial values of dry weight (W$_i$), leaf area (L$_i$), specific leaf area (SLA$_i$) and leaf weight as a proportion of total plant weight (leaf weight ratio, LWR$_i$). A constant daily incident solar radiation integral (I$_0$) is partially intercepted by leaves and converted to the chemical energy of dry matter with a fixed efficiency. The influence of doubling CO_2 concentration on model plant growth is solely by increasing the efficiency of conversion of intercepted radiation into dry matter (radiation use efficiency, RUE) by a fixed multiplier, r. A 30%

Table 13.3 *Parameters and initial values (i.e. day 0) of the standard configuration of the heuristic plant growth model to illustrate the inevitable disappearance of the initial CO_2 effect on relative growth rate with time after the start of CO_2 enrichment*

Constants

Ground area per plant	A	cm^{-2}	100
Incident solar radiation	I_0	$MJ\ m^{-2}\ day^{-1}$	15
Radiation use efficiency	RUE_{con}	$g\ DW\ MJ^{-1}$	1.8
Extinction coefficient	K	cm	0.33
Specific leaf area	SLA	$cm^{-2}g^{-1}$	250
Instantaneous leaf weight ratio	ILWR	$g\ g^{-1}$	0.5
CO_2 enhancement ratio	$r = RUE_{enr}/RUE_{con}$	dimensionless	1.3

Initial set values (day 0)

Total plant dry weight (live + dead)	W_i	g	0.1
Leaf weight ratio	LWR_i	$g\ g^{-1}$	0.5

Calculated values

Fractional light interception	$F = (1-e^{-k.LAI})$	dimensionless
Live plant dry weight increment	$\Delta W=F*RUE*I_0*A$	g
Live dry weight	$LIVW_d=LIVW_{d-1}+\Delta W$	
Relative growth rate	$RGR_i=\Delta W/LIVW$	$g\ g^{-1}$
Net assimilation rate (control)	$NAR=\Delta W/LA$	$g\ cm^{-2}day^{-1}$
Live leaf weight increment	$\Delta LW=(ILWR*\Delta W)-LWL$	g
Live leaf area formation	$\Delta LAF=\Delta LW*SLA$	cm^{-2}
Incremental live leaf area loss	$\Delta LAL=(LAI/10)^2*\Delta LAF$	cm^{-2}
Incremental live leaf weight loss	$\Delta LWL=\Delta LAL/SLA$	g
Green leaf area index	$LAI=LA/A$	dimensionless
Green leaf area ratio	$LAR=LA*LIVW$	$cm^{-2}g^{-1}$
Green live leaf area	$LA_d=LA_{d-1}+\Delta LAF-\Delta LAL$	cm^2

Subscript meanings are: con, control CO_2 level; enr, enriched CO_2 level; i, initial values; d, the current day of the simulation; d−1, the previous day of the simulation.

enhancement has been assumed as a typical experimental and theoretical value for C_3 plants (Gifford, 1992). Importantly, there is no photosynthetic down-regulation included in this model. As modelled leaf area develops over the allotted 100 cm^2 of ground, leaf area index (LAI) is calculated daily. The proportion of incident solar radiation (I_0) intercepted by leaves (F) is calculated by the Beer–Lambert Law relation: namely $F = I/I_0 = (^1 - e^{-k \cdot LAI})$ where k is the extinction coefficient of the canopy. Throughout growth the allocation of daily growth to leaf (instantaneous leaf weight ratio, ILWR) is held constant, as is the specific leaf area. As the canopy grows, lower leaves die and become leaf litter. Daily leaf death as a proportion of new leaf formation increases as $(LAI/10)^2$, thereby being slow at first and accelerating to be equal to daily leaf formation when $LAI = 10$. At that time all further leaf dry matter formation is matched by leaf litter formation. This particular choice of maximum LAI has no effect on the conclusions. The conclusions of the model concerning RGR are not sensitive to the detailed assumption about how leaf death occurs in relation to LAI. Indeed the qualitative conclusion about RGR is the same if no leaf death is assumed and if the LAI is unconstrained.

Fig. 13.1 shows the output of this model. The results for live dry weight, RGR, net assimilation rate (NAR) and green leaf area ratio (LAR) are also shown in the form of enriched/control ratios in Fig. 13.2. The important point here is that the enrichment ratio for the CO_2 effect on RGR quickly drops from the initial 1.3 to below 1.0 before stabilizing at unity (i.e. no CO_2 effect on RGR). Thus the model illustrates how it is possible for high CO_2 concentration to actually reduce RGR while increasing absolute growth rate (Fig 13.2), depending when the plants are harvested. Live dry weight and live plus dead dry weight both rise to very high levels of enrichment ratio, peaking at a time when the RGR effect is lost, before stabilizing at an enrichment ratio of 1.3. Thus, even without any metabolic acclimation or partitioning adjustment to CO_2 enrichment, the initial differential in RGR quickly disappears based just on the mathematics of growth that is departing from initial exponentiality. Even for isolated plants, which are departing from the initial exponential growth (for whatever reason, be it mutual shading, nutrient limits, pot size limits), the logic holds that the initial RGR advantage must disappear even though absolute growth rate is faster at high CO_2 concentration. The model results also suggest why it is that there is such a wide range in the magnitude of the CO_2 enrichment effect on plant size in the literature: many experiments are terminated during the transient shown in Fig. 13.2 where the plant size advantage is well above the long term effect of +30% in that example.

In addition to that mathematical inevitability of a declining response of

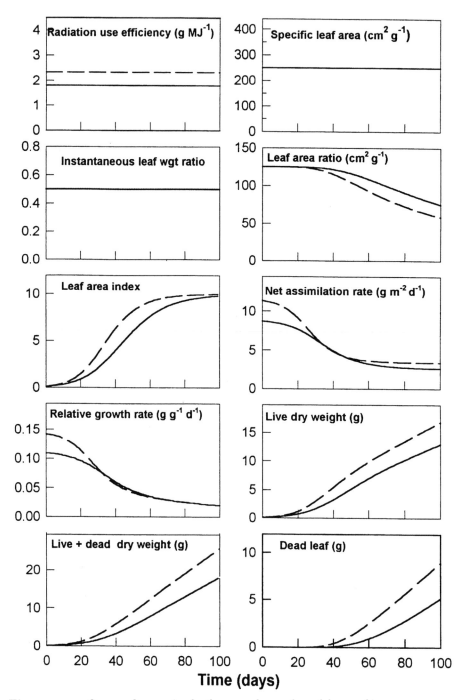

Figure 13.1 *Outputs from a simple plant stand growth model at ambient or 2× ambient CO₂ concentration, in which radiation use efficiency, specific leaf area, and incremental partitioning of dry matter increase into leaf (i.e. instantaneous leaf weight ratio), are constant inputs. The primary effect of elevated CO₂ concentration is to increase radiation use efficiency by 30%. Solid lines are for control and dashed lines are for CO₂-enriched conditions.*

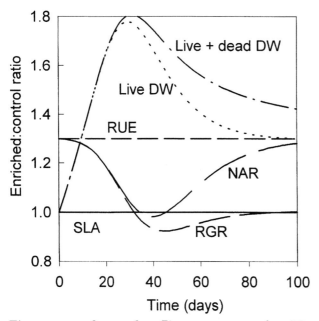

Figure 13.2 *Outputs from Figure 13.1 expressed as CO$_2$-enriched : control ratios through time.*

RGR to elevated CO$_2$, there may be, under some circumstances, accentuating biological effects such as photosynthetic down-regulation, or decrease in incremental partitioning of dry matter into leaf relative to root or stem under elevated CO$_2$ concentration.

Change in C-partitioning

The above model assumed that incremental partitioning of daily dry matter gain is not influenced by CO$_2$ concentration. The evidence on this is equivocal. Most reports deal with 'snapshot' measurements of the proportions of plants rather than incremental partitioning. From Fig. 13.1, even under the assumption of constant instantaneous partitioning of dry matter into leaf, the overall leaf area ratio (LAR) may decline under elevated CO$_2$ concentration (Fig. 13.1). That arose in the model from leaf death without concurrent stem or root death. Nevertheless instantaneous dry matter partitioning may change under elevated CO$_2$. Two aspects of partitioning are specific leaf area (SLA) and root : shoot ratio (R : S).

Specific leaf area has regularly been observed to decrease in C$_3$ plants at elevated CO$_2$ concentration. This effect may be partially connected with accumulation of higher levels of non-structural carbohydrates but also to increased thickness owing for example, to more palisade layers (Acock & Allen, 1985; Gunderson & Wullschleger, 1994). The effect occurs not only for plants grown with abundant nutrients but also for plants grown with strong N or P deficiencies (Fig. 13.3); it appears to be a fairly robust partitioning response of C_3 plants to growth in elevated CO_2 concentration.

If, in the above RGR model, the assumption about independence of SLA from CO_2 concentration is changed to the assumption depicted in Fig. 13.4 in which SLA declines by 30% over the first week as the first leaves emerge, then RGR quickly declines to be equal to the control value and stays equal from day 21 onwards. Nevertheless live and dead dry matter formation per unit ground area continue to be stimulated by the high CO_2 concentration. Thus, biomass accumulation in a community can be persistently enhanced by elevated CO_2 even though a RGR effect rapidly disappears; this effect is not simply a repercussion of an initial increase in capital.

The model was not set up to distinguish root versus shoot partitioning. However, if we supposed that all the incremental CO_2-stimulated growth were to be partitioned into new root, such that root : shoot (R : S) ratio continuously increased while R : S ratio of controls was constant, then the RGR of CO_2-enriched plants would be depressed relative to controls even though whole plant dry weight growth were increased by the CO_2. While early reviews suggested that an increase in R : S is a normal response of plants to CO_2 enrichment (Wittwer, 1978), later information suggests that in practice the R : S ratio is not very sensitive to CO_2 concentration for plants under non-stressed environmental conditions (Cure, 1985; Eamus &

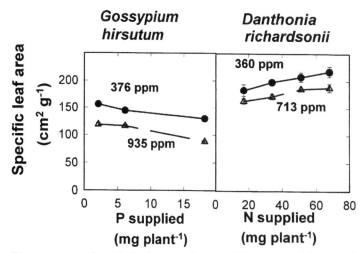

Figure 13.3 *Specific leaf areas of cotton (*Gossypium hirsutum*) and Wallaby grass (*Danthonia richardsonii*) grown with various growth restrictive levels of P or N supply at ambient and elevated CO_2 concentrations. (J. L. Lutze, D. J. Barrett & R. M. Gifford, unpublished data.)*

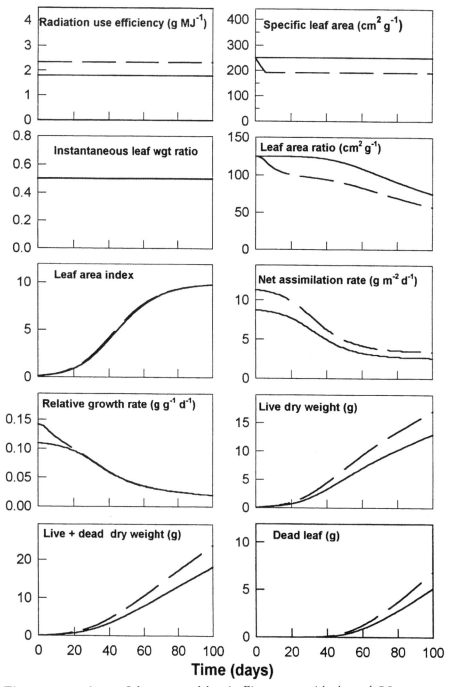

Figure 13.4 *A run of the same model as in Figure 13.1 with elevated CO₂ not only increasing radiation use efficiency by 30%, but also causing specific leaf area to decline by 30% over the first week.*

Jarvis, 1989). Evidence is conflicting as to whether nutrient stress, like low N supply, causes a positive response of R : S to CO_2 concentration (El Kohen *et al.*, 1992; Kuikman & Gorissen, 1993) or not (Hilbert *et al.*, 1991). However, the variability of the magnitude or even the direction of results for CO_2 effects on R : S may reflect a confounding of any direct CO_2 effects with those of other environmental variables which affect R : S ratio (Gifford, 1980), some of which may be affected by the plant's response to CO_2 concentration. For example, considering water and nutrient limitations, both of which tend to increase R : S, high CO_2 may decrease water stress, through stomatal closure, if there is little or no compensating leaf area increase (see below), while nutrient stress may increase via increased demand to meet the CO_2-enhanced growth. Another confounding effect is that root turnover and rhizo-deposition of exudates and other easily decomposed root material might be enhanced by CO_2 (Kuikman and Gorissen, 1993). Then R : S ratio may not be an adequate indication of the partitioning of C below-ground. For example Norby *et al.* (1992), reported that yellow poplar trees grown in elevated CO_2 concentration exhibited increased fine root production without any change in R : S ratio. Presumably there was increased fine root death and decomposition too.

Certainly there seems to be no consistent indication that elevated CO_2 concentration decreases the proportion of primary production going below-ground. If anything it increases it under some circumstances. Since net primary production is increased by elevated CO_2 concentration, we can expect therefore that the amount of C going below ground is also increased under elevated CO_2 concentration. This may have implications for nutrient acquisition in the longer term as discussed below.

Water use and soil water depletion

Water loss by transpiration is the 'operating cost' of carbon acquisition. Stomata continuously adjust to optimize the balance between water loss and C gain (Cowan, 1977) as a short term response to varying environmental variables including atmospheric CO_2 concentration (Wong *et al.*, 1979). At the single plant level in the short term, a CO_2-doubling typically reduces transpiration by about 30% via stomatal closure, while CO_2 uptake increases by a similar proportion (Morison, 1985). The expression of reduced evapo-transpiration at the community and regional scales can be attenuated by feedbacks involving increased leaf area index, increased leaf temperature, changed evaporation directly from the soil surface, and decreased leaf-to-air vapour pressure difference (Gifford, 1988; Dunin, 1991; McNaughton & Jarvis, 1991). Thus on balance, the impact on whole system evapotranspiration might be modest or, where there is complete leaf area compensation (Gifford, 1988), non-existent. But even a modest conservation of water could lead to substantial cumulative effects on seasonal average soil water content with biogeochemical and agronomic consequences (see below).

Species may differ with respect to the effect of elevated CO_2 on soil water status. Broadly speaking, for individual spaced plants growing in a regime of drying/re-wetting cycles, as may occur for arable crops before canopy closure, the increase of plant water use owing to the bigger leaf area under high CO_2 concentration tends to cancel out the water-conserving effect of stomatal closure (Morison & Gifford, 1984). However this is not always precisely true. Maize (C_4), having little photosynthetic effect of high CO_2, did reduce whole plant water use by about 30% because there was no leaf area increase but a reduction in stomatal conductance (Samarakoon & Gifford, 1995). By contrast, cotton plants, which have a strong leaf area response but weak stomatal response to high CO_2, actually used water more rapidly under high CO_2 concentration (Samarakoon and Gifford, 1995) thereby drying the soil faster (Fig. 13.5).

Once a full leaf canopy forms, the stomatal closure effect on transpiration cannot be offset by the enhanced leaf area effect on water use. Thus, in sufficiently well ventilated vegetation for which stomatal conductance influences evapotranspiration (McNaughton & Jarvis, 1991), high CO_2 may lead to less rapid soil water depletion and higher annual average soil water content. The same would be expected to apply in rangelands where low fertility may prevent greater leaf area index under elevated CO_2 (see below) and thus soil water may be higher. We have found this for controlled humidity greenhouse swards of native grass Wallaby grass *Danthonia richardsonii* under elevated CO_2 concentration with severe N restriction (J. L. Lutze & R. M. Gifford, unpublished data). Over each soil drying cycle the elevated CO_2 swards conserved water leading to higher average soil water contents (Fig. 13.6). This may, over time, tend to cause faster soil organic matter decomposition and hence net transfer of nutrients from soil organic matter to live plants and a secondary increase in productivity and LAI due indirectly to the CO_2 concentration increase via mineralization.

Changed C : N ratio, lignin content and decomposability of tissues and litter

Whether nitrogen is a limiting factor to seasonal growth or not, live plant tissues have usually been observed to show increased C : N ratio when grown at elevated CO_2 concentration. This seems to be true of stems and often roots as well as leaves. It also appears to be true whether one expresses the

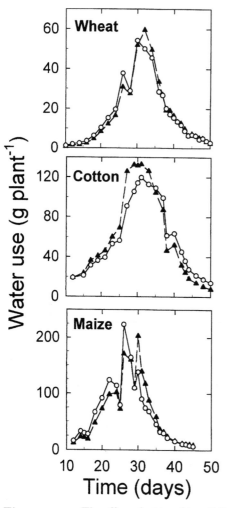

Figure 13.5   *The effect of 2 × ambient CO₂ concentration on whole plant water use of spaced plants growing in soil over a single drying cycle starting at field capacity and progressing to near death. The overall water use efficiencies were 5.6 ± 0.1 for wheat (control) and 9.2 ± 0.1 (enriched); 4.5 ± 0.1 for cotton (control) and 8.6 ± 0.2 (enriched); 4.5 ± 0.1 for maize (control) and 8.6 ± 0.1 (enriched). (From A. B. Samarakoon & R. M. Gifford, unpublished data.)*

C : N ratio on a total dry weight or on a structural dry weight basis (Conroy, 1992). Probably most agricultural and pastoral systems are N-limited in the sense that they show some growth response to inorganic N addition. For Australian native grass species grown with moderate, but not saturating, levels of N input we have found increases in C : N of stems (including sheaths) and roots as well as leaves (Table 13.4). This change in protein content of plant tissues has implications for forage quality for animals on pastures. It also has implications for pests and diseases (Teng *et al.*, this volume). C : N can increase in dead and live material at high

Table 13.4. *C : N of live leaf, root and stem material of native or naturalized Australian C_3 grasses after 73 days with and without CO_2 enrichment, expressed on a total carbon bases Nitrogen was supplied at a growth restricting rate of 1.6 mg plant^{-1} day^{-1}, with other nutrients supplied in non-growth limiting quantities (J. L. Lutze & R. M. Gifford, unpublished data)*

	C : N ratio			
	359 ppm CO_2	723 ppm CO_2	Ratio (723/359)	P
Danthonia richardsonii[a]				
Leaf	14·8	19·2	1·30	<0·001
Stem	23·4	30·1	1·29	<0·001
Root	20·7	23·5	1·14	<0·01
Microlaena stipoides[a]				
Leaf	12·1	18·6	1·54	<0.001
Stem	20·4	26·2	1·28	<0·01
Root	17·6	20·2	1·15	0·066
Vulpia bromoides[b]				
Leaf	20·2	30·1	1·49	<0·01
Stem	-	-	-	-
Root	43·4	64·5	1·49	<0·001
Vulpia ciliata[b]				
Leaf	16	24·3	1·52	<0·01
Stem	-	-	-	-
Root	29·5	42·8	1·45	<0·01

[a]Leaf is leaf lamina only, stem is true stem and leaf sheath.
[b]*Vulpia* sp. produced no true stem, leaf is leaf lamina plus leaf sheath.

CO_2. We have found that elevated CO_2 increased the C : N ratio of all tissues of swards of *Danthonia richardsonii*, including naturally senesced shoots, over several cycles of defoliation and regrowth (J. L. Lutze & R. M. Gifford, unpublished data). However, El Kohen *et al.* (1992) found for N-restricted sweet chestnut seedlings that despite the decreased N content of all live tissues after 1 year's growth under elevated CO_2 concentration, the dead leaf litter had a *higher* N content than from the control plant litter. This result, however, contrasted with another result for sweet chestnut litter (Couteaux *et al.* 1991) in which the N content decreased following growth in high CO_2 concentration.

The classical textbook notion that the higher the C : N ratio of organic matter the slower the decomposition has been shown to be an inadequate generalization (Fog, 1988). Whether or not an increase in litter C : N causes

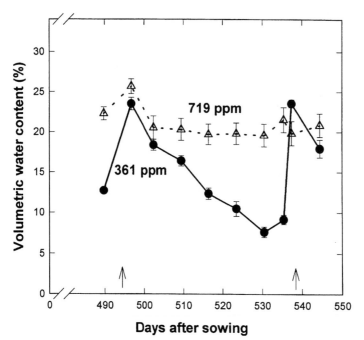

Figure 13.6   *The time-course of soil water content over a drying cycle for swards of Wallaby grass (*Danthonia richardsonii*) grown for 2 years with fixed allocations of soluble N of 6 g m⁻²year⁻¹ with and without continuous CO₂ enrichment to an average of 719 ppm CO₂. Arrows mark the times when the soil was returned to field capacity by supplying water to the point of drainage. In between the arrows water was supplied 2 times per week at a rate equivalent to 28% of an estimate of free water evaporation. (J. L. Lutze & R. M. Gifford, unpublished data.)*

faster or slower C mineralization or N mineralization may depend on concurrent changes in lignin, cellulose, non-structural carbohydrate contents, phenolics and probably other aspects of composition of the material, rather than just the C : N ratio (Melillo *et al.*, 1989). Thus the longer-term feedback through increased or decreased N mineralization on plant productivity will vary according to the composition of CO_2-induced litter formed. The feedback will be further modified by the effects on decomposition of the changed soil water content (see above), and also the effects of concurrent climate change involving warmer and wetter (or in some places, drier) conditions. Similarly with tissue quality for herbivory, features other than C : N ratio are important.

Data on responses of carbohydrate, cellulose and lignin contents to elevated CO_2 are, unfortunately, sparse. A study with the Australian native grass *Danthonia richardsonii* grown at different temperatures showed that elevated CO_2 concentration decreased the lignin content of both shoot and

Table 13.5 *The lignin content (% of dry weight) of* Danthonia richardsonii *plants grown in various (day/night) temperature regimes in the Canberra phytotron (R. M. Gifford, W. Doherty and J. L. Lutze, unpublished data).*

	Temperature (°C)	Ambient CO_2	High CO_2
Shoots	13/8	13.6	10.9
	18/13	12.7	11.9
	23/18	18.7	17.2
	28/23	22.7	22.1
	33/28	17.8	17.5
Roots	13/8	22.4	20.8
	18/13	26.4	21.2
	23/18	27.3	19.9
	28/23	23.3	23.5
	33/28	25.4	26.7

The CO_2 effects on decreasing the lignin content of both shoots and roots are significant ($P<0.001$); the effect diminishes, however, with increasing temperature for both shoot ($P<0.05$) and root ($P<0.001$).

root tissues in the lower part of the temperature range examined (Table 13.5). Similarly soybean plants grown under elevated CO_2 concentration had lower percentage lignin content in all tissues (leaves, enriched 6.0%, ambient 6.6%; stems, enriched 9.8%, ambient 11.8%; roots, enriched 11.2%, ambient 13.1%), (R. M. Gifford, W. Doherty and J. Lutze, unpublished data).

Increased mineral nutrient acquisition from the environment

For intensively managed agro-ecosystems it is likely that if the rising atmospheric carbon dioxide concentration causes increases in productive potential of crops then inputs like fertilizer will be used to match the increased nutrient requirement. For low output extensive agro-ecosystems like rangelands and rough grazing, however, the rate at which nutrients, especially N and P, are solubilized or mineralized from soil organic matter places limits on how much productivity and biomass can respond to the potential increase under elevated CO_2 concentration in any one season. Nevertheless it is probable that, over time, the pace or size of the entire nutrient cycle can change in response to a changed configuration of environmental inputs. It has been hypothesized (Gifford, 1992) that, over the time-scale appropriate to the rate of global atmospheric change, ecosystems may acquire N to meet the requirements of net primary production as paced by the primary drivers of productivity, namely solar radiation, water supply,

CO_2 concentration, and temperature. That hypothesis is reminiscent of, but apparently contradictory to, another hypothesis that in natural terrestrial ecosystems: 'given sufficient time, the supply of [biologically available] nitrogen comes into balance with the supply of phosphorus' (Cole & Heil, 1981). Thus plant communities on soils derived from parent material of high phosphorus content may gradually accumulate high organic P, C and N contents. At shorter time-scales the seasonal demand for P is largely met by mineralization of soluble organic P. Turnover of the organic pool is principally by microbial action but solubilization of inorganic P by root exudates may also be an important contributor to plant uptake. Thus, under the Cole and Heil hypothesis, both accumulation of organic carbon in the longer term and associated fixation of N are dependent on P availability, itself dependent on processes of mineralization of organic matter, solubilization of insoluble P salts, and weathering of rock fragments. Competition for soluble P by various biological and chemical sinks including root uptake, microbial growth, and precipitation as insoluble inorganic phosphates determine P availability to plants (Attiwill & Adams, 1993).

Resolving the apparent conflict between these hypotheses probably again involves consideration of time-scale: while N availability is often seen as the primary nutrient limitation to productivity in many ecosystems, this might in fact be a short-term representation of a long-term phosphate limitation in disguise (Vitousek & Howarth, 1991). Yet this idea is apparently incompatible with the observation that spatial variation in primary production of terrestrial ecosystems, and hence their correlated standing biomass (Gifford *et al.*, in press), is determined primarily by spatial variation in the aerial inputs of radiation and rainfall (Seino & Uchijima, 1992); incompatible, that is, unless the latter are, directly or indirectly, determining the amount of organic P circulating in an ecosystem. Is it possible that both N and P limitations are expressions of a long-term limitation of available energy (carbohydrate) used by various competing ecosystem processes including nutrient acquisition? As Attiwill and Adams (1993) have observed, Australian *Eucalyptus* forests growing on soils that are notoriously impoverished in available P have productivities that equal or exceed those of Northern hemisphere forests on richer soils but which have a similar climate. Furthermore, their regrowth following clear-felling was unresponsive to added nutrients such as N and P (Attiwill, 1994).

In attempting to resolve the apparent conflict between these hypotheses we suggest that three things need to be borne in mind. First, in complex systems like ecosystems, the concept of single limiting factors rarely applies (Gifford, 1992), rather multiple co-limiting resources interact in complex ways often involving feedbacks (Bloom *et al.*, 1985). Second, the time-scale of concern must be taken into account as it is possible that what seems to be a master limiter on a short time-scale may be a slave to another limiter on a longer time-scale. And thirdly, correlative evidence must be used with considerable caution in a situation of multiple interactive resource limitations with different characteristic time-scales: it is difficult to deconvolve chicken-and-egg problems.

For nitrogen, a mechanism is obvious whereby the element might gradually move into an ecosystem that has its productivity potential increased by elevated CO_2 concentration: biological fixation of the massive atmospheric dinitrogen pool could increase owing to the greater availability of chemical energy as carbohydrates. This cannot happen quickly to a substantial degree, but evidence and arguments have been advanced (Gifford, 1992, 1994; Gifford *et al.*, in press) that for the time-scale of decades over which atmospheric CO_2 concentration is increasing the N cycle may be able to track the C cycle. In some parts of the world such as Australia, phosphate can be an important limiter of N fixation and productivity. While there is no doubt that the phosphate status of the soil's parent material plays a substantial role in determining the distribution (Beadle, 1954) and productivity (Cole & Heil, 1981) of ecosystems, the looseness of the relationship (see below) means that other factors are involved. Dominant among these other factors is rainfall. Available evidence is poor, but an attempt to explore this for diverse ecosystems is illustrated in Figs. 13.7 and 13.8.

For a diverse range of Australian plant communities, from low productivity desert shrub vegetation to tall eucalypt forests of southeastern Australia, above-ground biomass, a rough indicator of productivity (Gifford *et al.*, 1995), correlates with annual precipitation, an approximate indicator of water availability ($r = 0.78$, $P<0.05$; Fig. 13.7). Since there is very little likelihood that annual rainfall correlates positively with the P content of the parent material, one may conclude that, on mineral soils, rainfall rather than the P content of the parent material drives productivity variation across this wide range of Australian ecosystems. Yet rainfall also correlates with its total P content ($r^2 = 0.34$, $P<0.05$; Fig. 13.7) but not with its total N content (Fig. 13.7) over this range of ecosystems. Thus it appears that in diverse Australian native plant communities, including rangelands, long-term ecosystem accumulation of P is related to the above-ground biomass, and consequently to ecosystem productivity, which in turn is dependent on water availability.

Similar relationships apply, over a narrower productivity range, to arid and semi-arid rangelands in the Sahel (Fig. 13.8). As for Australia, Sahelian soils are also P-deficient, yet the seasonal biomass production along a north–south transect is related to infiltrated rainfall ($r^2 = 0.69$, $P<0.05$;

Bremen & Krul, 1991). The P content of this above-ground biomass also correlates with infiltrated rainfall (r^2 = 0.45, $P<0.05$; Fig. 13.8), though less strongly than does biomass production. The N content of the biomass correlates even more weakly with the infiltrated rainfall (r^2 = 0.40, $P<0.05$; Fig. 13.8). Thus again standing biomass is correlated with rainfall and we may expect that phosphate and nitrogen accumulation over time matches rainfall-paced demand, be it with variability. The data of Fig. 13.7 and 13.8 confirm that the mineral content of vegetation is highly variable.

Presumably when the P content of the parent material is low the total amount of P potentially available to weathering is still very large compared with other P pools in the soil. Consequently it will take longer, all else equal, for an ecosystem on low P content rocks to accumulate P into its organic cycle to meet the productive potential determined by rainfall and radiation. Thus soon after land-clearing, where biomass is taken off-site, P might appear to be, in some ecosystems, the master manipulable limiting resource (i.e. regrowth would respond most strongly to its artificial addition), N the second most limiting, followed by water input. But such an observation would not invalidate the notion that the first order role of rainfall is as the primary driver of productive potential with P and N gradually being mobilized into the organic cycle over decades, centuries and, on very P-poor rocks, even millennia, until that potential is met.

As discussed above, water use and CO_2 uptake are two sides of the same coin in that transpiration is the operating cost that plants pay for CO_2 uptake by diffusion. That is why water use efficiency is increased in the long term by elevated CO_2 concentration at all levels of water limitation (Gifford, 1979; Samarakoon & Gifford, 1995). Consequently wherever productivity increases with increased water supply on some time-scale, it should also increase with increased CO_2 concentration on the same time-scale. (The converse is not necessarily true since well-watered C_3 crops can respond to CO_2 through its photosynthetic effect alone.) Thus where productivity and biomass of phosphate-deficient crops, pasture and native communities show rainfall-dependence on some time-scale they should also show CO_2-dependence on that time-scale. We therefore suggest that the correlations in Figs. 13.7 and 13.8 indicate that productivity of natural ecosystems should increase with gradual increase of atmospheric CO_2 concentration.

Consolidating this hypothesis we have identified five possible mechanisms whereby increased productivity could occur at elevated CO_2 when P is scarce. The first relates to the direct effect of increased CO_2 concentration on efficiency of within-plant nutrient utilization whereas the other four are concerned with processes leading to increased P acquisition.

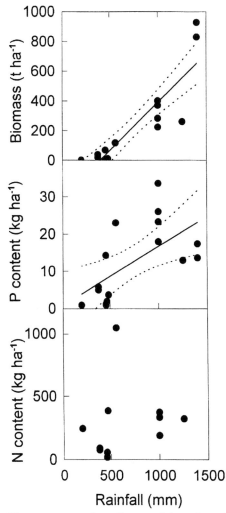

Figure 13.7 *Compilation of data from the literature on above-ground biomass, organic N and organic P for a range of Australian ecosystems, namely: tall eucalypt forest (Ashton, 1976; Attiwill, 1980; Baker & Attiwill, 1985); eucalypt forest on sandy soils (Westman & Rogers, 1977a,b; Hingston et al., 1980/81); arid woodland (Curtis, 1975; Burrows, 1976); arid shrubland (Charley & Cowling, 1968; Burrows, 1972); heath shrubland (Specht et al., 1958; Specht, 1966); Brigalow (N-fixing* Acacia) *woodland (Moore et al., 1967). The lines are the linear regressions with 95% confidence limits. Regression correlation coefficients were 0.78 and 0.34 for biomass and phosphorus models which were significant at* $P<0.05$ *in both cases. No linear relationship was found for nitrogen.*

Elevated CO_2 can increase the utilization efficiency of mineral nutrients including phosphate such that greater biomass increase is achieved for a given uptake of nutrient or tissue nutrient concentration. For example, in cotton plants grown under extremely low P supply (Barrett & Gifford,

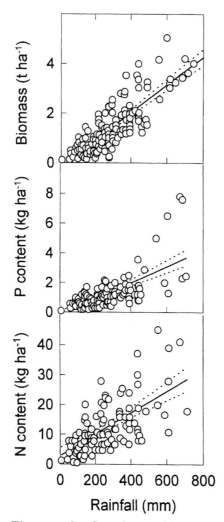

Figure 13.8 *Compilation of data on the end of season above-ground biomass for rangeland communities along a N–S transect across the Sahel, N. Africa (1976–9). Data were extracted from Bremen & Krul (1991). The lines are the linear regressions with 95% confidence limits. Regression coefficients were 0.69, 0.45 and 0.40 for biomass, P and N models, respectively. All regression models were significant at P<0.05.*

1995*a*), increased P utilization efficiency (biomass produced per unit tissue concentration) was observed at elevated CO_2. While this effect is relatively small in terms of plant growth, it may provide the energetic basis for the plant acquiring more phosphate from the environment by one or more of the following mechanisms.

As for N acquisition, we see the increased supply of available energy in the vegetation as being the key to increased ecosystem P acquisition. Increased P acquisition might be achieved in four ways: (1) by an increase in surface area to volume ratio through enhanced branching of lateral roots which may contribute to greater capture of soluble P in competition with microbial populations; (2) by enhanced biological mineralization of soluble organic P and increased solubilization of insoluble organic P in the rhizosphere; (3) by increased solubilization of inorganic P from Al and Fe precipitates and desorption of P from Al and Fe hydroxides or clay particles via organic acid exudation; and (4) by greater rates of weathering of parent material.

Soluble phosphates are taken up by roots and associated mycorrhizae. Since phosphate ions are relatively immobile in the soil, root and hyphal proliferations through the soil medium is important for plants to compete effectively with microorganisms for the minute fraction of available P. Thus, as elevated CO_2 concentration enhances root absorptive surface area by elongation of lateral branches (Rogers *et al.*, 1992) and stimulated mycorrhizal growth (O'Neill *et al.*, 1987), it may increase P uptake (Barrett & Gifford, 1995a).

Roots and mycorrhizae actively secrete organic acids (especially during phosphate stress (Tadano *et al.*, 1993)), which solubilize insoluble aluminium and iron phosphates (Marschner & Dell, 1994). It is also possible, but remains to be evaluated, that conditions which foster high potential productivity, such as high rainfall or CO_2 levels, will foster high specific rates of organic acid and phosphatase production by roots as these may be secreted in relation to the plants' needs for P, though moderated by competition with other energy-requiring processes.

Where nutrient supply tends to restrict plant growth, elevated CO_2 concentration does not necessarily increase leaf area or may even decrease it (Barrett & Gifford, 1995a; Lutze & Gifford, 1995) as illustrated in Fig. 13.9. Therefore the reduction in stomatal aperture at elevated CO_2 concentration, uncompensated by leaf area increase (see above), tends to conserve water, potentially producing higher average soil water contents as illustrated for N-restricted native grass growth (Fig. 13.6). Higher soil water contents foster faster decomposition and hence faster mineralization of organic P as well as organic N. This process may redistribute P from dead organic pools (having relatively low C : P ratios) to live tissues (having relatively high C : P ratios) thereby increasing carbon productivity and increasing the rate of C and mineral cycling. It is possible that the increased rate of turnover of trees in tropical rainforests over the last 2–3 decades (Phillips & Gentry, 1994) implies an increase in the rate of C and mineral cycling in response to past increases in atmospheric CO_2 concentration.

Over decades both soil moisture and the activity of plant roots are critical to rock weathering which is the main process by which new P is

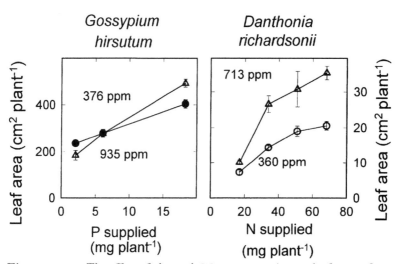

Figure 13.9   *The effect of elevated CO_2 concentration on leaf area of cotton plants grown for 77 days with growth restrictive allocations of phosphate, and wallaby grass grown for 41 days with growth restrictive supplies of nitrogen. (adapted from Barrett and Gifford, 1995a, and J. L. Lutze and R. M. Gifford, unpublished).*

brought into the biological P cycle. Roots increase weathering physically by penetrating cracks thereby increasing the weathering surface area and chemically by increasing the carbonic and organic acid concentration in the soil solution, by increasing the respiratory CO_2 output of both roots and micro-organisms and by exuding organic acids. It is believed that vegetation has increased weathering rates as much as 10- to 1000-fold above that which would occur without the Earth's vegetation (Lovelock & Whitfield, 1982). Hence over decades and longer time-scales weathering may be correlated with atmospheric CO_2 concentration (Gifford, 1992). Although annual P input through weathering is minute compared with P turnover in ecosystems, typically amounting to much less than 1 kg ha^{-1} year^{-1}, over the time-scale of decades to centuries appropriate to global change considerations, it would add up to a significant contribution.

Conclusions

The thrust of the hypothesis being put forward here is that with carbohydrates being the energy currency of ecosystem metabolism, the carbon input to ecosystems sets the limits for many other aspects of ecosystem function including resource capture. The abundance of available forms of each resource certainly has a major influence on the productivity and carbon stocks of ecosystems, but there is always more to be gained by an ecosystem providing there is more energy to gain it given enough time. By increasing the efficiency of solar energy conversion and of water use, elevated CO_2 concentration increases the energy content of an ecosystem and over time this energy supply may foster the capture of more mineral resources, notably N and P. For intensive agricultural ecosystems man is likely to add fertilizer to maximize the expression of the CO_2 fertilizing effect. For extensive agro-ecosystems where fertilizer is not an option, it seems that the increased productive potential, which the ever increasing atmospheric CO_2 concentration is fostering, will gradually mobilize unavailable forms of N and P from soil, air and rock, albeit with substantial lags. A question to be resolved concerns how big those time lags are.

This chapter highlights the concept that it is meaningless to attempt to specify dominant limiting factors to primary production without at the same time indicating the time-frame to which that specification applies. In the context of global change the relevant time-frame is decades to centuries. That time-frame is similar to the time for significant change in the terrestrial ecological carbon cycle.

Acknowledgements

This work is a contribution to the CSIRO Climate Change Programme and is partially supported by the Australian National Greenhouse Research Grant Scheme and by the Electric Power Research Institute, Palo Alto. It contributes to the GCTE Core Project of IGBP. We are grateful to Dr S. C. van de Geijn for drawing our attention to Bremen & Krul (1991) and to Wendy Doherty who conducted the lignin determinations (Table 13.5).

References

Acock, B. & Allen, L. H. Jr. (1985). Crop responses to elevated carbon dioxide concentrations. In *Direct Effects of Increasing Carbon Dioxide on Vegetation*, ed. B. R. Strain & J. D. Cure, pp. 53–98. Washington, DC: US Department of Energy, DOE/ER-0238.

Arp, W. J. (1991). Effects of source–sink relations on photosynthetic acclimation to elevated CO_2. *Plant, Cell and Environment*, 14, 869–75.

Ashton, D. H. (1976). Phosphorus in forest ecosystems at Beenak, Victoria. *Journal of Ecology*, 64, 171–86.

Attiwill, P. M. (1980). Nutrient cycling in a *Eucalyptus obliqua* (L'Herit.) forest. IV. Nutrient uptake and nutrient return. *Australian Journal of Botany*, 28, 199–222.

Attiwill, P. M. (1994). Ecological disturbance and the conservative management of eucalypt forests in Australia. *Forest Ecology and Management*, 63, 301–46.

Attiwill, P. M. & Adams, M. A. (1993). Tansley Review No. 30. Nutrient cycling in forests. *New Phytologist*, 124, 561–82.

Badger, M. (1992). Manipulating agricultural plants for a future high CO_2 environment. *Australian Journal of Botany*, **40**, 421–9.

Baker, T. G. & Attiwill, P. M. (1985). Above-ground nutrient distribution and cycling in *Pinus radiata* D. Don and *Eucalyptus obliqua* L'Herit. forests in southeastern Australia. *Forest Ecology and Management*, **13**, 41–52.

Barrett, D. J. & Gifford, R. M. (1995a). Acclimation of photosynthesis and growth by cotton to elevated CO_2: interactions with severe phosphate deficiency and restricted rooting volume. *Australian Journal of Plant Physiology*, **22** 955–63.

Barrett, D. J. & Gifford, R. M. (1995b). Photosynthetic acclimation to elevated CO_2 in relation to biomass allocation in cotton. *Journal of Biogeography*, **22** 331–9.

Baxter, R., Ashenden, T. W., Sparks, T. H. & Farrar, J. F. (1994). Effects of elevated carbon dioxide on three montane grass species. I. Growth and dry matter partitioning. *Journal of Experimental Botany*, **45**, 305–15.

Beadle, N. C. W. (1954). Soil phosphate and the delimitation of plant communities in Eastern Australia. *Ecology*, **35**, 370–5.

Bloom, A. J., Chapin, F. S. III & Mooney, H. A. (1985). Resource limitation in plants – an economic analogy. *Annual Review of Ecology and Systematics*, **16**, 363–92.

Bremen, H. & Krul, J. M. (1991). La pluviosite et la production de fourrage sur les paturages naturels. In *La productivite des paturages saheliens: Une etude des sols, des vegetations et de l'exploitation de cette ressource naturelle*, ed. F. W. T. Penning de Vries & M. A. Djiteye, pp. 304–22. Agricultural Research Report 918. Wageningen: Pudoc.

Bunce, J. A. (1992). Light, temperature and nutrients as factors in photosynthetic adjustment to an elevated concentration of carbon dioxide. *Physiologia Plantarum*, **86**, 173–9.

Burrows, W. H. (1972). Productivity of an arid zone shrub (*Eremophila gilesii*) community in south-western Queensland. *Australian Journal of Botany*, **20**, 317–29.

Burrows, W. H. (1976). Aspects of nutrient cycling in semi-arid mallee and mulga communities. PhD Thesis, Australian National University.

Cave, G., Tolley, L. C. & Strain, B. R. (1981). Effect of carbon dioxide enrichment on chlorophyll content, starch content and starch grain structure in *Trifolium subterraneum* leaves. *Plant Physiology*, **52**, 171–4.

Charley, J. L. & Cowling, S. W. (1968). Changes in soil nutrient status resulting from overgrazing and their consequences in plant communities of semi-arid areas. *Proceedings of the Ecology Society of Australia*, **3**, 28–38.

Cole, C. V. & Heil, R. D. (1981). Phosphorus effects on terrestrial nitrogen cycling. In *Terrestrial Nitrogen Cycles* (SCOPE 33), ed. F. E. Clark & T. Rosswall, Ecological Bulletin (Stockholm), **33**, 363–74.

Coleman, J. S., McConnaughy, K. M. D. & Bazzaz, F. A. (1993). Elevated CO_2 and plant nitrogen use: is reduced tissue nitrogen concentration size dependent? *Oecologia*, **93**, 195–200.

Conroy, J. (1992). Influence of elevated atmospheric CO_2 concentrations on plant nutrition. *Australian Journal of Botany*, **40**, 445–56.

Couteaux, M. M., Mousseau, M., Celerier, M. L. & Bottner, P. (1991). Atmospheric CO_2 increase and litter quality: decomposition of sweet chestnut leaf litter with animal foodwebs of different complexities. *Oikos*, **61**, 54–64.

Cowan, I. R. (1977). Stomatal behaviour and environment. *Advances in Botanical Research*, **4**, 117–227.

Cure, J. D. (1985). Carbon dioxide doubling responses: a crop survey. In *Direct Effects of Increasing Carbon Dioxide on Vegetation*, ed. B. R. Strain & J. D. Cure, pp. 99–116. Washington, DC: US

Department of Energy, DOE/ER-0238.

Curtis, Y. (1975). Soil-vegetation relationships in the cypress pine forests of the Pilliga region. Honours Thesis, University of New England.

Dunin, F. X. (1991). Extrapolation of 'point' measurements of evaporation: some issues of scale. *Vegetatio*, **91**, 39–47.

Eamus, D. & Jarvis, P. G. (1989). The direct effects of increase in the global atmospheric CO_2 concentration on natural and commercial temperate trees and forests. *Advances in Ecological Research*, **19**, 1–54.

Ehleringer, J. R. & Field, C. B. (eds.) (1993). *Scaling Physiological Processes: Leaf to Globe*. San Diego: Academic Press.

El Kohen, A., Rouhier, H. & Mousseau, M. (1992). Changes in dry weight and nitrogen partitioning induced by elevated CO_2 depend on soil nutrient availability in sweet chestnut (*Castanea sativa* Mill). *Annales des Sciences Forestières*, **49**, 83–90.

Evans, L. T. (1993). *Crop Evolution, Adaptation, and Yield*. Cambridge: Cambridge University Press.

Fog, K. (1988). The effect of added nitrogen on the rate of decomposition of organic matter. *Biological Reviews*, **63**, 433–62.

Gifford, R. M. (1979). Growth and yield of CO_2-enriched wheat under water limitations. *Australian Journal of Plant Physiology*, **6**, 367–78.

Gifford, R. M. (1980). Carbon storage by the biosphere. In *Carbon Dioxide and Climate: Australian Research*, ed. G. I. Pearman, pp. 167–81. Canberra: Australian Academy of Science.

Gifford, R. M. (1988). Direct effects of higher carbon dioxide concentrations on vegetation. In *Greenhouse: Planning for Climate Change*, ed. G. I. Pearman, pp. 506–19. Leiden: E. J. Brill.

Gifford, R. M. (1992). Interaction of carbon dioxide with growth-limiting environmental factors in vegetation productivity: Implications for the global carbon cycle. *Advances in Bioclimatology*, **1**, 24–58.

Gifford, R. M. (1994). The global carbon cycle: A viewpoint on the missing sink. *Australian Journal of Plant Physiology*, **21**, 1–15.

Gifford, R. M., Barrett, D. J., Lutze, J. L. & Samarakoon, A. B. (in press). The CO_2 fertilising effect: Relevance to the global carbon cycle. In *The Global Carbon Cycle*, ed. T. Wigley. Cambridge: Cambridge University Press.

Gifford, R. M. & Morison, J. I. L. (1993). Crop responses to the global increase in atmospheric carbon dioxide concentration. In *International Crop Science I*, ed. D. R. Buxton, R. Shibles, R. A. Forsberg, B. L. Blad, K. H. Assay, G. M. Paulsen & R. F. Wilson, pp. 325–31. Madison, Wisconsin: Crop Science Society of America.

Goudriaan, J. (1994). Using the expolinear growth equation to analyze resource capture. In *Resource Capture*, ed. J. L. Monteith, R. K. Scott & M. H. Unsworth, pp. 99–110. New York: Springer.

Gunderson, C. A. & Wullschleger, S. D. (1994). Photosynthetic acclimation in trees to rising atmospheric CO_2: A broader perspective. *Photosynthesis Research*, **39**, 369–88.

Hilbert, D. W., Larigauderie, A. & Reynolds, J. F. (1991). The influence of carbon dioxide and daily photon-flux density on optimal leaf nitrogen and root : shoot ratio. *Annals of Botany*, **68**, 365–76.

Hingston, F. J., Dimmock, G. M. & Turton, A. G. (1980/1). Nutrient distribution in a Jarrah (*Eucalyptus marginata* Donn ex sm.) ecosystem in southwestern Australia. *Forest, Ecology and Management*, **3**, 183–207.

Hurd, R. G. (1968). Effects of CO_2-enrichment on the growth of young tomato plants in low light. *Annals of Botany*, **32**, 531–42.

Kuikman, P. J. & Gorissen, A. (1993). Carbon fluxes and organic matter transformations in plant-soil-systems. In *Climate Change; Crops and Terrestrial Ecosystems*, ed. S. C. van de Geijn, J. Goudriaan & F. Berendse, pp. 97–107. Wageningen: Agrobiologische Thema's 9, CABO-DLO.

Lovelock, J. E. & Whitfield, M. (1982). Life span of the biosphere. *Nature*, **296**, 561–3.

Lutze, J. L., Gifford, R. M. (1995). Carbon storage and productivity of a carbon dioxide enriched nitrogen limited grass sward after one year's growth. *Journal of Biogeography*, **22** (227–33).

Marschner, H. & Dell, B. (1994). Nutrient uptake in mycorrhizal symbiosis. *Plant and Soil*, **159**, 89–102.

Masle, J., Hudson, G. S. & Badger, M. R. (1993). Effects of ambient CO_2 concentration on growth and nitrogen use in tobacco (*Nicotiana tabacum*) plants transformed with an antisense gene to the small subunit of ribulose-1,5-bisphosphate carboxylase/oxygenase. *Plant Physiology*, **103**, 1075–88.

Mauney, J. R., Fry, K. E. & Guinn, G. (1978). Relationship of photosynthetic rate to growth and fruiting of cotton, soybean, sorghum, and sunflower. *Crop Science*, **18**, 259–63.

McNaughton, K. G. & Jarvis, P. G. (1991). Effects of scale on stomatal control of transpiration. *Agricultural and Forest Meteorology*, **54**, 279–301.

Melillo, J. M., Aber, J. D., Linkins, A. E., Ricca, A., Fry, B. & Nadelhoffer, K. J. (1989). Carbon and nitrogen dynamics along the decay continuum: Plant litter to soil organic matter. *Plant and Soil*, **115**, 189–98.

Moore, A. W., Russell, J. S. & Coaldrake, J. E. (1967). Dry matter and nutrient content of a subtropical semiarid forest of *Acacia harpophylla* F. Meull. (Brigalow). *Australian Journal of Botany*, **15**, 11–24.

Morison, J. I. L. (1985). Sensitivity of stomata and water use efficiency to high CO_2. *Plant, Cell and Environment*, **8**, 467–74.

Morison, J. I. L. & Gifford, R. M. (1984). Plant growth and water use with limited water supply in high CO_2 concentration. I. Leaf area, water use and transpiration. *Australian Journal of Plant Physiology*, **11**, 361–74.

Neales, T. F. & Nicholls, A. O. (1978). Growth response of young wheat plants to a range of ambient CO_2 levels. *Australian Journal of Plant Physiology*, **5**, 45–59.

Norby, R. J., Gundeson, C. A., Wullschleger, S. D., O'Neill, E. G. & McCracken, M. K. (1992). Productivity and compensatory responses of yellow-poplar trees in elevated CO_2. *Nature*, **357**, 322–4.

Norby, R. J. & O'Neill, E. G. (1991). Leaf area compensation and nutrient interactions in CO_2-enriched seedlings of yellow poplar (*Liriodendron tulipifera* L.). *New Phytologist*, **117**, 515–28.

O'Neill, E. G., Luxmoore, R. J. & Norby, R. J. (1987). Increases in mycorrhizal colonisation and seedling growth in *Pinus echinata* and *Quercus alba* in an enriched CO_2 atmosphere. *Canadian Journal of Forest Research*, **17**, 878–83.

Phillips, O. L. & Gentry, A. H. (1994). Increasing turnover through time in tropical forests. *Science*, **263**, 954–8.

Rogers, H. H., Peterson, C. M., McCrimmon, J. N. & Cure, J. D. (1992). Response of plant roots to elevated atmospheric carbon dioxide. *Plant, Cell and Environment*, **15**, 749–52.

Rowland-Bamford, A. J., Baker, J., Allen, L. H. Jr. & Bowes, G. (1991). Acclimation of rice to changing atmospheric carbon dioxide concentration. *Plant, Cell and Environment*, **14**, 577–83.

Sage, R. F., Sharkey, T. D. & Seeman, J. R. (1989). Acclimation of photosynthesis to elevated CO_2 in five C_3 species. *Plant Physiology*, **89**, 590–6.

Samarakoon, A. B. & Gifford, R. M. (1995). Soil water content under plants at high CO_2 concentration and interactions with the direct CO_2 effects: a species comparison. *Journal of Biogeography*, **22** 193–202.

Seino, H. & Uchijima, Z. (1992). Global distribution of net primary productivity of terrestrial vegetation. *Journal of Agricultural Meteorology*, **48**, 39–48.

Sionit, N., Hellmers, H. & Strain, B. R. (1982). Interaction of atmospheric CO_2 enrichment and irradiance on plant growth. *Agronomy Journal*, **74**, 721–5.

Specht, R. L. (1966). The growth and distribution of Mallee–Broombush (*Eucalyptus incrassata–Melaleuca uncinata* association) and heath vegetation near Dark Island Soak, Ninety-Mile Plain, South Australia. *Australian Journal of Botany*, **14**, 361–71.

Specht, R. L., Rayson, P. & Jackman, M. E. (1958). Dark Island heath (Ninety-Mile Plain, South Australia). *Australian Journal of Botany*, **6**, 59–88.

Stitt, M. (1993). Enhanced CO_2, photosynthesis and growth: What should we measure to gain a better understanding of the plant's response? In *Design and Execution of Experiments on CO_2 Enrichment*, ed. E.-D. Schulze & H. A. Mooney, pp. 3–28. Ecosystems Research Report No. 6. Brussels: European Community Directorate-General for Science, Research and Development.

Tadano, T., Ozawa, K., Sakai, H., Osaki, M. & Matsui, H. (1993). Secretion of acid phosphatase by the roots of crop plants under phosphorus deficient conditions and some properties of the enzyme secreted by lupin roots. *Plant and Soil*, **155/156**, 95–8.

Thomas, R. B. & Strain, B. R. (1991). Root restriction as a factor in photosynthetic acclimation of cotton seedlings grown in elevated carbon dioxide. *Plant Physiology*, **96**, 627–34.

Vitousek, P. M. & Howarth, R. W. (1991).

Nitrogen limitation on land and at sea: How can it occur? *Biogeochemistry*, **5**, 7–34.

Watson, J. & Graves, J. D. (1993). Effect of elevated carbon dioxide on the performance of nine coexisting grassland species. In *Proceedings of the XVII International Grassland Congress*, vol. 2, pp. 1145–7. New Zealand Grassland Association.

Webber, A. N., Nie, G.-Y. & Long, S. P. (1994). Acclimation of photosynthetic proteins to rising atmospheric CO_2. *Photosynthesis Research*, **39**, 413–25.

Westman, W. E. & Rogers, R. W. (1977a). Nutrient stocks in a subtropical eucalypt forest, North Stradbroke Island. *Australian Journal of Ecology*, **2**, 447–60.

Westman, W. E. & Rogers, R. W. (1977b). Biomass and structure of a subtropical eucalypt forest, North Stradbroke Island. *Australian Journal of Botany*, **25**, 171–91.

Wittwer, S. H. (1978). Carbon dioxide fertilisation of crop plants. In *Crop Physiology*, ed. U. S. Gupta, pp. 310–33. New Delhi: Oxford and I.B.H. Publ. Co.

Wong, S. C. (1990). Elevated atmospheric partial pressure of CO_2 and plant growth. II. Non-structural carbohydrate content in cotton plants and its effect on growth parameters. *Photosynthesis Research*, **23**, 171–80.

Wong, S. C., Cowan, I. R. & Farquhar, G. D. (1979). Stomatal conductance correlates with photosynthetic capacity. *Nature*, **282**, 424–6.

Xu, D.-Q., Gifford, R. M. & Chow, W. S. (1994). Photosynthetic acclimation in pea and soybean to high atmospheric CO_2 partial pressure. *Plant Physiology*, **106**, 661–71.

Yelle, S., Beeson, R. C., Trudel, M. J. & Gosselin, A. (1989). Acclimation of two tomato species to high atmospheric CO_2. *Plant Physiology*, **90**, 1465–72.

⑭ Predicting crop yields under global change

J. Goudriaan[1]

Introduction

Climate and soil constitute the main environmental factors that determine the growth of plants, both in natural vegetation and in crops. The growth cycle of vegetation is adapted to the seasonal cycle at its specific geographic location, being determined by factors such as temperature in temperate and cool climates or precipitation as summer rains in the dry tropics, and winter rains in mediterranean-type climates. Soil factors determine root growth and the availability of water and nutrients to the plants. On a decadal time scale, changes in vegetation in their turn modify the soil properties, such as organic matter content, acidity and soil structure. Agriculture, however, changes the vegetation type entirely, and usually very rapidly. The farmers will weed out competing plant species or at least suppress them, and they will try to protect the crops against pests and diseases. Also they will try to alleviate limitations to crop growth imposed by shortage of water and nutrients. Adverse temperature conditions are avoided by proper timing of sowing and by using varieties that have a suitable crop phenology. In the long term agriculture has a large effect on the soil under cultivation.

Global change means that the environmental limitations to crop growth will be modified, and it will therefore alter farm management decisions. A scientific understanding of the changes that might occur will require a sound knowledge of the basic processes. Eventually, the complex interaction between management and environment calls for a multi-level approach, ranging from detailed research at the physiological process level, through farm systems analysis, up to agricultural economies and sociology. Because of the complexity of the systems, precise quantitative prediction will be extremely difficult and probably beyond the reach of our scientific capacity. What should be done is to explore feasible routes, and to make plausible assessments of the most likely impacts using the best tools we can develop.

1. With contributions from L. A. Hunt, J. Ingram, J. W. Jones, S. C. van de Geijn, M. J. Kropff, J. R. Porter, M. A. Semenov, P. D. Jamieson, T. Kartschall, T. R. Sinclair, R. Grant, J. T. Ritchie, F. Miglietta, J. I. L. Morison, E. Triboi, M. H. Jueffroi, H. van Keulen, W. Stol, D. W. Lawlor, R. Mitchell, F. Wang

Exploring global change effects on crop growth and yield

'Global change' is a rich term, encompassing several factors. Experimental programmes have explored various aspects of global change, both those acting independently and interactively. The empirical study of the effects of, for instance, temperature and CO_2 on crop growth and development has been of enormous value. It has shown that the primary effect of temperature is on developmental and morphogenetic processes, and that the primary effect of CO_2 is on photosynthesis and dry matter accumulation. Yet, many more factors need to be investigated and the way they exert their effects, not for just one plant species or variety but for many different ones. Although this process of knowledge accumulation and uncertainty reduction is very necessary, it is not sufficient by itself. Crop growth is a dynamic process, influenced by interaction of many factors. It is almost impossible to transform the multitude of experimental data directly into useful statements about their effect for the variable real-world conditions, let alone to extrapolate them into conditions that are different from the present ones.

To date, simulation modelling has the potential and promise to provide a scientifically sound framework for the combination of experimental data and scientific hypotheses. In a simulation model the hypotheses are implemented, and the processes are dynamically evaluated during the simulated crop growth process. The outcome of the simulation should then be compared with real-world data.

In this phase, which is in fact model validation, experiments are needed again. This time the experiments are done on a higher integration level than in the experiments for parameter determination, and they often pertain to applications. The emphasis is now not on process investigation, or parameter estimation, but rather on model validation under a wide range of conditions. The classical requirement of precise control of the experimental conditions is replaced by another requirement, that of careful monitoring of the unique time pattern of environmental and crop characteristics. The altered environmental conditions are simply made a part of the model input. This way the inevitable divergence between field conditions and the artefacts of our experimental set-up can at least be partly solved. Open-top chambers (OTCs) are a popular experimental device, in spite of their relatively large disturbance of the open-air conditions. Their popularity may well be explained by the fact that they do not have a very high threshold of initial investment, and that more units can be easily added on. Another, much more economic method that has recently become popular is the gradient-tunnel method (Horie *et al.*, 1991; GCTE Focus 3 Office, 1994). In this method the crop is even more sheltered from field conditions, but the vegetated experimental area is much larger. The light climate in OTCs and gradient tunnels differs profoundly from a homogeneous field crop, as does the wind and turbulence environment. Factors such as lateral irradiation, partial shading, and an increased fraction of diffuse radiation are all accessible to modelling (Jetten, 1992). It is usually better to accept that these factors are modified, and then to include them explicitly in the models, than to wage a costly battle for real-world resemblance of environmental conditions. Yet, it is justifiable that least-disturbance methods are demanded for at least a few cases, so that the validity of the chamber method, even if improved by modelling, can be tested.

Such a method is the Free Air Carbon dioxide Enrichment (FACE) of field crops, that is at the extreme end of a whole range of experimental methods, from enclosing individual leaves in cuvettes through growth chambers and OTCs (Schulze & Mooney, 1993). In this sequence the resemblance with the real world situation is improved. Also the duration of the exposure is extended, which offers the possibility that adaptations to the changed conditions can express themselves. Next to immediate effects, there will be an accumulation of secondary effects which will become noticeable only much later. After a prolonged exposure these immediate and accumulated effects will be simultaneously present, but for a correct interpretation they should be separated. Modelling can help to highlight the dynamic interaction of the different processes, that usually have widely different time coefficients. Leaf photosynthesis, for instance, may react immediately and its reaction will be reversible, whereas a change in leaf area growth continues to have an effect on light interception during several weeks after the change occurred.

Complexity of models

Depending on its purpose, modelling can be done at various levels of complexity. In many situations, the approach of describing accumulated dry matter as the integral of intercepted radiation times a radiation conversion coefficient (Russell *et al.*, 1989) is sufficient. This approach is followed in the simpler models such as SIMRIW (Horie *et al.*, 1992) or LINTUL (Spitters, 1990). Rossing *et al.* (1992) and Van Oijen (1991) successfully applied this approach to explain the main route of damage of potato late blight (*Phytophtora infestans*). Such a simplified summarizing approach is just as valid as a detailed reductionist approach in which a complex simulation model is used. Jones *et al.* (1994) showed that this simplifying approach is warranted as long as temperature and radiation are the only factors responsible for the diurnal course of assimilation. Some caution is needed when water shortage occurs. The distribution of rainfall intensity determines infiltration of water into the soil. Plant factors such as stomatal closure, leaf rolling and loss of turgor cause a variable response of net assimilation to radiation. Almost certainly a more detailed time resolution will then be needed to correctly capture these processes.

Feedback and slow state variables

In crop growth models, one of the most difficult processes to handle is the formation of new leaf area. A positive feedback loop between leaf area, intercepted radiation and canopy assimilation can amplify early effects on leaf photosynthesis by altering light interception later on. Separation of immediate effects and accumulated feedback effects is often difficult, even for the modeller himself, let alone that he could credibly convey his explanation to others. In such a situation modelling is best done in stages, introducing feedback processes step by step. An example of this step-wise approach was given by Kropff et al. (1993b). They made an assessment of the simulation of the effect of nitrogen fertilization on crop yield, in three steps of simulation. First the model was run with the climate data as input, and without any allowance in the simulation for observed values of crop characteristics. In the low nitrogen situation the formation of leaf area was overestimated, leading to a subsequent overestimation of dry matter formation. The second step was to reduce the leaf area formation in the low-N case, which raised the r^2 between observed and simulated yield from 0.56 to 0.86. Partitioning between shoot and root was simultaneously adapted. This step achieved a good fit between simulated and observed leaf area index. Finally the effect of low-N leaf status on the CO_2 assimilation rate was included, which raised r^2 further to 0.95.

The choice of the sequence of the constraints is not always obvious. It is usually a good idea to begin with phenology, then have the leaf area pattern correct, and finally internal plant characteristics, if available. This procedure of separation of feedbacks in subsequent runs is particularly powerful when soil water effects are involved. Leaf area is a major factor determining transpiration rate, soil water is a slow state variable, and delayed effects of leaf area formation can be cumbersome. For instance, excessive early irrigation may cause lush vegetation growth, with unduly large transpiration and water stress later on. These effects that are well known from agricultural practice also occur in models for crop growth.

Crop growth models worldwide

A number of crop growth models that have been used in recent climate change studies are listed in Table 14.1, based on Jones *et al.* (1994) and GCTE Focus 3 Office (1994). Although this list is not complete, its size is indicative of the amount of work that is going on. One important role of GCTE is to bring different groups in contact with each other; those that are active in modelling with other modellers, those that are active in experimental work with other experimentalists, and, most of all, these two groups with each other. We believe that modelling can help in interpretation and synthesis of experimental results. Science aims at the reduction of the dazzling amount of experimental results to as few and consistent rules as possible. This is, of course, the traditional task of scientific research, and the role of modelling brings the science then further to summarization and assessment of what has been found. In models the basic rules are put together again, and one then hopes to be able to reproduce the richness of behaviour observed in the real world.

GCTE provides a unique platform for interaction between modellers and experimentalists who might not have interacted otherwise. One of the mechanisms for attaining this goal has been the creation of Crop Networks. The members of a Crop Network have expressed their willingness to exchange both experimental data and source model codes on the basis of a Policy Document that each member has signed (GCTE Focus 3 Office, 1994). Under the regulations of this policy each member can make an appeal to another member for material. The policy implies that the material should not be further distributed, without consent of the original owner, and that the original ownership should be expressed by co-authorship of any manuscript that might result. It was thought to be most practicable to concentrate initially on a few major crops of worldwide importance, and to group the experimental and modelling research into a few networks, one for each crop. The reason for a division among crops is that on the technical level many definitions are crop-dependent, and it is crucial that interacting scientists agree on the definitions used. The concept of a crop network is well encapsulated in the GCTE Wheat Network, which was launched in Saskatoon in 1992. Currently the Wheat Network includes 31 formal members contributing models and datasets strictly pertinent for model validation and development. The Wheat Network has held two meetings so far, the latest one in November 1993, in Lunteren, The Netherlands. A similar network was launched for rice in March 1994, and those for potato and cassava are still in their planning phase.

14.1 *Crop growth models used in climate change studies*

Number	Model	Crop	References
1	SINCLAIR*	Wheat, soybean	T. R. Sinclair*
	CROP system*	Wheat, other crops	M. A. Semenov*
2	Sirius wheat*	Wheat	P. D. Jamieson*
3	SUCROS*	Wheat, other crops	J. Goudriaan*, Nonhebel (1993)
	IATAsubmodel*	Wheat	F. Miglietta*
5	CROPSIM*	Wheat	L. A. Hunt*
7	CERES-Wheat2.10*	Wheat	J. T. Ritchie*
6	AFRCWHEAT2*	Wheat	J. R. Porter*
8	SWHEAT*	Wheat	H. van Keulen*
9	DEMETER2.0*	Wheat	T. Kartschall*
10	ECOSYS*	Wheat	R. Grant*
4	SIMMCROG*	Wheat	F. Wang*
	WOFOST	Wheat, maize	Van Diepen et al. (1988), Wolf (1993)
	CERES-maize	Maize	Ritchie et al. (1991)
	SOYGRO	Soybean	Wilkerson et al. (1983) Jones et al. (1989)
	SIMRIW	Rice	Horie et al. (1992)
	ORYZA1	Rice	Kropff et al. (1993c)
	CERES-Rice	Rice	Singh et al. (1993)

*Models reported in GCTE Focus 3 Office, 1994.
The model number is used in Fig. 14.1 for identification of model results.

Initial results

Comparison with field data is not the only available method of model
testing and validation. This external validation must be done, but it is not
enough. Models should also be thoroughly tested internally. Such tests
include several aspects such as consistency of units, correct balances of mass
and energy, correct theoretical behaviour under extreme circumstances.
Transportability of models means that the models should be able to reproduce experimental results for a wide range of environmental conditions.

Porter et al. (1993) compared three simulation models for the wheat
crop, and tested their performance by comparison with observed crop yields
in a number of different growing seasons. They not only considered final
yield, they also included time courses of several crop attributes in their
comparisons, such as leaf area index (LAI) and absorbed radiation during

Table 14.2 *Maximum and minimum values of model outputs for the unperturbed standard datasets*

	Date of anthesis (day of year)	Date of maturity (day of year)	Maximum LAI (-)	Total Above-ground dry weight (t/ha)	Grain dry weight (t/ha)
Crookston, Minnesota, North America (spring wheat)[a]					
Prescribed	183	233			
Minimum	182	216	2.2	6.6	2.5
Maximum	187	233	8.5	20.6	8.0
Lelystad, The Netherlands, Western Europe (winter wheat)[b]					
Prescribed	166	207			
Minimum	164	204	6.3	14.2	5.4
Maximum	169	209	9.5	27.0	10.3

The models were run without any water stress. The dates of phenology were prescribed, but not all models were parameterized to meet exactly the prescribed dates.
[a]Results of 10 models. [b]Results of 8 models.

growth. Only by including various aspects of crop growth can a sound scientific analysis of model performance be made. Porter *et al.* (1993) were able to pinpoint a number of processes in which the models differed, and in some cases, in which the models should be improved. These processes included phenology and formation of new leaf area.

In the GCTE Wheat Network, a similar testing operation has been started on a larger scale. Quite to our surprise, at the latest workshop in Lunteren a considerable divergence in modelling approaches and results emerged, much more so than was anticipated. Each of the models presumably had been rigorously tested for local and regional conditions, and was expected to be able to handle the potential production under other climate conditions. The modellers were asked to present their simulated results for two typical climatic conditions: those of North America (Minnesota) and those of Western Europe (the Netherlands). The results of the models varied to a surprising degree, even though the dates of emergence, of anthesis and of maturity had been prescribed (Table 14.2, Fig. 14.1a). It should be emphasized here that Fig. 14.1 is not a scatter diagram of simulated versus measured results, but merely a plot of simulated grain dry matter versus simulated total above-ground biomass. If all model results fell on one line through the origin, it would only mean that all models simulate

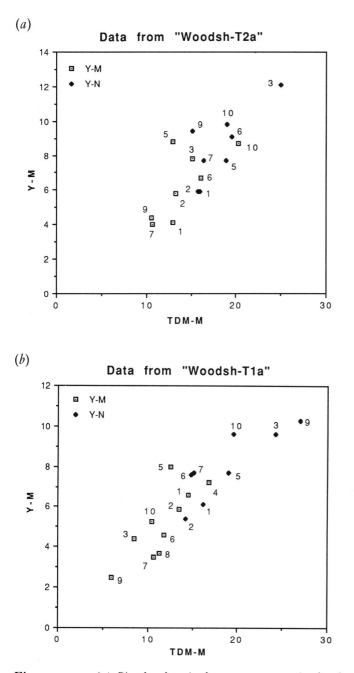

Figure 14.1 *(a) Simulated grain dry matter versus simulated total above-ground dry matter for wheat, as predicted by eight different simulation models for wheat growth and production. All models had the same climatic input (a 30 year mean for the site), and it was assumed that there was no constraining shortage of water or nutrients. The sites were at Crookston, Minnesota, North America (squares) and Lelystad, the Netherlands, North-Western Europe (diamonds). The model numbers have been listed in Table 14.1. (b) As in (a), but after adoption of the same time course for LAI in all models.*

the same harvest index. Here the simulated harvest index ranged from a low 0.3 to a high 0.6.

If all models were identical, their estimates should have collapsed onto one point for the North American site, and onto one other point for the European site. In reality the model estimates for total dry matter (above-ground) and grain dry matter deviated considerably and covered a range of almost a factor of 3 for North America and a factor of 2 for Europe. The models showed an approximately 50% higher production for the European site, this basically being the difference between a spring wheat and a winter wheat. The source of these differences between the models cannot be deduced in a straightforward way. The modellers felt that the method of comparison was satisfactory and that the method did not, in itself, account for the wide disparity. A more likely cause is that the models themselves are insufficiently robust to changed environmental conditions; i.e. those for which the model may not have been designed. This means that their parameter values implicitly allowed for some growth reduction, typical of the conditions for which the model was developed. For Europe in Fig. 14.1a, the two highest dry matter simulations came from European models, the two lowest from non-European models, and for America likewise the two lowest from European models and the two highest from non-European models. Even after imposing the same time-dependence of leaf area index (Fig. 14.1b) this pattern was still present, although the individual models shifted places. In calibration to the local conditions in some simple models the radiation use efficiency (RUE) was adapted, in other models adapted mechanisms were an imposed leaf senescence, or an imposed time pattern of the respiration coefficient as a function of development. Also, the models differed greatly in complexity. The simplest model consists of only 90 lines of code in GWBASIC, and the most complex one needs a CRAY to run it. However, there was no clear relationship between model output values and model complexity.

Models contain many feedbacks, making it difficult to trace the precise causes of deviation. From our experience of modelling crop growth we know that a major feedback route occurs through growth of leaf area. Once leaf area in early growth is overestimated, its high value will tend to be amplified through a higher rate of photosynthesis. For this reason, we wanted to force the models into line by imposing the same prescribed time course of leaf area index on all models. This artificial move enables exclusion of morphological model differences (specific leaf area, leaf area formation, tillering, leaf appearance) leaving nothing but functional model differences (light interception, photosynthesis, respiration). Internal model feedbacks may, however, have caused unintended imbalances (leaf area ratio,

Table 14.3 *As in Table 14.2, but all models were obliged to use the same time course of LAI (as generated by the model CROPSIM). For North America the date of maturity was advanced to a more realistic day of year of 216*

	Date of anthesis (day of year)	Date of maturity (day of year)	Maximum LAI (-)	Total above-ground dry weight (t/ha)	Grain dry weight (t/ha)
Crookston, Minnesota, North America (spring wheat)[a]					
Prescribed	183	216			
Minimum	182	215	4.5	10.6	4.0
Maximum	187	216	4.5	16.1	8.8
Lelystad, The Netherlands, Western Europe (winter wheat)[b]					
Prescribed	166	207			
Minimum	164	204	7.5	15.2	5.9
Maximum	166	207	7.5	25.0	12.1

[a]Results of 6 models. [b]Results of 6 models.

nitrogen content and photosynthetic capacity, etc.). For North America in the first exercise the date of maturity (Day of Year 233) was considered to be unduly late, and it was therefore advanced to Day of Year 216. The simulated time course of leaf area index in the model CROPSIM was adopted as a standard and used as input to the other models. (This choice does not imply any quality judgement.) The results are shown in Table 14.3 and in Fig. 14.1b. The differences shrank, but they did not disappear. In particular the North American simulations of dry matter and yield became higher, due to a larger leaf area duration. Not all models were able to handle the isolated change in leaf area index because of problems with, for instance, the balance of nitrogen. The more complex a model, and the more feedbacks it contains, the more difficult it is to impose a time course of just one variable without disturbing the delicate balance in the model.

Our goal was to test if the crop growth models are in such a stage of development that they can be used in generalized conditions just as they come. Clearly the answer is that this is not possible. If used in a different environment, each model will at least need calibration or tuning of its parameter set. Indeed, a comparable exercise on a smaller scale in rice (Kropff *et al.*, 1995) showed that an initial model divergence could be substantially reduced by removing the almost unconscious pre-adaptations of the various models to local growth constraints (Fig. 14.2). The considerably more homogeneous result of this rice modelling exercise is the result of a considerable time investment in calibration of each model separately.

Although only a few rice models were involved they differed almost as much as the wheat models did. They were certainly not just slightly modified versions of the same parent model. However, homogeneity of growing environment may have contributed to the better uniformity of the rice results. These models were all for irrigated rice, which is grown under very similar conditions that are much closer to potential production than those for most wheat crops. The wheat models, on the other hand, have been developed for a much wider range of conditions, covering both winter and spring wheat, from semi-arid to humid conditions, from low-input extensive cultivation to high-input intensive cultivation. This variety of environments has left its mark on the models.

Development of a single crop growth model is not the objective of GCTE, but model comparison does highlight critical features in models. It will stimulate self-criticism among modellers, and unsuccessful approaches will be more easily identified so that they can be replaced.

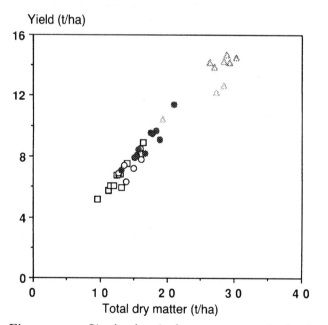

Figure 14.2 *Simulated grain dry matter versus simulated total above-ground dry matter for rice, as predicted by four different rice growth models for two cultivars on four sites. All models had the same climatic input (for actual years), and it was assumed that there was no constraining shortage of water or nutrients. The sites were at Los Banos, Philippines, 1991 (squares) and 1992 (filled circles), Maebashi, Japan (open circles), and Yanco, NSW, Australia (triangles).*

Sensitivity to temperature and CO_2

We also requested the effect of a temperature and a CO_2 perturbation, under both standard and drought conditions. There was much more agreement among the modellers on the relative response to CO_2 than there was on the response to temperature. In all models the CO_2 response followed a rising curve, gradually tapering off. There was, however, a difference in the steepness of the response, varying between a 20% and a 40% increase of dry matter upon doubling of the CO_2 concentration.

For temperature the situation is clearly much more complicated. The dominating factor in the temperature response is the rate of development through the various life stages. In most models this rate increases steadily with temperature, and so the growth duration decreases. As a result, the accumulated biomass, being a function of intercepted radiation, is mostly reduced as well. Sometimes more complex responses were found due to limitation by reduced photosynthesis at low temperatures or enhanced respiration at high temperatures. For North-Western Europe the predicted temperature response varied from a slight increase of yield with temperature to a strong decrease. Likewise the predicted optimal temperature shift varied between +3°C and −4°C.

In the situation where water shortage was permitted to occur, the models equivocally predicted a negative effect of a temperature rise. Even if higher temperatures do not always lead to more water stress *during crop growth* (Wolf, 1993), because of earlier emergence and earlier maturation that result in some avoidance of the dry season, there is practically always a reduction of total intercepted radiation.

Discussion

Unlike in nature, the selection of plant species that will grow and survive in the field is the result of management considerations. These human decisions will certainly be affected by climatic change, as the competitive position of a crop type may be undermined in comparison with that of another crop. Understanding single species performance may help to predict such choices and thereby the effect on land use. However, the farming system as a whole should be considered, especially when it comes to land use. The farming system may include factors such as the use of crops for fodder, the use of land for manure disposal, and timing of precious labour.

As an example, the history of the cultivation of maize in North-Western Europe tells an interesting story. At present the climate of North-Western Europe is not suitable for grain maize. Yet, the cultivation of maize has become much more extended in the past 30 years, not so much because of climate change, but because of the suitability of maize as silage fodder even when the cobs are not yet mature. Other promoting factors are its apparent resistance against excessive manuring, and its relative freedom from plant diseases in this cool climate. It is very hard to see how a single crop growth model could have led to the prediction of this large change in land use, without including crop management and general farming considerations.

The expected changes in CO_2 and temperature will have relatively small effects compared with the impact of crop management practices and genetic improvements in rice (Kropff *et al.*, 1993*a*) and in maize (Paruelo & Sala, 1994). This is not meant to belittle the climatic effects, but we are facing very rapid changes in terms of the dynamics of the human population, technology and economy. At their average rate of change of 2% per year those processes will probably have a much larger impact on food security and crop production than the dynamics of climate change. However, climate change effects will add on to the other changes that will occur anyway, and regionally critical thresholds may be passed, such as for sterility in rice (Matthews *et al.*, 1994).

According to Uehara and Tsuji (1993), systems simulation in agriculture may replace field trials as the primary means to generate information with which to identify new practices and policies for change. On a moderate scale such a trend is beneficial and cannot be refuted, but simulation can never really replace field experiments. Rather, the role of simulation is complementary to the role of experiments. Simulation can enhance their value by a more efficient analysis of field trials, and it can also help to guide further experimentation. Model development is a cumbersome, slow and expensive process, and Uehara and Tsuji (1993) recommended that the slow pace of model development should be accelerated by an integrated effort mounted on an international scale. At the same time, they strongly recommended avoidance of conformity and regimentation. We cannot agree more, and we hope that in IGBP/GCTE the benefit of international cooperation will be combined with continued pluriformity in the scientific process of model development.

The exercise reported here has shown the divergence in calibration of models as they tend to come from their authors. This report should not be taken as proof of poor reliability of models in themselves, nor as an attack on the usefulness of the modelling approach. It does underline the importance of continued testing of models with the real world. Modelling has shown its usefulness for a number of different and well-defined purposes (Penning de Vries *et al.*, 1993) such as studies of agro-ecological zoning, decision-support systems, yield gap analysis, multiple goal programming (Rabbings *et al.*, 1992), scientific analysis of experiments and data analysis. The technique of modelling has become an integrated part of the scientific methodology, but contrary to what the ease of using models suggests, their application requires skill, criticism and caution.

Acknowledgement

Part of this paper was prepared while the author was on study leave as a recipient of a CSIRO McMaster Fellowship, at the Centre for Environmental Mechanics of CSIRO, Canberra, Australia.

References

Connor, D. J. & Wang, Y. P. (1993). Climatic change and the Australian wheat crop. Proceedings of the Third Symposium on the Impact of Climatic Change on Agricultural Production in the Pacific Rim. Taiwan, ROC: Central Weather Bureau.

Diepen, C. A. van, Rappoldt, C., Wolf, J. & van Keulen, H. (1988). Crop growth simulation model WOFOST. Documentation version 4.1. Wageningen: Centre for World Food Studies.

GCTE Focus 3 Office (1994). GCTE Report No. 2. GCTE Focus 3 Wheat Network: 1993 Model and Experimental Meta Data. Canberra, Australia.

Gijzen, H. & Goudriaan, J. (1989). A flexible and explanatory model of light distribution and photosynthesis in row crops. *Agricultural and Forest Meteorology*, **48**, 1–20.

Goudriaan, J. & Monteith, J. L. (1990). A mathematical function for crop growth based on light interception and leaf area expansion. *Annals of Botany*, **66**, 695–701.

Horie, T. J., Nakano, H., Nakagawa, K., Wada, H. Y. & Kim, Seo T. (1991). Effects of elevated CO_2 and high temperature on growth and yield of rice. I. Development of temperature gradient tunnels. *Japanese Journal of Crop Science*, **60**, 127–8.

Horie, T., Yajima, M. & Nakagawa, H. (1992). Yield forecasting. *Agricultural Systems*, **40**, 211–36.

Jetten, T. H. (1992). Physical description of transport processes inside an open-top chamber in relation to field conditions. PhD thesis. ISBN 90-5485-011-6. Agricultural University, Wageningen, The Netherlands.

Jones, J. W., Boote, K. J., Hoogenboom, G., Jagtap, S. S. & Wilkerson, G. G. (1989). SOYGRO VS. 42: Soybean crop growth simulation model: user's guide. Florida Agricultural Experiment Station Journal no. 8304. Gainesville, Florida: Institute of Food and Agricultural Sciences, University of Florida.

Jones, J. W., Pickering, N. B., Rosenzweig, C. & Boote, K. J. (1994). Simulated impact of global climatic change on crops. Proceedings Rice Ecosystems Workshop, IRRI.

Kropff, M. J., Cassman, K. G., Penning de Vries, F. W. T. & van Laar, H. H. (1993a). Increasing the yield plateau in rice and the role of global climate change. *Journal of Agricultural Meteorology*, **48**, 795–8.

Kropff, M. J., Cassman, K. G., van Laar, H. H. & Peng, S. (1993b). Nitrogen and yield potential of irrigated rice. *Plant and Soil*, **155/156**, 391–4.

Kropff, M. J., van Laar, H. H. & ten Berge, H. F. M. (eds.) (1993c). ORYZA1, a basic model for irrigated lowland rice production. Los Baños, The Philippines: IRRI.

Kropff, M. J., Williams, R. L., Horie, T., Angus, J. F., Singh, U., Centeno,

H. G. & Cassman, K. G. (1995). Predicting the yield potential of rice in different environments. In *Temperate Rice: Achievements and Potential.* Proceedings of the Temperate Rice Conference, Yanco, 1994, Australia, ed. E. Humphreys *et al.*, pp. 657–64.

Matthews, R. B., Kropff, M. J., Bachelet, D. (1994). Climate change and rice production in Asia. *Entwicklung + Ländlicher Raum*, 1/94, 16–19.

Nonhebel, S. (1993). The importance of weather data in crop growth simulation models and assessment of climatic change effects. PhD thesis, ISBN 90-5485-114-7. Agricultural University, Wageningen, The Netherlands.

Oijen, M. van (1991). Light use efficiencies of potato cultivars with late blight (*Phytophtora infestans*). *Potato Research*, 34, 123–32.

Paruelo, J. M. & Sala, O. E. (1994). Effect of global change on maize production in the Argentinean Pampas. *Climate Research*, 3, 161–7.

Penning de Vries, F. W. T., Teng, P. S. & Metselaar, K. (1993). *Systems Approaches for Agricultural Development.* Dordrecht: Kluwer.

Porter, J. R., Jamieson, P. D. & Wilson, D. R. (1993). Comparison of the wheat simulation models AFRCWHEAT2, CERES-wheat and SWHEAT for non-limiting conditions of crop growth. *Field Crops Research*, 33, 131–57.

Rabbinge, R. & van Latesteijn, H. C. (1992). Ground for choices, ISBN 90-399-0367-0. The Hague: Netherlands Scientific Council for Government Policy, SDU.

Ritchie, J. T., Singh, U., Godwin, D. & Hunt, L. (1991). A user's guide to CERES-Maize- V2.10, 2nd edition. Muscle Shoals, Alabama: International Fertilizer Development Center.

Rossing, W. A. H., van Oijen, M., van der Werf, W., Bastiaans, L. & Rabbinge, R. (1992). Modelling the effects of foliar pests and pathogens on light interception, photosynthesis, growth rate and yield of field crops. In *Pests and Pathogens*, ed. P. G. Ayers. Oxford: Bios Scientific Publishers.

Russell, G., Jarvis, P. G. & Monteith, J. L. (1989). Absorption of radiation by canopies and stand growth. In *Plant Canopies*, ed. G. Russell, B. Marshall & P. G. Jarvis, pp. 21–39. Society for Experimental Biology Seminar Series, No. 31. Cambridge: Cambridge University Press.

Schulze, E.-D. & Mooney, H. A. (1993). Design and execution of experiments on CO_2 enrichment. Ecosystem Research Report 6. Brussels: CEC.

Singh, U., Ritchie, J. T. & Godwin, D. C. (1993). A user's guide to CERES-Rice-V2.10, 2nd edition. Muscle Shoals, Alabama: International Fertilizer Development Center.

Spitters, C. J. T. (1990). Crop growth models: their usefulness and limitations. *Acta Horticulturae*, 267, 349–68.

Uehara, G. & Tsuji, G. Y. (1993). Dealing with climate change and food security through systems analysis and simulation. Proceedings of the Third Symposium on the Impact of Climatic Change on Agricultural Production in the Pacific Rim. Taiwan, ROC: Central Weather Bureau.

Wilkerson, G. G., Jones, J. W., Boote, K. J. & Ingram, K. T. (1983). Modeling soybean growth for management. *Transactions of ASAE*, 27, 129–44.

Wolf, J. (1993). Effects of climate change on wheat and maize production potential in the EC. In *The Effect of Climate Change on Agricultural and Horticultural Potential in Europe*. Research Report No. 2, ed. G. J. Kenny, P. A. Harrison & M. L. Parry. Oxford: Environmental Change Unit, University of Oxford.

⑮ Global change impacts on managed forests

S. Linder, R. E. McMurtrie and J. J. Landsberg

Background

Forests cover more than one-third of the land surface of the Earth and account for 80% to 90% of plant and 30% to 40% of soil carbon (cf. Melillo *et al.*, 1990). Forests and other wooded land thus constitute the largest areal component of current and future land use and play an important role in the global carbon balance. Because forests are long-lived communities, current rapidly changing atmospheric carbon dioxide concentrations [CO_2], the likely rising temperatures and other associated changes in climate will have significant impacts not only on the forests of the future but also on forests currently growing (e.g. Eamus & Jarvis, 1989; Jarvis, 1989; Bazzaz, 1990).

From an economic point of view it is an obvious imperative to secure long-term wood supply to timber and pulp industries. Biomass production in less intensively managed forest ecosystems may, however, be equally important since they provide fuel, fodder and other utilities for a large proportion of the world's population.

To provide the information needed for accurate predictions of the effects of global change on forest ecosystems and to provide the knowledge required as the basis for appropriate and adaptive management programmes, we need focused, coordinated forest research programmes throughout the world. Research is also required on the role of managed forests as carbon sinks or sources, including the possibility of increasing carbon sequestering in forest ecosystems by means of silvicultural practices. Research programmes should aim to predict both short-term forest responses to altered climate, disturbance and silvicultural practices, and effects on long-term sustainable productivity and biodiversity.

Studies of forest ecosystems occur in all of the four foci within the core project 'Global Change and Terrestrial Ecosystems' (GCTE) of IGBP (Steffen *et al.*, 1992). The broad aim of this forest-related research is to understand and evaluate potential effects of global change on future structure, composition, and cycling of carbon, water and mineral nutrients, in natural and managed forests and to identify resource management strategies for sustainable forestry and ecosystem management.

Within GCTE there is one 'Activity' (3.5) specially designated to study the 'Impacts of global change on managed forests'. The primary aim of Activity 3.5 is to understand and evaluate potential effects of global change on future structure, biomass production, and yield of managed forests and to identify resource management strategies for sustainable forestry under changed climatic conditions. The term 'managed forest' is defined as any forest that is or has been subject to some form of management action, i.e. manipulation with the view of achieving some specified objective. Where the primary and overriding objective of forest management is wood production exploitation may lead to loss of species diversity and alteration of the structure of the forests (e.g. from mixed to even-aged). However, there is now a strong move towards the management of forests as sustainable ecosystems. Sustainability may be defined as the long-term maintenance or enhancement of the productive capacity and biodiversity of forests. Evaluation of the effects of management actions (in the broad sense used here) on sustainability requires evaluation of the resilience of the forest systems, where resilience refers to the ability of an ecosystem to return to its pre-disturbance state.

A strategic plan for research on managed forests in a globally changing environment has recently been prepared (Landsberg *et al.*, 1995*a*). The intention is that the plan will function as a supporting document to the GCTE Operational Plan (Steffen *et al.*, 1992). The objective is to develop a framework for planning and implementing scientific research concerned with forests so that it is consistent in terms of concepts, procedures and data recording and produces results that can be compared and used to test and validate models and assess the impacts of global change on forest growth and production.

It this chapter the main features of the strategic plan are presented and examples given of some ongoing and planned experimental and modelling activities. No attempt is made, however, to cover or review all aspects and current activities within this rapidly growing field of research.

Strategic plan for research on managed forests

The strategic research plan outlined by Landsberg *et al.* (1995*a*) provides the basis for co-ordination of international research: a framework within which research institutions, as well as individual scientists, can set their work, with reasonable expectation that it will be consistent and comparable with that being done elsewhere by other groups. It is axiomatic that the research done in a particular region should also provide information pertinent to the problems of management in the region. The plan clearly does not cover all aspects of the research needed to evaluate the likely effects of global change on forests, but identifies the main area of concern and the information needed as a basis for both management and policy decisions.

The working definition of global change that underlies the plan is that the following variables, associated with global change, are expected to affect forests: (i) increasing $[CO_2]$; (ii) increasing average temperatures, particularly at higher latitudes; (iii) changes in amount and distribution of precipitation; (iv) increasing UV-B; and (v) change in the frequency and severity of extreme events (e.g. extreme hot or cold periods, hurricanes, floods, droughts, fires, pest and disease outbreaks). Some of the variables mentioned have specially designated 'Activities' or 'Tasks' within GCTE (cf. Steffen *et al.*, 1992).

The plan is divided into five main parts: (1) Study sites and transects; (2) Studies and experimental treatments; (3) Measurements and observations; (4) Simulation models; and (5) Data bases and geographic information systems (GIS).

The research strategy underlying the plan is to sample a representative range of forest types, growing under different conditions in different parts of the world. Within these the establishment of identical, or very similar, experiments along transects of varying climatic conditions will allow evaluation of the variation of, and constraints on, the productivity and biological diversity of forests. The basic minimum requirement, at each experimental or observation site, is a series of baseline (Level I) measurements and observations, preferably carried out over a long period. Where resources allow, these baseline observations should be supplemented by more detailed studies on community dynamics (Level II) and, at some sites, measurements of the physiological processes governing forest growth and productivity (Level III).

All research programmes should be based on appropriate process-based models. These are expected to converge towards a universally applicable suite of forest models, for various purposes, that may have to be parameterized for particular areas but will handle most problems and analyses of forest growth in relation to climate and global change. Models should be incorporated into geographic information systems, to deal with the heterogeneity of large areas of forest, and as a framework for the recording and analysis of remotely sensed information.

Study sites and transects

Selected study sites should, in terms of structure and species composition, be representative of large areas. For assessing the impact of environmental conditions, and hence global change, on the growth and performance of forests it is important that study sites be located in areas where significant change is expected, where there is reason to believe that the forests are likely to be susceptible to change and where the forest ecosystems are of socio-economic importance. Based on these criteria, the following systems have initial priority:

- Boreal forests, which contain much of the world's available softwood and have high stores of organic matter. It is predicted that the boreal areas will experience considerable warming as a result of climate change. In those regions physiological responses to enhanced $[CO_2]$ are likely to be limited by current low temperatures and infertile soils. Recommended study sites include southern Siberia, southern Canada and Scandinavia.
- Temperate coniferous and mixed coniferous–deciduous forests, which form the basis for most of the world's present timber and pulp industry. Recommended sites include northern Europe, and northern USA, and conifer plantations in the southern hemisphere.
- Humid tropical and subtropical forests, which have the greatest biodiversity. These forests are under intense land-use pressures and may be highly responsive to $[CO_2]$ except where nutrient availability is low. Recommended sites include tropical rainforests in Central and South America, South-East Asia and deciduous broadleaf and coniferous forests in the southern USA.
- Semi-arid forests, which are expected to be particularly sensitive to altered moisture availability, are globally extensive, and are under considerable land-use pressure. Recommended sites include woodlands in Asia and Africa, eucalypt and softwood plantations in mediterranean climate zones, managed native eucalypt stands in Australia, and semi-deciduous forests in the southern Amazon.

The primary goal of the transect experiments is to quantify ecophysiological responses to current environmental variability, and the selection of sites along transects and design of experiments will, whenever possible, be closely coordinated with other IGBP/GCTE efforts (Koch *et al.*, 1995). Transects may be relatively short, spanning steep climate gradients – for example, from coast to mountain to rain shadow – or long, across countries and continents. In either case the objective is to set up similar (if possible identical) experiments or sets of observations. Transects in humid tropical and subtropical forests should cover the range of land-use pressure, the major threat to these forests. The developing network of modelling centres within GCTE's 'Long-term Ecological Modelling Activity' (LEMA; Steffen *et al.*, 1992) may provide a mechanism for linking modelling efforts to the development of particular transects.

Studies and experimental treatments

Experimentation and modelling is required to develop quantitative indices characterizing sustainability of forest ecosystem processes on specific sites subjected to treatments manipulating CO_2, climate or supply of water and nutrients. Indices should characterize the rate and extent of post-disturbance recovery of plant and soil pools of carbon and nutrients and may be either based on mechanistic models incorporating the closure of carbon, water and nutrient budgets, or empirically based (cf. Landsberg, 1986). The experiments should monitor both short- and long-term responses to disturbances such as soil warming, altered rainfall, CO_2 enhancement, fertilization, forest logging, and conversion of land use from agriculture or grassland to forestry.

Experimental work should be aimed not only at providing empirical information, but also at understanding the physiological control mechanisms and key processes underlying forest response to environmental factors, at both the tree and stand levels. The emphasis on process-level understanding acknowledges the lack of fundamental understanding about forest ecosystem carbon budgets, recognizes the impracticability of studying all major forest types and supports the development of general models of forest growth.

Observational studies

Considerable information can be obtained from observational/measurement studies along transects, where a wide range of environmental conditions can be obtained naturally, but in an uncontrolled manner. This information can be greatly enhanced by conventional forestry field experiments. These may involve testing several species or provenances, manipulating tree density (either initially or by later thinning) and fertilization treatments to vary nutrient availability.

Manipulation experiments

The effect of manipulating water and nutrition can be illustrated by results from some ongoing forest experiments where the availability of nutrients

Table 15.1. *The average current annual increment (CAI), during a 4 year period, in five forest stands subjected to different treatments in terms of water and nutrient supply. The treatments were: irrigation combined with a complete liquid fertiliser (IL), irrigation (I), annual supply of solid fertilizers (F), and untreated controls (C). The values are given in* m^3 ha^{-1} $year^{-1}$ *and per cent (bold type) of the untreated controls and represent the last 4 years of the treatment period. The age of the stands represents the last year of the treatment period and the total years of treatment are given within brackets.*

Species	Location	Treatment IL	I	F	C	Stand age	Source
Australia							
Monterey pine	35°20'S	50.4	33.0	20.5	20.5	15	1
(*Pinus radiata*)	148°55'E	**246**	**161**	**100**	**100**	(4)	
Portugal							
Tasmanian blue gum	35°02'N	52.3	43.6	33.7	28.3	5	2, 3
(*Eucalyptus globulus*)	9°15'W	**185**	**154**	**119**	**100**	(5)	
Sweden							
Scots pine	60°49'N	11.5	5.7	11.8	5.1	31	4, 5
(*Pinus sylvestris*)	16°30'E	**225**	**112**	**231**	**100**	(17)	
Norway spruce	57°08'N	22.7	12.2	16.9	9.4	19	5
(*Picea abies*)							
	14°45'E	**242**	**130**	**180**	**100**	(7)	
	64°07'N	11.6	2.8	10.3	2.7	31	5
	19°27'E	**430**	**104**	**382**	**100**	(8)	

Source: 1, Snowdon & Benson (1992); 2, Pereira *et al.* (1989); 3, Pereira *et al.* (1994); 4, Linder & Flower-Ellis (1992); 5, S. Linder & J. G. K. Flower-Ellis (unpublished data).

and water was controlled (Table 15.1). In all the stands in which the availability of nutrients and soil water was optimized by means of combined fertilization and irrigation, the yield of stemwood was similar to or surpassed the best yields obtained by conventional silvicultural means. In Sweden, the response to irrigation was small and 'high' yields were obtained by fertilization alone. In Australia and Portugal, however, irrigation was essential for obtaining a response to fertilization. In Portugal, the stands responded to both irrigation and fertilization, but the best response was obtained when the treatments were combined. The results emphasize the fact that, since yield in most forest stands is limited by the availability of nutrients and water, these factors must be included when interpreting responses to CO_2

and temperature. Therefore, each main site should include treatments which manipulate availability of water and mineral nutrients. Where possible experiments should incorporate high CO_2 and soil warming treatments and follow the same protocols. Three forest experiments along those lines have been established in conifer plantations in Sweden and southeastern USA, respectively, and will be extended to a network of sites representing major managed forest ecosystems of the world.

Nutrition

Nutrient treatments should include all essential nutrient elements and repeated fertilization to maintain 'non-limited' nutrient conditions. Wherever possible current understanding of optimal nutritional requirements should be utilized both to minimize nutrient excesses and to target tree requirements accurately (cf. Linder & McDonald, 1993; Linder, 1995). Untreated stands will generally be used as the nutrient-limited condition for comparison. Whenever possible nutrient $\times$ CO_2 interactions should be included in the experimental design.

Water

Supplementation to maintain field capacity should be imposed at all water-limited sites to provide a baseline for comparison of other water-related treatments. Natural (rainfall supplied) water levels will generally provide the other treatment conditions. For forests in regions where general circulation model (GCM) estimates of temperature and precipitation predict reductions in soil water, treatments might include reductions in water availability by means of throughfall diversion.

[CO₂]

It is recommended that elevated [CO_2] treatments should be ambient + 350 μmol mol^{-1}, applied in a step change. This level is sufficiently high to detect a response in susceptible systems, but is not outside the range of predictions for future CO_2 concentrations. CO_2 should be applied day and night, but may be discontinued during certain periods of the year when tree activity is minimal due to low temperatures or deciduous habit. Elevated [CO_2] can be achieved in open-topped chambers, branch bags or by 'Free Air CO_2 Enrichment' (FACE; see Schulze & Mooney, 1993). Branch bags are the simplest to use, although even these involve considerable technology to control and monitor (Barton et al., 1993). A European network of high CO_2 experiments along these lines on a number of tree species has been established, and could be a useful model for future similar research (Lee & Barton, 1993).

Temperature

Air temperature may be varied in enclosures and soil temperature by heating. Elevated temperature treatments should be based on the current consensus prediction of GCMs for a greenhouse forcing equivalent to a doubling of $[CO_2]$; this will range from relatively large (4–8oC) increases at higher latitudes to minor temperature changes at low latitudes. The temperature treatments should also attempt to match the predicted seasonality of a $2 \times [CO_2]$ climate because changes in seasonality may biologically be just as or more important than changes in means. A number of forest experiments including elevated soil temperatures have been initiated following the design described by Peterjohn *et al.* (1993). Within the CLIMEX project not only soil warming, but also controlled elevated air temperature is planned within a large enclosure (Jenkins & Wright, 1993).

Measurements and observations

It is proposed that measurements in forest research should be based on nested experiments characterized as either: extensive level – baseline measurements (level I); medium level – community dynamics (level II); or intensive level – process measurements (level III).

Level I: baseline measurements

Level I measurements should be made at all sites. These would include normal mensurational measurements that can be made with simple equipment but provide essential baseline information about the state of forests and their long-term growth patterns. Level I measurements accumulated over long periods at many well-characterized sites will provide an extremely valuable database from which considerable information about forest growth and performance, and the effects of weather and climate, may be obtained. The sites should be selected to represent the most common forest type(s) in the region. In many countries there exist established networks of permanent forest plots and repeated forest surveys, which can be used for initial estimates of the carbon balance within the forest sector (cf. Kurz *et al.*, 1992; Nabuurs, 1994).

Level II: community dynamics

Level II measurements are aimed at providing additional information about the factors that cause changes in species composition. Such studies are important for all natural stands since species composition, and hence biodiversity, is considered likely to be significantly affected by many aspects of global change. The structural characteristics of stands should be described as part of any study on species composition/community dynamics; the measurements include the factors to be measured at level I, but should include also leaf area index (LAI) and leaf litterfall.

Level III: process measurements

To allow estimates of diurnal, seasonal, and annual fluxes of carbon, water and nutrients in relation to environmental influences, level III models require measurements with a high temporal and spatial resolution of driving variables and studied processes. In addition to measurements and observations required at level I and II, respectively, measurements should include the physiological processes underlying and driving the growth and production of trees. These are primarily: photosynthesis, respiration, carbon allocation, litterfall, transpiration, and uptake and utilization of nutrients.

Simulation models

Models provide a framework for organizing and focusing experiments. They provide hypotheses to test, and allow pre-testing of hypotheses to help suggest critical experimental tests. It is not the primary role of models to 'integrate' the observations made in experiments. Models should be used as experimental tools, and sensitivity analyses conducted to identify key variables and weaknesses in understanding. Models should, therefore, be used *a priori* rather than *a posteriori*.

There are many stand-level, process-based, physiological models, written for various purposes at varying degrees of complexity (cf. Ågren *et al.*, 1991; Ryan *et al.*, 1996). There is now considerable convergence in the general structure of many of these models and it would not be advisable for new research programmes to set out to develop new models at this level. Careful assessment of extant models is likely to show that there is one or more available that will suit the purpose.

Process-based model (e.g. BIOMASS: McMurtrie *et al.*, 1990; MAESTRO: Wang & Jarvis, 1990; FOREST-BGC: Running & Gower, 1991) are used both for detailed analysis of stand responses to current environmental conditions (Fig. 15.1) and to assess the likely impact of climate change (Rastetter *et al.*, 1991; McMurtrie *et al.*, 1992; Comins & McMurtrie, 1993; Jarvis, 1993; McMurtrie & Wang, 1993).

The detailed process-based models should provide the basis for the development of less complex, large spatial scale models driven by remotely

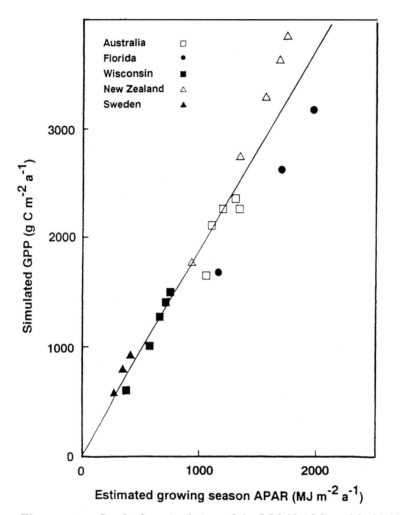

Figure 15.1   *Results from simulations of the BIOMASS model (McMurtrie et al., 1990) comparing the gross primary production (GPP) in relation to absorbed photosynthetically active radiation (APAR) during the growing season in pine forests growing in contrasting environments. (From McMurtrie et al., 1994.)*

sensed observations (cf. Running, 1993). The principle underlying these models is that net primary production in a forest (NPP) is proportional to the amount of absorbed photosynthetically active radiation (APAR), a function of LAI and incoming solar radiation (cf. Jarvis & Leverenz, 1983; Linder, 1987; Potter *et al.*, 1993; Ruimy *et al.*, 1993; Landsberg *et al.*, 1995*b*). Climate, nutrients, soils and topography all affect the relationship between NPP and APAR, and this must be determined from first principles for altered climate and [CO_2]. By nesting estimates of productivity within a region, derived from detailed (fine-scale) process models, with regional scale estimates of LAI and APAR, the influence of climate change on regional productivity can be estimated (cf. Running, 1993; Waring *et al.*, 1993).

Models also provide the best means of evaluating the likely consequences of various management actions, particularly in relation to the uncertainties arising from global change. It should be noted that regression-based models – which include most of the models derived as part of traditional forest biometrical research and management – being derived from historical observations, without mechanistic basis, are unable to predict the effects on growth of altered climatic conditions. Process-based submodels could be employed in this context to show how existing regression-based models can be modified to incorporate climate change effects (Hunt *et al.*, 1991).

The level of detail incorporated in a particular model will vary depending on modelling objectives, and scientific hypotheses to be addressed. Where experiments are nested, a nested modelling approach is also appropriate and will make it possible to extrapolate in both time and space. Theories validated at the micro-environmental scale can be extrapolated to the stand scale using detailed canopy models which can be used to derive simplified relationships predicting annual biomass production from absorbed photosynthetically active radiation (Fig. 15.1). The latter relationship can be applied to simulate growth on sites with less comprehensive databases and to predict growth at the regional scale. Data and models at each level of the modelling hierarchy are used to parameterize and validate models at the next level. Process formulations become less complex at each successive hierarchical level. However, new processes and feedbacks become important as we scale up in time and space, e.g. soil feedbacks (Fig. 15.2), which are irrelevant in simulating diurnal photosynthetic production, but critical at longer time-scales.

Databases and geographic information systems

The technology now exists to record data in computer databases, in formats that make them accessible and understandable to all who need them. There are enormous benefits in this; the problems lie not with the technology but in the willingness of scientists and managers to agree on the data to be collected, formats and labelling, and data documentation procedures. However, the IGBP Data and Information System (IGBP-DIS) has taken a pro-active role in coordinating international activities to ensure that the necessary data sets are produced and made available (cf. Townshend, 1992).

In the absence of international agreement it is recommended that all data be recorded in standard databases and labelled accurately and

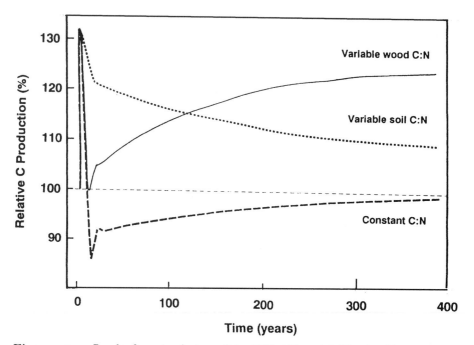

Figure 15.2   *Results from simulations of the G'DAY model (Comins &*
McMurtrie, 1993) showing the effects of an instantaneous doubling of [CO₂]
from 350 to 700 ppmv on the relative net primary production (NPP) in a
stand of Pinus radiata *growing near Canberra, Australia. Initial conditions were*
set by running the model to equilibrium with [CO₂] at 350 ppmv, giving an
initial NPP of 0.73 kg C m⁻² a⁻¹. The doubling of [CO₂] occurs at time 2
years. Simulations are illustrated under various assumptions about the plasticity
of carbon to nitrogen ratios (C : N) in woody tissue and soil organic matter;
large responses occur if C : N ratios decline significantly at high CO₂, with little
or even a negative response occurring if C : N ratios are inflexible. For further
details see McMurtrie et al. *(1992).*

comprehensively. For example, level I measurements, to be of value to
people other than those who collected the data, should include clear statements about number of (individual) measurements made, methods used and
estimated errors. Local classifications should be avoided and information
should, whenever possible, be stored in its primary form. It is seldom possible to disaggregate classifications into primary data, but it is usually simple
to derive classifications from primary data, if the classifications are needed
for local purposes. Meteorological data are generally recorded in standard
forms, but in the case of forests it is important to know where the data
were recorded, in relation to the forest.

The long-term objective must be to develop databases for each forest
type which provide accurate (geo)references for site location, long-term
climatic data, properly documented site descriptions and standardized

records of research data and observations. The scientific requirements and availability of remotely sensed data was discussed in Townshend (1992).

Geographic information systems (GIS) are spatially referenced databases and they serve not only as repositories for information about, and a means of mapping, forest types, soil types, topography, and any other spatially referenced information, but also as tools for analysis. The GIS packages available allow manipulation of data, so that, for example, it is possible to determine immediately, from data in the computer, the relative areas of different categories, their ratios, or similar information. More importantly, where the input data are held as layers for each polygon/cell, it is possible to apply models to evaluate the effects of change, or particular conditions, across heterogeneous landscapes. This is being done on a global scale for NPP, using satellite-derived estimates of LAI and photosynthetically active radiation. It can be done for any variable for which values or estimates of the input data are available for regions. Other examples are water balance and mineralization (from organic matter and climate data). Weather data, recorded at discrete points, can be interpolated across landscapes using digitized terrain data.

Given these data in GIS, in compatible formats, for many regions of the world, it will be possible to compare productivity and evaluate the consequences of global change at sites around the world. Hence GIS provide the means for scaling up from process-based models to the large-scale, simple models essential for wide area evaluation of forest growth and response to climate.

References

Ågren, G. I., McMurtrie, R. E., Parton, W. J., Pastor, J. & Shugart, H. H. (1991). State-of-the-art of models of production–decomposition linkages in conifer and grassland ecosystems. *Ecological Applications*, **1**, 118–38.

Barton, C. V. M., Lee, H. S. J. & Jarvis, P. G. (1993). A branch bag and CO_2 control system for long-term CO_2 enrichment of mature Sitka spruce (*Picea sitchensis* (Bong.) Carr.). *Plant, Cell and Environment*, **16**, 1139–48.

Bazzaz, F. A. (1990). The response of natural ecosystems to the rising global CO_2 levels. *Annual Review of Ecology and Systematics*, **21**, 167–96.

Comins, H. N. & McMurtrie, R. E. (1993). Long-term response of nutrient-limited forests to CO_2 enrichment: Equilibrium behavior of plant-soil models. *Ecological Applications*, **3**, 666–81.

Eamus, D. & Jarvis, P. G. (1989). The direct effect of increase in the global atmospheric CO_2 concentration on natural and commercial trees and forests. *Advances in Ecological Research*, **19**, 1–55.

Hunt, E. R. Jr., Martin, F. C. & Running, S. W. (1991). Simulating effects of climatic variation on stem carbon accumulation of a ponderosa pine stand:

comparison with annual growth increment data. *Tree Physiology*, **9**, 161–71.

Jarvis, P. G. (1989). Atmospheric carbon dioxide and forests. *Philosophical Transactions of the Royal Society of London series B*, **324**, 369–92.

Jarvis, P. G. (1993). MAESTRO: A model of CO_2 and water vapor exchange by forests in a globally changed environment. In *Design and Execution of Experiments on CO_2 Enrichment*, ed. E.-D. Schulze & H. A. Mooney, pp. 107–16. *CEC, Ecosystem Research Report* No. 6, ISBN 2-87263-112-7.

Jarvis, P. G. & Leverenz, J. W. (1983). Productivity of temperate, deciduous and evergreen forests. In *Physiological Plant Ecology IV. Ecosystem Processes, Mineral Cycling, Productivity and Man's Influence*, ed. O. L. Lange, P. S. Nobel, C. B. Osmond & H. Ziegler, pp. 233–80. Berlin: Springer.

Jenkins, A. & Wright, R. F. (1993). The CLIMEX project – Raising CO_2 and temperature to whole catchment ecosystems. In *Design and Execution of Experiments on CO_2 Enrichment*, ed. E.-D. Schulze & H. A. Mooney, pp. 211–19. *CEC, Ecosystem Research Report*, No. 6, ISBN 2-87263-112-7.

Koch, G. W., Scholes, R. J., Steffen, W. L., Vitousek, P. M. & Walker, B. H. (1996). *The IGBP Terrestrial Transects: Science Plan*. IGBP report no. 36. (In press).

Kurz, W. A., Apps, M. J., Webb, T. M. & McNamee, P. J. (1992). *The Carbon Budget of the Canadian Forest Sector: Phase 1*. Forestry Canada, Information Report NOR-X-326, ISBN 0-662-19913-8.

Landsberg, J. J. (1986). Experimental approaches to the study of the effects of nutrients and water on carbon assimilation by trees. *Tree Physiology*, **2**, 427–44.

Landsberg, J. J., Linder, S. & McMurtrie, R. E. (1995*a*). Effects of global change on managed forests. A strategic plan for research on managed forest ecosystems in a globally changing environment. GCTE Report No. 4. Canberra: IGBP/GCTE and IUFRO, ISSN 1024-414X.

Landsberg, J. J., Prince, S. D., Jarvis, P. G., McMurtrie, R. E., Luxmoore, R. & Medlyn, B. (1995*b*). Energy conversion and use of in forests: the analysis of forest production in terms of radiation utilization efficiency. In *The Use of Remote Sensing in the Modelling of Forest Productivity at Scales from Stand to the Globe*, ed. R. L. Gholz & K. Nakane. Dordrecht: Kluwer (in press).

Lee, H. S. J. & Barton, C. V. M. (1993). Comparative studies on elevated CO_2 using open-top chambers, tree chambers and branch bags. In *Design and Execution of Experiments on CO_2 Enrichment*, ed. E.-D. Schulze & H. A. Mooney, pp. 239–59. *CEC, Ecosystem Research Report* No. 6, ISBN 2-87263-112-7.

Linder, S. (1987). Responses to water and nutrition in coniferous ecosystems. In *Potentials and Limitations of Ecosystem Analysis*, ed. E.-D. Schulze & H. Zwölfer, pp. 180–202. Berlin: Springer.

Linder, S. (1995). Foliar analysis for detecting and correcting nutrient imbalances in Norway spruce. *Ecological Bulletins (Copenhagen)*, **44**, 178–90.

Linder, S. & Flower-Ellis, J. G. K. (1992). Environmental and physiological constraints to forest yield. In *Responses of Forest Ecosystems to Environmental Changes*, ed. A. Teller, P. Mathy & J. N. R. Jeffers, pp. 149–64. Amsterdam: Elsevier.

Linder, S. & McDonald, A. J. S. (1993). Plant nutrition and the interpretation of growth response to elevated concentrations of atmospheric carbon dioxide. In *Design and Execution of Experiments on CO_2 Enrichment*, ed. E.-D. Schulze & H. A. Mooney, pp. 73–82. *CEC, Ecosystem Research Report* No. 6. ISBN 2-87263-112-7.

McMurtrie, R. E., Comins, H. N., Kirschbaum, M. U. F. & Wang, Y.-P. (1992).

Modifying existing forest growth models to take account of effects of elevated CO_2. *Australian Journal of Botany*, **40**, 657–77.

McMurtrie, R. E., Gholz, H. L., Linder, S. & Gower, S. T. (1994). Climatic factors controlling the productivity of pine stands: A model-based analysis. *Ecological Bulletins (Copenhagen)*, **43**, 173–88.

McMurtrie, R. E., Rook, D. A. & Kelliher, F. M. (1990). Modelling the yield of *Pinus radiata* on a site limited by water and nitrogen. *Forest Ecology and Management*, **30**, 381–413.

McMurtrie, R. E. & Wang, Y.-P. (1993). Mathematical models of the photosynthetic response of tree stands to rising CO_2 concentrations and temperatures. *Plant, Cell and Environment*, **16**, 1–13.

Melillo, J. M., Callaghan, T. V., Woodward, F. I., Salati, E. & Sinha, S. K. (1990). Effects on ecosystems. In *Climate Change. The IPCC Scientific Assessment*, ed. J. T. Houghton, G. J. Jenkins & J. J. Ephraums, pp. 283–310. Cambridge: Cambridge University Press.

Nabuurs, G. J. (1994). State-of-the-art in the field of forest sector carbon balance studies – with reference to the European situation. European Forest Institute, Joensuu, Finland, Working Paper 2, ISBN 952-9844-01-8.

Pereira, J. S., Linder, S., Araújo, M. C., Pereira, H., Ericsson, T., Borallho, N. & Leal, L. (1989). Optimization of biomass production in *Eucalyptus globulus* – A case study. In *Biomass Production by Fast-growing Trees*, ed. J. S. Pereira & J. J. Landsberg, pp. 101–21. Dordrecht: Kluwer.

Pereira, J. S., Linder, S., Araújo, M. C., Tomé, M., Madeira, M. V. & Ericsson, T. (1994). Biomass production with optimized nutrition in *Eucalyptus globulus* plantations. In *Eucalyptus for Biomass Production. The State of the Art*, ed. J. S. Pereira & H. Pereira, pp. 13–30. Brussels: CEC.

Peterjohn, W. T., Melillo, J. M., Bowles, F. P. & Steudler, P. A. (1993). Soil warming and trace gas fluxes: Experimental design and preliminary flux results. *Oecologia*, **93**, 18–24.

Potter, C. S., Randerson, J. T., Field, C. B., Matson, P. A., Vitousek, P. M., Mooney, H. A. & Kloster, S. A. (1993). Terrestrial ecosystem production: A process model based on global satellite and surface data. *Global Biogeochemical Cycles*, **7**, 811–41.

Rastetter, E. B., Ryan, M. G., Shaver, G. R., Melillo, J. M., Nadelhoffer, K. J., Hobbie, J. E. & Aber, J. D. (1991). A general biogeochemical model describing the response of C and N cycles in terrestrial ecosystems to changes in CO_2 climate and N deposition. *Tree Physiology*, **9**, 101–26.

Ruimy, A., Saugier, B. & Dedieu, G. (1993). Methodology for the estimation of net primary production from remotely sensed data. *Journal of Geophysical Research*, **99**, 5263–83.

Running, S. W. (1993). Modeling terrestrial carbon cycles at varying temporal and spatial resolutions. In *The Global Carbon Cycle*, ed. M. Heimann, pp. 201–17. *NATO ASI Series I*, vol. 15. Berlin: Springer.

Running, S. W. & Gower, S. T. (1991). FOREST-BGC, a general model of forest ecosystem processes for regional applications. II. Dynamic carbon allocation and nitrogen budgets. *Tree Physiology*, **9**, 147–60.

Ryan, M. G., Hunt, E. R. Jr., McMurtrie, R. E., Ågren, G. I., Aber, J. D., Friend, A. D., Rastetter, E. B., Pulliam, W. J., Raison, R. J. & Linder, S. (1996). Comparing models of ecosystem function for coniferous forests. I. Model description and validation. In *Effects of Climate Change on Production and Decomposition in Coniferous Forests and Grasslands*, ed. J. M. Melillo, G. I. Ågren & A. Breymeyer, pp. 00–00. Scientific Committee on Problems in the Environment (SCOPE). Chichester: John Wiley and Sons. (In press).

Schulze, E.-D. & Mooney, H. A. (eds.)

(1993). *Design and Execution of Experiments on CO₂ Enrichment*. CEC, Ecosystem Research Report No. 6, ISBN 2-87263-112-7.

Snowdon, P. & Benson, M. L. (1992). Effects of combinations of irrigation and fertilisation on the growth and above-ground biomass production of *Pinus radiata*. *Forest Ecology and Management*, **52**, 87–116.

Steffen, W. L., Walker, B. H., Ingram, J. S. I. & Koch, G. W. (eds.) (1992). Global change and terrestrial ecosystems. The operational plan. *IGBP Report* **21**, ISSN 0284-8015.

Townshend, J. R. G. (ed.) (1992). Improved global data for land applications. A proposal for a new high resolution data set. *IGBP Report* **20**, ISSN 0284-8015.

Wang, Y.-P. & Jarvis, P. G. (1990). Description and validation of an array model: MAESTRO. *Agricultural and Forest Meteorology*, **51**, 257–80.

Waring, R. H., Runyon, J., Goward, S. N., McCreight, R., Yoder, B. & Ryan, M. G. (1993). Developing remote sensing techniques to estimate photosynthesis and annual forest growth across a steep climatic gradient in western Oregon, USA. *Studia Forestalia Suecica*, **191**, 33–42.

Linked pest-crop models under global change

P. S. Teng, K. L. Heong, M. J. Kropff, F. W. Nutter
and R. W. Sutherst

Introduction

One of the first effects of changes in global climate, atmospheric composition or land use on agriculture is likely to be on the distribution, abundance and dynamics of pests (insects, weeds, pathogens) and natural enemies or antagonists, and pest–plant interactions. In GCTE Activity 3.2, pests are anticipated to react in a variety of ways: (i) the distribution and abundance of current key pests may be affected and hence change their effects on yield and consequently alter demands for mitigation techniques such as applications of pesticides; (ii) the outcome of competition between weeds and crops may be affected through changes in ecophysiology; and (iii) pests currently of minor significance may become more important, thereby causing increased losses. In addition, the behaviour of migratory or polyphagous pests may alter. GCTE Activity 3.2 will develop networks to (i) standardize experimental and analytical methods; (ii) conduct comparative trials in selected areas; and (iii) share data and models, in support of the broad goal of providing assessments of global change on pests, and from that, the socio-economic cost of such impacts and their mitigation. The focus of GCTE Activity 3.2 in the medium term is the development of models and their application to key pests of six agronomic species (rice, cassava, peanut, wheat, maize and potato) and to this end, strong links are envisaged between pest scientists and crop modellers. In this paper, we review progress in the use of models to predict the effects of global change, firstly, on pests and pest communities, and secondly, on pest–crop interactions, with particular attention to rice and rice pests.

Models for predicting the distribution, dynamics and abundance of pests under global change

There are a number of models that could be used for predicting the distributions, dynamics and abundance of pests and diseases under global change.

Generally, available models are either inductive, deductive or a combination of both approaches. They range from static, statistical models to dynamic climate-matching models and process-based population models. The advantages and disadvantages of each approach vary depending on the application.

Certain features are required in models to deal with global change:

- the ability to handle spatial and temporal variance in climate data;
- improved reliability in the extreme range of parameter values;
- linkages with non-climatic tools to handle qualitative data and quantitative data on the physical environment; and
- vertical integration with models of different trophic levels to account for host plant–pest–predator interactions.

The priority tasks to be addressed by IGBP-GCTE in order to meet the Task 3.2.2 objective of refining existing models to predict the responses of key pests and pathogens to predicted changes in the physical environment, climate and atmospheric composition are:

- identification of the key global changes that need to be taken into account so that these effects can be included and described in the models;
- agreement on the roles and limitations of the existing models, so that a degree of standardization can be reached to facilitate more concentrated effort on improving those tools;
- development of linkages between models at different trophic levels; and
- improved methods of automated data collection and interpretation to develop a sound factual basis for developing and testing the models.

Effects of climate change on pest species *per se* have been estimated using climate matching (for example, CLIMEX) and population dynamic models (for example, BLASTSIM2) in conjunction with geographic information systems (GIS) to identify geographic zones of potentially high impact.

Climate matching algorithms

Approaches for investigating geographical distributions in relation to changing climates were recently reviewed by Sutherst *et al.* (1995). BIOCLIM produces a 'climate profile' for a species based on 16–24 input meteorological variables. The core of BIOCLIM is a high-resolution 'climate surface', used to interpolate temperature and precipitation between meteorological stations for any given grid point. BIOCLIM-based approaches have been particularly useful in conservation biology and forestry. BIOCLIM is suitable for climate change studies where there is no immediate interest in understanding the mechanisms involved in changes to the geographical distributions of organisms.

CLIMEX is a dynamic climate-matching model which attempts to emulate the process by which species integrate the effects of climate in different times and places and reflect this in their response through geographical distribution. The model is based on an integrated inductive and deductive approach as appropriate to the given situation. It is designed to describe the climatic requirements of a species with a minimum of data on biological processes. An example of the use of CLIMEX to investigate the response of a species to climate change is given in Fig. 16.1.

There is a large number of climate-driven pest population models suitable for investigating the impact of climate change on phenology and abundance of pests. In the case of insects, these population models are generally not accurate enough for predictions of abundance and are mostly useful for estimating the timing of specific events in life cycles. Nevertheless, changes in timing and the length of the growing season are likely to be important effects of climate change.

Pest models may be classified by various functional groups (Pinnschmidt *et al.*, 1994). It is important to determine the objective of model building before commencement of the exercise as this would influence strongly the type of model to build. The world's scientific environment is such that it is unlikely that the data required to build detailed model descriptions of many species will be obtainable (Sutherst *et al.*, in press). Therefore, pragmatic approaches offer the best prospects of progress (Worner, 1991). While the CLIMEX model provided a tool to approach such problems in the cases where minimal data are available, the statistical model described by Hudes and Shoemaker (1988) provides a method for modelling insect phenology based on field observations. Similarly, the algorithm of Dallwitz and Higgins (1978) is useful for estimating development rates from observations under ambient temperature conditions. The PEST package of J. E. Doherty (personal communication) also provides a ready means of optimizing parameter values in a simulation model using a least-squares statistical routine. Generic approaches to building process-based simulation models (Larkin *et al.*, 1988; Logan, 1988) provide means of reducing coding costs.

Quantifying effects of global change on arthropod communities

Climatic factors limit the distribution and abundance of insect pest species and their associated communities. Temperature, for example, affects insect survival, reproduction and development. In the rice Brown Plant Hopper

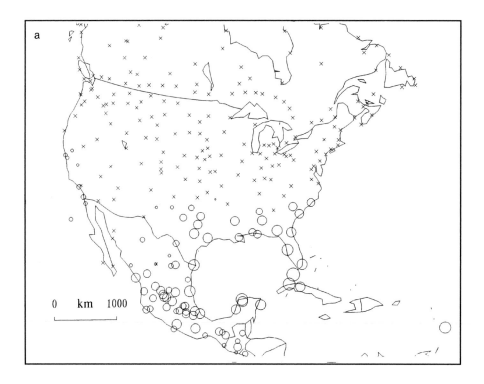

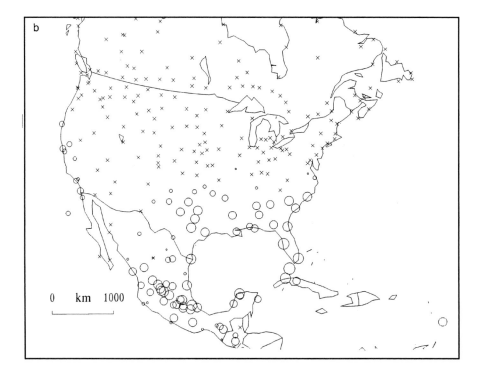

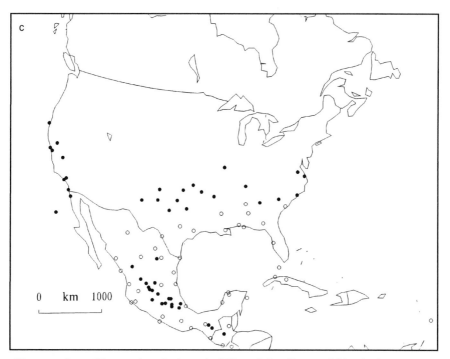

Figure 16.1 *Changes in relative suitability of the climate of North America for the Medfly,* Ceratitis capitata *given a change in the climate of 30°C as a result of the enhanced greenhouse effect. a, current scenario; b, greenhouse scenario; c, difference between greenhouse and current. Solid circles represent increase in suitability, open circles represent decrease.*

(BPH), adult survival remains relatively unchanged between 25 °C and 35 °C, but is drastically reduced at 40 °C (Heong *et al.*, 1994). Oviposition of BPH females at 35 °C and 40 °C are known to be higher than that at 25 °C and 30 °C, although egg survival is reduced at 35 °C relative to the lower temperatures.

In regions frequently subjected to higher temperatures, there is also likely to be higher tolerance by pest species to high temperatures. This phenomenon may be quantified using probit analysis, as has been done for the egg predator, *Cyrtorhinus lividipennis*. The median lethal exposure time, LT_{50}, for *Cyrtorhinus* was found to be very low relative to that for BPH (Heong & Domingo, 1992). This suggests possible dissociation in species range overlap between *Cyrtorhinus* and BPH which in turn implies reduced egg predation. The wolf spider, *Pardosa pseudoannulata*, another predator of BPH, on the other hand, is more adaptable. In Khon Kaen, Thailand, where BPH is frequently subject to higher temperatures, another egg predator belonging to the same family, *Tytthus*, was found to be more common (Heong *et al.*, 1994). This further implies that there is great variability in rice arthropod communities in their response to abiotic stresses, especially if the stress changes gradually. The impact of global warming appears to depend on genetic variability within the populations of arthropods and the distribution of tolerant phenotypes, migration patterns and competition.

Modelling and estimating effects of global change on pest–crop interactions

A conceptual framework for generating assessments of impact has been proposed for ICBP/GCTE Activity 3.2, in which pest effects are linked directly to crop models, and through appropriate input data, spatial effects are also estimated (Fig. 16.2). This framework is proposed for application to six crops – rice, potato, wheat, maize, peanuts and cassava – for which crop modelling networks have or are being created under the auspices of GCTE. The intention, however, is not to exclude other crop–pest combinations from consideration but to provide a focus on these crops.

Much progress has been made to identify potential damage mechanisms that link pest effects to crop growth and development at the physiological level. The damage mechanism (Rabbinge *et al.*, 1989) is a quantitative equation that links a pest variable (examples include population number, proportion injured, and amount of plant tissue) to a crop variable (for example, leaf area and photosynthetic rate). A pest can influence the crop via one or more damage mechanisms. Pest effects which have been linked to crop models and potentially may be used to predict effects of changes in climate and UV-B, are those of rice (ORYZA1, CERES-RICE), peanuts (PEANUTGRO), soybean (SOYGRO) and wheat (MACROS) (Pinnschmidt *et al.*, 1994).

Crop models available for linking pest effects

Two major types of crop models can be distinguished based on how they simulate dry matter accumulation. The first type involves models that simulate daily dry matter production based on leaf photosynthesis and respiration, using detailed routines for light absorption at different heights in the canopy and for different times during the day. Effects of pests and diseases on physiological processes at the organ level (for example, leaf photosynthesis) can be introduced. Examples are the SUCROS models (generic model that has been parameterized for many crops such as wheat, potato, maize: Spitters *et al.*, 1989), MACROS models (Penning de Vries *et al.*, 1989) and the ORYZA models (rice) that are derived from those of Kropff *et al.* (1993) and the GRO models for crops such as soybean, peanut and

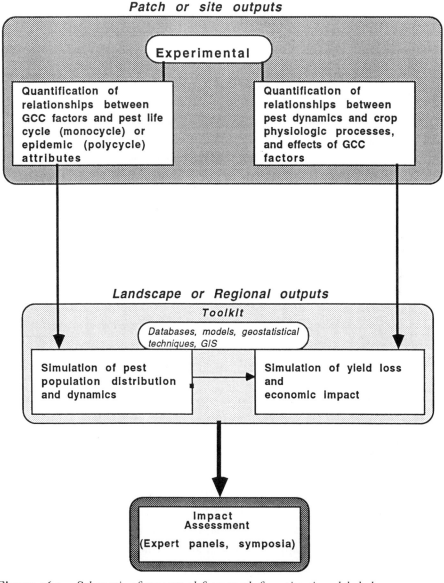

Figure 16.2 *Schematic of conceptual framework for estimating global change effects on pests using pest–crop models and associated databases. (Source: GCTE Activity 3.2 Operational Plan.)*

other leguminous crops (Jones *et al.*, 1989). The second type involves models that use light use efficiency to estimate dry matter accumulation directly from the daily absorbed radiation by the canopy, for example, Simulation Model for Rice–Weather relations (SIMRIW) for rice (Horie, 1987) and GUMCAS for cassava (Matthews & Hunt, 1994). Other features such as morphological development, phenological development, water stress effects, and nitrogen effects are simulated in varying detail by the different models. Some models only simulate potential yield and/or water limited yields. In many cases, more than one model is available for a crop species, and it is then necessary to evaluate the different models for their utility in pest coupling. The following sections describe models for rice which may have pest effects incorporated into their structures.

Three models of varying complexity have been developed for rice: ORYZA1 (developed from the Wageningen models SUCROS and MACROS), CERES-RICE, and SIMRIW. CERES-RICE is a dynamic model for simulating plant growth and development in different production situations (Singh *et al.*, 1993). The model simulates growth and development of a rice crop which has been established by transplanting, wet and dry broadcast seeding or row seeding, from transplanting/seeding to maturity on the basis of physiological processes determined by the crop's response to soil and canopy environmental conditions. It simulates water balance under fully irrigated conditions, rain-fed conditions with fluctuating water regime, and/or in fully upland conditions where the soil is never flooded. Crop development in CERES-RICE has two distinct features: phasic development and morphological development. The rice model predicts the following phenological events to mark the end or the beginning of a growth stage: dates of emergence, panicle initiation, anthesis, and physiological maturity. CERES-RICE simulates the number of leaves, tillers, and grains. The principal environmental factor affecting crop morphogenesis is temperature. Potential tiller formation and growth are controlled by rate of leaf appearance while actual growth is dependent on assimilate availability, temperature, and water and nutrient stresses. The rate of leaf appearance in the CERES-RICE model is thermal time driven. Leaf area index (LAI) is simulated as a function of leaf tip appearance rate (which in turn is temperature-driven) and leaf expansion growth rate. CERES-RICE uses this approach to leaf area prediction until the plant becomes source-limited.

The growth submodel in CERES-RICE simulates plant mass accumulation, expansion growth, assimilate partitioning, and senescence. Potential mass accumulation is dependent on absorption of photosynthetically active radiation, the radiation use efficiency, LAI, and light extinction coefficient as affected by row spacing. The plant's development phase dictates assimilate partitioning for the growth of roots, leaves, culms, panicles, and grains. Partitioning of assimilates is not only growth stage-dependent but is also under genotypic influence. Calculation of sink size or the grain number is the most critical step in accurate yield prediction. In the CERES-RICE model, the sink size is determined by the culm biomass at anthesis and genetic coefficients for potential spikelet number and single grain weight. Senescence of leaves is dependent on stage of growth, shading, non-optimal temperatures, and water and N stresses.

SIMRIW is a simplified process model for simulating growth and yield of irrigated rice in relation to weather. The model was developed by a rational simplification of the underlying physiological and physical processes of the growth of the rice crop (Horie, 1987). Because of this, it requires only a limited number of crop parameters which can easily be obtained from well-defined field experiments, and hence is applicable to a wide range of environments.

Crop dry matter production is proportional to the amount of photosynthetically active radiation (PAR) or short-wave radiation absorbed by a crop canopy. For rice, the proportional constant – the conversion efficiency from solar radiation to biomass – is constant until the middle of the ripening stage but thereafter decreases curvilinearly. These functions and parameters have been determined from field experiment data for eight major cultivars of both *japonica* and *indica* rice varieties grown under widely different environmental conditions (Horie, 1987). Moreover, Horie concluded from both simulations and experiments that the conversion efficiency is practically unaffected by climatic conditions in a wide range of environments. Full details of the derivation of each functional relationship and parameters of the model are given in Horie (1987).

The IRRI-Wageningen ORYZA models simulate rice growth and development in irrigated and rain-fed rice production systems. The basic model ORYZA1 simulates potential growth of a rice crop, based on detailed physiological calculations of CO_2 assimilation and respiration (Kropff *et al.*, 1993). It uses a calculation scheme for the rates of daily dry matter production of the plant organs, the rate of leaf area development and the rate of phenological development. By integrating these rates over time, dry matter production of the crop is simulated throughout the growing season. The total daily rate of canopy CO_2 assimilation is calculated from the daily incoming radiation, temperature and LAI. The model contains a set of subroutines that calculate the daily rate of canopy CO_2 assimilation by integrating instantaneous rates of leaf CO_2 assimilation. On the basis of the photosynthesis characteristics of single leaves, which depend upon the N concentration, the photosynthesis profile in the canopy is obtained. Integration over the LAI of the canopy and over the day gives the daily CO_2 assimilation rate. After subtraction of respiration requirements, the net daily growth rate in kilograms of dry matter per hectare per day is obtained. In contrast to the other models, ORYZA1 simulates the light use efficiency (LUE) from photosynthesis and respiration, whereas the LUE is input or built-in in the other two models. The dry matter produced is partitioned among the various plant organs following a partitioning pattern that is determined by the developmental stage of the crop.

Phenological development rate is tracked in the model as a function of ambient daily average temperature and photoperiod. When the canopy is open, leaf area increment is calculated from daily average temperature (sink-limited), because carbohydrate production does not limit leaf expansion. When the canopy closes, the increase in leaf area is obtained from the increase in leaf weight (source-limited). Integration of daily growth rates of the organs and leaf area results in dry weight increments during the growing season. The model simulates the formation of spikelets as a function of the growth between panicle initiation and flowering. A simple procedure was built in the model to simulate sink limitation as a result of spikelet sterility at high or low temperatures. These functions were derived from Horie (1987).

Generic damage mechanisms: linking pest effects to crop models

There have been many efforts recently to use crop growth models to simulate the effects of pest damage on crop growth and yield. One attempt is to link pest population models to crop growth models. This pest–crop modelling approach has been discussed in detail by Boote *et al.* (1983), Loomis and Adams (1983), Rouse (1988), Teng and Johnson (1988), Rabbinge *et al.* (1989), and Kropff and van Laar (1993). Table 16.1 summarizes research related to modelling pest damage on rice using crop growth models. Most of the work has been conducted using either MACROS (Penning De Vries *et al.*, 1989) or CERES-RICE (Alocilja & Ritchie, 1988). Pinnschmidt *et al.* (1994) surveyed the literature and found that in most studies (Table 16.1), empirical pest levels were introduced into the crop simulation, with few studies on pest–crop models, where pest development is driven by crop variables and vice versa (examples include Benigno *et al.*, 1988, Graf *et al.*, 1990*a*). In most cases, damage effects of single pest species were simulated separately. There have been several attempts to incorporate effects of multiple pest populations on simulated crop yield (Haque, 1991; Pinnschmidt, 1991; Pinnschmidt *et al.*, 1990, 1991, 1994). These typically involve incorporation of interactive effects of a limited number of damage types such as reduction of photosynthesis, leaf feeding, removal of assimilates, detillering, respiration increase, and competition for light, nitrogen, and water. In many cases, parameterization of the pest damage mechanisms was done using 'educated guesses', due to a lack of quantitative data.

The simulation of pest effects with pest-coupled crop models requires knowledge of the mechanisms of pest damage. Table 16.2 from Pinnschmidt *et al.* (1994) shows a detailed list of pests and their associated physiological coupling points and effects on rice. In some cases, pest effects could be measured quantitatively and in other cases, damage characterization was qualitative. Only quantitative data is useful for simulation. For example, daily leaf consumption rates for leaffolders (Heong, 1990) or the functional relationship between leaf blast severity and leaf photosynthesis (Bastiaans, 1991*a*) can directly be used for formalizing and parameterizing pest damage effects in crop simulations. Where direct observations of damage mechanisms are difficult, researchers have used indirect methods, like honeydew production, to estimate the assimilate consumption of brown plant hoppers (Sogawa, 1970; Kim & Kim, 1986). Factors such as plant age, larval age and development stage, varietal resistance, and temperature may affect feeding activities and have to be accounted for (Kim & Kim, 1986; Cheng, 1987; Heong, 1990). An alternative to 'mechanistic' studies (Table 16.2) may consist of estimating damage coefficients by means of multi-criteria parameter estimation and optimization techniques based on field data (Alocilja & Ritchie, 1988; Klepper & Rouse, 1991).

Most physiologically based techniques for linking pest-induced injury to crop models use area-under-disease-progress-curves (AUDPC), healthy-area-duration (HAD), or healthy-area-absorption (HAA) to quantify pest effects on yield and economic thresholds (Pinnschmidt *et al.*, 1994). These methods are based on the biological and physiological aspects of the plant–pest interaction and measure either the cumulative diseased host fraction (AUDPC) or the cumulative amount of healthy host material (HAD) and PAR intercepted by healthy host tissue (HAA) over time. Although various applications have been proposed (Johnson, 1987; Waggoner & Berger, 1987; Teng & Johnson, 1988), these have only been recently employed to quantify pest-induced yield losses or yields under pest stress for rice. Their applicability may be limited to the varieties that were used to develop particular equations, because varieties differ in leaf photosynthetic rate (IRRI, 1968; Bastiaans & Roumen, 1993) and other characteristics of crop productivity (Yoshida, 1981).

A generic approach for simulating multiple pest effects was recently proposed by Pinnschmidt *et al.* (1994) for rice but has general applicability. The authors used CERES-RICE, in which biomass at harvest (Y_h) is the cumulative outcome of interactions between physiological processes (M_{mt}) as influenced by environment (E_t) and crop management (C) inputs. Each M_{mt} is a potential coupling point where pest effects can change crop growth. In CERES-RICE, 20 coupling points were identified (Pinnschmidt *et al.*, 1994): leaf, root, stem, panicle and grain consumption, leaf shading,

Table 16.1 *A survey of literature on pest effects simulation in rice*

Pest or damage type	Damage effects simulated	Method of quantifying pest level	Crop model	Authors
Bacterial leaf blight	Photosynthesis reduction	Empirical	MACROS	Narasimhan et al., 1991; Reddy et al., 1991
Brown Planthoppers	Assimilate consumption	Empirical	No name	Kenmore, 1980
Defoliators	Leaf consumption	Empirical	CERES	Teng et al., 1990
Leaf blast	Leaf consumption respiration increase, leaf covering, photosynthesis reduction, shading	Empirical	MACROS	Bastiaans, 1991a, b; Chen et al., 1991
Leaf blast and Leaffolders	Leaf covering Leaf consumption	Simulation model Analogue model	CERES	Alocilja & Ritchie, 1988
Leaffolders	Leaf consumption, Leaf folding	Simulation model Empirical Simulation model Empirical Simulation model Simulation model	MACROS RICESYS MACROS	Benigno et al., 1988 De Jong, 1991 De Kraker, 1991 Ma Jufa et al., 1991 Graf et al., 1992 Fabellar et al., 1994
Malayan black bug	Assimilate consumption	Empirical	MACROS	Hamid & Selamat, 1991
Multiple pests	Various	Empirical/ simulation model	CERES	Pinnschmidt, 1991 Pinnschmidt et al., 1990, 1991

Sheath blight	Photosynthesis reduction, respiration increase	Empirical	MACROS	Singh & Das, 1991
Sheath rot and ufra	Panicle exsertion rate reduction	Empirical	MACROS	Haque, 1991
Stem borers	Stand reduction, (shading)	Empirical	MACROS	Rubia & Penning De Vries, 1990; Pathak & De Jong, 1991; Singh et al., 1991; Xu Zhihong & Wang Zhaoqian, 1991
Weeds	Competition for light, N, (water)	Simulation model	RICESYS	Graf et al., 1990a
		Simulation model	INTERCOM	Rajan et al., 1991; Kropff et al., 1993

From Pinnschmidt et al. (1994).

Table 16.2 Survey of studies on damage mechanisms and damage activities of rice pests

Pest	Affecting through	Damage mechanism respectively: affected crop physiological process	Data[a] type	Authors
Bacterial leaf blight	Disease severity	Photosynthesis	2	Reddy et al., 1991
Brown plant hoppers	No. of pest individuals, instar, sex	Assimilate consumption	2	Suenaga, 1959, Sogawa, 1970, Kenmore, 1980, Kim & Kim, 1986
		Leaf senescence, phloem blockage, 'hopperburn', plant biochemistry	1	Santa, 1959; Hisano, 1964; Sogawa, 1971; 1973a; b; Sogawa & Cheng, 1979; Cagampang et al., 1974
Leaf blast	Disease severity	Respiration	2	Sun et al., 1986
		Photosynthesis, shading, respiration	3	Bastiaans, 1991a, b
Leaf folder	No. of pest individuals, plant age, temperature	Leaf consumption	3	Heong, 1990
		Leaf consumption	2	
Malayan black bug	No. of pest individuals, plant age	Assimilate consumption	3	Hamid & Selamat, 1991

Sheath blight	Disease incidence	Photosynthesis, transpiration, respiration, plant biochemistry	1 / 1	Roy, 1982
Stem borers	Pest incidence	Detillering, shading	2	Rubia & Penning De Vries, 1990; Pathak & De Jong, 1991; Singh *et al.*, 1991; Xu Zhihong & Wang Zhaoqian, 1991

From Pinnschmidt *et al.* (1994).

[a]1, qualitative ('soft') data – 3, quantitative ('hard') data.

photosynthesis reduction, assimilate consumption, respiration increase, phloem blockage/translocation reduction, growth rate reduction of leaf, root, stem and grain, light competition, leaf senescence acceleration, xylem blockage, altered transpiration, stand reduction, and blockage of leaf and stem reserves. The 20 coupling points corresponded to damage sites incurred by the most important rice pests. In order to implement the equations associated with each coupling point and damage mechanism, a structure had to be developed to define pest damage and apply pest damage to coupling points in the crop model. A pest module, consisting of two primary components, was developed to read pest data into the model and compute effects of pest damage on coupling points (Pinnschmidt *et al.*, 1994). The first module creates, edits, and retrieves pest input data and the second module computes pest damage effects on coupling points. Module 1 is executed on the first day of simulation before crop growth begins, while module 2 is embedded in the daily simulation loop of the crop model and allows a dynamic simulation of pest damage effects via the coupling points. When simulating specific examples from the literature, the authors obtained excellent fit between simulated and observed yield losses for simple damage, such as that caused by detillering (Rubia & Penning De Vries, 1990) and for complex damage, such as that caused by sheath blight (Singh & Das, 1991).

Predicting global change effects using linked pest–crop models

In order to estimate global change impact on pests and crops, both components need to be modelled using climate as a driving variable. Most insect–crop models do not yet meet this criterion. Further, few models include the interactions of insect pests with natural enemies within a community. Diseases and weeds seem to provide relatively simple two species interaction systems for modelling global change.

Diseases

The estimation of global change effects on a specific disease, leaf blast, caused by the fungus *Pyricularia grisea*, is discussed here to illustrate the approach proposed in GCTE Activity 3.2. Two simulation models (CERES-RICE for rice growth and development, and BLASTSIM for leaf blast) were coupled based on quantitative effects of leaf blast on rice leaf photosynthesis and biomass production (Luo *et al.*, 1995). The CERES-RICE model can be used to estimate yield through the process of grain formation, filling, and maturity. The effects of N, water and crop management practices on rice growth are also considered in the model. In the BLASTSIM model (Calvero & Teng, 1992), the disease cycle of blast is described by subroutines such as spore production, dispersion, deposition, and infection, latency period, lesion formation, lesion development, and infection process. Rates of change between the processes are dictated by environmental factors (e.g. temperature, rainfall, relative humidity) quantified empirically. CERES-RICE provides leaf growth data for blast development and simulates final yield with or without blast epidemics. The quantitative relationship between reduction of photosynthesis and blast severity was:

$$BL\text{--}COEF_1 = (1 \cdot 0 \text{--} DISSEV)^b$$

where $BL\text{--}COEF_1$ is a parameter describing effects of disease on photosynthesis in terms of proportion of reduction of photosynthesis, $DISSEV$ is disease severity, and b is a parameter with the value of 3.74 (Bastiaans, 1991a). The effects of disease on rice leaf photosynthesis were then calculated using the equation:

$$CARBO = CARBO \cdot BL\text{--}COEF_1$$

where $CARBO$ is daily biomass production (g/plant) calculated in CERES-RICE.

$DISSEV$ was calculated by:

$$DISSEV = SIZEOL/(PLA\text{--}LA\text{--}DEAD)$$

where $SIZEOL$ is diseased leaf area simulated from $BLASTSIM$, PLA is total plant leaf area calculated from CERES-RICE, and $LA\text{--}DEAD$ is dead leaf area estimated by the newly constructed subroutine. Disease severity is the diseased leaf area divided by total leaf area.

Historical daily weather data were collected from 53 locations in five Asian rice-growing countries. From the DSSAT system (Jones, 1993), the weather generators WMARK and WGEN were used to produce estimated weather data. These generators need at least 5 years of daily weather data including solar radiation, maximum temperature, minimum temperature, and rainfall. The generators produce the relevant coefficients for each weather factor from historical data, which are then used to produce the estimated weather data. Thirty years of estimated weather data were produced for each location and utilized for each simulation treatment.

To study the effects of global climate changes on rice blast epidemics, two weather factors (temperature and rainfall) were considered in simulations, increasing or decreasing daily data on the basis of estimated current, 'normal' weather data. Different levels of temperature change were defined in different locations in countries belonging to a specific agroecological zone. Simulations were conducted for each rice-growing season in each location driven by 30 years' weather data. From each combination of climate change conditions, the outputs generated were: frequency distributions of AUDPC, maximum disease severity, and simulated yield loss compared with simulated yield without disease under the same weather conditions.

Statistical analysis of simulation results showed that changes in rainfall levels had no significant effect on variation of disease epidemics in all cases. This was because rainfall changes led to very slight change of dew period simulated in the BLASTSIM model. The changes in temperature levels had significant effects on blast epidemics in most cases, but effects varied with the agroecological zone, as follows (Luo *et al.*, 1995):

1. In most parts of the cool subtropical zones, increasing temperature increased maximum blast severity and AUDPC. Elevated temperatures will therefore increase the risk of blast epidemics in these zones.
2. In the humid tropics and warm humid subtropics (Philippines and Thailand), elevation of temperature will inhibit blast development (i.e. lower risk).

Weeds

Ecophysiological models for interplant competition for light, water, and nutrient resources have been developed in the past decade by linking plant growth models for crop and weed species (review by Kropff & van Laar, 1993). These models are based on the principle that competition is a dynamic process that can be understood from the distribution of growth-determining (light) or growth-limiting (water and nutrients) resources to the competing species and the efficiency with which each species uses the resources available. Ecophysiological models that simulate these processes provide insight into competition effects observed in (field) experiments and may help in seeking ways to manipulate competitive relations, such as those between crop and weeds, by determining the most important factors in crop–weed competition.

The ecophysiological models consist of coupled crop growth models equal to the number of competing species. The model INTERCOM is the most widely tested to date (Kropff & van Laar, 1993). Experiment-specific input requirements of the ecophysiological model include geographical latitude, standard daily weather data, soil physical properties, dates of crop and weed emergence, and crop and weed density. A detailed description of the ecophysiological simulation model is given by Kropff and van Laar (1993). Under favourable growth conditions, light is the main factor determining the growth rate of the crop and its associated weeds. From the LAI of the species, the vertical distribution of their leaf area and their light extinction properties, the light profile within the canopy is calculated. Based on the species characteristics for the photosynthetic light response of single leaves, the vertical photosynthesis profile of each species in the mixed canopy is obtained. Integration over the height of the canopy and over the day gives the daily assimilation rate for each species. After subtracting the respiration requirements for maintenance, the net daily growth rate in kilograms of dry matter per hectare per day is obtained using a conversion factor for the transformation of carbohydrates into structural dry matter. The dry matter produced is partitioned among various plant organs, using partitioning coefficients that are introduced as a function of the phenological development stage of the species. Phenological development rate is tracked in the model as a function of ambient daily average temperature. When the canopy is not closed, leaf area increment is calculated from daily average temperature, since leaf expansion is sink-limited. When the canopy closes, increase in leaf area is obtained from increase in leaf weight using specific leaf area (SLA m^2 leaf/kg leaf) since leaf expansion is source-limited at this stage. Integration of daily growth rates of the organs and leaf area results in the time course of LAI and dry weight during the growing season. Height growth rate is calculated as a function of temperature.

Nutrient competition has not yet been included in the model because nutrients were always in ample supply in the validation experiments. For situations where N is limiting, an extended version of the model can be used, according to the principles described by Kropff and van Laar (1993). The model INTERCOM has been tested with data from competition experiments with the following combinations: maize *Sinapis arvensis* L., maize *Echinochloa crus-galli* L. and sugarbeet *Chenopodium album* L. in the Netherlands; tomato *Amaranthus retroflexus* L. and tomato *Solanum americana* in Canada; and rice *Echinochloa crus-galli* in the Philippines (Kropff & van Laar, 1993). The model INTERCOM was evaluated in most detail using a wide range of data sets on competition between *Echinochloa* species and transplanted or direct-seeded rice (Kropff & van Laar, 1993). The yield loss response surface was accurately predicted by the model over this wide range of competition situations. This means that the differences in yield loss between the experiments can be explained by the ecophysiological model based on crop density, weed density and the period between crop and weed emergence and establishment method of rice.

Another ecophysiological simulation model for rice–weed competition was developed by Graf *et al.* (1990*a*,*b*). This model for nitrogen and light competition in rice is based on a general crop growth model developed by Graf *et al.* (1990*a*,*b*). Many aspects of their model are similar to the approaches used in INTERCOM. They divided the weed flora into six groups based on differences in leaf shape, growth form, height and phenology. Graf *et al.* (1990*a*,*b*) also found a close correspondence between

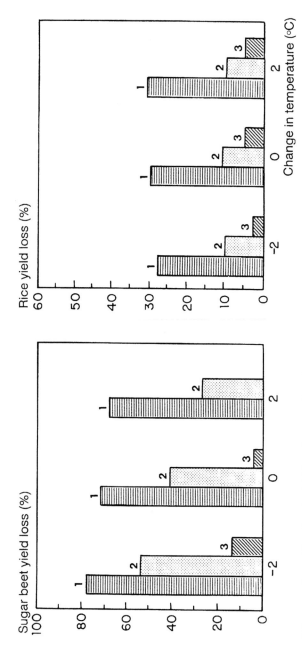

Figure 16.3 *Simulated yield loss in sugar beet by competition with C. album (5.5 plants m⁻², 1985, Wageningen, Netherlands) and rice (700 plants m⁻² direct-seeded) by competition with E. crus-galli (100 plants m⁻², Los Banos, Philippines, 1992) at different periods between crop and weed emergence at normal temperatures, and a temperature change of +2°C and −2°C for the whole growing season. Periods between crop and weed emergence 0 (1), 10 (2) and 20 days (3). (From Kropff & van Laar, 1993.)*

simulated and observed dry matter production of crop and weeds. The effect of different weeding treatments was simulated accurately.

The INTERCOM model was used to analyse the effect of temperature changes on rice *E. crus-galli* competition (tropical conditions) and sugarbeet *Chenopodium album* competition (temperate conditions) (Fig. 16.3). Yield loss in sugarbeet decreased at increasing temperatures because the life cycle of *C. album* shortens, thereby reducing the period of competition for light. The effect of temperature changes on competition between rice and *E. crus-galli* had little effect on yield losses by *E. crus-galli*. Yield losses slightly increased at a higher temperature, because *E. crus-galli* is a C_4 species, with higher optimum temperatures for several growth processes.

The effect of changes in CO_2 concentration can also be analysed by the model by introducing the effect of CO_2 on the initial light use efficiency and the maximum rate of leaf photosynthesis. It is well known that C_4 species show much less photosynthetic response to CO_2 enrichment than C_3 species, as a result of differences in the biochemical pathways. So, increased CO_2 may be more beneficial to C_3 species. Differences in morphological responses to CO_2 between species like changes in the leaf area ratio have been observed (Bazzaz *et al.*, 1989). Such responses will strongly affect the competitive ability of species.

Conclusions

Dynamic pest-coupled crop models are an exciting new way to quantify the effects of pests on crop development and yield. Their great potential lies in the capacity for adjusting crop simulations to different levels of pest stress (injuries) and subsequently simulating the corresponding yield. Their potential application is not limited to specific pest situations or cropping conditions, since the approach is strictly mechanistic. Thus, yield losses resulting from single as well as multiple pest scenarios can be simulated for any period of pest occurrence and for any cropping condition. Due to their physiologically based structure and dynamic nature, pest-coupled crop models allow mimicry of yield loss more realistically than other approaches. If the crop model employed is sensitive to the effect of physiological crop age, then its pest-coupled version will respond by recognizing the differential effect of crop age on yield loss caused by pests. Thus, the yield loss caused by a given level of injury or disease severity will depend on the time of its occurrence.

Because of their flexibility, pest-coupled crop models might become the ideal tool for developing control tactics and strategies and thus improving decision making in integrated pest management as emphasized by Teng (1988). Such models provide the means to incorporate the crop and its growing conditions as a component in yield loss predictions and to estimate pest-free and pest-affected yields under variable conditions. Certainly, with the trend towards smaller and lower cost portable microcomputers, increasing hardware should not be a constraint to their use.

Isoloss-curves and isodamage-curves are some of the decision tools that have been developed using pest-coupled crop models. These reflect the different levels of injury that may cause equal yield loss or the same level of injury that may cause different yield loss, depending on the physiological crop age. By including economic values such as yield goal, price per unit of yield, control costs, and benefit from control in the consideration, damage thresholds can be suggested at which actions to control specific pests are economically justifiable. Least-loss strategies can thus be developed and pesticide application schemes be optimized (Teng, 1988). All this might be extremely useful especially when dealing with multiple pests, since the level of complexity involved in economically guided decision making increases under multiple pest situations. An additional advantage of pest-coupled crop models over conventional methods is that dynamic rather than static damage thresholds can be developed that account for variability in the phenological patterns of insect pest population, disease severity or weed population development.

References

Alocilja, E. C. & Ritchie, J. T. (1988). Upland rice simulation and its use in multicriteria optimization. Research Report Series No. 1, IBSNAT project. Honolulu: University of Hawaii.

Bastiaans, L. (1991a). The ratio between virtual and visual lesion size as a measure to describe reduction in leaf photosynthesis of rice due to leaf blast. *Phytopathology*, **81**, 611–15.

Bastiaans, L. (1991b). Quantifying the effect of rice leaf blast on leaf photosynthesis, crop growth, and final yield using an ecophysiological model of crop growth. In *Rice Blast Modeling and Forecasting*, pp. 89–99. Los Baños, Philippines: International Rice Research Institute.

Bastiaans, L. & Roumen, E. C. (1993). Effect on leaf photosynthetic rate by leaf blast for rice cultivars with different types and levels of resistance. *Euphytica*, **66**, 81–7.

Benigno, E. A., Shephard, B. M., Rubia, E. G., Arida, G. S., Penning De Vries, F. W. T. & Bandong, J. P. (1988). Simulation of rice leaffolder population dynamics in lowland rice. *IRRI Research Paper Series*, **135**, 1–8.

Boote, K. J., Jones, J. W., Mishoe, J. W. & Berger, R. D. (1983). Coupling pests to crop growth simulators to predict yield reductions. *Phytopathology*, **73**, 1581–7.

Cagampang, G. B., Pathak, M. D. & Juliano, B. O. (1974). Metabolic changes in the rice plant during infestation by the brown planthopper,

Nilapar lugens Stal (Hemiptera: Delphacidae). *Applied Entomology and Zoology*, **9**, 174–84.

Calvero, S. B. & Teng, P. S. (1992). Validation of BLASTSIM.2 model in IRRI blast (BI) nursery and Cavinti, Lafuna, Philippines. *International Rice Research Newsletter*, **17**(5), 20–1.

Chen Zhongxiao, Hu Guowen & Ma Jufa (1991). Modelling rice growth dynamics and the damage caused by leaf blast (*Pyricularia oryzae* Cavara). In *Simulation and Systems Analysis for Rice Production* (SARP), ed. F. W. T. Penning De Vries, H. H. Van Laar & M. J. Kropff, pp. 305–9. Wageningen: Pudoc.

Cheng, C. H. (1987). Investigation on bionomics of the rice leafroller, *Cnaphalocrocis medinalis* (Guenee) in the south of Taiwan. *Plant Protection Bulletin*, (Taiwan), **29**, 135–46.

Dallwitz, M. J. & Higgins, J. P. (1978). User's guide to DEVAR. A computer program for estimating development rate as a function of temperature. CSIRO Division of Entomology Report No. 2.

De Jong, P. (1991). Simulation of yield loss by the rice leafroller *Cnaphalocrocis medinalis* under different growth conditions. SARP internal report.

De Kraker, J. (1991). Modelling a crop–pest–natural enemy system: leafrollers in rice. Poster presented at the SAAD symposium, AIT, Bangkok, Thailand, 2–6 Dec. 1991.

Fabellar, L. T., Fabellar, N. G. & Heong, K. L. (1994). Stimulating rice leafroller feeding effects on yield using MACROS. *International Rice Research Newsletter*, **19**(2), 7–8.

Graf, B., Gutierrez, A. P., Rakotobe, O., Zahner, P. & Delucchi, Y. (1990*a*). A simulation model for the dynamics of rice growth and development. II. The competition with weeds for nitrogen and light. *Agricultural Systems*, **32**, 367–92.

Graf, B., Lamb, R., Heong, K. L. &

Fabellar, L. T. (1992). A simulation model for the population dynamics of rice leafrollers and their interactions with rice. *Journal of Applied Ecology*, **29**, 558–70.

Graf, B., Rakotobe, O., Zahner, P., Delucchi, Y. & Gutierrez, A. P. (1990*b*). Simulation model for the dynamics of rice growth and development: the carbon balance. *Agricultural Systems*, **32**, 341–65.

Hamid, M. N. & Selamat, A. (1991). Simulation of yield reduction due to Malayan black bug infestation on several rice varieties. In *Simulation and Systems Analysis for Rice Production (SARP)*, ed. F. W. T. Penning De Vries, H. H. Van Laar & M. J. Kropff, pp. 310–19. Wageningen: Pudoc.

Hague, M. A. (1991). Effect of sheath rot and ufra diseases. A report of a case study of the SARP Project. Los Banos: IRRI.

Heong, K. L. (1990). Feeding rates of the rice leafroller, *Cnaphalocrocis medinalis* (Lepidoptera: Pyralidae) on different plant stages. *Journal of Applied Entomology*, **7**, 81–90.

Heong, K. L. & Domingo, I. (1992). Shifts in predator–prey ranges in response to global warming. *International Rice Research Newsletter*, **17**(6), 29–30.

Heong, K. L., Song, Y. H., Pimsamarn, S., Zhang, R. & Bao, S. D. (1994). Global warming and rice arthropod communities. Paper presented at International Conference on Global Change and Rice, March 1994, International Rice Research Institute, Los Baños, Philippines.

Hisano, E. (1964). Occurrence and injury of white-backed and brown planthopper and control measures. *Agriculture Horticulture*, **39**, 141–4 [in Japanese].

Horie, T. (1987). A model for evaluating climatic productivity and water balance of irrigated rice and its application to Southeast Asia. *Southeast Asian Studies*, **25**, 62–74.

Hudes, E. S. & Shoemaker, C. A. (1988).

Inferential method for modeling insect phenology and its application to the spruce budworm (Lepidoptera: Tortricidae). *Environmental Entomology*, **17**, 97–108.

International Rice Research Institute (IRRI) (1968). *Annual Report for 1968*. Los Baños, Philippines: IRRI.

Jones, J. W. (1993). Decision support systems in agriculture. In *Systems Approaches for Agricultural Development*, ed. F. W. T. Penning de Vries, P. Teng & K. Metselaar, pp. 459–72. Amsterdam: Kluwer.

Jones, J. W., Boote, K. J., Hoogenboom, G., Japtap, S. S. & Wilkerson, C. G. (1989). *SOYGRO V5.42, Soybean Crop Growth Simulation Model: User's Guide*. Florida Agricultural Experiment Station Journal No. 8304. Gainsville, Florida: Agricultural Engineering and Agronomy Department, University of Florida.

Johnson, K. B. (1987). Defoliation, disease, and growth: A reply. *Phytopathology*, **77**, 1495–7.

Kenmore, P. E. (1980). Ecology and outbreaks of a tropical insect pest of the green revolution, the rice brown planthopper, *Nilaparvata lugens* (Stal). PhD thesis. Berkeley: University of California.

Kim, J. W. & Kim, D. H. (1986). Studies on the resistance of rice varieties to biotypes of the brown planthopper, *Nilaparvata lugens* Stal. *Korean Journal of Plant Protection*, **24**, 209–17.

Klepper, O. & Rouse, D. I. (1991). A procedure to reduce parameter uncertainty for complex models by comparison with real system output illustrated on a potato growth model. *Agricultural Systems*, **39**, 375–95.

Kropff, M. J. & van Laar, H. H. (eds.) (1993). *Modelling Crop–Weed Interactions*. Wallingford, UK: CAB International, and Los Baños, Philippines: International Rice Research Institute.

Kropff, M. J., van Laar, H. H. & ten Berge, H. F. M. (1993). *ORYZA1, A Basic Model for Irrigated Rice Production*. Los Baños, Philippines: International Rice Research Institute.

Larkin, T. S., Carruthers, R. I. & Soper, R. S. (1988). Simulation and object-oriented programming: the development of SERB. *Simulation*, **52**, 993–1000.

Logan, J. A. (1988). Toward an expert system for development of pest simulation models. *Environmental Entomology*, **17**, 359–76.

Loomis, R. S. & Adams, S. S. (1983). Integrative analysis of host–pathogen relations. *Annual Review of Phytopathology*, **21**, 341–62.

Luo, Y., Teng, P. S., Fabellar, N. G. & TeBeest, D. O. (1995). Simulation of rice blast epidemics under global change in several Asian countries. *Agricultural Systems* (in press).

Ma Jufa, Hu Guowen, Zhu Defeng & Chen Zhongxiao (1991). Simulation of damage caused by leaffolder on rice crops. In *Simulation and Systems Analysis for Rice Production (SARP)*, ed. F. W. T. Penning De Vries, H. H. Van Laar & M. J. Kropff, pp. 328–31. Wageningen: Pudoc.

Matthews, R. B. & Hunt, L. A. (1994). GUMCAS: a model describing the growth of cassava (*Manihot esculenta* L. Crantz). *Field Crops Research*, **36**, 69–84.

Narasimhan, V., Mohandass, S., Abdul Kareem, A. & Palanisamy, S. (1991). Simulation of the effect of bacterial leaf blight infection at different growth stages on rice yield. In *Simulation and Systems Analysis for Rice Production (SARP)*, ed. F. W. T. Penning De Vries, H. H. Van Laar & M. J. Kropff, pp. 332–9. Wageningen: Pudoc.

Pathak, P. K. & De Jong, P. (1991). The damage relation of stem borer: simulation and experiment. Poster presented at the SAAD symposium, AIT, Bangkok, Thailand, 2–6 Dec. 1991.

Penning De Vries, F. W. T., Jansen, D. M., Ten Berge, H. F. M. & Bakema, A. (1989). *Simulation of Ecophy-*

siological Processes of Growth in Several Annual Crops. Wageningen: Pudoc.

Pinnschmidt, H. (1991). Effects of multiple pests on crop growth and yield. *Philippine Phytopathology*, **27**, 1–11.

Pinnschmidt, H. O., Batchelor, W. D. & Teng, P. S. (1994). Simulation of multiple species pest damage in rice using CERES-RICE. *Agricultural Systems*, **48**, 193–222.

Pinnschmidt, H. O., Teng, P. S. & Batchelor, W. D. (1991). Simulation of damage effects of multiple pests and diseases in rice. Poster presented at the SAAD symposium, AIT, Bangkok, Thailand, 2–6 Dec. 1991.

Pinnschmidt, H., Teng, P. S., Yuen, J. E. & Djurle, A. (1990). Coupling pest effects to the IBSNAT CERES crop model for rice. Paper presented at the APS meeting in Grand Rapids, Michigan, USA, August 1990. *Phytopathology*, **80**, 997, abstract no. A311.

Rabbinge, R., Ward, S. A. & van Laar, H. H. (eds.) (1989). *Simulation and Systems Management in Crop Protection*. Wageningen: Pudoc.

Rajan, A., Surjit, S. & Ibrahim, Y. B. (1991). Simulation of rice–weed competition in direct seeded rice. Poster presented at the SAAD symposium, AIT, Bangkok, Thailand, 2–6 Dec. 1991.

Reddy, P. R., Nayak, S. K. & Bastiaans, L. (1991). Simulation of the effect of bacterial blight disease on crop growth and yield of rice. *Simulation and Systems Analysis for Rice Production (SARP)*, ed. F. W. T. Penning De Vries, H. H. Van Laar & M. J. Kropff, pp. 340–7. Wageningen: Pudoc.

Rouse, D. J. (1988). Use of crop growth models to predict the effects of disease. *Annual Review of Phytopathology*, **26**, 183–201.

Roy, A. K. (1982). Effect of sheath blight infection on respiration and transpiration of rice plants. *International Rice Research Newsletter*, **7**(1), 20.

Rubia, E. G. & Penning De Vries, F. W. T. (1990). Simulation of yield reduction caused by stemborers in rice. *Journal of Plant Protection in the Tropics*, **7**(2), 87–102.

Santa, H. (1959). Damages of rice plants caused by planthoppers [in Japanese]. *Shokibutsu-Boeki [Plant Protection]*, **13**, 307–10.

Singh, R. A. & Das, B. (1991). Simulation of yield loss due to sheath blight in rice crops. In *Simulation and Systems Analysis for Rice Production (SARP)*, ed. F. W. T. Penning De Vries, H. H. Van Laar & M. J. Kropff, pp. 348–52. Wageningen: Pudoc.

Singh, R. A., Pathak, P. K. & Misra, B. (1991). Simulation of yield losses due to major diseases and insect pests. Report on a case study of the SARP project. Los Baños: IRRI.

Singh, U., Ritchie, J. T. & Godwin, D. C. (1993). *A User's Guide to CERES-RICE*, v2.10. Mussel Shoals, Alabama: International Fertilizer Development Center.

Sogawa, K. (1970). [Studies on the feeding habits of the brown planthopper. II. Honeydew excretion.] *Japanese Journal of Applied Entomology and Zoology*, **14**, 134–9 [In Japanese, English summary].

Sogawa, K. (1971). Effects of feeding of the brown planthopper on the components in the leaf blade of rice plants. *Japanese Journal of Applied Entomology and Zoology*, **15**, 175–9.

Sogawa, K. (1973*a*). Feeding behavior of the brown planthopper and brown planthopper resistance of indica rice Mudgo. Laboratory of Applied Entomology, Faculty of Agriculture, Nagoya University Bulletin no. 4.

Sogawa, K. (1973*b*). Feeding of the rice plant- and leafhoppers. *Review of Plant Protection Research*, **6**, 31–43.

Sogawa, K. & Cheng, C. H. (1979). Economic thresholds, nature of damage, and losses caused by the brown planthopper. In *Brown Planthopper: Threat to Rice Production in Asia*, pp. 125–44. Los

Baños, Philippines: International Rice Research Institute.

Spitters, C. J. T., van Keulen, H. & van Kraalingen, D. W. G. (1989). A simple and universal crop growth simulator: SUCROS87. In *Simulation and Systems Management in Crop Protection*, ed. R. Rabbinge, S. A. Ward & H. H. van Laar, pp. 147–81. Wageningen: Pudoc.

Suenaga, H. (1959). Damage caused by the plant- and leafhopper and its assessment. Abstract of the 3rd Symposium on the Assessment of Insect Damage. Tokyo: Japanese Society of Applied Entomology and Zoology.

Sun, S. Y., Jin, M. Z., Zhang, Z. M., Tao, X. L., Tao, R. X. & Fang, D. F. (1986). Rice blast disease and its control. Shanghai: Shanghai Scientific and Technology Press [in Chinese].

Sutherst, R. W., Maywald, G. F. & Skarratt, D. B. (1995). Predicting insect distributions in a changed climate. In *Insects in a Changing Environment*, ed. R. Harrington & N. E. Stork, pp. 59–91. London: Academic Press.

Teng, P. S. (1988). Pests and pest-loss models. *Agrotechnology Transfer*, 8,(1), 5–10.

Teng, P. S., Calvero, S. & Torres, C. Q. (1989). The CERES-RICE-blast simulation model. In IBSNAT symposium: The decision support system for agrotechnology transfer. Part 2. Poster presentation. American Society of Agronomy Meeting, Oct. 1989, Las Vegas, Nevada.

Teng, P. S. & Johnson, K. B. (1988). Analysis of epidemiological components in yield loss assessment. In *Experimental Techniques in Plant Disease Epidemiology*, ed. J. Kranz & J. Rotem, pp. 179–90. New York: Springer.

Teng, P. S., Torres, C. Q., Nuque, F. L. & Calvero, S. B. (1990). Current knowledge on crop losses in tropical rice. In *Crop Loss Assessment in Rice*, pp. 39–53. Los Baños, Philippines: International Rice Research Institute.

Waggoner, P. E. & Berger, R. D. (1987). Defoliation, disease, and growth. *Phytopathology*, **77**, 393–8.

Worner, S. (1991). Use of models in applied entomology: the need for perspective. *Environmental Entomology*, **20**, 768–73.

Xu Zhihong & Wang Zhaoqian (1991). Simulation of reduction in rice yields due to striped stem borer. In *Simulation and Systems Analysis for Rice Production (SARP)*, ed. F. W. T. Penning De Vries, H. H. Van Laar & M. J. Kropff, pp. 353–60. Wageningen: Pudoc.

Yoshida, S. (1981). *Fundamentals of Rice Crop Science*, p. 269. Los Baños, Philippines: International Rice Research Institute.

Zuber, M. & Rao, K. M. (1983). Physiology of host–pathogen interaction in rice sheath blight disease with reference to changes in carbohydrate and nitrogen contents. *Indian Journal of Botany*, **6**, 117–24.

17 Soil erosion under global change

C. Valentin

Introduction

In recent years, increasing attention has been given to the potential impacts of climate and land-use changes on soils (Arnold *et al.*, 1990; Bouwman, 1990; Varallyay, 1990). Despite the imprecision of general circulation models, the uncertainty of the effect of warming on global precipitation and doubts about the validity of assumptions on land-use changes, it is crucial to anticipate the consequences of future scenarios on soil, which is one of the most essential non-renewable natural resources. Soil erosion has been identified as the major type of human-induced land degradation in a global perspective, nearly one-sixth of the world's usable land having already been degraded (Oldeman *et al.*, 1991). As shown diagrammatically in Fig. 17.1, climatic changes are expected to influence soil erosion both directly through changes in wind and rainfall regimes, and indirectly through changes in vegetation and surface cover. Also changes in atmospheric composition may have some impact on soil erosion through changes in vegetation and soil organic matter. However, the intensity and extent of water and wind erosion are likely to be primarily affected by land-use changes, the effects of which are more present, immediate and extensive.

To assess what will happen if climate and land use change we need:

1. to review what is known about the relationships between climate, land use and soil erosion;
2. to analyse the potential impacts of global change on soil erosion;
3. to identify the major gaps in the present knowledge and infer the future research needed.

These questions are addressed with specific reference to West Africa, a region which has been experiencing a definite downward trend in rainfall amount for 25 years and where the near doubling of population over the same period has severely exacerbated erosion problems. Moreover, West Africa constitutes a large continuous mass, with a fairly homogeneous terrain, distributed across a strong and clear moisture gradient. Such conditions seem, therefore, most propitious for examining the possible

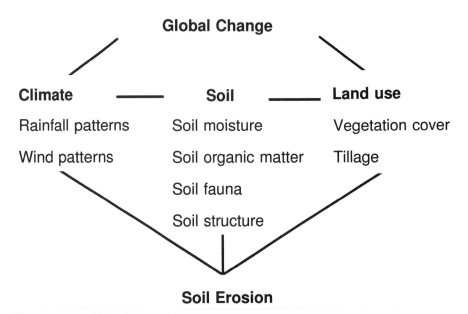

Figure 17.1   *Main factors of soil erosion under global change.*

impacts of global change on soil erosion and its potential feedbacks on climate in tropical regions. Some of these issues will also be considered in other areas, including mid- and high-latitudes areas as well as steeplands, coastal lowlands and urban and peri-urban areas.

Past data

Climate

One way to predict the effect of changes, whether climatic or man-induced, is to look back at previous events.

Reliable information can be drawn from the earliest records of rainfall in West Africa since the beginning of the twentieth century. Whatever the statistical approaches, three main periods can be clearly distinguished (Lamb, 1982; Olivry, 1983; Ojo, 1987; Sircoulon, 1989): (1) from the beginning of the century to the late 1940s, annual variability of rainfall is high. The higher or below average periods are limited to 2 or 3 consecutive years. In particular, three severe droughts occurred in 1911–1914, 1941–1942 and 1947–1949. (2) From 1950 to 1968, the time-series was continuously exceeding average. (3) Since 1969, annual rainfall remains below average with two major droughts in 1972–1974 and 1983–1985. The recent drought has resulted in a pronounced southward shift of the isohyets, particularly in the already driest zone (nearly 500 km for the 100 mm isohyet and 100 km for the 1000 mm isohyet; Le Borgne, 1990).

Scientific and historical evidence attests that persisting droughts occurred also in the fifteenth, sixteenth, eighteenth and nineteenth centuries (Maley, 1981). Moreover, abundant geological and palaeoecological evidences revealed that West Africa has undergone major climatic changes during the late Quaternary. The Sahara encroached with wind-driven sand far southwards between 20 000 and 12 000 BP. African savannahs were wetter than now 12 000–7000 BP. At 5000–4500 BP, the environmental saharization started stretching southwards.

While past climate changes are fairly well documented, climatic forecasts for the next decades remain highly controversial. Some authors postulate that the Sahel's northern limit will still be regressing southwards (Petit-Maire, 1990), arguing that the latest wet and dry periods coincided with sea-level transgressive and regressive phases and matched the Milankovitch-type orbital variations. Human influence upon these processes is believed to be very limited since the scant neolithic groups could by no means be blamed for the deterioration of the environment which started nearly 5000 years ago. This astronomically and geologically based arid scenario (Kukla, 1980; Berger, 1981; Petit-Maire, 1990) would run counter to the potential increase in precipitation due to the greenhouse effect in the region.

While annual rainfall is one of the most relevant variables to consider in the context of climate change, high-intensity rainfall is more crucial as far as erosion is concerned. As shown by Albergel (1987) using the records from 20 stations in Burkina Faso over a period of nearly 50 years, daily rainfall higher than 40 mm shows a substantial decrease since 1969. However, no concurrent decrease was observed either for high intensity rainfall or for the depth of the 10 year frequency rainfall. In some locations, this rainfall depth has even increased (Fig. 17.2). Similar results were obtained in Nioro du Rip, Senegal, over the period 1931–92 (Perez, 1994). Likewise, Yu & Neil (1993) showed that in southwest Western Australia, where annual rainfall is known to have decreased significantly over the last 70–80 years, there has been no corresponding decrease in high-intensity rainfall for the same period.

Population growth and land use

Assuming the maintenance of the present rate of population growth of 3%, the population in the Sahel should double by 2011. Migration from rural areas to cities is expected to be insufficient to maintain the rural density below supporting capacities, especially in the driest zones of West Africa where these capacities are already exceeded (Le Houérou, 1989).

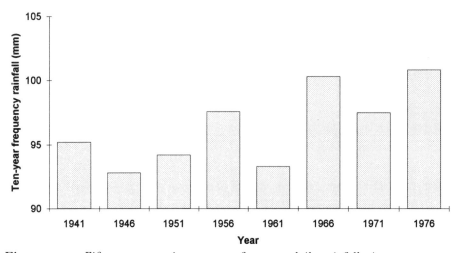

Figure 17.2   *Fifteen year running average of 10-year daily rainfalls, in 5 year increments, over the period 1934–83 for Ouagadougou, Burkina Faso. (From Albergel, 1987.)*

Where arable farming is generally not viable, i.e. in regions with <350 mm year[-1] rainfall, pastoralism is the most appropriate response to the space- and time-variability of precipitation. Better than average rains in the 1960s encouraged expansion of grazing into some marginal land and an increase in stock numbers. When rainfall decreased these herds damaged rangeland through overgrazing (Le Houérou, 1989). Apart from the most severe drought period, the increase in livestock has roughly followed the increase in human population during these four last decades, suggesting a similar trend in the near future. Sedentarization of pastoralists, which has been promoted by the authorities, tends also to accelerate soil degradation through overgrazing and trampling, especially in the vicinity of the recently drilled deep wells (Valentin, 1985).

In the semi-arid zone, a number of surveys using remote sensing techniques indicate a rate of increase in cultivated land almost equal to the growth of human populations (Le Houérou, 1989). Such an expansion of cultivated surface, mainly by clearing marginal and grazing land, is a response of farmers to decreasing yields subsequent to drought and gradual loss of soil fertility. Furthermore, the fallow period was drastically shortened to meet the increasing demand for agricultural land.

The wettest zone of West Africa suffered a dramatic deforestation. While the rate of loss of rainforest is about 4.0% for the whole region (Harrison, 1987), it attains 7% in Ivory Coast where only 22% of the 1957 cover remained 30 years later (Bourke, 1987). In more densely populated countries, such as Nigeria, there is virtually no rainforest left.

Consequences on soil structure and crusting

Organic matter is one of the most important aggregate stabilizing agents in soil (Combeau & Quantin, 1964; Charreau & Nicou, 1971, among others). Since it is difficult to maintain high carbon levels during cultivation of tropical soils (Feller *et al.*, 1991), the aggregate stability declines under cultivation and increases under fallow (Combeau & Quantin, 1963; Valentin *et al.*, 1990). However, carbon content should not be considered separately from mineral contribution to soil texture. Minor decreases in carbon content may have a more detrimental effect upon structural stability of sandy than clayey soils (Pieri, 1989). The increased organic matter decomposition induced by global warming could thus favour soil crusting, notably in the sandy soils of the Sahelian zone. Crusts are thin soil surface layers more compact and hard, when dry, than the material directly beneath. They hamper seedling emergence, reduce infiltration rate and favour runoff generation.

Rainfall simulation tests conducted in the Sahel (Hoogmoed, 1986; Valentin, 1991; Ambouta, 1994) indicated that the texture most prone to crusting consists of approximately 90% sand and 10% silt + clay. Below 5% of silt + clay, the amount of fine particles is apparently not sufficient to clog the pores (Ambouta, 1994). As shown in West Africa (Casenave & Valentin, 1989; Valentin & Bresson, 1992; van der Watt & Valentin, 1992), structural degradation of sandy soils involves the successional development of various types of crusts (Fig. 17.3). When soil is protected by natural or cultivated cover, no severe form of crusting is observed and surface roughness and porosity remain high, so that the rates of water infiltration into the soil remain high (>30 mm hour^{-1}; Casenave & Valentin, 1989, 1992). When the soil surface is denuded owing to drought or land overuse, the direct impact of energetic raindrops results in the formation of micro-craters. The walls of these micro-craters present a clear, vertical sorting of particles (Valentin, 1986). When coalescing these micro-craters constitute a 'sieving crust' which consists of a layer of loose sand overlaying a thin, dense and hard layer of fine material. In its most advanced form, the crust can consist of three well-sorted layers. The uppermost layer is composed of loose, coarse sand; the middle one consists of fine, densely packed grains with vesicular pores; and the lower layer shows a higher content of fine particles with considerably reduced porosity. The lower fine-textured layer is responsible for the low infiltrability of such crusts (0–5 mm hour^{-1}; Casenave & Valentin, 1989, 1992). Loose particles of the sandy micro-layers of the sieving structural crusts can be readily removed by wind and by runoff, favouring thereby a high soil detachment rate (Fig. 17.3). Wind-

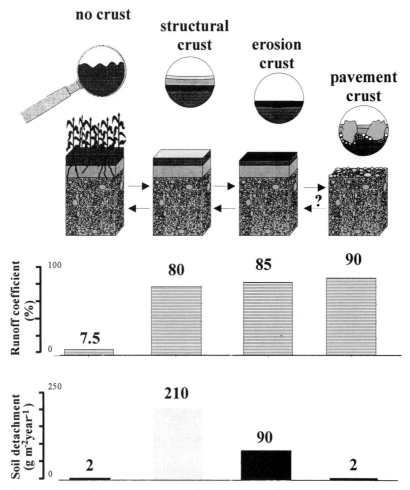

Figure 17.3 *Time-sequence of surface crusting and removal by erosion of the top layers in the Sahelian zone and the consequences for runoff production and soil detachment. Runoff coefficient and soil detachment have been measured using rainfall simulation on 1 m² plots. Standard conditions for slope angle (2.5%) and rainfall aggressivity index (R = 250) have been used to normalize soil detachment values. (From Valentin, 1981; Collinet & Valentin, 1985; Chevallier & Valentin, 1985; Valentin, 1991; Casenave & Valentin, 1992; Valentin & Bresson, 1992.)*

drifted sands are entrapped by surrounding vegetation and can evolve in turn into sieving structural crusts if vegetation decays due to drought and/ or overgrazing (Valentin, 1985).

When the sandy upper layers are removed, the exposed thin-textured layer evolves into an erosion crust. The erosion crusts are built up with one smooth and hard layer made of thin particles. Porosity is restricted to few cracks and vesicles so that infiltrability is very low (0–2 mm hour^{-1}; Casenave & Valentin, 1989, 1992), as well as evaporation (1–2 mm day^{-1};

Le Fevre, 1993). These erosion crusts promote runoff but are resistant to wind and water erosion compared with sieving crusts (Fig. 17.3). Erosion crusts cannot usually be colonized by vegetation because of their resistance to seedling emergence, the very dry pedoclimate they produce, and primarily because seeds that deposit on the soil surface are invariably removed by wind and overland flow. Such processes lead to positive feedbacks which favour the expansion of these crusted patches under drying (and/or overused) conditions. This vicious spiral evolution has been recorded as 'sahelization' to indicate the occurrence of such a typical Sahelian feature within desertified Sudanian degraded areas as a result of drought and/or a shortened fallow system (Albergel & Valentin, 1988).

The erosion crusts and the top layers can in turn be eroded as the consequence of sheet flow generated upslope or regressive rill erosion. In West Africa, many soils contain coarse fragments. When gravel-free top layers are removed by erosion, these layers are exposed. In the semi-arid and arid zones, these surface coarse fragments are embedded in a crust the microstructure of which is very similar to the sieving crust with three layers. Such a crust is referred to as 'pavement crust'. Vesicular porosity is much pronounced especially below the coarse fragments. Infiltrability is extremely low (0–2 mm hour^{-1}; Casenave & Valentin, 1989, 1992) and decreases when the size of the coarse fragments increases (Valentin & Casenave, 1992). However, these pavement crusts tend to protect the soil underneath, playing the role of a mulch, and thereby limit further erosion (Collinet & Valentin, 1985).

This sequence of degradation raises two main questions: (1) how quickly the pavement crust stage is reached, and (2) whether such processes are reversible. Many observations in the Sahelian zone (Casenave & Valentin, 1989) show that under natural conditions, the soil surface consists of a two-phase mosaic made of erosion-crusted barren spots surrounded by vegetated micro-dunes. Depending on the rainfall, the relative proportions of these two phases can be altered. Crusted patches can stretch out due to drought irrespective of any human or animal influence (Valentin, 1985). However, such a system is highly resilient; crusted patches regress when precipitation returns to higher levels. Though the development of pavement crust can also result from natural processes, it is greatly accelerated by severe overgrazing. In some cases observed in northern Senegal and northern Burkina Faso, large patches of pavement crusts have developed within three decades. At this stage, the ability of the soil to revert in the foreseeable future to the near-original state seems highly questionable.

Table 17.1 *Variations of main hydrological parameters as influenced by natural surface conditions along a climatic gradient in West Africa.*

Ecological zone	Mean annual rainfall (mm)	Faunal cover (%)	Vegetation cover (%)	Type of crust	Krdry (%)	Krwet (%)	Prd (mm)	Prw (mm)	FIR (mm hr^{-1})
Guinean	>1200	>20%	>50%	No	4	5	30	15	23
Sudanian	600–1200	<20%	>50%	Structural	25	30	19	8	14
Sahel	200–600	<20%	<50%	Structural	50	59	8	4	6
Desert	<200	0	0	Erosion + Gravel	83	87	4	3	1

After Valentin *et al.* (1990), Casenave & Valentin (1992)

Krdry, runoff coefficient under dry initial conditions; Krwet, runoff coefficient under moist initial conditions; Prd, pre-ponding rainfall under dry initial conditions; Prw, pre-ponding rainfall under moist initial conditions; FIR, final infiltration rate at saturation.

Consequences for runoff and water erosion

Moisture conditions prior to rainfall largely control hydrological parameters. In particular, the pre-ponding rainfall represents one of the threshold conditions for bringing about overland flow and rainwash erosion. However, extensive rainfall simulation experiments throughout West Africa (Collinet, 1988; Casenave & Valentin, 1992) clearly showed that structural surface conditions and surface cover were even more important controlling factors of infiltration than antecedent soil moisture conditions. The ranges of variations between dry and wet antecedent soil moisture conditions are much lower than those resulting from surface features. Table 17.1 shows that antecedent moisture conditions have more influence upon runoff in the Guinean savannah zone than in the drier areas. In all regions where surface crusts prevail, runoff conforms to the Horton model of overland flow generation.

In the Guinean savannah zone, runoff is primarily controlled by rainfall amount and is consistent with the saturation or storage control model of runoff generation. In such regions, runoff should therefore respond positively to increasing precipitation. By contrast, in the drier regions, possible changes in soil water regime should alter runoff production much more indirectly through faunal activity and vegetation cover than directly, so that this effect should be the reverse of what is normally expected. Any increase in precipitation should result in activated faunal activity and protective vegetation cover (Langbein & Schumm, 1959), thus in lower runoff production. Under natural conditions, at the regional scale, the effect of annual rainfall on runoff coefficient (runoff depth/rainfall depth expressed in percentage terms) is totally offset by the protective effect of vegetation and litter cover and by the destruction of surface crusts by faunal activity (Fig. 17.4). Therefore, the dryer the climate is, the higher the runoff coefficient. Even more interesting in the context of climatic change is the recorded evolution of runoff between the 1950s and the 1980s on a watershed basis (Fig. 17.5). The combination of drought, surface degradation, with no change in exceptional precipitation, has led to a pronounced increase in runoff coefficient in the Sahelian zone, a slight increase in the Sahelo–Sudan zone, and a minute increase in the Sudan zone (Albergel, 1987). Such results suggest that any further decrease in precipitation in the most arid zone would induce a further increase in runoff coefficient.

In contrast to runoff, a less surprising general trend appears for water erosion measured under bare conditions. The potential erosion gradually increases from less than 2 tonnes hectare^{-1} year^{-1} for an annual rainfall of 150 mm in the fringe of the Sahara to nearly 80 tonnes hectare^{-1} year^{-1} under 2000 mm in southern Ivory Coast (Collinet & Valentin, 1985).

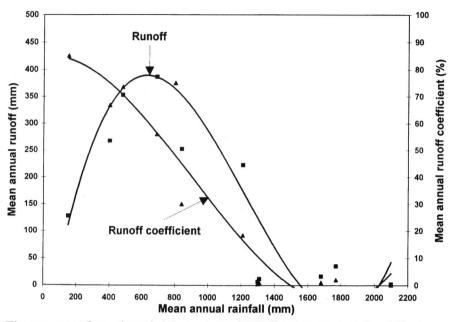

Figure 17.4   *General trends in mean annual runoff and mean annual runoff coefficient (runoff depth/rainfall depth, expressed as a percentage) under natural cover for the various ecological zones of West Africa as measured under natural and simulated rainfall. (From Roose, 1973; Valentin, 1981; Collinet, 1988.)*

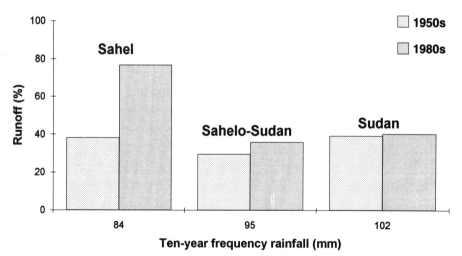

Figure 17.5   *Changes in runoff production between the 1950s and the 1980s in three watersheds of Burkina Faso of few 10 km² given a steady 10 year frequency daily rainfall. (From Albergel, 1987.)*

However, the large temporal variability of rainfall, particularly that of high intensity rainfall, as well as the relatively small dimensions of the erosion plots of generally a few tens of square metres make it difficult to draw a firm conclusion as to the likely influence upon erosion of changes in rainfall regime on a regional basis. It is the incidence of above-average years rather than any increase in average erosivity that will be significant not only on inter-rill but also on rill and gully erosion. Furthermore, it is well known that vegetation cover plays a key role in protecting soil from erosion. Compared with bare soil conditions, dense herbaceous cover was found to reduce the erosion rate 100-fold in the Sahel, and 500-fold in the Sudano-Guinean zone (Collinet & Valentin, 1985). Consequently the decay of vegetation greatly enhances erosion even though precipitation declines. The soil denudation due to drought, overgrazing and expansion of cultivated land resulted in a marked increase of eroded land in the semi-arid zone. By way of example, Albergel & Valentin (1988) showed in a Burkinabe watershed that within less than 25 years the surface of cultivated land had doubled, fallow land was halved, shrub savannah was reduced from 80% to 45% while severely eroded patches increased 20-fold. In a less densely populated region of Burkina Faso bare and crusted patches increased 6-fold within 28 years (Serpantié et al., 1991).

Consequences on wind erosion

It is well known that wind erosion rates significantly increase with increasing aridity. Besides the higher erodibility of dry soil compared with wet soil, the removal of vegetation associated with drought greatly enhances soil loss. Wind erosion increases exponentially when soil cover is decreased (Fryrear, 1985, among others). For instance, when the percentage of soil surface cover decreases from 40% to 20%, soil loss due to wind erosion is increased four-fold. Diminishing grass cover as the dry season progresses can favour therefore dust-entrainment, as described in northern Nigeria by McTainsh (1980). Also in northern Nigeria, in Sokoto, as well as in Nouackchott, Mauritania, a strong anticorrelation was established between the frequency of days of dust haze since the late 1960s and the annual rainfall during the same period (Middleton et al., 1986). Similarly, while the annual rainfall decreased steadily between 1952 and 1987 in Gao (Mali) from 300 mm to 130 mm the number of days of dust haze passed from nearly zero to more than 250 (N'Tchayi, 1988). There is thus much evidence that the drought in the Sahel has produced a marked increase in dust storms (Middleton, 1985). Dust events involve such broad-scale processes that trans-Atlantic export of dust from West Africa currently occurs throughout the year. The severe drought in the Sahel increased the quantity of dust emanating from West Africa as detected through the records of dust concentration in the West Indies: in 1967–1968 8 micrograms per m³, in 1972 15, and in 1973 42 (Prospero & Nees, 1977; Mann, 1987).

Besides climate, land use also influences dust processes. A detailed study of the local dune soil-profile characteristics in the vicinity of deep drilled water-holes in northern Senegal revealed an impoverishment in fine particles in the top layers which had been subject to heavy trampling (Valentin, 1985). Weeding is also blamed in the Sahel for depriving topsoils of fine particles (Ambouta, 1994). A beneficial effect of this loss lies in the reduced hazards of crusting that can occur only when silt + clay content exceeds 5%. By contrast, soils left under fallow are subject to severe crusting. This surprising result can be ascribed to the gradual trapping of dust by recovering vegetation. The dust is transported from source areas in the desert, in periodic pulses over the West African savannah, and is deposited at a decreasing rate and with a decreasing particle size with transport distance (McTainsh, 1986). The total dustfall in the Sahel represents 200 g m² year⁻¹ which represents a deposition rate of 0.2 mm year⁻¹ (Orange & Gac, 1990). Considering that silt + clay constitute between 82% (in northern Nigeria; Moberg *et al.*, 1991) and 97.5% (in Dakar, Senegal; Orange & Gac, 1990) of the dust samples, dustfall is a major source of fine particles in the sandy soils of the Sahel.

While the influence on wind erosion of climate and human activities are well documented, the feedback effects of wind-blown dust in the atmosphere on the climate have not been thoroughly investigated. In the size range of the dust particles – diameter usually 0.1–20 μm for the particles transported over a long distance – atmospheric aerosols modify the radiation budget by absorption and scattering of solar and telluric radiations. High concentrations of mineral aerosols could produce a warming of the upper atmospheric layers and a cooling at surface level, reducing the formation of clouds and changing their spatial distribution, thus affecting the climate (Bergametti, 1992).

Although water and wind erosion are generally studied separately, the two processes are often intimately linked. At the same location, disaggregation of soil clods into individual particles during the rainy season due to the impact of raindrops and particle sorting induced by sheetflow greatly encourage wind erosion processes during the subsequent dry season. On larger temporal and spatial scales, large quantities of sediments accumulated during the late Quaternary wet periods within river and lacustrine systems, as in the case of Lake Chad, 5000–6000 BP (Maley, 1981). At present these dry depressions constitute the major source areas for dust storms in West Africa (McTainsh *et al.*, 1992).

Potential impacts of global change

Current climatic models, especially predictions of future wind and rainfall regime, lack precision. Similarly, political, economical, cultural and socio-logical processes can alter substantially the land-use change projections based on recent trends. However, it is essential to analyse the implications of climate change scenarios coupled with land-use change hypotheses on the severity, frequency and intensity of soil erosion. This section concentrates primarily on a transect in West Africa which will help to consider such potential impacts of global change in the warm regions.

Wet forest zone

Potential water erosion is very high in the wet forest region especially on the frequent hillslopes (Collinet & Valentin, 1985). However, the erodibility of soils is lower than in the dryer regions due to better structural stability and higher permeability of the soils. Moreover, the maintenance of a complete soil cover, that can be readily achieved in the humid regions, can reduce erosion to zero. As a whole, increased water erosion might result from change in land management that would not maintain forest cover conditions. At present, the prevailing tree-based farming systems with perennial crops (cocoa, coffee, tea, rubber, oil palm, etc.) associated with cover crops and mulch include sufficiently effective conservation practices to resist higher rainfall intensities. Any increase in precipitation should certainly lead to more severe water erosion on bare soils, or relatively unpro-tected soils as for cassava. Major hazards could also stem from an extended use of heavy machinery for land clearing and land preparation (Lal, 1987).

Wet savannah zone

The wet savannah still has a relatively low density of population because of river blindness that has only recently been more or less eradicated. Most suitable soils for agriculture are generally located on the upper parts of the catena. Such gravelly soils are highly resistant to erosion. Whilst wind erosion is not an acute problem in the region, water erosion is likely to increase only if important migration of the Sahelian population increases the clearing of lower soils of the catena where aggregate stability is very low. In that case, soil erosion might become a major problem particularly at the onset of the rainy season when the bare (due to late bush-fires) and tilled soils are exposed to high-intensity storms (Planchon et al., 1987). The combination of an expected increase in high-intensity storms and the exposure of larger bare areas should thus increase water erosion.

Semi-arid zone

In the Sahel, the greenhouse effect is expected to favour a higher potential evapotranspiration rate (PET). Assuming an increase of 3.0 °C uniformly spread over the year, one may estimate an annual increase of 210 mm in PET (Le Houérou, 1993). Possible rainfall increase subsequent to global warming will allow more vegetation cover, at least in some areas, but is likely to remain insufficient to counterbalance this effect. Consequently water shortages for vegetation and crop production should increase, thereby favouring soil denudation and salinization. In addition, warming should slightly increase the rates of organic matter mineralization, depleting soil organic matter content, whether precipitation increases or not. Assuming the maintenance of the present rate of population growth of 3% in the Sahelian zone, the population should double by 2011. Migration from rural areas to cities will be insufficient to maintain the rural density below supporting capacities. The increase in livestock should roughly follow the increase in human population. As a result, soil denudation, decline in soil organic matter, coupled with an increased pressure on pasture and arable land should substantially increase the risks of soil erosion either by wind or by water, whatever the changes in wind and rainfall regime. Increased runoff coefficient due to more severe soil crusting should foster rill and gully development. However, uncertainty about the scenarios make it difficult to make firm predictions.

Mid-latitude areas

Whilst in the warmer regions both water and wind erosion will often be accelerated by land-use change, especially the clearing of vegetation cover as a consequence of rapid population growth, in the temperate countries, temperature rise is likely to become the major global change driver. This can affect the winter and summer rainfall regimes. Runoff and erosion are primarily controlled by rainfall and soil saturation in winter (as in the wet tropics), by rainfall intensity and surface conditions in summer (as in the semi-arid regions). Higher temperature may be expected to lead to more intense rainfall events in summer, hence to more severe erosion (Kwaad, 1991). Warmer and drier summers may also be accompanied by an increase in wind erosion (Boardman *et al.*, 1990). Increased evapotranspiration due to warming would induce water shortages for agriculture. As a result, land use, the type of crops and timing of farming operations will have to change in response to changed climate, enhancing the erosion hazards. Nevertheless because a large part of the mid-latitude areas belong to the industrialized world, it can be postulated that sensible precautions will be taken to mitigate the effects of such changes upon erosion.

High latitude areas

The majority of global climatic models predict warming and increased precipitation in high latitude areas. While runoff would respond to higher rainfall on an annual basis, glacier, permafrost, vegetation and organic matter responses would be measured in decades to centuries. In these areas, sediment systems might take centuries to millennia to reflect climate change (Slaymaker, 1990). However, while water erosion may not change significantly in the next decades, wind erosion might increase (Dregne, 1990).

Steeplands

In the tropics, outward population movement from the steeplands to the lowlands will no longer relieve the pressure sufficiently to reduce the rate of destroying pastures and forests, hence of soil losses. In cold mountain regions, predicted climatic warming suggests an upward migration of natural vegetation zones. Similarly, longer growing season and fewer frosts will favour a shift in the mountain agroecological zones with the production of new crops which require higher temperatures. As a result, cultivation and grazing are likely to encroach on higher and still preserved lands, thus increasing the risks of erosion. As for the other areas, change in land use due to population growth (in the tropics) and warming (in colder areas) is expected to have a greater impact on erosion than the direct effect of possible change in rainfall patterns.

Coastal lowlands

Rising sea-level due to a warmer climate presents the potential for increased coastal erosion, loss of highly productive wetlands, and thus for an intensified pressure on the remaining land. Inundation and salt water intrusion may swallow up extensive areas, especially in Casamance, southern Senegal, where beach erosion might severely curtail the now flourishing tourism industry. Serious consequences for the cultivation of coastal lowlands may also be expected in the Gulf of Bengal, the Mekong Delta and the Guyanas. These effects might further exacerbate the consequences of human activity. For example, embanking rivers to channel flood flows directly to the sea tends to deprive basin areas of the sediments which formerly offset land subsidence (Greenland *et al.*, 1994).

Urban and peri-urban areas

Owing to the burgeoning urban population, especially in the developing world, one can anticipate that sheet erosion, gullying, mass movement and landslides would all seriously increase in the next decades in cities and their surroundings, irrespective of climate change. Any increase in high-intensity rainfall would further worsen the situation, provoking more frequent disruption of communication and other networks, along with pollution damage and off-site health problems.

Research needs

Past data

To anticipate the consequences of possible future scenarios it is crucial to examine data currently available on present erosion. In this respect one must be aware that the already large archive of aerial photographs and satellite data has not yet been sufficiently used as a retrospective basis for predicting models. New techniques such as radioactive fallout ^{137}Cs measurements could also provide information on recent erosion rates (Lance *et al.*, 1986; Richtie & McHenry, 1990; Walling & Quine, 1991; Bernard & Laverdière, 1992). Additional information on longer time-span erosion, related to climatic changes, could be derived from the numerous data on the Late Quaternary. Pollen, charcoal, chemical, physical, magnetic mineral and radiocarbon dating (^{14}C) along with archaeological records and local documentary evidence could be interpreted in the light of models of climatic change. Even though they are of value, the use of paleoclimatic analogues should be pursued with care as it is not clear that they are realistic for mid twenty-first century climatic and land-use rates of change.

Long-term monitoring and experimental programmes

Since erosion processes are commonly related to exceeding thresholds, emphasis should be put on the determination of the critical thresholds, on reversibility of processes, and on soil resilience. Low-frequency climatic events, such as heavy storms and typhoons, can trigger severe erosion unpredictable from short-term records. There is also a need to monitor the effect of climatic events in relation to their frequency, to allow weighted integration over the frequency spectrum. Monitoring erosion in the long run is therefore essential to observe possible transient as well as equilibrium responses to climatic and land-use changes.

Despite the numerous current and past experiments conducted on the factors affecting soil erosion, some uncertainties remain. An international set of collaborative experiments should, therefore, be specifically designed to cope with changing environmental conditions. This research could concentrate on the impact upon soil losses of rain and wind storm profiles, and of soil surface structure as related to soil organic matter status and faunal activity. Another pivotal issue should be the integration of erosional processes, i.e. crusting, wind and water erosion. Such field experiments, conducted at various scales from the plot to the watershed, and possibly at larger scales, should be combined with remotely sensed data and integrated into GIS. Field programmes should be fully integrated with model development and validation.

Modelling

The broad range of data covering the space–time domain can be used to calibrate, initialize, and validate models. Models should concurrently be highly sensitive to climatic and land-use changes and sufficiently flexible to be relevant under the largest range of conditions. The recent soil erosion process-based model, WEPP (Water Erosion Prediction Process; Lane & Nearing, 1989), is highly sensitive to precipitation (Nearing et al., 1990). However, considerable developments are needed to link global climatic models with more detailed erosion models accounting for interactions between physical and human factors. Some attempts have been made in this direction. Physically based simulation models such as the erosion productivity impact calculator (EPIC; Williams et al., 1984) allow us to link climate with other factors to estimate erosion and to examine the impact on crop yields. Skidmore and Williams (1991) have developed a wind erosion interface with EPIC. However, such models are data-intensive in nature and thus seem rather difficult to use in developing countries. The use of expert systems may be an effective alternative (Boardman et al., 1990). The rules generated by the expert system may be used to predict erosion rates at difficult scales across the landscape under varying climatic conditions, but, like other models, require an adequate and robust underlying physical basis.

Conclusion

Even though our climatic models are still inaccurate, sufficient information has been collected to make some assessment of the likely effect of climate and land-use-induced changes on soil erosion. This requires a better integration of long-term monitoring, experimental and modelling programmes at different scales in both time and space. Experimental programmes need to improve process-based understanding, incorporate different scales and integrate process-scale issues. These experiments will be based on clearly defined working hypotheses to be derived from the current scenarios pertaining to global changes. For model comparison and evaluation, three spatial scale categories were identified: patch, landscape and watershed. Larger scales should also be considered to establish some linkages with the Global Circulation Models. Another goal to meet the dual GCTE objectives is the assessment of potential feedbacks of soil erosion to the physical climate system. In this context, two main lines of research can be identified: (1) the interactions between denudation, crusting, albedo and climate changes, and (2) the interactions between windblown dusts and climate changes. This requires a close collaboration with other programmes such as BAHC (Biospheric Aspects of the Hydrologic Cycle).

Soil erosion results from complex interactions of physical and human factors, but the latter are potentially more damaging than the direct effects of climate change on soils. Strong links should therefore be established with HDP (International Human Dimensions Programme), notably with LUCC (Land Use/Cover Change).

As a response to these needs, a GCTE network was inaugurated in Paris, in March 1994. Its first objectives are:

- To design and undertake experimental and monitoring programmes to provide a predictive understanding of the impacts of changes in climate and land use on soil erosion.
- To refine and adapt current erosion models for use in global change studies from the plot to the region.

Experimental and modelling programmes must strongly interact. Due to the food security issue, the network will initially concentrate on the impacts of global change on water erosion in the humid tropics, and water and wind erosion in semi-arid regions. In this respect, the IGBP transects (NATT, Northern Australia Tropical Transect; SALT, Savannahs in the Long Term, from Ivory Coast to Niger) appear to be the most relevant.

References

Albergel, J. (1987). Genèse et prédétermination des crues au Burkina Faso. Du m² au km² étude des paramètres hydrologiques et de leur évolution PhD, University of Paris.

Albergel, J. & Valentin, C. (1988). Sahélisation d'un petit bassin versant soudanien: Kognere-Boulsa au Burkina-Faso. In *Les hommes face aux Sécheresses, Nordeste brésilien – Sahel africain*, ed. B. Bret, pp. 179–91.

EST/IHEAL, Collection Travaux et Mémoires no. 42.

Ambouta, J. M. K. (1994). Etude des facteurs de formation d'une croûte d'érosion et de ses relations avec les propriétés internes d'un sol sableux fin au Sahel. PhD thesis, University of Laval, Québec.

Arnold, R. W., Szabolcz, I. & Targulian, O. (1990). *Global Soil Change.* Laxenburg, Austria: IIASA, ISSS, UNEP.

Bergametti, G. (1992). Atmospheric cycle of desert dust. *Encyclopedia of Earth Systems Science*, **1**, 171–82.

Berger, A. (1981). The astronomical theory of paleoclimates. In *Climatic Variations and Variability: Facts and Theories*, ed. A. Berger, pp. 501–25. Dordrecht: Reidel.

Bernard, C. & Laverdière, M. R. (1992). Spatial redistribution of Cs-137 and soil erosion on Orléans Island, Québec. *Canadian Journal of Soil Science*, **72**, 543–54.

Boardman, J., Evans, R., Favis-Mortlock, D. T. & Harris, T. M. (1990). Climate change and soil erosion on agricultural land in England and Wales. *Land Degradation & Rehabilitation*, **2**, 95–106.

Bourke, G. (1987). Forests in the Ivory Coast face extinction. *New Scientist*, **114**, 22.

Bouwman, A. F. (1990). *Soils and the Greenhouse Effect*. Chichester: Wiley.

Casenave, A. & Valentin, C. (1989). *Les états de surface de la zone sahélienne. Influence sur l'infiltration*. Paris: ORSTOM, Collection 'Didactiques'.

Casenave, A. & Valentin, C. (1992). A runoff capability classification system based on surface features criteria in the arid and semi-arid areas of West Africa. *Journal of Hydrology*, **130**, 213–49.

Charreau, C. & Nicou, R. (1971). L'amélioration du profil cultural dans les sols sableux et sablo-argileux de la zone tropicale sèche Ouest Africaine et ses incidences agronomiques. *Agronomie Tropicale*, **26**, 903–78, 1183–247.

Collinet, J. (1988). Comportements hydrodynamiques et érosifs de sols de l'Afrique de l'Ouest. PhD thesis, University of Strasbourg.

Collinet, J. & Valentin, C. (1985). Evaluation of factors influencing water erosion in West Africa using rainfall simulation. In *Challenges in African Hydrology and Water Resources*, pp. 451–61. Wallingford, UK: International Association of Hydrological Sciences, Publication no. 144.

Combeau, A. & Quantin, P. (1963). Observations sur les variations dans le temps de la stabilité structurale des sols, en région tropicale. *Cahiers ORSTOM, série Pédologie*, **1**, 17–26.

Combeau, A. & Quantin, P. (1964). Observations sur les relations entre stabilité structurale et matière organique dans quelques sols d'Afrique Centrale. *Cahiers ORSTOM, série Pédologie*, **2**, 3–11.

Dregne, H. E. (1990). Impact of climate warming on arid region soils. *Developments in Soil Science*, **20**, 177–84.

Feller, C., Fritsch, E., Poss, R. & Valentin, C. (1991). Effet de la texture sur le stockage et la dynamique des matières organiques dans quelques sols ferrugineux et ferrallitiques (Afrique de l'Ouest, en particulier). *Cahiers ORSTOM, série Pédologie*, **26**, 25–36.

Fryrear, D. W. (1985). Soil cover and wind erosion. *Transactions of the American Society of Agricultural Engineers*, **28**, 781–4.

Greenland, D. J., Bowen, G., Eswaran, H., Rhoades, R. & Valentin, C. (1994). Soil, water, and nutrient management: a new agenda. Position paper. Bangkok: International Board for Soil Research and Management.

Harrison, P. (1987). Trees for Africa. *New Scientist*, **114**, 54–7.

Hoogmoed, W. B. (1986). Crusting and sealing problems on West African soils. In *Assessment of Soil Surface Sealing and Crusting*, ed. F. Callebaut, D. Gabriels & M. De Boodt, pp. 48–55. Ghent: Flanders Research Centre for Soil Erosion and Soil Conservation.

Kukla, G. (1980). End of the last Interglacial: a predictive model for the future? *Palaeoecology of Africa*, **12**, 395–408.

Kwaad, F. J. P. M. (1991). Summer and winter regimes of runoff generation and soil erosion on cultivated loess soils (the Netherlands). *Earth Surface Processes and Landforms*, **16**, 653–62.

Lal, R. (1987). Managing the soils of sub-

Saharan Africa. *Science*, **236**, 1069–76.

Lamb, P. J. (1982). Persistence of subsahara drought. *Nature*, **299**, 46–7.

Lance, J. C., McIntyre, S. C., Naney, J. W. & Rousseva, S. S. (1986). Measuring sediment movement at low erosion rates using caesium-137. *Soil Science Society of America Journal*, **50**, 1303–9.

Lane, L. J. & Nearing, M. A. (eds.) (1989). USDA-Water Erosion Prediction Project: profile model documentation. Report no. 2. West Lafayette: NSERL.

Langbein, W. B. & Schumm, S. A. (1958). Yield of sediment in relation to mean annual precipitation. *Transactions of the American Geophysical Union*, **39**, 1076–84.

Le Borgne, J. (1990). La dégradation actuelle du climat en Afrique, entre Sahara et Equateur. In *Dégradation des paysages en Afrique de l'Ouest*, ed. J. F. Richard, pp. 17–36. University of Dakar.

Le Fevre, J. (1993). L'évaporation au sein de sols sableux du Niger au cours de la saison humide. MSci thesis. Rennes: ENSA.

Le Houérou, H. N. (1989). *The Grazing Land Ecosystems of the African Sahel*. Ecological Studies no. 75. Berlin: Springer.

Le Houérou, H. N. (1993). Changements climatiques et désertisation. *Sécheresse*, **4**, 95–111.

Maley, J. (1981). Etudes palynologiques dans le bassin du Tchad et paléoclimatologie de l'Afrique nord-tropicale de 30 000 ans à l'époque actuelle. Collection Travaux et Documents no. 129. Paris: ORSTOM.

Mann, R. (1987). Development and the Sahel disaster: the case of Gambia. *The Ecologist*, **17**, 84–90.

McTainsh, G. H. (1980). Harmattan dust deposition in northern Nigeria. *Nature*, **286**, 587–8.

McTainsh, G. H. (1986). A dust monitoring programme for desertification control in West Africa. *Environmental Conservation*, **13**, 17–25.

McTainsh, G. H., Rose, C. W., Okwach, G. E. & Palis, R. G. (1992). Water and wind erosion: similarities and differences. In *Erosion, Conservation, and Small-scale Farming*, ed. H. Hurni & K. Tato, pp. 107–19. Merceline, USA: Wadsworth.

Middleton, N. J. (1985). Effect of drought on dust production in the Sahel. *Nature*, **316**, 431–4.

Middleton, N. J., Goudie, A. S. & Wells, G. L. (1986). The frequency and source areas of dust storms. In *Aeolian Geomorphology*, ed. W. G. Nikling, pp. 237–59.

Moberg, J. P., Esu, I. E. & Malgwi, W. B. (1991). Characteristics and constituent composition of Harmattan dust falling in Northern Nigeria. *Geoderma*, **48**, 73–81.

Nearing, M. A., Deer-Ascough, L. & Laflen, J. M. (1990). Sensitivity analysis of the WEPP hillslope profile erosion model. *Transactions of the American Society of Agricultural Engineers*, **33**, 839–49.

N'Tchayi, M. G. (1988). Etude satistique des brumes sèches au-dessus du Sahara à partir des réductions de visibilité au sol. MSci thesis, University of Abidjan, Côte d'Ivoire.

Ojo, O. (1987). Rainfall trends in West Africa. In *The Influence of Climate Change and Climatic Variability on the Hydrologic Regime and Water Resources*, ed. S. I. Solomon, M. Beran & W. Hogg, pp. 37–44. Wallingford, UK: International Association of Hydrological Sciences, publication no. 168.

Oldeman, L. R., Hakkeling, R. T. A. & Sombroek, W. G. (1991). World map of the status of human-induced soil degradation: an explanatory note. Wageningen: ISRIC, Nairobi: UNEP.

Olivry, J. C. (1983). Le point en 1982 sur l'évolution de la sécheresse en Sénégambie et aux îles du Cap-Vert. Examen de quelques séries de longues durées (débits et précipitations). *Cahiers ORSTOM, série Hydrologie*, **20**, 47–69.

Orange, D. & Gac, J. Y. (1990). Bilangéochimique des apports atmosphériques en domaines sahélien et soudano-guinéen d'Afrique de l'Ouest (bassins supérieurs du Sénégal et de la Gambie). *Géodynamique*, **5**, 51–65.

Perez, P. (1994). Genése du ruissellement sur les sols cultivés du Sud Saloum (Sénégal). Du diagnostic à l'aménagement de la parcelle. PhD thesis, ENSAM, Montpellier.

Petit-Maire, N. (1990). Natural aridification or man-made desertification? A question for the future. In *Greenhouse Effect, Sea Level and Drought*, ed. R. Paepe *et al.*, pp. 281–5. Amsterdam: Kluwer.

Pieri, C. (1989). Fertilité des terres de savane. Bilan de trente ans de recherche et de développement agricoles a sud du Sahara. Paris: Ministère de la Coopération/Cirad.

Planchon, O., Fritsch, E. & Valentin, C. (1987). Rill development in a wet savannah environment. *CATENA*, Suppl., **8**, 55–70.

Prospero, J. M. & Nees, R. T. (1977). Dust concentration in the atmosphere of the equatorial North Atlantic: Possible relationship to the Sahelian Drought. *Science*, **196**, 1196–8.

Ritchie, J. C. & McHenry, J. R. (1990). Application of radioactive fallout cesium-137 for measuring soil erosion and sediment accumulation rates and patterns: a review. *Journal of Environmental Quality*, **19**, 215–23.

Serpantié, G., Tezenas du Montcel, L. & Valentin, C. (1991). La dynamique des états de surface d'un territoire agropastoral soudano-sahélien sous aridification climatique: Conséquences et propositions. In *L'aridité: une contrainte pour le développement*, Collection 'Didactiques', pp. 419–47. Paris: ORSTOM.

Sircoulon, J. (1989). Bilan hydropluviométrique de la sécheresse 1968–1984 au

Sahel et comparaison avec les sécheresses des annes 1910–1916 et 1940–1949. In *Les hommes face aux sécheresses*, pp. 107–14. Paris: EST IHEAL.

Skidmore, E. L. & Williams, J. R. (1991). Modified EPIC wind erosion model. In *Modeling Plant and Soil Systems*. Agronomy monograph no. 31, pp. 457–69. ASA-CSSA-SSSA.

Slaymaker, O. (1990). Climate change and erosion processes in mountain regions of Western Canada. *Mountain Research and Development*, **10**, 171–82.

Valentin, C. (1985). Effects of grazing and trampling on soil deterioration around recently drilled water-holes in the Sahelian zone. In *Soil Erosion and Conservation*, ed. S. A. El Swaïfy, W. L. Moldenhauer & A. Lo, pp. 51–65. Ankeny, USA: Soil Conservation Society of America.

Valentin, C. (1986). Surface crusting of arid sandy soils. In *Assessment of Soil Surface Sealing and Crusting*, ed. F. Callebaut, D. Gabriels & M. De Boodt, pp. 9–17. Ghent: Flanders Research Centre for Soil Erosion and Soil Conservation.

Valentin, C. (1991). Surface crusting in two alluvial soils of northern Niger. *Geoderma*, **48**, 201–22.

Valentin, C. & Bresson, L. M. (1992). Morphology, genesis and classification of surface crusts in loamy and sandy soils. *Geoderma*, **55**, 225–45.

Valentin, C. & Casenave, A. (1992). Infiltration into sealed soils as influenced by gravel cover. *Soil Science Society of America Journal*, **56**, 1167–73.

Valentin, C., Chevallier, P., Fritsch, E. & Janeau, J. L. (1990). Le fonctionnement hydrodynamique aux échelles ponctuelles. In *Structure et fonctionnement hydropédologique d'un petit bassin de savane humide*, Collection Etudes et Thèses, pp. 147–63. Paris: ORSTOM.

Van der Watt, H. V. H. & Valentin, C. (1992). Soil crusting: the African view. In *Soil Crusting. Chemical and Physical*

Processes. Advances in Soil Science, pp. 301–38. Boca Raton, USA: Lewis.

Varallyay, G. (1990). Consequences of climate induced changes in soil degradation processes. 14th International Congress of Soil Science, Kyoto, Japan, 12–18 August. V:265–270.

Walling, D. E. & Quine, T. A. (1991). Use of ^{137}Cs measurements to investigate soil erosion on arable fields in the UK: potential applications and limitations. *Journal of Soil Science*, **42**, 147–65.

Williams, J. R., Jones, C. A. & Dyke, P. T. (1984). A modeling approach to determining the relationship between erosion and soil productivity. *Transactions of the American Society of Agricultural Engineers*, **27**, 129–44.

Yu, B. & Neil, D. T. (1993). Long-term variations in regional rainfall in the south-west of Western Australia and the difference between average and high intensity rainfalls. *International Journal of Climatology*, **13**, 77–88.

Part five

Ecological complexity

18 Global change and ecological complexity

O. E. Sala

The aim of the Global Change and Terrestrial Ecosystems (GCTE) core project has been to assess the effects of global change on the functioning of ecosystems and how these changes feed back to the atmosphere and the physical climate system. The drivers of global change, which are changes in land use, atmospheric composition, and climate, also directly affect ecological complexity which in turn affects ecosystem functioning (Fig. 18.1). How important is this indirect effect? Focus 4, Global Change and Ecological Complexity, is a new programme launched by GCTE to answer this question. The objective of this new Focus is to assess the effects of global change on ecological complexity and on the relationship between ecological complexity and ecosystem function (the dotted arrow in Fig. 18.1).

Ecological complexity represents biological diversity but in a broad sense, including not only species diversity but also diversity of ecosystems and landscapes, as well as genetic diversity within species. In addition, ecological complexity involves the diversity of trophic pathways and inter-actions. We can envision systems with similar diversity but contrasting complexity as a result of different organizational structures. Ecosystem func-tioning represents the collection of processes including primary production, decomposition, and nutrient cycling and their interactions.

Ecologists are intrigued by the diversity of organisms which inhabit the earth and, therefore, have studied the mechanisms that may account for this wealth of diversity. There is currently evidence supporting several available hypotheses which explain diversity as a function of ecosystem properties. However, less effort has been concentrated in understanding the effect on the opposite direction (Fig. 18.1): the effects of ecological complexity (or changes in ecological complexity) on ecosystem functioning. SCOPE (Scien-tific Committee on Problems of the Environment) has just finished a project led by H. A. Mooney which synthesized our current understanding of effects of biodiversity on ecosystem functioning (Schulze & Mooney, 1993).

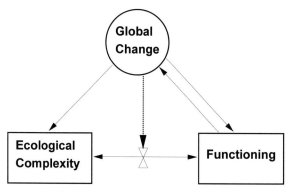

Figure 18.1　*The relationship between global change, ecological complexity and ecosystem functioning.*

This project consisted of a series of parallel workshops around the globe for different biomes and a final synthesis conference which attempted to identify similarities and differences among biomes regarding this issue. The reports from each biome and the cross-biome comparisons helped identify gaps in our understanding. There is a close connection between the ending SCOPE project and the starting GCTE Focus 4 Global Change and Ecological Complexity. The SCOPE project synthesized our knowledge and identified gaps which are the base of the research project which will be carried out by the new Focus 4. This agrees with the missions of SCOPE (synthesis) and of IGBP (fostering and coordinating research).

Several models of the relationship between ecological complexity and ecosystem functioning have been suggested, ranging from one which proposes that each species plays a unique role in the functioning of ecosystems and that therefore deletion of any species results in a change in ecosystem functioning, to those which consider that most species are redundant and that changes in ecological complexity should not result in changes in functioning (Vitousek & Hooper, 1993). A recently developed model relates previous diversity–function models with rank dominance models (Sala *et al.*, 1996). The effects on ecosystem functioning depend not only on changes in complexity but how these changes occur, and on which species are added or deleted. The model suggests a way of identifying those species which will have maximum effect on ecosystem processes.

Experiments scattered around the world provide evidence to support or reject the different diversity–function models. These include a range of studies from field to controlled environment conditions (McNaughton, 1993; Naeem *et al.*, 1994). Primary production and its relationship to plant species diversity has been one of the best-studied relationships. For example, in the Serengeti grasslands, removal of grasses with different contributions to total productivity shows the limits of ecosystems to compensate for the deletion of different species (McNaughton, 1983). Removal of rare plant species resulted in full compensation of production by remaining species, removal of species of intermediate abundance resulted in partial compensation, whereas removal of dominant species resulted in a significant reduction in production.

Ecological complexity may affect not only average ecosystem functioning but also the system response to extreme conditions. The diversity–stability hypothesis suggests that perturbations will result in a larger change in ecosystem functioning in simple systems than diverse systems. In old fields in New York, McNaughton (1993) described the response of ecosystems with different diversity resulting from being at different successional stages to a perturbation caused by fertilization. In the USA tall grass prairie, Tilman and Downing (1994) analysed the effects of a severe drought along a diversity gradient created as a result of an experimental fertilization where diversity was maximum in the native system and decreased as fertility increased. In both cases, the effect of perturbation on production was maximum in simple systems and minimum in the most diverse systems. Some of the conclusions drawn from these experiments have been criticized because fertilization simultaneously modified diversity and selected for species with lower root–shoot ratio, higher leaf conductance, and greater photosynthetic capacity, which are characteristics which result in lower drought resistance (Givnish, 1994). Consequently, the lower productivity during the drought year in low-diversity plots could have been the result of those plots being dominated by drought-sensitive plants. The critical experiment to disentangle the effect of diversity from the effect of individual species has not been performed yet.

The chapters presented in this section provide an excellent overview of the current understanding of global change and ecological complexity. Heal *et al.* note that a major part of the world's biodiversity lies unseen in the soil. However, it is important for many key ecosystem functions, such as organic matter decomposition, nutrient transformation and translocation, greenhouse gas formation and breakdown, and pedogenesis and soil morphology. They explore the following questions in their chapter: How diverse is the soil biota? How is it structured? How does diversity affect individual ecosystem functions? What are the research priorities?

Chapin *et al.* provide a useful link between below-ground processes and above-ground diversity by focusing on the direct and indirect effects of individual species on ecosystem processes, such as litter decomposition. They note that it is individual organisms that carry out essential ecosystem processes, and thus the traits of the individuals and their abundances should be important in determining the pool sizes and rates of energy and material fluxes in ecosystems. Their chapter addresses two general questions about the ecosystem significance of species traits and species diversity. If we know the traits of key organisms in an ecosystem, can we predict (1) the ecosystem impacts of species invasions or losses, and (2) the rates and patterns of processes in intact ecosystems? How does species diversity influence ecosystem processes?

Focus 4 is also concerned with diversity at the landscape level and its impacts on ecosystem functioning. The chapter by Holling *et al.* highlights recent advances in understanding the spatial and temporal functioning of terrestrial ecosystems at the landscape scale. It shows how contagious processes such as disturbances (e.g. fire and insect outbreaks) can self-organize vegetation patterns to create mosaics that are persistent over time and are resilient to broad ranges in variation in vegetation species composition, topography, and climate. These patterns in vegetation entrain discontinuous patterns in the body sizes of resident mammal and bird communities. The chapter concludes that when change in vegetation pattern occurs, perhaps as a result of global change, it will be sudden and extensive. Such dramatic change will probably have significant effects on ecological complexity and ecosystem functioning.

Chapters in this section describe our current understanding of the effects of ecological complexity on ecosystem function at different scales and from different standpoints. At the same time as reviewing what is known, they have highlighted gaps in our understanding. In order to fill those gaps the starting Focus 4 of GCTE will seek problems which are best solved collectively. It will avoid tasks which can be accomplished individually by investigators or groups and will concentrate on those studies which yield more than the sum of individual experiments.

References

Givnish, T. J. (1994). Does diversity beget stability? *Nature*, **371**, 113–14.

McNaughton, S. J. (1983). Serengeti grassland ecology: the role of composite environmental factors and contingency in community organization. *Ecological Monographs*, **53**, 291–320.

McNaughton, S. J. (1993). Biodiversity and function of grazing systems. In *Biodiversity and Ecosystem Function*, ed. E.-D. Schulze & H. A. Mooney, pp. 361–83. Berlin: Springer.

Naeem, S., Thompson, L. J., Lawler, S. P., Lawton, J. H. & Woodfin, R. M. (1994). Declining biodiversity can alter the performance of ecosystems. *Nature*, **368**, 734–7.

Sala, O. E., Lauenroth, W. K., McNaughton, S. J., Rusch, G. & Xinshi Zhang (1996). Biodiversity and ecosystem function in grasslands. In *Functional Roles of Biodiversity: A Global Perspective*, ed. H. A. Mooney, J. H. Cushman, E. R. Medina, O. E.

Sala & E.-D. Schulze. Chichester: Wiley (in press).

Schulze, E.-D. & Mooney, H. A. (eds.) (1993). *Biodiversity and Ecosystem Function*. Ecological Studies Edition, vol. 99. Berlin: Springer.

Tilman, D. & Downing, J. A. (1994). Biodiversity and stability in grasslands. *Nature*, **367**, 363–5.

Vitousek, P. M. & Hooper, D. U. (1993). Biological diversity and terrestrial ecosystem biogeochemistry. In *Biodiversity and Ecosystem Function*, ed. E.-D. Schulze & H. A. Mooney, pp. 3–14. Biodiversity and Ecosystem Function. Berlin: Springer.

⑲ Self-organization in ecosystems: lumpy geometries, periodicities and morphologies

C. S. Holling, G. Peterson, P. Marples, J. Sendzimir,
K. Redford, L. Gunderson and D. Lambert

Introduction

Adult body masses of mammals and birds clump into a small number of size categories in the forest, prairie and aquatic ecosystems of boreal biomes of North America (Holling, 1992). That is, there are gaps in body mass distributions that, in effect, are forbidden zones where there are no species with those adult sizes. This paper describes recent work designed to determine how such patterns, and related discontinuous vegetation patterns, are caused by, and in turn control, processes of ecosystem dynamics. These are the self-organizing processes that mediate the interactions among plants, animals and abiotic variables. The goal is to understand how species diversity might contribute to the robustness of the organizing processes and hence affect the resilience of ecosystems to possible changes in climate and land use.

Background

The observation that adult body masses are lumpily distributed might only be an intriguing curiosity, except that its discovery came from testing the postulate that landscapes and ecosystems are structured by a small set of self-organizing processes, each of which produces spatial patterns and temporal frequencies over a particular range of scales unique to each process. Each of those scale ranges, therefore, has unique distributions of the sizes of and spacing between objects and of the time intervals for their renewal. In short, there will be lumpy distributions of geometric and temporal attributes at least over the range from centimetres and days to hundreds of kilometres and millennia. If so, then the 'echo' of those persistent, lumpy patterns on landscapes should also be seen in equally lumpy patterns of behavioural and morphological attributes (such as adult size) of the communities of organisms that occupy the landscapes.

That turned out to be the case (Holling, 1992). Alternative taxonomic, historical, biomechanical or trophic hypotheses were disproved as primary generators of the discontinuous morphological patterns. The only hypothesis that consistently resisted disproof was that discontinuous landscape patterns generate discontinuous body mass patterns. For example, in one test, species assemblages from different biomes showed different patterns of adult body mass lumps if they resided in different biomes with different spatial lumps and textures (boreal latitude forests, grasslands, marine pelagic). In contrast, species assemblages from different taxa with fundamentally different organizations and habits, like birds and mammals, or different trophic status, like carnivores and herbivores, showed basically the same body mass lump structure if they resided within the same biome. In effect, the body mass clump structure of species of any taxon is an ecoassay of the structure of the landscape within which they reside.

The underlying postulate that ecosystem and landscape dynamics are controlled by scale-specific clusters of self-organizing processes emerged after comparing the similarities among a set of 23 examples of ecosystems dominated by four different classes of landscape scale disturbance. These included forests dominated by insect outbreaks, forests controlled by fire, grassland and savannah maintained by grazing pulses and aquatic systems organized by predation (summarized in Holling, 1986). In each example, it was observed that most of the dynamics in space and time could be ascribed to three or four sets of variables, each of which had a spatial extent and periodicity that typically differed from one another by one or more orders of magnitude (Table 19.1). Moreover, there clearly are still smaller and faster variables and even slower and larger ones than the ones chosen in these studies, and those also shape patterns, but over other smaller and larger scale ranges.

In natural forests, the small, fast processes are soil and plant physiological processes that determine structures from the scale of leaves to tree crowns. The next larger and slower processes are ones of inter-specific plant competition for light, nutrients and moisture that control structure up to scales of the patch or gap. The still larger and slower contagious abiotic and zootic processes of fire, wind, insect outbreak and large mammal herbivory organize structure up to scales of forest stands. Finally, still larger and slower regional geomorphological and climate processes characterize ecosystem types on landscapes.

Although this postulate emerged from an inductive search for common causes in a comparison of the dynamics of a number of different ecosystems, it is consistent with a hierarchical representation (Allen & Starr, 1982; O'Neill et al., 1986). An example is shown in Figs. 19.1 and 19.2 for the Everglades system of South Florida where the hierarchies of landscape

Table 19.1 *Key variables and speeds in ecosystems with four classes of disturbance*

Disturbance	Fast and fine variables	Intermediate variables	Slow and coarse variables	Reference
Forest insect	Insect needles	Foliage crown	Trees	McNamee *et al.*, 1981; Holling 1991
Forest fire	Intensity	Fuel	Trees	Holling, 1980
Savannah grazing	Annual grasses	Perennial grasses	Shrubs	Walker *et al.*, 1969
Aquatic predation	Phytoplankton	Zooplankton	Fish	Steele, 1985

objects are compared with those of atmospheric phenomena (Fig. 19.1) and to those of behavioural choices of a typical large wading bird such as the White Egret (Fig. 19.2). Any one of these structures form a set of nested objects, each progressing from lower levels of fast small objects to higher levels of progressively slower and larger objects. But each structure – ecosystem, atmosphere and animal choice – overall occupies a different part of the time/space domain. The atmospheric structures are the fastest, the ecosystem/landscape are the slowest, and animal choice lies between the two. Since the dynamic interactions between elements of these different structures are strongest when there are similar speeds, such representations provide clues as to which elements of the ecosystem, atmosphere and animal choice are most immediately coupled together.

The image of a small set of self-organizing processes forming dynamic spatial and temporal patterns in vegetation and in animal communities provides one direction to search for a simplified way to understand the great complexity of and diversity within ecosystems. It could open ways to set priorities for policies to create, restore and protect critical habitats, ecosystem functions, and species in a world of intensifying and expanding global and regional change.

This possibility can be explored by posing four questions:

▪ What characterizes the processes that produce lumpy patterns on the landscape? The brief answer, at this stage, is that the processes that form these patterns are reinforced by the pattern, i.e. they are self-organizing and the patterns they produce are repeated and persistent within broad, but not unlimited, ranges of climate change or land transformation.

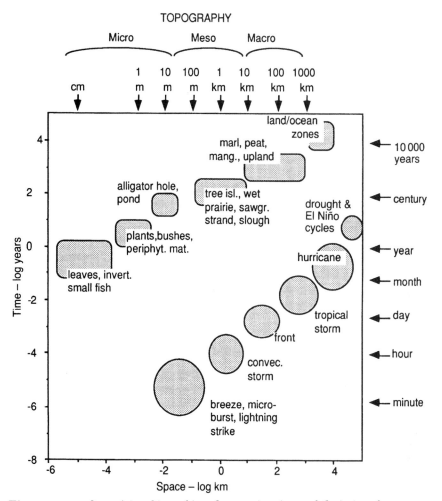

Figure 19.1 *Space/time hierarchies of vegetation (upper left tier) and climatic features (lower right tier) in the Everglades ecosystem.*

■ How do such self-organized patterns affect animal communities and biodiversity? The brief answer, so far, is that persistence of the landscape patterns encourages adaptation to them. Thus lumpy patterns in landscapes generate lumpy categories of body size and scales of choice. Such structures diagnose which species are most affected by vegetation transformation – as ones vulnerable to extinction or open for invasion. The lumpy patterns provide an ecoassay that can set one priority for policy attention on those species most sensitive to the impact of climate or land-use change.

■ How does biodiversity contribute to ecosystem function? The brief answer is that some species not only exist because of the niches

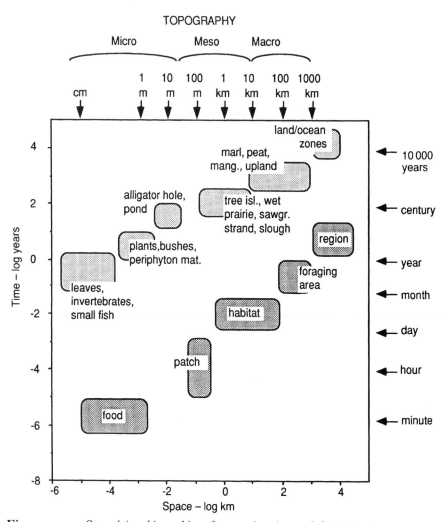

Figure 19.2 *Space/time hierarchies of vegetation (upper left tier) and foraging opportunities for wading birds (lower right tier) in Everglades ecosystem.*

provided by the lumpy structures, they also contribute to the formation of that structure as variables in the self-organizing processes. When several species are involved, they provide a functional diversity that contributes to the robustness of the self-organizing processes. This functional diversity provides a second priority for policy attention on those species most important in maintaining the structure and dynamics of ecosystems.

■ What causes the entrainment of body mass clump structures to landscape lump structures? The brief, and still very tentative, answer is that the gaps in body masses represent forbidden zones of landscape scale where the resources are insufficient for utilization by animals of those body sizes. Finer scale or coarser scale resources can be in abundance but can only be exploited by smaller or larger animals. When these resources fluctuate seasonally, generalists can bridge those gaps, but specialists still reflect their existence. The clumps themselves represent regimes of existence within which various strategies and objectives for optimization of resource acquisition and of fitness can express themselves.

This chapter will concentrate on recent results obtained in answering the first three questions.

Self-organizing ecosystem processes

The question 'What characterizes the processes that produce lumpy patterns on the landscape?' can be approached by modelling processes that control space/time dynamics and evaluating the resulting patterns against empirical measurements of boreal forest ecosystems. Experience in modelling global climate change and its impacts has covered a restricted range of scales. Global circulation models resolve the earth's surface into cells in the order of 100 km, though efforts continue to develop finer regional scale variants that couple with the global models (Lean & Warrilow, 1989; Pitman *et al.*, 1990). At a finer scale, modelling of forest dynamics has successfully aggregated knowledge of small-scale vegetative ecological processes up to the scale of a gap in the forest (about 30 m) where species composition and growth are dominated by processes of plant competition for light, water and nutrients (Pastor & Post, 1986; Smith & Urban, 1988; Leemans & Prentice, 1989; Bonan *et al.*, 1990; Shugart & Prentice, 1992). It is the range of scales between the tens of metres encompassed by gap models and the tens of kilometres resolved by global climate models, where the least modelling of space/time dynamics has occurred.

Species differences between ecosystems are partly explained by gap models as the consequence of local competitive interrelationships. But the structural differences between ecosystems from the tundra to the tropics is dominated more by larger-scale disturbance processes which are initiated locally and spread across landscapes. These contagious processes include abiotic processes such as fire, storm and water, and zootic processes such as insect outbreaks and large mammal herbivory – from beavers to ungulates (McNaughton, 1988; Naiman, 1988). Their roles are sufficiently unknown that they can equally equate to the mysteries of the biblical forces of fire, plague, flood and pestilence.

Nevertheless, these processes, interacting with topography and regional climate, form the ecosystem-specific lumpy patterns on the landscape that shape the morphology and diversity of animal communities. They also generate spatial and temporal variation which increase the diversity of plant species by periodically overriding the competitive dominance relations occurring at the gap scale (Holling, 1991).

The role of fire in the mid-continental North American boreal forest is examined as an initial, representative, example of a contagious meso-scale process. A family of models of forest disturbance has been developed to explore the role of various ecological processes and structures and to explain the fire patterns observed in nature. The range of scales cover the cross-hatched region in Fig. 19.3, essentially the region between gap scale and regional scale phenomena.

Other work has also examined the role of disturbance in structuring landscapes. Work done at the Oak Ridge National Laboratory (Turner *et al.*, 1989; Turner, 1989) has explored the structuring influence of self-propagating disturbance upon different types of landscape structure. Green (1989) has examined interactions between fire processes, seed dispersal and vegetative pattern in forests.

Fire is one of the dominant meso-scale process in the North American boreal forest (Payette, 1989). At the local scale, fire is a homogenizing force in the boreal forest. Fire burns large areas of forest which regenerate as even-aged stands. At a regional scale, fire introduces heterogeneity, because different fires fragment a landscape into many different patches. Fire mediates between large, fast atmospheric processes, which influence the combustibility of the forest, and smaller slowly changing forest features, which provide the fire's fuel (see Fig. 19.3).

Although fire occurs quickly, compared with forest growth, fire has a persistent effect on forest structure. By changing the forest landscape, and therefore constraining the properties of the future forest landscape, fire influences the spread of future fires. It is a self-organizing process.

The extent of this self-organizing influence is determined by the degree to which the patterns in the landscape induced by a fire, influence the spread of a future fire. The strength of this feedback in fire–forest interaction depends upon the forest's species composition, the region's topography and climatic conditions. These interrelationships have been explored using a variety of simulation models, each designed to identify and explain the specific features of the patterns produced by a typical meso-scale disturbance process such as fire.

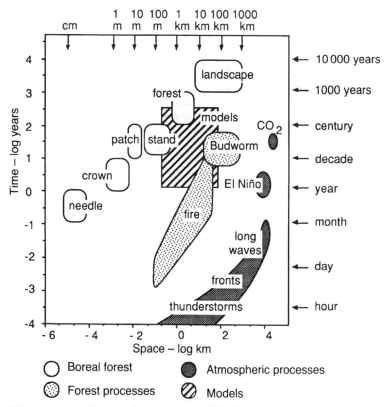

Figure 19.3 *Time and space scales of the boreal forest and their relationship to some of the processes which structure the forest. Contagious meso-scale disturbance processes provide a linkage between macro-scale atmospheric processes and micro-scale landscape processes.*

Models

A family of four meso-scale landscape simulation models were constructed to examine these spatial-temporal relationships between fire and forest structure. These models, the Null, Prototype, Fuel and Tree Species models, incorporate a range of ecological processes and structures as noted in Table 19.2. The behaviour of each of these models was compared with both other models and empirical data from southeastern Manitoba. These models share a common architecture, and this commonality makes it simplest to describe the structure of the basic Prototype model first and then detail the ways in which the other three models differ from it.

Model organization

The model represents a region of forest as a rectangular matrix. Each element in the matrix, termed a site, represents an area of forest

Table 19.2 *Alternative models and the defining differences between models. The models are of increasing complexity with the models at the top of the table encompassing the functioning of those at the bottom*

Model	Key process
Null	Fire spread independent of forest age
Prototype	Fire spread function of forest age
Fuel	Fire spread function of fuel and site dryness
Tree Species	Landscape composed of 2 tree species

approximately 150 m on edge which is equivalent to an area of 2.25 ha. At each site forest succession processes occur, while fire processes occur across sites.

Each site is described by the age of the forest at that site. Site age is considered to be analogous to the physical state of the site. Every simulated year the site ages. The modelled forest region experiences a fixed number of fire initiations/unit area at random locations every year. The probability of fire spreading from a burning site to an unburned site is a function of the age of the unburned site. A fire will continue to spread until it fails to spread to any unburned cells. The entire process of fire initiation and spread occurs within a simulated year (see Fig. 19.4).

No external disturbance enters the area from beyond the model's edges, and fire cannot spread beyond the edges of the simulated forest area. These constraints produce an edge effect, which results in the cells near the edges of the simulated area burning less frequently than the cells near the centre of the matrix. The extent of the area experiencing edge effects depends upon the relative size of the simulated fires compared with the total extent of the simulated area. The size of the landscape in the model is arbitrary, but we examined landscapes composed of 256 by 256 and 512 by 512 sites (about 150 000 ha and 600 000 ha respectively) which typically exhibited minimal edge effects over the range of conditions modelled.

The two main features which control the dynamics of the model are the probability of fire spread as a function of tree age and the rate of fire initiations/unit area/year. The behaviour of a model run is also influenced by the initial structure of the landscape and the extent of the simulated forest, but by appropriate experimental design these temporal and spatial edge effects can be reduced.

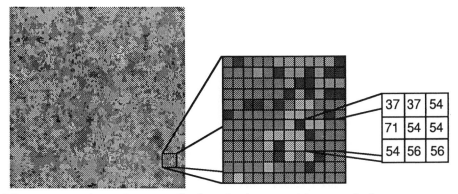

Figure 19.4 *A sample output of a forest landscape model composed of a matrix of sites. Fire organizes and maintains the landscape as a mosaic of patches composed of even-aged sites. The sites making up the landscape are each shaded by age. Young sites are lightly shaded while old sites are black. At the right of the figure the ages, in years, of a set of sites are shown.*

Alternative models

The Prototype model provides a template from which simplified or more complex models can be produced. For example, the Null model is a simplified model of boreal forest/fire interactions which assumes that fire spreads independently of stand age. This model provides a baseline, or null, against which to compare other models/hypotheses of boreal forest organization. In order to determine the roles of fuel and tree species, the Prototype model was also made more realistic in one alternative version by introducing fuel dynamics and in another alternative by including fuel and tree species differences. The behaviour of this family of models is summarized below.

Model behaviour

The analysis of the set of four models (Null, Prototype, Fuel and Tree Species) revealed a great commonality of behaviour and some key differences. The pre-eminent result is that when past fires influence future fire spread, the forest-fire system self-organizes into a self-maintaining, heterogeneous, patchy landscape even in the absence of external patchiness, such as topographic variation. Such self-organization does not occur in the 'memoryless' Null model, while it does occur in all the other models. Compared with this difference, other differences between the models are superficial (see Figs. 19.5 and 19.6).

Landscape self-organization occurs when fire produces a heterogeneous landscape of persistent, homogeneous age patches. This self-organization, and the memory that produces it, occurs whenever fire-spread is a function

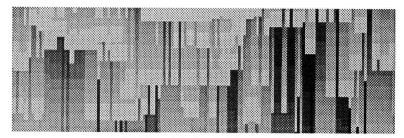

Figure 19.5 *Space/time transect of a Null model. The vertical axis is time, and the horizontal axis is distance along a transect of the spatial array of model cells. Time progresses from the bottom to the top. Young sites are lightly shaded, while older sites are darkly shaded. Fires burn over many different patches, overlapping with one another.*

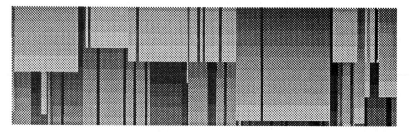

Figure 19.6 *Space/time transect of a Prototype model. Burns are more likely to burn an entire patch. The transect exhibits much more order than transects from the Null model.*

of site age. As a consequence, landscape pattern greatly influences where fires burn in the future. The homogeneity within patches produces a high probability of either none or most of a patch burning if part of it is ignited, and the differently aged patches bordering a burning patch reduce the probability of the fire spreading beyond a burning patch. These two processes encourage patch maintenance and patch cohesiveness.

This landscape self-organization occurs in the models only when the forest system is provided with memory, such as fire spreading as a function of stand age. System memory increases with the rate at which the probability of fire-spread increases with age. Increasing either the abruptness or the magnitude of the relationship between the probability of fire-spread and age increases the 'memory' of the forest landscape and its degree of self-organization.

The second major result of comparing model behaviour is that this self-organization process is very robust. Despite many changes in system parameterization the particular patterns persist. This suggests that the question to ask when seeking the applicability of these results to an actual forest is not what the fuel dynamics are, but how age affects the probability of fire-spread compared with other factors, such as climate and terrain.

Table 19.3. *Key behaviour of each of the four model types*

Models	Key behaviour
Null	Produces randomly distributed overlapping fires
Prototype	Landscape self-organizes into persistent patches
Fuel	Landscape self-organizes into persistent patches and fires can burn same area during several successive years
Tree Species	Same as Fuel model

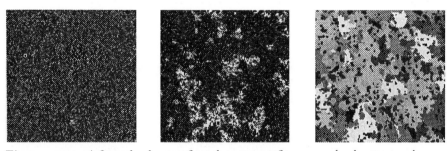

Figure 19.7 *A forest landscape of patches emerges from a randomly structured landscape composed of many small differently aged patches. Left, forest at 1 year; middle, forest at 10 years; right, forest after 52 years.*

The sequence of models were developed along an axis of progressively greater detail and greater complexity. With these increases in detail came greatly increased problems of model parameterization and analysis, and, after a point, little benefit in terms of producing different output patterns. The more complex models produced qualitatively more realistic behaviour, when compared with fire data from nature, but they also failed to produce any major novel behaviours. The key behaviours of each of the model types are described in Table 19.3.

Empirical comparison

The principal result of comparing the Null and Prototype models with the empirical Manitoban data is that the Prototype model is a much better approximation of the Manitoban data than the Null model. Furthermore, the Prototype model could be transformed in several ways both by including new processes or simply changing model parameters to produce a burn frequency map quantitatively very similar to the Manitoba data. It is much less possible to perform similar parameter tuning for the Null model (see Fig. 19.8).

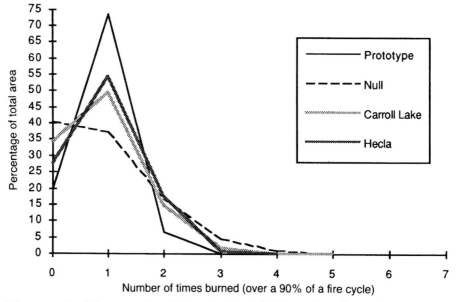

Figure 19.8   *The proportion of sites experiencing different burn frequencies within the Hecla and Carroll Lake regions of southeastern Manitoba (1929–89), and the Null and Prototype fire simulation models. These models have not been configured to mimic the Manitoba region, so that what is important is the form of the graphs. The peak of the Manitoba fire data and output of the Prototype model contrasts with the declining curve of the Null model.*

Conclusions

The Prototype model produces landscape features and fire properties similar to those actually found in the boreal forest. It also shows that the fire/forest system can produce a discontinuous patchy structure which persists and maintains itself in the absence of any external heterogeneity. The increasingly realistic models produce more detailed models of the forest/fire interactions, but at the cost of increased complexity.

The ability of the forest landscapes to self-organize suggests that fire and forests have an ability to adapt to variation, but when this resilience is exceeded, the landscape will change in a punctuated style in response to changes in climate, topography, or anthropogenic landscape modification. As a consequence, changes in forest structure, fire regimes and boundaries will occur suddenly. This clearly has important consequences for the management of forest resources, the prediction of the effects of global climate change and the design of international policies for moderating or adapting to climate change.

Although these models were developed as generalized meso-scale forest fire models, the fundamental conclusions they reveal have broader significance. Essentially, any contagious process with memory will have the self-organizing character described, generating a distribution of patch sizes, which, once established, will tend to persist in similar positions. There are two major consequences.

First, a large number of meso-scale processes have just those characteristics: insect outbreaks, tree diseases, large ungulate grazing, storm damage, seed dispersal and disturbance by vertebrates. All are contagious processes, i.e. they are initiated at one location and spread to others. All have systems memory, i.e. the likelihood of impact on another site is a function of the age and successional stage at that site. All are robust processes yielding patterns that are resilient. We will explore this last property in more detail later in the chapter since it bears strongly on the role of biodiversity in maintaining ecosystem patterns in space and time.

Second, as a consequence of the robust, persistent patterns, species living on those landscapes will have the opportunity to adapt to the persistent patch structures or utilize their presence. The imprint of that structure should therefore appear on a variety of attributes, including body mass structure. That is the subject we now address.

Lumpy body mass patterns and lumpy vegetation patterns

We have approached the question, 'How do lumpy patterns of vegetation, ecosystems and landscapes affect animal communities and biodiversity?' by comparing differences in body mass lump structures of animals residing in different sites where vegetations have been determined or transformed by different climates, or by changes in human activities. These examples include:

1. Comparison of body mass and vegetation lump structures of different taxa on sites chosen from an east–west transect through the boreal regions of continental North America and from a north–south transect from the boreal region into the tropical regions of South America. The goal is to have representations from the major biomes of the New World (J. Sendzimir, unpublished observations).

2. Lump analysis of pre- and post-agricultural bird and mammal fauna in a site near the Savanna River that has moved from intensive agriculture, principally cotton, to a mixed species temperate forest (J. Sendzimir & L. Gunderson, unpublished observations).

3. Comparison of body mass structure of bird communities and vegetation structures in eight paired comparisons of urban and non-urban sites in North America (Hostetler & Holling, in press).

4. Analysis of the body mass lump structure of fossil and extant mammals in southern California and northern Florida before and after the extinctions of the Pleistocene (Lambert & Holling, 1996).

5. Analysis of the relation of endangered, invasive and persistent species to the body mass lump structures of herps, birds and mammals in the transforming landscape of Florida (R. C. Allen, E. Forys & C. S. Holling, unpublished observations).

Much of that work is still in progress, but the analysis has been completed of body mass lump structures for 75 data sets drawn from specific biomes and ecosystems in Africa, Central America, Europe, North America and South America. All data sets show strong and statistically significant patterns of gaps and clumps. The lumpiness of body mass distribution for animals in terrestrial ecosystems is clearly real and universal.

Nevertheless, considerably more tests are needed to establish unambiguously that lumpy body mass patterns are caused by ecosystem-specific lumpiness within vegetated landscapes. In the original paper, three separate tests point in that direction and fail to support alternative hypotheses (Holling, 1992). One test compared body mass structure of species in a single taxon (birds or mammals) across ecosystem/biome types (grassland, forest and pelagic). One compared body mass structure within one biome type across major taxa (birds and mammals) and one across trophic categories (carnivores, herbivores and omnivores). The new projects described above are all designed to develop additional multiple evidence to define the causes. One set emphasizes comparison among landscapes with extremes of vegetation, different spatial size units, and different taxa. Another set approaches the question from a behavioural direction, asking how size-dependent attributes of animals can generate discontinuous size distributions on landscapes with various kinds of discontinuities. Here we will only review the work that is largely completed, particularly the results from nine sites established in the tropical regions of the north–south transect described in (1) above.

Biome scale structure

As a point of departure, Fig. 19.9 presents two graphs that show the body mass clump structure in different ways. Fig. 19.9a is a plot of the changes in a so-called gap detector as a function of the rank order of ascending body masses of birds of the boreal region prairies: the grasslands and savannas of Alberta (Holling, 1991). High values in the gap statistic represent jumps in the values of body masses between neighbouring members arranged in an ascending order of body masses. Fig. 19.9b translates the peaks (the gaps) and valleys (the clumps) revealed by the gap

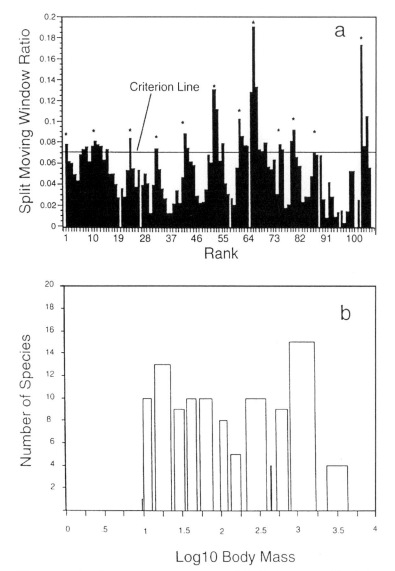

Figure 19.9 *Gap and clump structure of the body masses of birds in the boreal region prairies. (a) Value of the gap statistic (split-moving window ratio; see Appendix) as body mass increases. Gaps in body mass are indicated with an asterisk and are those that exceed the criterion line. (b) Body mass clump structure, i.e. the number of species in body mass clumps.*

detector, into a plot of number of species in each body mass clump, by selecting gaps that cross the criteria line shown. Details of the statistic and technique of lump analysis are described in the Appendix.

Other regional scale assemblages of animals drawn from distinct biomes also show lump structures whose complexity reflects the complexity of the ecosystems within the biome. The most complex body mass lump

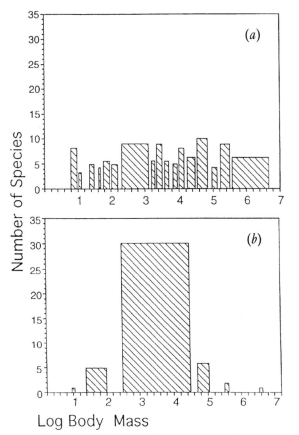

Figure 19.10   *Body mass clump structure for mammals of the (a) African savannas and (b) forests of Sierra Leone.*

structures we have discovered to date are within mammal data sets from the savannahs of Africa (Fig. 19.10a), a biome containing a complex set of discontinuous structures at all scales from the rock piles inhabited by dik diks to forests utilized by elephants. The simplest are from the tropical forests of Sierra Leone in Africa (Fig. 19.10b) which are notably homogeneous over meso- to macro-scales.

Ecosystem scale structure

The above examples represent fauna of whole biomes covering over hundreds of thousands of square kilometres containing mixed ecosystems within them. We were interested in seeing whether lump structure would also be found in a faunal assemblage from specific ecosystems covering a few tens of square kilometres. Here we will only review the work that is largely completed, particularly the results concerning mammal masses from nine ecosystem-specific small sites established in the tropical regions of the

Table 19.4 *Physical characteristics of nine Neotropical forest ecosystems*

Place	Latitude	Longitude West	Annual rainfall (mm)	Forest type[a]	Area (km$_2$)[b]
Masaguaral, Venezuela	8,5 N	67,5	1500	Seasonally wet savanna (Llanos) with forested areas (higher ground and river gallery)	<30
Chamela, Mexico	19,5 N	105,1	748	Dry deciduous forest	<7
Barro Colorado, Panama	9,9 N	79,51	2656	Rainforest: island, montane	>15
Guatopo, Venezuela	10,0 N	66,5	1500	Mix deciduous and non-deciduous montane forests with dry season, lower elevation in rain shadow	<10
Belem, Brazil	1,0 S	48,5	2337	Rainforest: eastern edge of Amazon Basin	<5
Lacandona, Mexico	16,5 N	91	2500	Rainforest: corridor within isthmus	>15
La Selva, Costa Rica	10,5 N	84	3994	Rainforest: corridor within isthmus	>15
Cuzco Amazonico, Peru	12,5 S	70,4	2387	Rainforest: continental	>20
Manu National Park, Peru	11,54 S	71,22	2028	Rainforest: continental	>15

[a]Ecosystems arranged roughly in ascending order of landscape complexity.
[b]The approximate collection area from which the species list was generated.

north–south transect described in (1) above. These include a subtropical wet savannah site, one neotropical dry deciduous forest site and seven neotropical rain forest sites (Table 19.4).

The significance of these sites is that they extend the tests made in the original study (Holling, 1992) in two ways. First, that study, as well as the additional data presented in Fig. 19.10, dealt with assemblages of species over large, biome-scale areas, whereas the nine tropical sites represent small sites of a few square kilometres containing one ecosystem type. In the original study, the persistence of lumpy structure in the data from large biome-scale aggregations was interpreted as evidence for the existence of some geometric landscape and vegetation attributes that were universal across biomes and others that were constrained to more site-specific conditions. For example, the geometry of foliage and trees is essentially the same wherever they occur, whereas the geometries generated by meso-scale disturbances such as fire or herbivory are unique to specific ecosystems. This combination of attributes with different degrees of universality would reveal persistent and similar clumps in the body mass distributions of different large-scale faunal assemblages and dissimilar clumps in distributions drawn from different small sites of the kind described here. The choice of small sites is therefore another way to test the different hypotheses originally proposed. If the body clump structures from different small sites are similar, particularly at intermediate scales where unique disturbance processes function, the lumpy vegetation/landscape geometry hypothesis is unlikely.

The second way these sites differ from the original study is that they represent southern tropical forested systems as compared with northern boreal coniferous ones. The latitudes of the boreal region were chosen originally as places where there were sufficient species to provide adequate data sets and potentially enough simple lump structure to reveal patterns. The question now is whether the higher productivity of tropical forests and their different scales of disturbance masks discontinuities or not.

Fig. 19.11 presents the results from the data set with the most complex structure. It shows the body mass structure of non-volant mammals residing in an approximately 15 km^2 portion of Manu National Park, Peru. In this example, at least, there is strong lump structure as predicted by the basic hypothesis. Neither the small size of the spatial unit nor the tropical latitudes eliminated or masked body mass clump structure.

In contrast, note that Wiens & Rotenberry (1981), in testing a different set of hypotheses concerning the causes of animal community structure, found that the conclusions depended upon the scale of the area sampled. They demonstrated that habitat structure correlates well with measures of bird community structure, like species diversity, at continental scales but

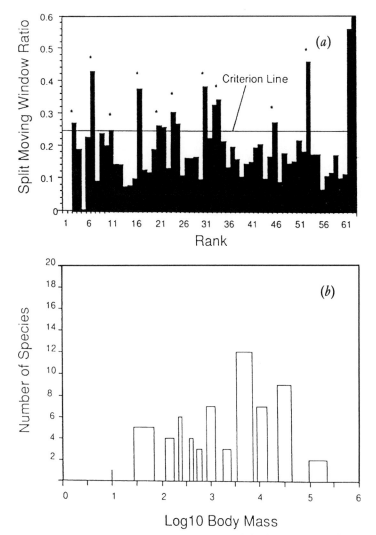

Figure 19.11 *Gap and clump structure of the body masses of non-volant mammals of Manu National Park. (a) Value of the gap statistic (split-moving window ratio) as body mass increases. Gaps in body mass are indicated with an asterisk and are those that exceed the criterion line. (b) Body mass clump structure, i.e. number of species in body mass clumps.*

not within local subsets of habitat. That is, the size of the sampling area affected the results of the tests of their hypotheses. Here, however, as expected, the existence of community body mass structure did not depend on the size of the area. The difference is that their hypotheses concerned one guild of species rather than all species within a major taxon, and their measures of community structure were species measures rather than morphological ones. Here we emphasize a much coarser set of measures of discontinuities in landscape geometry and discontinuities in animal morphology. Both measures are of form, and are therefore more likely to be connected.

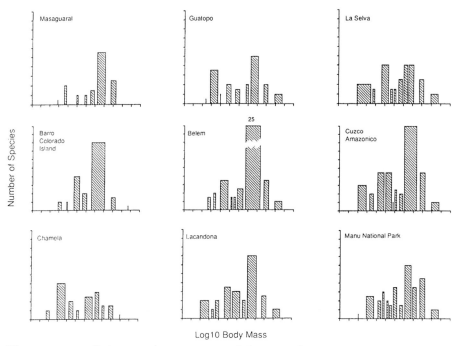

Figure 19.12   *Body mass clump structure of the non-volant mammals on all nine neotropical sites.*

Fig. 19.12 presents the data from all nine sites. First, as expected, there is variation in clump structure between sites. The number of clumps range from six (Masaguaral) to eleven (Manu). Second, the position of the clumps and gaps varies between sites over the intermediate size ranges, but less so over small and large ranges. For example, at the upper range, the four largest clumps of the Manu data sets are repeated in most others, with occasionally the upper one dropping out or the third and fourth largest combining into one. That is to be expected, because these are all low elevation, largely forested ecosystems in the tropics. Over the larger scale ranges, similar geomorphological processes should produce similar structure on all sites. Similarly, each of the two smallest clumps are either absent or repeated in all sites. Again that is expected because the fine-scale structure of foliage and trees is the same in any system containing trees. It is over intermediate ranges of size that clump structure varies between sites. That, also, is expected, since meso-scale disturbance processes are more location- and ecosystem-specific.

Are there any suggestions that the differences between sites, if they occur, are correlated with the architectural complexity of the sites? A measure of architectural complexity is the number of scale ranges that have unique geometric properties such as fractal dimension, object type and size

and inter-object distance. As scale of resolution increases in an architec-
turally simple ecosystem, for example, few breaks in fractal dimension occur
and few gaps in object types, sizes and inter-object distances are perceived.
In contrast, as scale of resolution increases in a complex ecosystem, many
breaks and gaps in spatial geometry occur.

We are still collecting the data to measure these attributes of complexity
directly, but we can at least infer some differences between sites. As noted
earlier, for example, Masaguaral at one extreme, and Cuzco Amazonico and
Manu at the other, show the greatest differences in number of body mass
clumps, suggesting they represent the greatest extremes in a spectrum of
landscape complexity.

Masaguaral, as a seasonally flooded savannah with some forest, is
predominantly covered with grasses, and most of the forest is relatively
short with no continuous canopy cover. Cuzco Amazonico and Manu, on
the other hand, are multiple-canopy rainforests, with much greater diversity
in terms of vegetative structures than the grassland/forests of Masaguaral.

The evidence presented here continues to reinforce the argument that
discontinuities in the geometry of vegetated landscapes impose discontinuities on the morphology of animals. That is, gaps in body mass distributions of resident species of animals correlate with scale-dependent
discontinuities in the geometry of vegetated landscapes. This relationship
has yet to be 'cross-calibrated' quantitatively in a way that turns gaps in
body mass measured in grams into breaks in geometry of landscape
elements measured in metres. That is the goal of the ongoing set of specific
studies mentioned at the beginning of this section. But even at this point
the more qualitative conclusion has consequences for questions of biodiversity loss.

For example, the pattern of Pleistocene extinctions that occurred about
11 000 years ago has been examined by comparing the body mass structures
of fossil and extant mammal fauna in northern Florida and southern California. In both data sets the impacts were specific to body mass clumps.
Whole clumps disappeared above body masses of 40 kg in both data sets
(Lambert & Holling, 1996), while the clumps and species numbers in
clumps below that weight remained unchanged. In both the Florida and
California cases, the extinctions can be explained in part by changes in the
mosaic patterns of vegetative productivity at particular scale ranges.

The patterns at those scales are likely to have been both generated and
maintained by grazers with large body masses, operating as self-organizing
disturbances similar to those described earlier for fire. As a consequence,
once the pattern of self-organization was disturbed beyond a certain point
the whole structure interrelating large herbivores and vegetation would

precipitously unravel. The initiating agents are likely to have been a combination of climate change and kills of large herbivores by peoples of the newly arrived Clovis hunting culture, but the suddenness of the disappearance of mega-herbivores suggests they played a role in self-organizing patterns. A somewhat similar explanation has been proposed by Morton (1990) for the extinction of middle-sized mammals in Australia subsequent to European settlement. Morton (1990) attributed this to transformation of the vegetation mosaic at intermediate scales associated with changes in burning practices and the influence of introduced rabbits. Short and Turner (1994), however, have challenged that conclusion with empirical evidence from an island fauna still containing middle-sized mammals. Their evidence comes from such a short period when rainfall was abundant that the issue is still very much open.

The evidence discussed in this section reinforces the possibility that the gaps in morphology and geometry could provide a predictive ecoassay useful for designing landscape and ecosystem restoration policies and managing impacts caused by possible climate and land-use change.

An ecoassay proposal

An ecoassay could be used in two ways. First, changes in landscape structure at defined scale ranges caused by land-use practice or by climate change will have predictable impacts on animal community structure (e.g. animals of some body masses can disappear or become rare if an ecosystem structure over a predicted scale range is changed). Therefore, observed (using remote imagery) or predicted (from models or from land-use development plans) impacts of changing climate or land use on vegetation can be used to predict the impacts on the α-diversity of animal communities.

Such an application uses vegetation attributes to predict changes in animal community diversity. But the reverse could also be done. That is, the critical scales where ecosystems are transforming could be predicted or diagnosed by analysing which body mass clumps contain species whose populations are declining. Before either of these ecoassay possibilities becomes a reality, however, considerably more work is necessary to explain the precise way that lumpy landscape patterns create lumpy body mass patterns and to cross-calibrate the two. That work is being carried out by continuing the process of hypotheses formation and tests using landscape and animal community data sets, and by developing an individual-based model to reveal the way simple size-specific rules of animal reaction interact with landscapes having different continuous and discontinuous patterns.

The role of biodiversity in self-organization of ecosystems

The preceding section explained the body mass clumps as being adaptations to lumpy vegetation and lumpy landscape structure. But some of those species not only exist because of the niches provided by the lumpy landscape structures, they also contribute to the formation of that structure as variables in the self-organizing processes. Using Walker's terms (Walker, 1992), they are not simply passengers going along for the ride, the structuring species are drivers. When a single species dominates in this structuring role, as in Paine's elegant demonstration of the role of starfish in structuring some intertidal communities (Paine, 1974), we call it a keystone species. When several species are involved, they provide a functional diversity that gives robustness to the self-organizing processes. That is the topic to be addressed in dealing with the third question posed in the introduction: How does biodiversity contribute to ecosystem function?

An example of ecosystem dynamics

One of the earliest interdisciplinary studies of space/time dynamics of an ecosystem provides an example to analyse. That was Morris's classic examination of epidemic spruce budworm–forest interactions in the central eastern portions of North America (Morris, 1963). That study exposed many of the processes in nature that controlled ecosystem dynamics. A subsequent project integrated that understanding and added process knowledge of forest growth within a space/time model of regional forest dynamics (Holling *et al.*, 1977; Jones, 1979).

Morris's study and the subsequent expanded model clarify the basic dynamics and the role of species diversity in generating and maintaining the temporal and spatial patterns. In particular, these studies have been used to demonstrate the structuring effect of a set of relations among three tree species, the budworm and 35 species of insectivorous birds over different scale ranges from local to continental (Holling, 1988).

For centuries prior to forest management, spruce budworm was the principal variable that structured the forest by setting the timing of the successional cycle and by forming the spatial attributes of the forest (Baskerville, 1976). In New Brunswick, for example, an outbreak would occur in a quasi-periodic cycle of 35–40 years. In any one local site of a few hundred hectares, the outbreak in most cases would last 3–5 years, killing, in the process, a large proportion of the mature balsam fir trees and damaging a smaller proportion of spruce. The outbreaks would sweep across the region as outbreaks in local stands spread to other locations, with a speed of spread such that the outbreaks in the region as a whole would be completed in about 15 years.

The causes of the boom–and–bust behaviour can be summarized by a zero-isocline surface of the kind shown in Fig. 19.13. This surface represents the conditions where budworm population densities are unchanging from one time period to another, assuming all the other slower variables (foliage and trees) are held fixed. The surface therefore separates regions of insect population increase from regions of decrease. The arrows indicate a typical cycle. Starting with a young stand of trees at point A, the budworm is scarce and in a region of decrease in its populations. That remains the case as the trees mature to point B in the figure because of the strongly recurved nature of the lower isocline surface. At that point, the low density regulation of budworm collapses and an insect outbreak is triggered. Foliage is then reduced sufficiently that trees die and the stand returns to one with younger trees and reduced tree crown volume.

The critical timing of that cycle is set by speed of tree growth and the depth of the 'pocket' in the lower recurved portion of the isocline surface. In effect, the speed of tree growth represents the ticking of a clock and the depth of the pocket the setting of the alarm that triggers the point of outbreak. It is this alarm setting or the depth of the pocket that makes the probability of outbreaks a function of stand age. It is, therefore, the

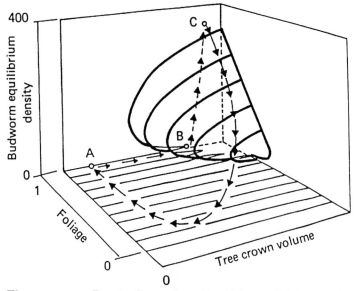

Figure 19.13 *Zero–isocline surfaces identifying equilibrium conditions for spruce budworm densities as a function of foliage and tree crown volume. The trajectory shows a typical natural outbreak sequence.*

property that gives the budworm outbreaks a memory, earlier described as one of the two attributes of a self-organizing process.

The particular form of the pocket – its breadth and depth – is caused by predation by a suite of insectivorous birds. As budworm densities increase from low levels, predation rate increases until the summed 'appetite' of the whole suite of predators saturates. At still higher densities, the now fixed level of attacks simply becomes diluted by progressively larger numbers of prey. Hence the major impact of predation is confined to a range of low budworm densities. That determines the breadth of the pocket.

The depth of the pocket is determined by rates of tree growth and foliage accumulation. That is, the impact of predation weakens as the trees in the stand grow, because attacks have to be distributed over an ever-expanding volume of foliage. On the one hand, that expanding volume provides more habitat for budworm larvae to feed. On the other hand, it weakens the impact of avian predation. The result is eventual collapse of regulation in older stands and the generation of an outbreak whose exact timing is often set by weather. The probability of budworm outbreaks therefore becomes a function of stand age, and that memory is introduced by the interaction among the suite of insectivorous birds and the rate of growth of volume of foliage in a stand.

The second requirement for self-organization described earlier was contagion. That is provided by both larval and adult moth dispersal. Dispersal of larval budworm is measured in tens to hundreds of metres and transfers a local impact within a patch of the forest (the 'gap' scale), to stands of trees measured in hundreds of metres. Dispersal of adult budworm moths is more extensive and is measured in tens to hundreds of kilometres. It transfers outbreaks in stands to whole forested landscapes of hundreds of kilometres. Together with the geomorphological processes that determine site conditions, the budworm-induced interactions produced a meso-scale pattern of stands of even-aged trees of tens to hundreds of hectares over regional landscapes of tens of thousands of square kilometres.

Thus the complex of variables that control the dynamics of this self-organizing process includes budworm and its dispersal, susceptible trees species and their rate of growth of foliage volume, and the suite of avian predators that, with tree growth, defines the memory for self-organization. That provides the focus now to explore how biotic diversity contributes to the robustness of the self-organization and the resilience of the resulting patterns.

Functional diversity in self-organizing processes

The foundations for the dynamics described above are determined by a set of interacting processes: tree growth and interspecific competition among three species of tree (birch, spruce and balsam) at the scale of a patch in the forest, the differential susceptibility of those tree species to budworm defoliation, budworm intraspecific competition and dispersal, and predation of budworm by a suite of 35 species of insectivorous birds. It is that cluster of processes and of the species that mediate them that determines the patterns in space and time, the scale-range over which those patterns persist and the resilience of the behaviour of the system.

Note that those processes operate over several scales. The finest and fastest scales are dominated by the physiological determinants of susceptibility of trees to insects, individual attack rates by birds (functional responses) and feeding competition by budworm. Growth of tree crowns is coarser and slower as are tree competitive relations, which function at the scale of the gap over years. Budworm dispersal, while a fast process, is still coarser spatially, operating at stand and inter-stand scales. All those processes, at all those scales, could potentially be affected by globally induced climate change and accumulation of CO_2 in the atmosphere.

If the system was only structured by small scale inter-tree competition, the patterns would be fine-scaled and tree species would be dominated by balsam, the superior interspecific competitor. But because balsam is most sensitive to budworm attack, budworm outbreaks periodically shift advantage from balsam to spruce so that both species persist in a dynamic inter-relationship. The addition of budworm contagious disturbance shifts the scale of the patterns from ones with a resolution of tens of metres in gaps, to ones with a resolution of thousands of metres in even-aged stands.

Now we can turn to a more precise analysis of the role of functional diversity by examining the way predation by the suite of insectivorous birds interacts with tree growth at the level of a stand. Predation by birds fluctuates from year to year and place to place because of differences in the species mix, in their population densities and in their attack rates. There is considerable evidence establishing ranges for all these attributes (Clark *et al.*, 1979; Holling, 1988). We can, therefore, make an extreme estimate of the range of fluctuation by assuming that the absolute minimum impact will occur when all species are simultaneously at their observed lowest population densities and lowest attack rates. That is truly an extreme, since minima of those attributes are unlikely to occur simultaneously for all species. Similarly, an absolute maximum can be set by assuming that all species simultaneously achieve their maximum population densities and attack rates. The resulting range is shown in Fig. 19.14.

The predation impact shown in the figure can be summarized by two parameters: the prey density at which the maximum predation occurs and the value of that maximum. When we used the full budworm/forest simulation model to test the sensitivity of the pattern to those two parameters,

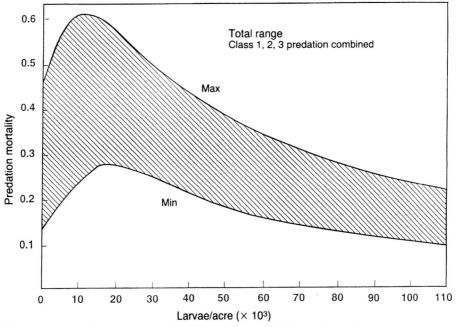

Figure 19.14 *Extreme range of total predation of budworm larvae by 35 species of insectivorous birds. Range is calculated from extrema estimated for each of three size groupings of birds.*

we essentially were testing the sensitivity to the depth and breadth of the 'predation pocket' that establishes the recurved portion of the isocline surface of Fig. 19.13.

Three basic patterns were generated in the parameter space (Fig. 19.15). The first pattern was a persistent 13 to 15 year cycle of budworm in which tree mortality was low. The timing of the cycle is set by regeneration rates for the foliage crowns and the spatial scale of resolution by the size of those crowns. The second pattern generated the classic 30–40 year boom-and-bust cycle where the timing was set by the interaction between the rate of tree growth and the depth of the predation pocket. Spatial scale of resolution was set by dispersal distances of budworm and was measured in hundreds of metres. The third pattern was a 70–100 year cycle set by tree regeneration rates and the spatial resolution by gap-scale interspecific competition among plant species. Because the controlling processes themselves act on a variety of time and space scales, the overall patterns shift suddenly rather than gradually when parameter values change incrementally. That is, the separation between the regions shown in Fig. 19.15 is very sharp and well defined.

The shift in patterns is accompanied by a shift in the processes that

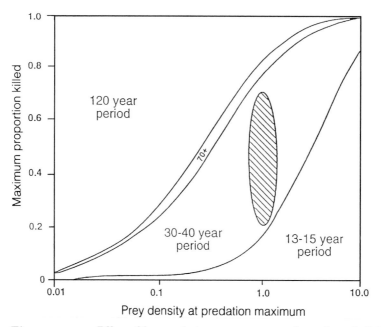

Figure 19.15 *Effect of key predation parameters on the cycle periodicity generated by the budworm/forest model.*

organize the patterns. The 13–15 year periodicity is controlled by crown dynamics and rates of tree recovery from defoliation. In the 30–40 year periodicity region, control shifts to the rate of tree growth and predation by the suite of birds. In the 70–120 year period region, control shifts to finer-scaled processes of interspecies plant competition.

   Now we can relate those patterns and their causes to the role of species diversity in maintaining the 30–40 year cycle. The oval hatched region in Fig. 19.15 represents the extreme range of those parameters in nature as earlier estimated in Fig. 19.14. Clearly, even though that range was developed to represent an exaggerated extreme, it still fits within the region of the classic 30–40 year cycle with its large even-aged stands. That is the pattern that dominates both in the model and nature. Occasionally, in local sites of the model, however, a persistent smaller-scale 15 year cycle takes hold by chance of initial conditions and weather. Such a cycle has been seen in aberrant stands in New Brunswick as well (Royama, 1984). More rarely, budworm populations remain persistently low in a site and gap-scale dynamics start to dominate. These features of a dominant pattern with a mosaic of smaller-scale patterns occur in all the examples of meso-scale disturbance we have modelled. They demonstrate that self-organizing processes of this kind generate lumpy distribution of patch sizes.

The robustness of the regulation provided by birds is partly the consequences of a contagious type of competition which makes predation very efficient when there are few birds and many prey and inefficient when the conditions are the reverse (Holling & Ewing, 1971; Holling, 1988). But more important, there is an overlapping redundancy provided by the very diversity of species that are functioning. First, the 35 species represent five of the body mass clumps identified in the 'lump paper' (Holling, 1992). The kinglets, chicadees and warblers are examples of the first two of the smallest size clumps. The Ovenbird and White-breasted Nuthatch are examples of the third size clump. The Slate-coloured Junco, White-throated Sparrow and Hermit Thrush are examples of the fourth clump size, and the grosbeaks are examples of the fifth (Holling, 1980). Moreover, other predators exist that were not represented in the model's parameter ranges. For example, still larger birds from larger clumps, like woodpeckers, are also known to attack budworm larvae, as are squirrels, ants and bats, all of which would have predation impacts of the same form as the predation curve in Fig. 19.14.

The significance is that each size clump represents a different set of values defining the attack responses. Therefore when the predation by birds of all clumps is combined, an overlapping set of influences is produced. Predation by birds of each clump dominate over different ranges of budworm density. Moreover, there are a small number of different searching styles among this suite of predators that establish a lumpy range of searching rates. Some species are more common in younger stands, some in older. Some search high in the tree canopy, some low. And in their Neotropical over-wintering areas, some are found in brush land, some in forests, some in clearings.

The consequence of all that variety is that the species combine to form an overlapping set of reinforcing influences that are less like the redundancy of engineered devices and more like portfolio diversity strategies of investors. The risks and benefits are spread widely to retain overall consistency in performance independent of wide fluctuations in the individual species. That is at the heart of the role of functional diversity in maintaining the resilience of ecosystem structure and function.

We chose to term this 'functional diversity' or 'functional redundancy' following the terms suggested by Schindler (1990) and by Holling *et al.* (1995). Such diversity adds great robustness to the process and, as a consequence, great resilience to the system behaviour. Moreover, this seems the way many biological processes are regulated: overlapping influences by multiple processes, each one of which is inefficient in its individual effect but together operating in a robust manner. For example, those are the features of the multiple mechanisms controlling body temperature regulation in endotherms, depth perception in animals with binocular vision and direction in bird migration.

Because of the robustness of this functional redundancy, and the non-linear way behaviour suddenly flips from one pattern to another and one set of controls to another, gradual loss of species involved in controlling structure initially would have little perceived effect over a wide range of loss of species. Then as loss of those species continued, suddenly, different behaviour would emerge more and more frequently in more and more places. To the observer, it would appear as if only the few remaining species were critical when in fact all add to the resilience. Although behaviour would change suddenly, resilience (measured as the size of stability domains, *sensu* Holling, 1973) would gradually contract. The system, in gradually losing resilience, would become increasingly vulnerable to perturbations that earlier could be absorbed without change in function, pattern and controls.

Conclusions

There is a debate whether every species is important in ecosystem dynamics and function or only a smaller subset of those involved in the self-organization (Baskin, 1994). On the one side is evidence from controlled experiments which show that declining generalized diversity reduces productivity (Naeem *et al.*, 1994) or reduced numbers of grass species reduces rates of recovery from drought (Tilman & Downing, 1994). But in such examples the physical limitations of the experiments limits the conclusions to small-scale interactions (plots ranged from 1 m to 4 m to the side), to short periods of time and to the set of structuring species and processes that can exist at those scales. Species involved in larger and slower processes are necessarily ignored so that attention shifts to the short term near equilibrium behaviour.

In contrast, those who argue that only a subset of species controls dynamics and function draw their evidence from large-scale manipulations of whole ecosystems like lakes (Schindler, 1990), from understanding of process function at different scales, from landscape and ecosystem scale models and field measures of disturbed and managed ecosystems – fresh water, marine, boreal forest, grassland and tropical forest. Those are summarized in a multi-authored synthesis by Holling *et al.* (1994) involving boreal forest, marine, freshwater and savannah ecosystems. There the argument is that it is the number of species involved in the structuring set of processes that determines the functional diversity (Schindler, 1990, 1993), and not all species. An example is the set of grass species and the set of ungulate grazers that maintain the productivity and resilience of savannahs (Walker *et al.*, 1969), or the tree species and suite of 35 species of insectivorous birds that mediate budworm outbreak dynamics in the eastern boreal forest, as described earlier.

Any ecosystem – a forest, a lake, a grassland or a wetland – contains hundreds to thousands of species interacting among themselves and their physical and chemical environment. But not all those interactions have the same strength or the same direction. That is, although everything might be ultimately connected to everything else if the web of connections is followed far enough, the first order interactions that structure the system increasingly seem to be confined to a small number of biotic and abiotic variables whose interactions form the 'template' (Southwood, 1977) or niches that allow a great diversity of living things, to, in a sense, 'go along for the ride' (Carpenter & Leavitt, 1991; Cohen, 1991; Holling, 1992). Those species are affected by the ecosystem but do not, in turn, notably affect the ecosystem, at least in ways that our relatively crude methods of measurement can detect. Hence at the extremes, species can be regarded either as 'drivers' or as 'passengers' (Walker, 1992), although this distinction needs to be treated cautiously. The driver role of a species may only become apparent every now and then under particular conditions that trigger their key structuring function.

The lumpy patterns generated by the models presented here and by the empirical measures of community structure highlight where the priority for global change or biodiversity policy should lie. The lumpy patterns indicate that change will not be incremental and local but sudden and extensive. If change does occur, there will be fundamental transformations from one ecosystem type to another, i.e. from forest to grassland, or grassland to a shrubby semi-desert, for example (Walker *et al.*, 1969; Holling, 1973). Then control of structure will shift from one set of organizing processes and variables to another set. And it is the diversity of overlapping influences within those controls that defines the resilience in those sudden shifts.

In summary, we currently believe that only a subset of species and physical processes are critical in forming the structure and overall behaviour of terrestrial ecosystems, although there are insufficient long-term studies to deduce whether the presence or absence of rare species may cause slow, subtle shifts in ecosystem structure or function.

The fundamental point is that only a small set of self-organizing processes made up of biotic and physical processes are critical in forming the structure and overall behaviour of ecosystems, and that these establish sets of relationships, each of which dominates over a definable range of scales in space and time. Each set includes several species of plants or animals, each species having similar but overlapping influence to give functional redundancy.

Regional and local climate and geomorphology sets the stage within which those sets of self-organizing processes establish persistent self-repeating patterns. The self-organized patterns uniquely define any specific ecosystem – forests in the maritime regions of Canada, in the mid-continental region of the Canadian shield or on the edge of the tundra – as well as temperate and tropical variants of forests, grasslands, savannahs and deserts.

Acknowledgements

The research reported on here was supported by NASA/EOS grant no. NAGW 2524, a part of the Interdisciplinary Scientific Investigations of the Earth Observing Systems programme, and NASA grant no. NAGW 3698, a part of the Terrestrial Ecology programme.

Appendix. Methods to detect gaps and lumps

The detection for discontinuities analyses structure in a univariate data set (such as adult animal masses). The data are rank ordered (lowest to highest value), then one of the specific techniques is applied to the rank-ordered set. The evaluation of methods has been done in two stages: first, a statistic to measure gaps or clumps is developed, then a series of randomization programs are compiled to test the significance of clumping found from analysing a data set against a series of random data from a known distribution.

A number of different gap and clump detection statistics have been applied to determine discontinuous structure in data sets. These include: the original and detrended versions of the Size Difference Index or SDI (Body Mass Index from Holling, 1992), the Silverman Difference Index (Silverman, 1986), a Kernel Estimator (Chambers *et al.*, 1983; Silverman, 1986), a Split-moving Window Index (Cornelius & Reynolds, 1991) and Hierarchical Clustering Analysis (Hartigan, 1975; Romesburg, 1984; Ludwig & Reynolds, 1988; Kaufman & Rousseeuw, 1990). All reveal the same gaps and clumps, subject to difference in the sensitivities of each method. Two of the techniques – the size difference index and the modified split-moving window method – appear to provide the simplest and most objective way to identify gaps for comparison across data sets. A brief description of these two techniques follows.

Size Difference Index

The original gap detector is expressed in Equation (1).

$$\text{Size Difference Index (SDI)} = (S_{i+1} - S_{i-1})/(S_i)\gamma \tag{1}$$

where S_i is the body mass of the ith species in ascending size rank order and γ is a detrending exponent. Holling (1992) found that $\gamma = 1.3$ was sufficient to detrend boreal bird data and $\lambda = 1.1$ to detrend mammal data. Applied to other data sets, S represents the size value used in the formula. Hence the size difference index is a more general formulation than the body mass difference index. The size index tends to emphasize differences at the low end of the value scale, that is, small absolute changes can be large relative changes. For example, an increase from 0.5 to 1.0 is an absolute change of 0.5, but a relative change of 1. This is corrected by the parameter γ; however, γ must be estimated. This technique creates a U-shaped artefact across the distribution of indices; values of the difference index at the low end of the distribution were approximately the same magnitude as those at the high end of the distribution. Another problem with this formulation is the population of index values do not have a normal distribution, and appear to follow a chi-square distribution. The non-normality may change the value of the confidence level used to determine the statistical significance of a measured gap size.

Split-Moving Window analyses

The modified split-moving window method is based in its initial form as a simple ratio of successive masses in an ascending order of body masses, where S_i is the body mass of the ith species in ascending size rank order. The technique passes two adjacent windows over the rank-ordered data set incrementally. The sum of all sizes within one window is divided by the sum of all sizes within the other. This technique's advantage is that the window size may be varied: increased window size examines coarser (i.e. low resolution) features within the data. However, this technique appears to create the same bias as the Size Difference Index.

We have made three adaptations. The first allows various levels of smoothing by changing the window size. The ratio ($SMWR_n$) is centred on one observation (M_n), and smoothing is increased by expanding window size, or number of observations, outward from the centre (Equation 2). Window width (w) usually starts at 2 and increases by doubling (i.e. 2, 4, 8). Larger body mass data sets ($n > 100$) sometimes required greater smoothing ($w = 4$) to reduce the 'noise' associated with numerous small gaps of questionable significance.

$$SMWr_n = \left(\sum_{i=n-w+1}^{n} Si \Big/ \sum_{i=n}^{n+w-1} Si \right) \qquad (2)$$

Where S_n is the size of the nth observed datum in a rank order of increasing size, and i denotes the relative position within a pair of windows of width w centred over that datum.

The second adaptation is made simply to highlight gaps rather than clumps, with the argument that clumps are defined in an ascending order of values as a set of similar values separated from each other by gaps in those values (Equation 3). Such an index ranges from 0 to 1, with the largest values representing the largest gaps, or largest jumps in values between members of a set of ascending values.

$$SMWr_n = I - \left(\sum_{i=n-w+1}^{n} Si \Big/ \sum_{i=n}^{n+w-1} Si \right) \qquad (3)$$

The third, and final, adaptation is to detrend the $SMWR$ index data (Equation 4). Algorithms to detect discontinuities in size distributions often overemphasize gaps at the extreme ends of a size distribution and underemphasize gaps in the middle. This can result either in U-shaped or generally increasing trends in index output when a $SMWR$ is applied to a size distribution ranked in ascending order. Such trends are diminished and usually eliminated by fitting a third order polynomial ($SMWR_{Fit}$) to the set of index values from the original body mass data set ($SMWR_{Orig_n}$) and subtracting the fitted curve from the original $SMWR$.

$$SMWR_{Detrend_n} = SMWR_{Orig_n} - SMWR_{Fit_n} \qquad (4)$$

The resulting set of split-moving window ratio indices ($SMWR_{Detrend_n}$) flatten the 'U-shaped' trend in that they exhibit a smaller range of index values with less variation between the beginning, middle and end portions of the body mass size distribution. These detrended SMWR values are standardized to range upward from 0 by subtracting the lowest $SMWR_{Detrend}$ value from all others. Large deviations from 0 in the detrended output are more likely to reflect significant gaps at any body size, i.e. the importance of a gap can be more equally assessed at any point along the range of body mass sizes.

Significance of gaps

Once the gap statistic is detrended, gaps can be defined by drawing criteria lines across the gap graph at various levels along the y-axis. Whenever the gap statistic crosses the line a gap is recorded provided that at least the

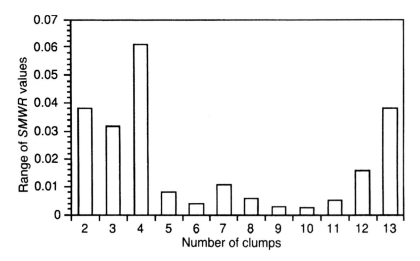

Figure 19.16 *Robustness Index for* SMWR *values of boreal prairie bird body masses.*

next two gap statistics are below the line – this is simply an arbitrary requirement so that a clump has at least two members. The higher the line drawn, the greater the level of significance of the gaps detected and the more conservative the estimate. But note that as the line is lowered, there are robust ranges where gap numbers do not change, separated by ranges where the number of gaps change frequently. Typically, as in the example in Fig. 19.16, there are two ranges where the estimate of gap number is robust. The lowest number of gaps can be chosen as a conservative estimate and the higher number as a liberal one. We have found that the liberal estimate provides a consistent way to reveal useful similarities and differences when comparing different data sets and it is those liberal estimates that are used in this paper.

References

Allen, T. F. H. & Starr, T. B. (1982). *Hierarchy: Perspectives for Ecological Complexity.* Chicago: University of Chicago Press.

Baskerville, G. L. (1976). Spruce budworm: super silviculturalist. *Forestry Chronicles*, **51**, 138–40.

Baskin, Y. (1994). Ecologists dare to ask: how much does diversity matter? *Science*, **124**, 202–3.

Bonan, G. B., Shugart, H. H. & Urban, D. L. (1990). The sensitivity of some high-latitude boreal forests to climate change parameters. *Climate Change*, **16**, 9–29.

Carpenter, S. R. & Leavitt, P. R. (1991). Temporal variation in paleolimnological record arising from a tropic cascade. *Ecology*, **72**, 277–85.

Chambers, J. M., Cleveland, W. S., Beat, K. & Tukey, P. A. (1983). *Graphical Methods for Data Analysis.* Boston: Duxbury Press.

Clark, W. C., Jones, D. D. & Holling, C. S. (1979). Lessons for ecological policy design: a case study of ecosystems management. *Ecological Modeling*, **7**, 1–53.

Cohen, J. (1991). Trophic topology. *Science*, **251**, 686–7.

Cornelius, J. M. & Reynolds, J. F. (1991). On determining the statistical significance of discontinuities within ordered ecological data. *Ecology*, **72**, 2057–70.

Green, D. G. (1989). Simulated effects of fire, dispersal and spatial pattern on competition with forest mosaics. *Vegetatio*, **82**, 139–53.

Hartigan, J. A. (1975). *Clustering Algorithms*. New York: Wiley.

Holling, C. S. (1973). Resilience and stability of ecological systems. *Annual Review of Ecology and Systems*, **4**, 1–23.

Holling, C. S. (1980). Forest insects, forest fires and resilience. In *Fire Regimes and Ecosystem Properties*, ed. H. Mooney, J. M. Bonnicksen, N. L. Christensen, J. E. Lotan & W. A. Reiners, pp. 445–64. USDA Forest Service General Technical Report. Washington, DC: USDA.

Holling, C. S. (1986). The resilience of terrestrial ecosystems, local surprise and global change. In *Sustainable Development of the Biosphere*, ed. W. C. Clark, & R. E. Munn, pp. 292–317. Cambridge: Cambridge University Press.

Holling, C. S. (1988). Temperate forest insect outbreaks, tropical deforestation and migratory birds. *Memoirs of the Entomological Society of Canada*, **146**, 21–32.

Holling, C. S. (1991). The role of forest insects in structuring the boreal landscape. In *A Systems Analysis of the Global Boreal Forest*, ed. H. H. Shugart, R. Leemans & G. B. Bonan, pp. 170–91. Cambridge: Cambridge University Press.

Holling, C. S (1992). Cross-scale morphology, geometry and dynamics of ecosystems. *Ecological Monographs*, **62**, 447–502.

Holling, C. S. & Ewing, S. (1971). Blind man's bluff: exploring the response space generated by realistic ecological simulation models. In *Statistical Ecology*, vol. 2, ed. G. P. Patil, pp. 207–29. Pennsylvania State Statistics Series. Philadelphia: Pennsylvania State University Press.

Holling, C. S., Jones, D. D. & Clark, W. C. (1977). Ecological policy design: a case study of forest and pest management. In *Pest Management*. Proceedings of an International Conference 25–29 October 1976, ed. G. A. Norton & C. S. Holling, pp. 13–90. Oxford: Pergamon.

Holling, C. S., Schindler, D. W., Walker, B. & Roughgarden, J. (1995). Biodiversity in the functioning of ecosystems: an ecological primer and synthesis. In *Biodiversity: Ecological and Economic Foundations*, ed. C. Perrings, K.-G. Mäler, C. Folke, C. S. Holling & B.-O. Jansson, pp. 44–83. Cambridge: Cambridge University Press.

Hostetler, M. & Holling, C. S. (in press). Using a new ecological technique to measure scale-specific impacts on landscape structure in urban environments. In *Proceedings of the National Symposium on Urban Wildlife*. Bellevue, Washington: National Institute for Urban Wildlife.

Jones, D. D. (1979). The budworm site model. In *Pest Management*. Proceedings of an International Conference, 25–29 October 1976, ed. G. A. Norton & C. S. Holling, pp. 91–159. Oxford: Pergamon.

Kaufman, L. & Rousseeuw, P. J. (1990). *Finding Groups in Data: An Introduction to Cluster Analysis*. New York: Wiley.

Lambert, W. D. & Holling, C. S. (1996). The cause of the terminal Pleistocene extinctions: evidence from mammal body mass distributions. *Paleobiology* (submitted).

Lean, J. & Warrilow, D. A. (1989). Simulation of the regional climatic impact of

Amazon deforestation. *Nature*, **342**, 411–13.

Leemans, R. & Prentice, I. C. (1989). FORSKA: a general forest succession model. *Meddelanden*, **2**, 1–45.

Ludwig, J. A. & Reynolds, J. F. (1988). *Statistical Ecology: a Primer on Methods and Computing*. New York: Wiley.

McNamee, P. J., McLeod, J. M. & Holling, C. S. (1981). The structure and behavior of defoliating insect/forest systems. *Research Population Ecology*, **23**, 280–98.

McNaughton, S. J., Ruess, R. W. & Seagle, S. W. (1988). Large mammals and process dynamics in African ecosystems. *BioScience*, **38**, 794–800.

Morris, R. F. (1963). The dynamics of epidemic spruce budworm populations. *Memoirs of the Entomological Society of Canada*, **21**, 1–332.

Morton, S. R. (1990). The impact of European settlement on the vertebrate animals of arid Australia: a conceptual model. *Proceedings of the Ecological Society*, **16**, 201–13.

Naeem, S., Thompson, J. L., Lawler, S. P., Lawton, J. & Woodfin, R. M. (1994). Declining diversity can alter the performance of ecosystems. *Nature*, **368**, 734–7.

Naiman, R. J. (1988). Animal influences on ecosystems dynamics. *BioScience*, **38**, 750–2.

O'Neill, R. V., DeAngelis, D. L., Waide, J. B. & Allen, T. F. H. (1986). *A Hierarchical Concept of Ecosystems*. Princeton: Princeton University Press.

Paine, R. T. (1974). Intertidal community structure: experimental studies on the relationship between a dominant competitor and its principal predator. *Oecologia*, **15**, 93–120.

Pastor, J. & Post, W. M. (1986). Influence of climate, soil moisture and succession on forest carbon and nitrogen cycles. *Biogeochemistry*, **2**, 3–27.

Payette, S. (1983). The forest tundra and present tree-lines of the northern Quebec–Labrador peninsula. In *Tree Line Ecology*. Proceedings of the Northern Quebec Tree Line Conference, ed. P. Morisset & S. Payette, pp. 3–23. Nordicana, Quebec.

Payette, S., Morneau, C., Sirois, L. & Desponts, M. (1989). Recent fire history of the northern Quebec biomes. *Ecology*, **70**, 656–73.

Pitman, A. J., Henderson-Sellers, A. & Yang, Z.-L. (1990). Sensitivity of regional climates to localized precipitation in global models. *Nature*, **346**, 734–7.

Romesburg, H. C. (1984). *Cluster Analysis for Researchers*. London: Wadsworth.

Royama, T. (1984). Population dynamics of the spruce budworm *Choristoneura fumiferana*. *Ecological Monographs*, **54**, 429–62.

Schindler, D. W. (1990). Experimental perturbations of whole lakes as tests of hypotheses concerning ecosystem structure and function. Proceedings of 1987 Crafoord Symposium. *Oikos*, **57**, 25–41.

Schindler, D. W. (1993). Linking species and communities to ecosystem management. Proceedings of the 5th Cary Conference, May 1993.

Short, J. & Turner, B. (1994). A test of the vegetation mosaic hypothesis: a hypothesis to explain the decline and extinction of Australian mammals. *Conservation Biology*, 8, 439–49.

Shugart, H. H. & Prentice, I. C. (1992). Individual-tree-based models of forest dynamics and their application in global change research. In *A Systems Analysis of the Global Boreal Forest*, ed. H. H. Shugart, R. Leemans & G. B. Bonan, pp. 313–33. Cambridge: Cambridge University Press.

Silverman, B. W. (1986). *Density Estimation for Statistics and Data Analysis*. London: Chapman and Hall.

Smith, T. M. & Urban, D. L. (1988). Scale and the resolution of forest structural pattern. *Vegetatio*, **74**, 143–50.

Southwood, T. R. E. (1977). Habitat, the template for ecological strategies?

Journal of Animal Ecology, **46**, 337–65.

Steele, J. H. (1985). A comparison of terrestrial and marine systems. *Nature*, **313**, 355–8.

Tilman, D. & Downing, J. A. (1994). Biodiversity and stability in grasslands. *Nature*, **367**, 363–5.

Turner, M. (1989). Landscape ecology: the effect of pattern on process. *Annual Review of Ecology*, **20**, 171–97.

Turner, M. G., Gardner, R. H., Dale, V. H. & O'Neill, R. V. (1989). Predicting the spread of disturbance across heterogeneous landscapes. *OIKOS*, **55**, 121–9.

Walker, B. H. (1992). Biological diversity and ecological redundancy. *Conservation Biology*, **6**, 18–23.

Walker, B. H., Ludwig, D., Holling, C. S. & Peterman, R. M. (1969). Stability of semi-arid savanna grazing systems. *Journal of Ecology*, **69**, 473–98.

Wiens, J. A. & Rotenberry, J. T. (1981). Habitat associations and community structure of birds in shrub steppe environments. *Ecological Monographs*, **51**, 21–41.

20 Diversity of soil biota and ecosystem function

O. W. Heal, S. Struwe and A. Kjøller

Introduction

A major part of the world's biodiversity lies unseen in the soil; unseen but critical to many ecosystem processes and vulnerable to human influence. Although it is rarely possible to attribute the process measurements to other than a broad group of organisms, a general link can be made to the key ecosystem functions performed by the soil biota:

- organic matter decomposition
- nutrient transformation and translocation
- greenhouse gas formation and breakdown
- pedogenesis and soil morphology

Whilst performed by the soil biota, many of these processes can be measured directly without study of the organisms responsible. However, understanding of the factors controlling these processes is therefore partial. The only major function that is dependent on the direct investigation of the organisms is the maintenance of food webs (and related energy transformation). Whether in the study of major ecosystem processes or food webs, research on soil biota suffers three major technical limitations: (1) taxonomic problems, particularly with bacteria, actinomycetes and fungi, but also with the wide range of fauna from protozoa through to mammals and including many temporary immature stages of insects; (2) an opaque medium which inhibits direct observation, even under experimental conditions; and (3) the specialized methods, some of which are grossly inefficient, that are required for isolation, extraction and quantification of the biota, with different methods required for different groups of organisms.

In examining the state of knowledge on soil biodiversity and ecosystem function we explore sequentially and briefly the following questions: How diverse is the soil biota? How is it structured? How does diversity affect individual ecosystem functions? What are the main conclusions and research priorities?

How diverse is the soil biota?

Taking number of species as the measure of biodiversity, there are no comprehensive lists even for individual sites; no single site has a complete inventory of soil biota and that may be impossible given the state of microbial taxonomy. Inventories of particular groups of microorganisms exist from many different ecosystems but more complete investigations are rare; for example, the investigation of an agroecosystem by Coleman (1994). One important reason is found in the classical work of Domsch (1975), where he demonstrated that not until 12 000 fungal strains were isolated from a soil sample was the frequency of new isolates dramatically reduced. Another reason is found in detailed taxonomic studies, where it is seen that a high diversity may be found within complex genera such as *Penicillium*. However, as an indication of the diversity of fauna, two sites which have been comprehensively studied are Moor House and Wytham Woods in the UK, an upland moorland and lowland deciduous woodland respectively. Extensive sampling both above and below ground has recorded, in the respective sites, about 1200 and 4200 species of higher invertebrate groups (omitting protozoa, nematodes and annelids) and vertebrates. Four major groups (Acarina, Collembola, Coleoptera and Diptera) are either permanent or temporary soil residents. These amount to *c*. 700 and *c*. 2000 species, i.e. about 50% of the fauna of these two ecosystems are soil-dwelling. In these two sites, amounting to a few hectares, the soil fauna represent 5–10% of the species recorded for the British Isles (Coulson & Whittaker, 1978).

Another study, that of the dominant faunal groups (nematodes and termites) in a moist tropical rain forest in Cameroon, further illustrates the challenge. Hodda *et al.* (in press) and Eggleton *et al.* (1995) recorded over 200 species of nematodes and 110 species of termites (57 genera) from a forest subject to a series of experimental management treatments. Some main features were:

- the number of nematode species is still rising
- species number for nematodes was lower in the primary forest compared with cleared and secondary forest
- for termites there were indications of a trend towards increased number of species and individuals with moderate disturbance, with a decrease under intense disturbance (Fig. 20.1)
- amongst the termites, only seven of the 110 species were major dominants in terms of numbers of individuals, but each treatment had a different dominant

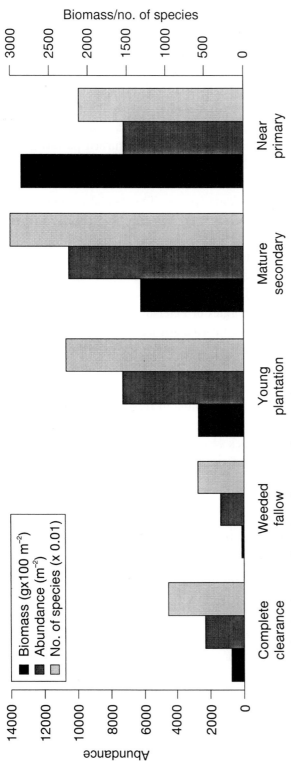

Figure 20.1 *Biomass, abundance and number of species of termites in a tropical forest subject to varying degrees of disturbance. (From Eggleton et al., 1995.)*

These preliminary results, part of the TIGER (Terrestrial Initiative in Global Environmental Research) programme, show some of the highest levels of nematode and termite diversity recorded, yet the Cameroon studies are only at the early stages of documenting soil biodiversity.

The impact of management on diversity is also shown amongst earthworm communities in the tropical network studies of Lavelle *et al.* (1994). The preliminary results summarized in Table 20.1 indicate that the higher diversity of species tends to occur in the moderately disturbed pastures, tree plantations and savannas which also have the highest earthworm biomass. The increase in species richness with disturbance results mainly from an increase in the number of exotic species. The data also show an increase in the richness of native species with increasing soil organic matter content.

Are there clear patterns in soil biodiversity related to particular environmental variation? What level of predictability exists? The most comprehensive compilations, resulting from the IBP research of the 1970s, are not rewarding despite the perceptive energies of Peterson and Luxton (1982) and Kjøller and Struwe (1982). More varied statistical approaches and theoretical considerations have subsequently begun to provide insights into the variability in composition of soil fauna communities. For example, Bengtsson (1994) explored data from various sources and showed that on the level of both species and higher taxa, fauna in reasonably stable deciduous and coniferous forests are quite predictable, i.e. constant over periods of decades. However, the lack of data on species interactions and functional groups still constrains understanding of biodiversity and structure of soil communities.

How is the biota structured in relation to ecosystem functions?

In the context of IGBP there are many formal representations or models of ecosystem function in organic matter, carbon and nutrient dynamics. These provide highly credible descriptions of the process dynamics and, in some cases, of the physical structure and species dynamics in ecosystems (e.g. Parton *et al.*, 1987; Stewart *et al.*, 1990). However, a recurrent feature of most ecosystem models is the minimal representation of the soil fauna and microflora. Some have a token compartment representation of the organisms associated with the well-defined dynamics of decomposition and nutrient mobilization.

This minimalist approach to the soil biota reflects the fundamental

Table 20.1 *Biomass abundance and species richness of earthworm communities in ecosystems in the humid tropics*

Ecosystem	Biomass (g m^{-2})	Abundance (no. m^{-2})	Number of species
Pastures	59.7 (0.6–153)	310 (93–740)	6.5 (2–9)
Tree plantations	28.6 (2.9–87)	170 (84–341)	9.5 (7–12)
Crops	1.1 (0.6–1.5)	19 (0–42)	2.5 (2–3)
Savannas	44.1 (38–50)	236 (187–286)	8 (8)
Tropical deciduous forest	4.1 (0.6–10)	37 (8–78)	2.7 (2–4)
Tropical rainforest	13.9 (0.2–72)	77 (4–401)	6.6 (5–11)

From Lavalle *et al.* (1994).
The range for sites within ecosystem types is shown in brackets.

distinction between process and population ecologists. Population dynamics have had remarkably little influence on the process models, perpetuating the separation which occurred in IBP between production and population ecologists. The problem of integration of process and population studies remains despite major efforts such as the Swedish project on Ecology of Arable Lands (Andren *et al.*, 1989). There are moves towards integration of the two approaches as seen in the Ecotron experiments described later and in Hendrix *et al.* (1986) in which the detritus food web of a managed grassland is represented. In this model (Fig. 20.2) a shift from a fungi-dominated (1) to a bacteria-dominated (2) food web results from an increase in tillage. The proximal factors which cause the shift are the changes in composition of the organic matter input to the soil and in the microclimate. In this model, the population dynamics are linked to the organic matter and nutrient dynamics represented in a simplified version of the CENTURY model (Parton *et al.*, 1987).

Within the Hendrix *et al.* (1986) model, the soil organisms are structured to represent their feeding relationships, grouped into five trophic levels described in more detail in Hunt *et al.* (1987) (Fig. 20.3). The trophic classification rests uneasily in soil biology because of the lack of specificity in many of the feeding relationships or possibly our lack of knowledge of them. The danger is in forcing information into an unrealistic paradigm. However, the concept does allow a first approximation to the dynamics of the microflora and fauna populations which can be related to processes of decomposition and nutrient transformation under the control of environmental factors and there are important developments in this area of research (Moore & Hunt, 1988; De Angelis, 1992; Bengtsson *et al.*, in press).

A brief summary of some research on the microbiology of decomposition

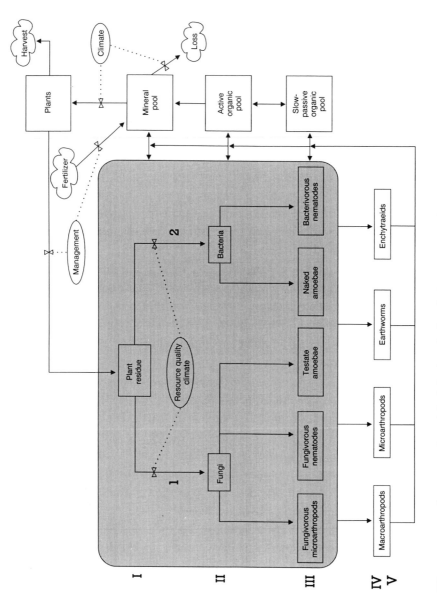

Figure 20.2 *A conceptual model of detritus food webs in conventional and no-tillage agroecosystems, showing a switch from a fungal-based (1) to a bacterial-based (2) food web in response to tillage. (From Hendrix et al., 1986.)*

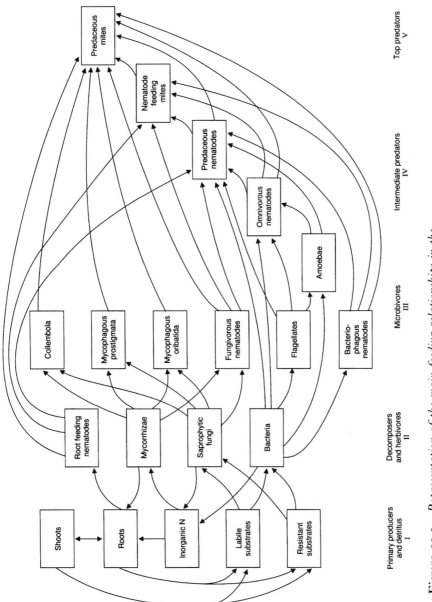

Figure 20.3 *Representation of the main feeding relationships in the detritus food web of a short grass prairie. (From Hunt et al., 1987, modified by Verhoef & Brussard, 1990.)*

illustrates the state of knowledge in one of the major processes. Succession of the fungal flora during decomposition has been investigated with many methods for many different kinds of plant litter. The stages of decomposition have been related to the activities of different taxonomic groups associated with depletion of the energy source (Kjøller & Struwe, 1982; Frankland, 1992). However, few studies have defined the functional roles of the taxa of decomposer fungi. Flanagan and Scarborough (1974), as part of the IBP tundra biome project, analysed the functional succession during decomposition in Arctic environments. They demonstrated that the fungi have different enzymatic potential at various temperatures due to different activity peaks, e.g. a pectinolytic ability at 20 °C and cellulolytic at 5 °C, indicating a high degree of functional flexibility within decomposer populations.

In decomposing litter from temperate deciduous forests, Kjøller and Struwe (1990, 1992) showed initially a maximum of pectinolytic fungi followed by cellulolytic fungi, but with ligninolytic species active throughout the entire decomposition period. These investigations were carried out at 10 °C, the mean temperature during decomposition in temperate forests. The extent to which taxonomic and functional diversity would change with increasing temperature remains an open question.

One line of research has focused on changes in the ratio between fungi and bacteria in litter and soil in relation to climatic conditions, pH and resource quality (Anderson & Domsch, 1975; Beare *et al.*, 1990; Neely *et al.*, 1991; Tate, 1991; Yang & Insam, 1991) as illustrated in Fig. 20.3. Determination of the metabolic quotient $Q CO_2$ (unit CO_2-C/unit biomass C/h^{-1}) has also been demonstrated to be a sensitive parameter to assess environmental conditions such as pH, land-use change or soil management, plant succession, changes in organic matter or soil texture (Insam & Domsch, 1988; Insam & Haselwandter, 1989; Anderson & Domsch, 1990, 1993; Insam, 1990).

The study of composition and function of the bacterial part of the microflora is complicated and time consuming. However, modern techniques such as the Biolog microtitre system can screen large numbers of bacteria for utilization of many carbon sources by direct metabolic measurement rather than detection of byproducts (Garland & Mills, 1991). Intensive use of these systems in ecological projects are needed to gain new knowledge about the composition of bacterial communities during decomposition. Molecular methods are increasingly applied in soil microbiology, including molecular probes for identification and DNA analysis to determine diversity. These techniques characterize the composition and biochemical potential of microbial populations and can demonstrate their physical position within the ecosystems. The challenge is to relate these characteristics to the observed rates of processes in the field.

How does diversity affect individual ecosystem functions?

The formulation of the food web model is based partly on the knowledge of the biology of the organisms but also on specific experiments. From the latter there is a gradual accumulation of quantitative information on the extent to which increasing trophic diversity affects particular process rates. Some examples illustrate this, recognizing that this is a highly selective sample from a large literature, including the valuable review by Verhoef and Brussard (1990), of the contribution of soil fauna to organic matter decomposition and nitrogen mineralization.

- Clarholm (1985, 1989), Elliott *et al.* (1979) and others have shown that, when protozoa were added to bacterial populations, the N uptake by plants was increased, probably as a result of grazing by the protozoa which increased the rate of turnover and hence availability of nitrogen. Kuikman and Van Veen (1989) also showed that there was no difference in plant uptake whether a single or mixed population of bacteria was present (Table 20.2).

- Kuikman *et al.* (1989, 1991) also showed an increase in plant uptake of ^{15}N from bacteria in soil when protozoa were present. What may be more significant is that mobilization of ^{15}N declined when bacteria alone were subjected to stress in the form of reduced or fluctuating moisture but the rate of mineralization was maintained when protozoa were also present (Fig. 20.4). To what extent does an increase in trophic diversity generally enhance resistance to stress?

- Verhoef and Meintser (1990) explored the effect of adding the collembolan *Tomocerus minor* to microflora in F layer organic matter from a *Pinus nigra* forest. Leaching of ammonia was increased by the collembola. The 'stimulus' was maintained under fluctuating temperatures but not under constant temperatures. Does physical heterogeneity or disturbance amplify the effect of trophic heterogeneity?

- Verhoef and Meintser (1990) also showed that addition of N to pine needle litter reduced the decay rate of the litter and reduced the degree of enhancement of N mobilization by the collembola (Table 20.3). This example illustrates a potentially more general phenomenon identified in the comprehensive review by Fog (1988). Contrary to general belief, Fog showed that N addition tends to retard rather than enhance decomposition of litters with a high C : N ratio, probably

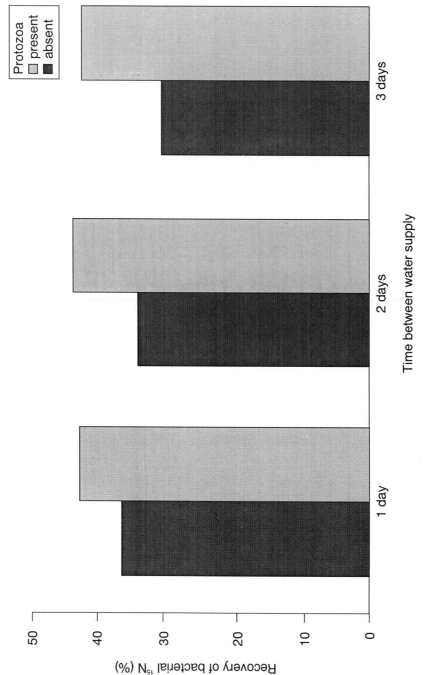

Figure 20.4 *The influence of protozoa on the recovery of bacterial*
^{15}N in plant nitrogen, as a percentage of amount inoculated, under
different moisture regimes. Water was added at 1, 2 or 3 day intervals.
(From Kuikman et al., 1991.)

Table 20.2 *The influence of pure and mixed inocula of bacteria on the uptake of nitrogen by wheat, in the presence and absence of protozoa*

	Plant mass (g mc^{-1})	Plant N uptake (mg mc^{-1})	^{15}N in plant (%)	Uptake of ^{15}N (%)	Bacteria numbers ($\times 10^8$ g^{-1})
Pseudomonas	5.6	66.5	0.86	28.6	4.7
Mixed bact. +*Ps.*	4.1	54.7	0.88	24.1	38.0
Ps.. + Protozoa	5.2	76.2	1.25	47.9	0.6
Mixed bact. + *Ps* + Pr.	4.0	65.3	1.17	38.2	5.2
LSD	0.53	4.67	0.06	3.33	18.35

From Kuikman & Van Veen (1989).
mc microsom; *Ps.*, *Pseudomonas*; Pr. protozoa; LSD, least squares differences at $P = 0.050$.

because nitrogen blocks production of ligninase by some basidiomycetes which may be replaced by cellulose decomposers leaving recalcitrant organic matter. This has implications for C sequestration given the global trend of increasing atmospheric N concentrations.

■ Couteaux *et al.* (1991) first demonstrated the now widely held view that enhanced atmospheric CO_2 results in an increase in C : N ratio of leaf litter and hence reduced decay rate. What is less widely recognized is that they showed that with increased trophic diversity (successive addition of protozoa, nematodes, collembola and isopods) the negative effect of CO_2 enhancement was obliterated and may even be turned into a positive effect (Fig. 20.5). Another example of increased trophic diversity enhancing resistance to disturbance?

■ The 'trophic effect' is less clearly shown where 'omnivores' such as earthworms are considered. Nevertheless their effect on ecosystem function has been clearly demonstrated (and often ignored) since the pioneer work of Charles Darwin. For example Lavelle *et al.* (1994) at Yurimaguas, Peru, showed enhanced crop production in four crops grown in rotation on ultisols when the earthworm *Pontoscolex corethrurus* was introduced.

■ In this *Pontoscolex* case, as with other fauna (Anderson, 1988), it is particularly their influence on soil physical properties rather than their trophic relationships which influence nutrient dynamics and availability to the crop. But the fauna also have a role in specific processes such as the complex process–population relationships related to nitrous oxide release from soils. Nitrous oxide emissions result from a

Table 20.3 *Effects of addition of nitrogen to pine needle litter*

Variable	Control (%)	+ NH$_3$ (150 ppb) (%)
Decomposition rate[a]	12.0	8.3
Animal-stimulated N mobilization[b]	54.0	32.0

From Verhoef & Meintser (1991).
[a]Dry mass loss over 12 weeks.
[b]Increase in presence of *Tomocerus minor* over total inorganic N mobilization with microorganisms.

> balance between microbially mediated nitrification and denitrification and are influenced by carbon, nitrogen and oxygen availability (Granli & Bøckman, 1994). Earthworm activity results in significant variation in substrate availability and microclimatic conditions for nitrifiers and denitrifiers, providing potential 'hot spots' for nitrous oxide emission from casts (Lavelle *et al.*, 1992).

To what extent do these isolated examples of relationships between diversity (both between and within functional groups) of soil microflora, fauna and ecosystem processes combine to form a predictable whole? Two pieces of evidence suggest that patterns are emerging. First, in an analysis of food webs and species diversity, Briand (1983) examined relationships between statistical 'connectedness' and 'diversity'. There was no detectable difference between food webs which were detritus-based and those which were not (i.e. above-ground). Four additional data sets, for below-ground food webs in the USA and the Netherlands, fell within the same general relationship (Moore & de Ruiter, 1991). Thus the indications are that the population structure theories developed in studies of above-ground fauna are applicable to the soil biota. Expansion of the relationship of community structure to function is coming from controlled multi-species experiments. For example, Allen-Morley and Coleman (1989) tested hypotheses of food web behaviour with specific microflora and nematodes representing three trophic levels in microcosms. The experiments indicated that addition of higher trophic levels reduced populations at lower levels and increased carbon release and nitrogen mineralization. However, recovery from stress (freezing) was more dependent on the species composition of the microbial and faunal community than on community structure.

Second, in an innovative experiment, Naeem *et al.* (1994) used a sophisticated controlled environment facility (Ecotron) in which to construct ecosystems with three levels of diversity, but similar trophic structures. The decomposers consisted of an earthworm and several collembolan species.

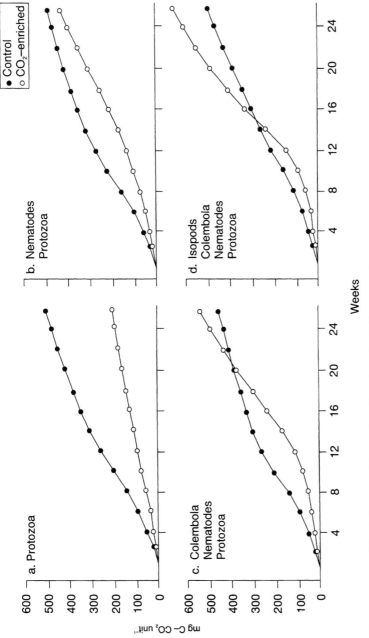

Figure 20.5 *The influence of increasing diversity of soil fauna on the rate of decomposition of sweet chestnut leaf litter grown under natural and enriched CO_2 atmosphere. Decomposition is represented as cumulative CO_2 emission. (From Couteaux et al., 1991.)*

Trends in ecosystem attributes (community respiration, net primary production, cover and canopy architecture) increased significantly with functional diversity. What may be important is that no trends were detected in the relationships between either decomposition or nutrient concentration and diversity. When combined with the specific examples quoted earlier, is there an indication that soil processes are less affected by changes in species diversity *within* a trophic group than they are by changes in the diversity *of* trophic groups? Is there a high degree of functional redundancy amongst soil biota within trophic groups, particularly the decomposer community? In this community, catabolic processes are largely non-specific and for every given reaction there is a cohort of organisms with the appropriate capacity, i.e. functional equivalent (Swift & Heal, 1986).

Conclusions

However tentative, the development of research on soil biodiversity in relation to ecosystem function raises some general hypotheses which provide foci for further studies.

- There is a potentially high degree of species redundancy amongst soil biota within functional groups. The diversity of species may result more from physical niche separation than from substrate or food utilization.
- Increased diversity of functional groups tends to dampen the effects of perturbation or stress.
- The direct response of soil organisms to climate change is likely to be small (slight changes in process rates) relative to change in land use and to variation between biomes which can modify species composition directly and strongly (Tinker & Ineson, 1990). There are, as yet, undefined effects of climate change on the migration and dispersal of soil species, especially macrofauna.

There are also some clear practical implications from plant response research to date:

- There is a danger in extrapolating results from experiments in which the soil biota have been simplified, e.g. lack of mycorrhizae or earthworms. Significantly different responses to the effect of a factor, e.g. moisture or CO_2, on a process can be obtained from simple or complex systems.
- The response of processes to environmental variables is significantly affected by soil type or conditions. Soil variability is particularly lacking as a factor in experiments and can lead to erroneous extrapolation.

■ The range and sensitivity of techniques for analysis of soil systems have improved dramatically in the last decade or so. Stable and radio-isotopes, genetic discrimination, microchemical analysis, experimental microcosms, etc. have greatly enhanced our ability to analyse the complexity of soil biota in relation to ecosystem processes.

■ A high priority is to develop systems to characterize soil organisms in terms of functional attributes, and to increase the effort in population/community–process relationships in theory and practice, particularly in microflora.

References

Allen-Morley, C. R. & Coleman, D. C. (1989). Resilience of soil biota in various food webs to freezing perturbations. *Ecology*, **70**, 1127–41.

Anderson, J. M. (1988). Invertebrate-mediated transport processes in soils. *Agriculture, Ecosystems and Environment*, **24**, 5–19.

Anderson, J. P. E. & Domsch, K. H. (1975). Measurement of bacterial and fungal contributions to respiration of selected agriculture and forest soils. *Canadian Journal of Microbiology*, **21**, 314–22.

Anderson, T.-H. & Domsch, K. H. (1990). Application of ecophysiological quotients (qCO$_2$ and qD) on microbial biomasses from soils of different cropping histories. *Soil Biology and Biochemistry*, **22**, 251–5.

Anderson, T.-H. & Domsch, K. H. (1993). The metabolic quotient for CO$_2$ (qCO$_2$) as a specific activity parameter to assess the effects of environmental conditions, such as pH, on the mocrobial biomass of forest soils. *Soil Biology and Biochemistry*, **25**, 393–5.

Andren, O., Lindbert, T., Paustian, K. & Rosswall, T. (1989). Ecology of arable land. *Ecological Bulletins*, **40**.

Beare, M. H., Neely, C. L., Coleman, D. C. & Hargrove, W. L. (1990). A substrate-induced respiration (SIR) method for measurement of fungal and bacterial biomass on plant residues. *Soil Biology and Biochemistry*, **22**, 585–94.

Bengtsson, J. (1994). Temporal predictability in forest soil communities. *Journal of Animal Ecology*, **63**, 653–65.

Bengtsson, J., Zheng, D. W., Agren, G. I. & Persson, T. (in press). Food webs in soil: an interface between population and ecosystem ecology. In *Linking Species and Ecosystems*, ed. C. G. Jones & J. H. Lawson. London: Chapman & Hall.

Briand, F. (1983). Environmental control of food web structures. *Ecology*, **64**, 253–63.

Clarholm, M. (1985). Interactions of bacteria, protozoa and plants leading to mineralization of soil nitrogen. *Soil Biology and Biochemistry*, **17**, 181–7.

Clarholm, M. (1989). Effects of plant–bacterial–amoebal interactions on plant uptake of nitrogen under field conditions. *Biology and Fertility of Soils*, **8**, 373–8.

Coleman, D. C. (1994). Compositional analysis of microbial communities: Is there room in the middle? In *Beyond the Biomass: Compositional and Functional Analysis of Soil Microbial Communities*, ed. K. Ritz, J. Dighton & K. E. Giller, pp. 201–20. Chichester: Wiley.

Coulson, J. C. & Whittaker, J. B. (1978). Ecology of moorland animals. In

Production Ecology of British Moors and Montane Grasslands, ed. O. W. Heal & D. F. Perkins, pp. 52–93. Berlin: Springer.

Couteaux, M. M., Mousseau, M., Célérier, M.-L. & Bottner, P. (1991). Increased atmospheric CO_2 and litter quality: decomposition of sweet chestnut leaf litter with animal food webs of different complexities. *Oikos*, **61**, 54–64.

De Angelis, D. L. (1992). *Dynamics of Nutrient Cycling and Food Webs*. London: Chapman & Hall.

Domsch, K. H. (1975). Distribution of soil fungi. In *Developmental Microbiology, Ecology*, ed. T. Hasegawa. Proceedings of 1st International Congress, IAMS, vol. 2, pp. 340–53, Tokyo: Science Council of Japan.

Eggleton, P., Bignell, D. E., Sands, W. A., Waite, B. & Wood, T. G. (1995). The species diversity of termites (Isoptera) under differing levels of forest disturbance in the Mbalmayo Forest Reserve, Southern Cameroon. *Journal of Tropical Ecology*, **11**, 85–98.

Elliott, E. T., Coleman, D. C. & Cole, C. V. (1979). The influence of amoebae on the uptake of nitrogen by plants in gnotobiotic soil. *The Soil–Root Interface*, ed. J. L. Harley & R. Scott Russell, pp. 221–9, London: Academic Press.

Flanagan, P. & Scarborough, A. (1974). Physiological groups of decomposer fungi on plant remains. In *Soil Organisms and Decomposition in Tundra*, ed. A. J. Holding, O. W. Heal, S. F. MacLean & P. W. Flanagan, pp. 151–8. Stockholm: Tundra Biome Steering Committee.

Fog, K. (1988). The effect of added nitrogen on the rate of decomposition of organic matter. *Biological Reviews*, **63**, 433–62.

Frankland, J. C. (1992). Mechanisms in fungal succession. In *The Fungal Community: Its Organisation and Role in the Ecosystem*, ed. G. C. Carroll & D. T. Wicklow, pp. 383–401. New York: Dekker.

Garland, J. L. & Mills, A. L. (1991). Classification and characterisation of patterns of community-level carbon-source utilization. *Applied and Environmental Microbiology*, **57**, 2351–9.

Granli, T. & Bøckman, O. C. (1994). Nitrous oxide from agriculture. *Norwegian Journal of Agricultural Sciences*, Supplement 12.

Hendrix, P. F., Parnelee, R. W., Crossley, D. A., Coleman, D. C., Odum, E. P. & Groffman, P. M. (1986). Detritus food webs in conventional and non-tillage agroecosystems. *BioScience*, **36**, 374–80.

Hodda, M., Bloemers, G. F., Wanless, F. R. & Lambshead, P. J. D. (in press). Nematodes and carbon flux in a tropical forest in Cameroon: preliminary results. In *Carbon and Nutrient Cycling in Forest Ecosystems*, ed. L.-O. Nilsson & P. Mathy. Brussels: Commission of the European Communities.

Hunt, H. W., Coleman, D. C., Ingham, E. R., Ingham, R. I., Elliott, E. T., Moore, J. C., Rose, S. L., Reid, C. P. P. & Morley, C. R. (1987). The detrital food web in a shortgrass prairie. *Biology and Fertility of Soils*, **3**, 57–68.

Insam, H. (1990). Are the soil microbial biomass and basal respiration governed by the climatic regime? *Soil Biology Biochemistry*, **22**, 525–32.

Insam, H. & Domsch K. H. (1988). Relationship between soil organic carbon and microbial biomass on chronosequences of reclamation sites. *Microbial Ecology*, **15**, 177–88.

Insam, H. & Haselwandter, K. (1989). Metabolic quotient of the soil microflora in relation to plant succession. *Oecologia*, **79**, 174–8.

Kjøller, A. & Struwe, S. (1982). Microfungi in ecosystems: fungal occurrence and activity. *Oikos*, **39**, 391–422.

Kjøller, A. & Struwe, S. (1990). Decomposition of beech litter: A comparison of fungi isolated on nutrient rich and nutrient poor media. *Transactions of the Mycological Society of Japan*, **31**, 5–16.

Kjøller, A. & Struwe, S. (1992). Functional groups of microfungi in decomposition. In *The Fungal Community*, ed. G. C. Carroll & D. T. Wicklow, pp. 619–30. New York: Dekker.

Kuikman, P. J., Jansen, A. G. & Van Veen, J. A. (1991). ^{15}N-nitrogen mineralization from bacteria by protozoan grazing in different soil moisture regimes. *Soil Biology and Biochemistry*, **23**, 193–200.

Kuikman, P. J. & Van Veen, J. A. (1989). The impact of protozoa on the availability of bacterial nitrogen to plants. *Biology and Fertility of Soils*, 8, 13–18.

Kuikman, P. J., van Vuuren, M. M. I. & Van Veen, J. A. (1989). Effect of soil moisture regime on predation by protozoa of bacterial biomass and the release of bacterial nitrogen. *Agriculture, Ecosystems and Environment*, **27**, 271–9.

Lavelle, P., Blanchart, E., Martin, A., Spain, A. V. & Martin, S. (1992). Impact of soil fauna on the properties of soils in the humid tropics. In *Myths and Science of Soils of the Tropics*, ed. R. Lal & P. A. Sanchez. SSSA Special Publication no. 29. Madison: Soil Science Society of America.

Lavelle, P., Dangerfield, M., Fragoso, C., Eschenbrenner, V., Lopez-Hernandez, P., Pashanasi, B. & Brussard, L. (1994). The relationship between soil macrofauna and tropical soil fertility. In *The Biological Management of Tropical Soil Fertility*, ed. P. L. Woomer & M. J. Swift, pp. 137–69. Chichester: Wiley.

Moore, J. C. & de Ruiter, P. C. (1991). Temporal and spatial heterogeneity of trophic interactions within belowground food webs. *Agriculture, Ecosystems and Environment*, **34**, 371–97.

Moore, J. C. & Hunt, H. W. (1988). Resource compartmentation and the stability of real ecosystems, *Nature*, **333**, 261–3.

Naeem, S., Thompson, L. J., Lawler, S. P., Lawton, J. H. & Woodfin, R. M. (1994). Declining biodiversity can alter the performance of ecosystems, *Nature*, **368**, 734–7.

Neely, C. L., Beare, M. H., Hargrove, W. L. & Coleman, D. C. (1991). Relationships between fungal and bacterial substrate-induced respiration, biomass and plant residue decomposition. *Soil Biology and Biochemistry*, **23**, 947–54.

Parton, W. J., Schimel, D. S., Cole, C. V. & Ojima, D. S. (1987). Analysis of factors controlling soil organic matter levels in Great Plains grasslands. *Soil Science Society of America Journal*, **51**, 1173–9.

Peterson, H. & Luxton, M. (1982). A comparative analysis of soil fauna populations and their role in decomposition processes. *Oikos*, **39**, 287–8.

Stewart, J. W. B., Anderson, D. W., Elliott, E. T. & Cole, C. V. (1990). The use of models of soil pedogenic processes in understanding changing land use and climate change. In *Soils on a Warmer Earth*, ed. H. W. Scharpenseel, M. Schomaker & A. Ayoub, pp. 121–31. Amsterdam: Elsevier.

Swift, M. J. & Heal, O. W. (1986). Theoretical considerations of microbial succession and growth strategies: intellectual exercise or practical necessity? In *Microbial Communities in Soil*, ed. V. Jensen, A. Kjøller & L. H. Sørensen, pp. 115–31. Amsterdam: Elsevier.

Tate, R. L. (1991). Microbial biomass measurement in acidic soil: effect of fungal/bacterial activity ratios and of soil amendment. *Soil Science*, **152**, 220–5.

Tinker, P. B. & Ineson, P. (1990). Soil organic matter and biology in relation to climate change. In *Soils on a Warmer Earth*, ed. H. W. Scharpenseel, M. Schomaker & A. Ayoub, pp. 71–87. Amsterdam: Elsevier.

Verhoef, H. A. & Brussard, L. (1990). Decomposition and nitrogen mineralization in natural and agroecosystems: the contribution of soil animals. *Biogeochemistry*, **11**, 175–211.

Verhoef, H. A. & Meintser, S. (1991). The role of soil arthropods in nutrient flow and the impact of atmospheric deposition. In *Advances in Management and Conservation of Soil Fauna*, ed. G. K. Verresh, D. Rajagopa, C. C. A. Viraktamath, pp. 497–506. New Delhi: Oxford and IBH Publishing Company.

Yang, J. C. & Insam, H. (1991). Microbial biomass and relative contributions of bacteria and fungi in soil beneath tropical rain forest, Hainan Island, China. *Journal of Tropical Ecology*, **7**, 385–93.

21 The functional role of species in terrestrial ecosystems

F. S. Chapin, III, H. L. Reynolds, C. M. D'Antonio
and V. M. Eckhart

Introduction

The expansion of human populations and their increasing access to technology have led to two general environmental concerns (NRC, 1994): (1) the increasing human impact on the earth's environment and ecosystems through changes in biospheric carbon pools, element cycling and climate; and (2) changes in the earth's biota, including species introductions and extinctions. These changes are occurring more rapidly than at any time in the last several million years. To date, these two areas have received largely separate research efforts, the first by ecosystem ecologists and the second by conservation biologists and community ecologists. One relatively new role of the research programme of GCTE has been to merge these efforts to explore the ways in which global change may alter biodiversity and the effects of altered biodiversity on ecosystem processes (Sala, this volume).

Biodiversity is important at levels ranging from diversity of landscapes to genetic diversity within species. Diversity within landscapes is clearly important in controlling landscape processes such as the transfer of nutrients from terrestrial to aquatic systems. At the opposite extreme, genetic diversity can be important to the persistence of small, isolated populations. However, most concern about global loss of biodiversity has focused at the species level because loss of species through changes in land use is well documented, frequent, and irrevocable (Solbrig, 1991). A major factor hindering the understanding of possible effects of loss of species diversity on ecosystem processes has been the lack of a clear conceptual framework. The purpose of this chapter is to present a framework for predicting strong species effects on ecosystem processes.

It is individual organisms that gain carbon and nutrients from the environment, transfer plant tissues to higher trophic levels, and decompose plant litter. It would, therefore, be surprising if the traits of individuals and their abundances did not determine the pool sizes and rates of energy and material flux in ecosystems. Species also have substantial indirect effects on ecosystem processes as a result of shading, deposition of faeces and urine, tissue-quality effects on decomposition, etc. For example, differences among ecosystems in decomposition rate are sometimes more strongly determined by species differences in litter quality than by ecosystem differences in climate (Flanagan & Van Cleve, 1983). In this chapter we address two general questions about the ecosystem significance of species traits and species diversity:

1. If we know the *traits* of key organisms in an ecosystem, can we predict (a) the ecosystem impacts of species invasions or losses and (b) the rates and patterns of processes in intact ecosystems? If key traits could be determined by remote sensing, this would improve our ability to estimate the global pattern of ecosystem processes such as productivity and carbon sequestration.
2. How does *species diversity* influence ecosystem processes?

Predicting ecosystem change from change in species traits

Conceptual framework

Species differ within and among communities in traits that affect ecosystem processes (Hobbie, 1992; Wilson & Agnew, 1992; Chapin, 1993; van Breemen, 1993), which in this chapter we define as the fluxes of energy or materials. Traditionally three major functional groups of species with clearly different roles in ecosystem processes have been recognized: producers, consumers, and decomposers, corresponding roughly with a broad phylogenetic classification of organisms into plants, animals and microbes. The cycling of energy and matter between soil, water, atmosphere and organisms results from the activities of these major functional groups, and removal of any of these three groups will dramatically change or break energy and material cycling processes. All but the very simplest ecosystems, however, contain many species of plants, animals and microbes which vary to some degree in physiological, morphological, and phenological traits. Predicting the impact of removing or adding any one of these species thus requires finer scale definition of species groupings in terms of traits relevant to ecosystem processes.

To predict the community and ecosystem consequences of a species invasion or extinction, we must know at least three things. First, which species traits have the largest effects on community and ecosystem processes? Vitousek (1990) suggested that invasion by a species would have large effects on ecosystem processes if the species greatly altered (a) soil resource supply pools and supply rates, (b) the rates of consumption of resources by plants (resource acquisition) or of prey by animals (trophic interactions), or (c) disturbance regime. In addition, we must know how strongly the species differs from other species in the community and how abundant the species was or could become.

For example, traits such as size and relative growth rate, which affect resource acquisition in plants, should generally have small effects on ecosystem processes. Such traits tend to be continuously distributed among plants (Grime & Hunt, 1975), so that it is unlikely that a single species will have highly distinct values for such traits (Fig. 21.1). In addition, to the extent that such traits are involved in interspecific competition for limited

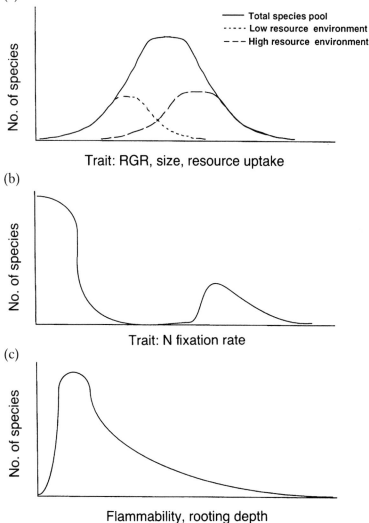

Figure 21.1 *Typical frequency of occurrence of (a) continuously varying traits and (b and c) discrete traits among species in a community.*

resources, species additions or deletions will tend to result in *compensatory* responses by the existing or remaining species. Thus, although community composition may be altered, resource consumption by the trophic level as a whole will be relatively unchanged.

By contrast, many traits are distributed discontinuously among species (that is species either have or lack the traits) and are uncommon, so that a single species is likely to be highly distinctive from others in a community. For example, if, as often occurs in upper trophic levels, a species differs strikingly from other community members in the type of resource it consumes, gain or loss of such a species is likely to have large effects on community and ecosystem processes ('top-down' controls). The same argument applies to traits which affect soil resource pools and supply rates ('bottom-up' controls; for example, nitrogen fixation, hydraulic lift) or disturbance frequency and intensity (the relative importance of equilibrium and non-equilibrium processes; for example, digging by animals, flammability in plants). In the real world, there is a spectrum from continuous to discontinuous traits, but the concept is useful in predicting whether a particular trait will have large or small ecosystem effects. Using examples, we will argue that uncommon traits can have *amplifying* effects on ecosystem processes.

The abundance of a species obviously affects its ecosystem impact. Even if a species is similar to other community members, its addition or deletion from an ecosystem may have a large initial impact. Conversely, a species may be quite dissimilar to other species in the community but, if it is rare, the ecosystem impact of its deletion may be small. However, some species have unexpectedly large ecosystem effects. Keystone species are those with effects that are disproportionate to their abundance (Mills *et al.*, 1993).

In summary, compensatory responses to a species introduction or loss are expected to occur when the species is similar to other species in its resource requirements. However, species introductions or losses are expected to have amplifying effects when big differences exist in traits that affect resource consumption, resource supply, or disturbance regime. In addition, abundant species often have greater effects than species with low abundance, although keystone species provide surprises.

Resource supply

The supply of soil resources is an important 'bottom-up' control of ecosystem processes (Jenny, 1980). Ecosystem processes are highly sensitive to changes in traits which influence the supply rates of soil resources because these traits are often discrete and usually have amplifying effects on ecosystem processes. In terrestrial ecosystems, introduction of exotic N fixers in Hawai'i greatly increased N availability, and, therefore, productivity and N cycling of forests (Vitousek *et al.*, 1987). Seeding with legumes increases the N inputs and productivity of pastures, particularly when combined with P fertilization (Ball *et al.*, 1979). Most plant species lack symbiotic N fixation, causing this trait to be discontinuous among species (Fig. 21.1). However, N fixation is not a perfectly discrete trait. There is considerable range in N-fixing capacity among N-fixing species, and some N fixation occurs in the leaf litter layer (Heath *et al.*, 1988) or in the rhizosphere of non-fixing species (van Berkum & Gohlool, 1980).

Deep-rooted species can increase the pool of soil resources available to support production. For example, species such as *Eucalyptus* or *Tamarix* have deep roots which access the water table, increasing both the amount of water absorbed and transpired (van Hylckama, 1974) and ecosystem productivity (Robles & Chapin, 1995). Deeply rooted species may also redistribute water from depth by hydraulic lift, making it available to other species in the ecosystem (Caldwell & Richards, 1989). Similarly, *Eriophorum vaginatum* is the only tussock-tundra species with roots that are deep enough to access nutrients in ground water that flows over the permafrost surface. By tapping nutrients at depth, its productivity increases 10-fold in sites with abundant ground water flow, whereas productivity of other species is unaffected by deep resources (Chapin *et al.*, 1988). In the absence of this species, ecosystem productivity and nutrient cycling would be greatly reduced. Differences in rooting depth can be important at the regional scale. Simulations suggest that conversion of the Amazon basin from forest to pasture would cause a permanent warming and drying of South America because the shallower roots of grasses would lead to less evapotranspiration and greater energy dissipation as sensible heat (Shukla *et al.*, 1990).

Nitrification, methanogenesis, and, consequently, the loss of trace gases to the atmosphere are controlled by relatively few species of microorganisms; thus these traits are uncommon and discontinuous. Rates of nitrification also influence the susceptibility of N to leaching loss. Changes in the abundance of microorganisms involved in these processes could, therefore, have large effects on loss of N and cations from ecosystems (Schimel, 1995). It is a common perception that all microbial functions are universally distributed and readily dispersed (Meyer, 1993) so that gain or loss of a microbial species might have minimal ecosystem impact. However, to our knowledge, little evidence exists to evaluate this perspective.

Animals can influence the resource base of the ecosystem by foraging in one area and depositing nutrients in another area (for example, concentration of nutrients in polygon troughs by lemmings (Batzli *et al.*, 1980), on hilltops by sheep, or beneath desert shrubs by rodents). Animals can also transport nutrients among ecosystems (for example, movement of ungulates to water holes).

Differences among plant species in tissue quality strongly influence litter decomposition rates (Melillo *et al.*, 1982; Flanagan & Van Cleve, 1983; Berg & McClaugherty, 1989). Litter from low-resource plants decomposes slowly because of the negative effects of low nitrogen and phosphorus and high lignin, tannins, waxes, and other recalcitrant or toxic compounds on soil microbes, reinforcing the low nutrient availability of these sites (Chapin, 1991; Hobbie, 1992; van Breemen, 1993). By contrast, species from high-resource sites produce rapidly decomposing litter with more N and P (Vitousek, 1982) and fewer recalcitrant compounds. Long-term field experiments suggest that the nutrient content of litter is more important than carbon quality in exerting these ecosystem effects (Berendse *et al.*, 1994). Forest trees which differ in carbon quality of litter and preferred form of nitrogen absorbed also influence nutrient supply rates through effects on soil acidity (Bormann & Likens, 1979). Litter nutrient concentration varies continuously among species (Chapin & Kedrowski, 1983) and is correlated with other traits governing competitive success in particular environments (Chapin, 1980). Therefore, we suggest that it is unlikely that a species with radically different litter quality will invade a stable community, unless it also differs in other important respects. For example, the invasion of Hawaiian forests by exotic grasses with low litter quality is also associated with increased fire frequency which is the trait that most directly increases tree mortality and alters ecosystem structure (D'Antonio & Vitousek, 1992). Invasion of early successional seral communities by species with lower litter quality is common (Van Cleve *et al.*, 1991; Wilson & Agnew, 1992). In summary, although invasions of communities by species with radically different litter quality does occur and has important consequences, this is most common in early successional communities or in situations where the change in litter quality is correlated with change in other species traits which strongly influence ecosystem processes.

Plants *indirectly* influence rates of nutrient supply through modification of the microenvironment (Wilson & Agnew, 1992; Hobbie, 1995). For example, arctic mosses, with their low rates of evapotranspiration (leading to water-logging) and effective insulation (preventing soil warming) indirectly inhibit decomposition. These species-specific effects could be important in determining both the pools of resources available to plants and the rate at which these pools turn over.

Resource consumption

Many animal species (and some plant species) differ strikingly from other community members in the types of resources or prey which they consume. Addition or deletion of these species, therefore, strongly influences the abundances of limiting resources or prey. These top-down controls are particularly well developed in aquatic systems, where removal of sea otters releases sea urchins which graze down kelp (Estes & Palmisano, 1974), or where addition or removal of a fish species can have large 'keystone' effects (that is effects much larger than would be expected from the biomass of the species) that cascade down the food chain (Carpenter et al., 1992; Power, 1992). Many non-aquatic ecosystems also exhibit strong responses to changes in predator abundance (Hairston et al., 1960; Strong, 1992). For example, removal of wolves can release deer populations which graze down vegetation (Rasmussen, 1941) or removal of elephants or other keystone mammalian herbivores leads to encroachment of woody plants into savannas (Owen-Smith, 1988; Wilson & Agnew, 1992). Similarly, disease organisms, such as rinderpest in Africa, can act as keystone species by greatly modifying competitive interactions and community structure (Bond, 1993), perhaps resulting in less energy flow to higher trophic levels.

Often these top-down controls by predators or pathogens have much greater effect on biomass and species composition of lower trophic levels than on the flow of energy or nutrients through the ecosystem (Carpenter et al., 1985) because declines in producer biomass are compensated by increased productivity and nutrient cycling rates by the remaining organisms. For example, intensely grazed grassland systems such as the southern and southeastern Serengeti Plains (McNaughton, 1985) have a low plant biomass but rapid cycling of carbon and nutrients due to treading and excretion by large mammals, which prevent accumulation of standing dead litter and return nutrients to soil in plant-available forms (McNaughton, 1988). Keystone predators or grazers thus alter the pathway of energy and nutrient flow, modifying the balance between herbivore-based or detritus-based food chains, but we know less about effects on total energy and nutrient cycling through the ecosystem.

Introduction or loss of a plant species is less likely than that of an animal species to have large ecosystem effects, as mediated by resource consumption. Most plant species are similar to one another in the types of resources used (light, water, and nutrients), although there can be some specialization by rooting depth (Cody, 1986), form of nitrogen utilized (Read, 1991; Chapin et al., 1993; Schulze et al., 1995), or level to which soil resources are depleted (Tilman, 1988). Traits of plant species that best predict rate of resource consumption are size and relative growth rate (RGR). Large size enhances resource capture in plants by allowing the plant to reach the top of the canopy where light is most available and to exploit a large soil volume. RGR is correlated with potential rates of carbon and nutrient acquisition (Olson & Lubchenco, 1990; Lambers & Poorter, 1992; Chapin, 1993). As with litter quality, size and RGR are traits which vary continuously among organisms and are correlated with other traits governing competitive success in particular environments. Therefore, it is unlikely that plant species with radically different size or RGR will be capable of invading stable communities.

In closed plant communities any reduction in abundance of one plant species should cause a compensatory increase in the abundance of other species due to release from competition, with little change in the total quantity of resources accumulated by vegetation at the ecosystem level (McNaughton, 1977; Chapin & Shaver, 1985). Consequently, we expect that gain or loss of a plant species will have little effect on biogeochemical cycles with the ecosystem under 'steady-state' conditions (Shaver *et al.*, in press), if plant species differ only in resource acquisition. There are three sources of evidence for this hypothesis.

First, the *pattern* of biomass distribution among plant species in closed communities fits a geometric model (Pastor, 1995), which theory suggests is best explained by competitive partitioning of resources among species (MacArthur, 1965; Whittaker, 1972). The dominant plant species pre-empts most resources, and the remaining resources are partitioned among species according to a competitive hierarchy. Second, *experimental manipulation* of resource supply causes much larger changes in the abundance of individual plant species than in biogeochemical pools or fluxes measured at the ecosystem level (Fig. 21.2) (McNaughton, 1977; Lauenroth *et al.*, 1978; Chapin & Shaver, 1985; Chapin *et al.*, 1995). When experimental manipulations alter the abundance of the dominant plant species, other plant species change their growth and resource acquisition to utilize the remaining resources. Biomass distribution in these experimental manipulations generally improves the fit of the community to the geometric distribution (Pastor, 1995), suggesting that competitive interactions are particularly important in explaining patterns of diversity under conditions of environmental change. Third, *natural variation* in weather causes greater change in abundance of individual plant species than in biogeochemical pools or fluxes (Table 21.1), again showing that biogeochemical processes are little affected by changes in species abundance. In summary, even large changes in species diversity and abundance have only moderate *direct, short-term* effects on pools and fluxes of carbon and nutrients, when species differ

Table 21.1 *Annual variation in production (% of 5 year mean) of major tussock-tundra species and total community above-ground production*

	Production (% of average)					Coefficient of variation (%)
	1968	1969	1970	1978	1981	
Eriophorum	77	58	148	101	116	35
Betula	30	52	55	248	121	88
Ledum	106	138	62	103	91	27
Vaccinium	135	172	96	28	71	56
Total production	93	110	106	84	107	11

From Chapin & Shaver (1985).

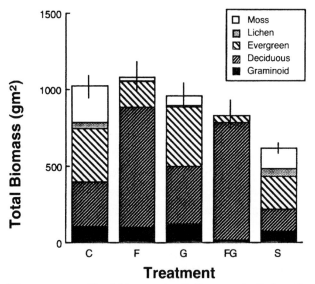

Figure 21.2 *Total biomass (excluding roots) of plant functional groups in arctic tussock tundra after 9 years of different environmental manipulations: control (C), fertilizer addition (F), greenhouse which raised air temperature by 3°C (G), fertilized greenhouse (FG), and shading to reduce irradiance by 50% (S) (from Chapin et al., 1995).*

only in resource consumption. Similarly, many animal species (especially grazers) have broad overlap in diet, such that changes in abundance of one species are compensated by changes in abundance of competing species with minimal effect on rate of consumption of their common prey species. We expect these generalizations to apply to closed communities where resource supply rather than colonization determine productivity and nutrient cycling.

Although we have argued that plants are less likely than animals to have large effects on ecosystem processes through their own resource consumption, there are cases in which they have large ecosystem impacts through their effects on consumption by animals. Some plants are so well defended against major herbivores that herbivores greatly change the competitive balance among plant types. For example, livestock grazing favoured invasion of Australian rangeland by *Opuntia*, a cactus with a pattern of plant defence distinctly different from that of other plants in the community. This created a large plant biomass that was inaccessible to herbivores and presumably shifted productivity to periods of lower soil moisture compared with ecosystems with more grass and less *Opuntia*. *Cactoblastis*, an insect specialist on *Opuntia*, was subsequently introduced and reduced the abundance of *Opuntia* (Dodd, 1959). Thus, animals, disease microorganisms, and sometimes plants, when added to or removed from ecosystems, can be 'keystone groups' that alter ecosystem processes through changes in trophic interactions much more dramatically than would be expected from their biomass and abundance.

We expect these generalizations about the importance of traits determining resource consumption to apply to closed communities where resource supply rather than colonization determine productivity and nutrient cycling and where multiple species use the same limiting resources. However, disturbance regime strongly influences the expression of these ecosystem effects (Menge *et al.*, 1994). As disturbance rate and intensity increase, patterns of resource supply and consumption become less important determinants of ecosystem processes, and the impact of strong biotic interactions and keystone predators is diminished (Menge *et al.*, 1994).

Disturbance regime

Animals or plants which greatly alter disturbance regime can also have large effects on ecosystems by altering the relative importance of equilibrium and non-equilibrium processes or by directly causing loss of nutrients from ecosystems. This is one of the most important ways in which animals affect ecosystem processes. For example, gophers and pigs create large areas of soil disturbance, creating sites for seedling establishment, and favouring early successional species (Hobbs & Mooney, 1991; Kotanen, 1995), generally leading to a lower biomass and a higher ratio of production to biomass (Shaver, 1995). Disturbance by feral pigs can promote ecosystem N loss (Singer *et al.*, 1984). At the regional scale, disturbance created by overgrazing can alter albedo of the land surface and change patterns of regional temperature and precipitation (Charney *et al.*, 1977; Schlesinger *et al.*, 1990).

Plants can also alter disturbance regime through effects on soil stability and flammability. For example, introduction of grasses into forest or shrubland ecosystems can increase fire frequency and cause a replacement of forest by savanna (D'Antonio & Vitousek, 1992). Similarly, boreal conifers are more flammable than deciduous trees because of their large leaf and twig surface area, low moisture content, and high resin content (Van Cleve *et al.*, 1991). For this reason, the invasion of the northern hardwood forests by hemlock in the early Holocene caused a change in fire frequency (Davis *et al.*, 1992), with associated changes in both plant and animal communities (Slobodchikoff & Doyen, 1977).

In early succession, plants are often critical in stabilizing soils and reducing wind and soil erosion. This allows successional development and retains the soil resources that determine the structure and productivity of late-successional stages. Introduced dune grasses have altered soil accumulation patterns and dune morphology in the western United States (D'Antonio & Vitousek, 1992).

Both plant and animal traits which alter disturbance regime tend to be discontinuous traits. Plants either have low or high flammability. Whether deletions or invasions of species which influence disturbance regime cause large changes in ecosystem processes depends on whether there are other species in the ecosystem sharing these traits.

Predicting ecosystem patterns from species traits

The same categories of species traits that govern ecosystem change (traits which influence resource supply, resource consumption and disturbance), are also useful in predicting patterns of ecosystem rates and structure. However, the species traits most useful in predicting patterns of ecosystem processes are plant traits related to resource consumption (plant size, RGR), precisely those traits that are least useful in predicting ecosystem change in response to changes in species composition. These traits are useful in predicting ecosystem patterns because (1) these and other correlated traits directly determine fluxes of energy and material through vegetation, as elaborated below, and (2) these traits can be readily measured by remote sensing (Chapin, 1993).

Biomass and carbon flux

Plants comprise >90% of live biomass in most non-aquatic ecosystems, so the size of plants and, to a lesser extent, their density directly determine the pools of carbon and nutrients in live biomass in terrestrial ecosystems. Therefore, in predicting patterns of biomass distribution, we emphasize

Table 21.2 *Above-ground biomass, production, and nitrogen flux in major temperate ecosystem types, maximum height, and relative growth rate of species typical of these ecosystem types.*

Parameter	Grassland	Shrubland	Deciduous forest	Ever-green forest
Above-ground biomass (kg m⁻²)	0.3+0.02	3.7+0.5	15+2	31+8
Above-ground production (kg m⁻² year⁻¹)	0.3+0.02	0.4+0.07	1.0+0.08	0.8+0.08
N flux (g m⁻² year⁻¹)	2.6+0.2	3.9+1.6	7.5+0.5	4.7+0.5
Canopy height (cm)	100	400	2200	2200
Field above-ground RGR (year⁻¹)	1.0	0.1	0.07	0.03
Laboratory above-ground RGR (week⁻¹)	1.3	0.8	0.7	0.4

From Chapin (1993).

plant traits. Plants also are the energetic base of the food chain, so the rate of carbon fixation by plants constrains energy flow through other components of the ecosystem. In equilibrium communities, where plant competition regulates density, stand biomass correlates positively with biomass of individuals (= b) and negatively with density (= $d^{-1/2}$) (Chapin, 1993). Thus, forests have larger vegetation pools of carbon and nutrients than do shrub lands than do grasslands because of larger plant size – despite lower plant density (Table 21.2) (Whittaker, 1975; Schlesinger, 1991). Within any one biome, communities dominated by larger plants generally have greater biomass (Schlesinger, 1991). Because satellites can detect various indices of plant size, remote sensing provides a promising mechanism of documenting global patterns of plant biomass and nutrient pools.

   Carbon flux between land and the atmosphere is determined by the balance between net primary production (NPP), which is determined directly by plant traits, and decomposition, which is indirectly influenced by plant traits (see above).

$$NPP = biomass \times RGR \times length\ of\ growth\ season$$

In other words, size, RGR, and phenology are the main traits determining productivity. In the field, RGR is constrained by many environmental and developmental factors. To support a high RGR, plants must have high rates of carbon and nutrient acquisition. Plants exhibit high rates of carbon gain because of some combination of high leaf allocation (leaf weight ratio), a large leaf area per unit leaf weight (specific leaf area), and high photosynthetic rate per unit leaf weight (Poorter & Remkes, 1990; Lambers & Poorter, 1992). Photosynthetic potential per gram leaf correlates well with RGR when growth forms are compared (Schulze & Chapin, 1987). In comparisons among genotypes or closely related species, leaf allocation and specific leaf area are generally more important than photosynthetic rate in explaining a high stand-level photosynthesis and RGR (Field & Mooney, 1986; Poorter & Remkes, 1990; Lambers & Poorter, 1992).

The RGR exhibited by a plant depends on both resource acquisition (see above) and tradeoffs between growth and alternative allocations. There is a consistent *negative* relationship between size and RGR across growth forms (Grime & Hunt, 1975; Field & Mooney, 1986; Tilman, 1988) (Table 21.2), among species or populations within a growth form (Chapin et al., 1989; Poorter & Remkes, 1990), and among different developmental stages of the same individual (Chapin et al., 1989; Lambers & Poorter, 1992). In general, high-resource sites are characterized by species with a high RGR and low allocation to woody support tissue in early succession and by large size (and correspondingly low RGR) in late succession. By contrast, low-resource sites are characterized by perennial species with lower RGR at any given size (for example evergreen shrubs or trees) and in extreme cases may not have sufficient resources to support trees. Thus, size is the major factor governing differences in NPP among biomes, and RGR accounts for differences in productivity of ecosystems of similar stature. Moreover, size varies by several orders of magnitude among plant species, whereas RGR varies by less than an order of magnitude (Grime & Hunt, 1975). Other allocation tradeoffs which compete with growth for resources include storage (Chapin et al., 1990; Lambers & Poorter, 1992), plant defence (Coley et al., 1985), and reproduction (Snow & Whigham, 1989). High allocation to these functions reduces instantaneous RGR and stand productivity.

Phenological specialization in the timing of plant activity can increase the total time available for plants to acquire resources from their environment. For example, evergreen trees can photosynthesize for a longer period of time than deciduous trees under favourable conditions (Schulze et al., 1977). Drought-tolerant species can extend the season of photosynthetic activity in dry environments. Spring ephemerals exhibit most of their growth before the tree canopy leafs out. However, in the few studies that have critically examined the importance of phenology, the extension of the growing season into unfavourable (dry or cold) seasons has surprisingly little influence on NPP (Schulze et al., 1977).

In most regions there is a plant species pool with a broad range of RGR, size and phenology. Ecological sorting (Vrba & Gould, 1986) causes species to dominate at those points along resource gradients where they have the greatest competitive advantage in acquiring and conserving the resources necessary for growth, survival, and reproduction (Whittaker, 1953). Thus, species with large size and high rates of resource acquisition tend to dominate nutrient-rich sites, and species with low rates of resource acquisition and turnover occupy infertile sites (Grime, 1977; Chapin, 1980; Berendse & Aerts, 1987; Lambers & Poorter, 1992; Aerts & van der Peijl, 1993). Within a given site it is the large-statured individuals that dominate resource capture and cycling, so, in ecosystems with a few dominant canopy species, relatively few species account for most of the biogeochemical cycling. Thus, differences among individuals and species in RGR, size, and phenology are extremely important in explaining (1) site differences in steady-state rates of biogeochemical cycling and (2) the identity of species in a given site that are responsible for most of the production and nutrient cycling. These traits are the basis of the life-form classification systems used in many global vegetation models (Holdridge, 1947; Woodward, 1987; Prentice *et al.*, 1992) and provide the physiological basis for vegetation distribution and activity.

Species effects on resource supply and disturbance regime (that is, the other categories of species effects on ecosystem processes) have a less predictable relationship to patterns of productivity than do species effects on resource acquisition. For example, biomass, productivity, and N uptake by N-fixing shrub communities are intermediate between values for successional stages which precede or follow this stage (Van Cleve *et al.*, 1971). Across communities in general there is little relationship between productivity and biomass of N-fixing species. Similarly, when averaged across the successional cycle, there is no clear relationship between patterns of productivity and disturbance regime.

Nutrient cycling

In a fashion analogous to carbon exchange, rates of nutrient cycling through plants are governed by rates of nutrient uptake and loss. There is a positive correlation between nutrient uptake and RGR which reflects species and growth-form differences in nutrient uptake potential per gram root. For example, rapidly growing tundra species (Chapin & Tryon, 1982, but see Kielland & Chapin, 1994) and taiga trees (Chapin *et al.*, 1983) have higher rates of phosphate uptake than do more slowly growing species. Within a genus, rapidly growing species of New Zealand tussock grasses (Chapin *et al.*, 1982) have high rates of phosphate uptake. By contrast, nutrient retention through increased leaf longevity is the major factor governing success in infertile sites (Chapin, 1980; Berendse *et al.*, 1987; Aerts & van der Peijl, 1993). Long-lived leaves have low photosynthetic rates and low specific leaf areas, thus reducing both RGR and rates of nutrient cycling. Size and total root length are also important contributors to nutrient uptake on a ground-area basis, so that annual nutrient flux is often greater in ecosystems dominated by large-statured plants (Chapin, 1993). Because of the correlations of nutrient uptake with RGR and size at the physiological level, there is also a correlation between productivity and nutrient cycling at the ecosystem level (Fig. 21.3).

Phenology can also strongly influence nutrient uptake. In ecosystems dominated by annual plants, early-season rains can leach nutrients from the ecosystem because there is little plant biomass present to absorb the nutrients (Jackson *et al.*, 1988), whereas in ecosystems with continuously active plant cover, plants absorb most nutrients that become available, so leaching loss of nutrients from the ecosystem is negligible (Vitousek & Reiners, 1975). In heavily fertilized agricultural systems, most nutrient leaching occurs during the fallow season and when seedlings are too small to fully exploit the soil volume or after crops have been harvested (Radke *et al.*, 1988).

Water and energy exchange

The major factors governing energy exchange between the land and the atmosphere are evapotranspiration (ET), albedo, and surface roughness, each of which changes with plant size or biomass. ET is strongly determined by leaf area (a function of plant size), which determines (1) the amount of precipitation that is intercepted by the canopy and quickly evaporates after a rain, (2) the size of the transpiring surface, and in some ecosystems (3) the amount of precipitation input from fog and rain (Wilson & Agnew, 1992). Annual ET is also influenced by phenology, with evergreen species having a transpiring surface present for more of the year. ET correlates with photosynthesis because of the dependence of both processes on leaf area and stomatal conductance. Thus, plants with a high RGR tend to have high rates of ET at both the leaf and canopy levels (Schulze *et al.*, this volume). Plant biomass indirectly influences ET because of its correlation with the quantity of litter on the soil surface, which influences the partitioning of water between surface runoff and infiltration into the soil. Surface runoff is negligible in forests and other communities with a well-developed litter layer (Running & Coughlan, 1988).

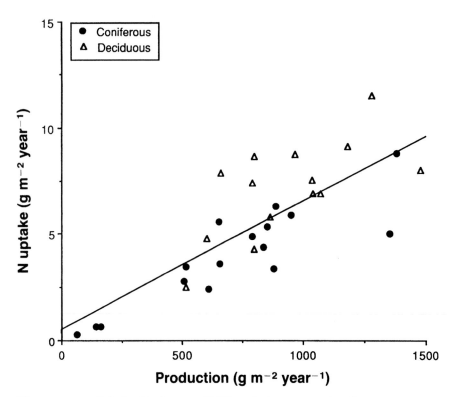

Figure 21.3   *Relationship between NPP and nitrogen uptake of temperate and boreal coniferous (●) and deciduous (Δ) forests (Chapin, 1993).*

In boreal regions, albedo (reflectance) is even more important than ET in regulating annual energy exchange. If plants are taller than the snow pack (boreal forest, for example), there is much greater energy absorption than in snow-covered tundra. Simulations in which the boreal forest was replaced by snow-covered short-statured vegetation demonstrated that this change in albedo could cause a large regional cooling that would be most pronounced in boreal regions but would extend to the tropics (Bonan *et al.*, 1992). Conversely, if trees expand northward into current tundra, this could act as a positive feedback to global warming (Foley *et al.*, 1994). Thus, plant size has large effects on albedo and annual energy exchange.

Plant size also strongly influences surface roughness in flat landscapes. Low-statured prairies have a well-developed boundary layer that is more important than stomatal conductance in governing water-vapour transport to the atmosphere, when stomates are open (McNaughton & Jarvis, 1991). By contrast, forest trees are large enough that turbulent eddies regularly penetrate their canopies, reducing the thickness of the boundary layer so that stomatal conductance consistently influences transpiration rate. Thus, plant size and stomatal conductance (a correlate of RGR) strongly influence surface roughness and ET.

Effects of diversity on ecosystem processes

The number of species in an ecosystem is functionally important because it (1) increases the rate or efficiency of resource capture under steady-state conditions and (2) provides insurance against large changes in ecosystem processes in response to disturbance or environmental change. When species diversity is extremely low (for example, in a crop monoculture), total nutrient uptake and productivity of a crop and its consumption by higher trophic levels is often less than in more complex ecosystems (Swift & Anderson, 1993; Vitousek & Hooper, 1993; Naeem et al., 1994), although forests dominated by a single tree species are not notably less productive than highly diverse forests (Rodin & Bazilevich, 1967). In artificial tropical communities, a variety of biogeochemical processes differed strikingly between plots with 0, 1 and 100 species but not among highly diverse plots of differing species composition (Ewel et al., 1991; Vitousek & Hooper, 1993). Thus, we know that species diversity affects ecosystem processes somewhere between 1 and 100 species, but we do not know where or for what reason this relationship saturates. Diverse grasslands maintained higher productivity in response to drought than did grasslands whose productivity had been reduced by experimental nutrient addition (Tilman & Downing, 1994). Artificial communities with differing numbers (1–4) of species per trophic level also differed in productivity (Naeem et al., 1994). The challenge in experimental studies is to separate the effects of species number from the effects of the traits of the component species. This represents an important area for future research.

Species often show unique responses to climate and resources (Whittaker, 1975; McNaughton, 1977; Chapin & Shaver, 1985), and for this reason, species diversity provides insurance against change in ecosystem function independent of their existing effects on ecosystem function via traits discussed above. Any change in climate or climatic extremes that is severe enough to cause extinction of one species is unlikely to eliminate all members from a functional group (Walker, 1995). Thus, the more species there are in a functional group, the less likely that any extinction event or series of such events will have serious ecosystem consequences (Holling, 1986; Chapin et al., 1995). For these reasons, genetic and species diversity is important to long-term maintenance of community and ecosystem structure and processes. This argues that no two species are ecologically redundant, even if they are similar in their ecosystem effects under a particular set of environmental conditions.

Species diversity is important to the maintenance of ecosystem integrity over a complete cycle of common disturbance events. Following disturbance, initial colonization by early successional species often stabilizes substrate or retains nutrients that are later used by other species. For example, riparian shrubs can stabilize stream banks sufficiently to allow colonization by forest trees (Van Cleve *et al.*, 1991). Rapid colonization by early successional species following fire or forest cutting retains nutrients that, in the long term, support the growth of late-successional forests (Stark & Steele, 1977; Bormann & Likens, 1979). The long-term stability and resilience of communities and ecosystems, therefore, probably requires a diversity of species whose ecosystem impact is minimal at most times but critical at certain phases of succession.

There is an unresolved debate as to whether each species is like a rivet in an aeroplane, such that any species loss makes the ecosystem more susceptible to catastrophic failure (Ehrlich & Ehrlich, 1981) or whether ecologically similar species can substitute functionally for one another, such that loss of a species has less effect if other species of the same functional group are retained (Lawton & Brown, 1993). For those characteristics that are continuous (size, RGR), the impact of the loss of a given species probably depends on the degree of ecological similarity of remaining species to those that have been lost from the community. Thus, loss or gain of a single species would probably have greater impact in the arctic than in a tropical rain forest and among a poorly represented (or missing) functional group than among a speciose functional group. Moreover, loss of a species that is extreme with respect to these traits could be particularly important. For example, loss of a canopy tree species (chestnut) might have greater direct impact on productivity, at least initially, than loss of a mid-canopy species (dogwood). For those characteristics that are discontinuous (for example, presence or absence of symbiotic N fixation) or where one or a few species differ strikingly from the rest of the community (for example, highly flammable grasses), gain or loss of a species will have large ecosystem consequences.

Ecosystem-level feedbacks of invasions and extinctions to biodiversity

In natural ecosystems, biodiversity may be of greater inherent interest to society than ecosystem processes, because many of the goods and services (direct benefits of species; Ehrlich & Ehrlich, 1981) that people derive from ecosystems relate to properties of species rather than to biogeochemical processes. However, the existence and quality of the earth's atmosphere, climate, water and soil ('indirect benefits' of species; Ehrlich & Ehrlich, 1981) depend on biogeochemical processes. We have established that species traits and biodiversity do have implications for ecosystem processes. We also know that species are quite sensitive to their environment and that subtle changes in environment can alter competitive balances, leading to changes in species composition and biodiversity. Perhaps the most important consequences of changes in species traits and biodiversity in natural ecosystems have to do with the largely unknown feedbacks of the altered environment to further changes in biodiversity. For example, invasion by non-indigenous grasses in Hawaiian forests resulted in increased fire frequency and a decline in diversity of fire-sensitive woody species (Hughes *et al.*, 1991; D'Antonio & Vitousek, 1992).

A second, largely unexplored consequence of changing biodiversity involves species-specific interactions with other species that have large ecosystem effects. For example, a seed disperser or pollinator which has little direct effect on ecosystem processes may be essential for persistence of a canopy species with greater direct ecosystem impact. Conversely, a seed disperser may indirectly promote changes in ecosystem processes by facilitating the dispersal of an N-fixing tree (Vitousek & Walker, 1989). At our present level of ignorance, these *indirect* effects of species on ecosystem processes are difficult to predict, suggesting that we should be conservative in drawing conclusions about the ecosystem impacts of loss of a given species or level of diversity.

Conclusions

Species have large and often predictable effects on ecosystem processes through alteration of soil resource supply, the consumption of resources or food, and disturbance regime. Gain or loss of a species will have greatest impact on ecosystem processes when it differs strongly from other species in the community. When species are similar to one another in their resource requirements, as often occurs with plants and generalist herbivores, gain or loss of a species has large effects on species composition of the community but less effect on ecosystem processes because of the compensatory responses of other species to the altered competitive environment. The traits which govern resource acquisition in these species are often continuously distributed among species, so that species differ quantitatively rather than qualitatively in their effects on ecosystem processes.

By contrast, gain or loss of species which consume unique food or soil resources or which alter resource supply or disturbance regime can have large effects on ecosystem processes, which propagate through the ecosystem through a chain of indirect effects. Traits which govern these processes are often qualitatively different among species, so that changes in the abundance of these species have widespread ecosystem impacts. Species effects on ecosystem processes are often as large as direct climatic effects and must be included in predictive models of the role of terrestrial ecosystems in global processes.

Species diversity is functionally important because it provides insurance against large changes in ecosystem processes and may enhance the efficiency with which resources are captured from the environment and transferred among species. Because each species shows a unique response to climate and resources, any change in climate or climatic extremes that is severe enough to cause extinction is unlikely to eliminate all members of a functional group. The more species there are in a functional group, the less likely that any extinction event or series of such events will have serious ecosystem consequences. Although species clearly differ in their magnitude of impact on community and ecosystem processes, the differences among species in their responses to disturbances and environmental extremes and their indirect ecosystem effects, mediated by species interactions, make it unlikely that there is much, if any, ecological redundancy in communities over time scales of decades to centuries, the time period over which environmental policy should operate.

References

Aerts, R. & van der Peijl, M. J. (1993). A simple model to explain the dominance of low-productive perennials in nutrient-poor habitats. *Oikos*, **66**, 144–7.

Ball, R., Brougham, R. W., Brock, J. L., Crush, J. R., Hoglund, J. H. & Carran, R. A. (1979). Nitrogen fixation in New Zealand pastures. *New Zealand Journal of Experimental Agriculture*, **7**, 1–5.

Batzli, G. O., White, R. G., MacLean, S. F. Jr, Pitelka, F. A. & Collier, B. D. (1980). The herbivore-based trophic system. In *An Arctic Ecosystem: The Coastal Tundra at Barrow, Alaska*, ed. J. Brown, P. C. Miller, L. L. Tieszen & F. L. Bunnell, pp. 335–410. Stroudsburg: Dowden, Hutchinson and Ross.

Berendse, F. & Aerts, R. (1987). Nitrogen-use efficiency: A biologically meaningful definition? *Functional Ecology*, **1**, 293–6.

Berendse, F., Oudhof, H. & Bol, J. (1987). A comparative study on nutrient cycling in wet heathland ecosystems. I. Litter production and nutrient losses from the plant. *Oecologia*, **74**, 174–84.

Berendse, F., Schmitz, M. & de Visser, W. (1994). Experimental manipulation of succession in heathland ecosystems. *Oecologia*, **100**, 38–44.

Berg, B. & McClaugherty, C. (1989). Nitrogen and phosphorus release from decomposing litter in relation to lignin. *Canadian Journal of Botany*, **67**, 1148–56.

Bonan, G. B., Pollard, D. & Thompson, S. L. (1992). Effects of boreal forest vegetation on global climate. *Nature*, **359**, 716–18.

Bond, W. J. (1993). Keystone species. In *Ecosystem Function and Biodiversity*, ed. E.-D. Schulze & H. A. Mooney, pp. 237–53. Berlin: Springer.

Bormann, F. H. & Likens, G. E. (1979). *Patterns and Process in a Forested Ecosystem*. New York: Springer.

Caldwell, M. M. & Richards, J. H. (1989). Hydraulic lift: water efflux from upper roots improves effectiveness of water uptake from deep roots. *Oecologia*, **79**, 1–5.

Carpenter, S. R., Fisher, S. G., Grimm, N. B. & Kitchell, J. F. (1992). Global change and freshwater ecosystems. *Annual Review of Ecology and Systematics*, **23**, 119–39.

Carpenter, S. R., Kitchell, J. F. & Hodgson, J. R. (1985). Cascading trophic interactions and lake productivity. *BioScience*, **35**, 634–9.

Chapin, F. S., III. (1980). The mineral nutrition of wild plants. *Annual Review of Ecology and Systematics*, **11**, 233–260.

Chapin, F. S., III. (1991). Effects of multiple environmental stresses on nutrient availability and use. In *Response of Plants to Multiple Stresses*, ed. H. A. Mooney, W. E. Winner & E. J. Pell, pp. 67–88. San Diego: Academic Press.

Chapin, F. S., III. (1993). Functional role of growth forms in ecosystem and global processes. In *Scaling Physiological Processes: Leaf to Globe*, ed. J. R. Ehleringer & C. B. Field, pp. 287–312. San Diego: Academic Press.

Chapin, F. S., III, Fetcher, N., Kielland, K., Everett, K. R. & Linkins, A. E. (1988). Productivity and nutrient cycling of Alaskan tundra: Enhancement by flowing soil water. *Ecology*, **69**, 693–702.

Chapin, F. S., III, Follett, J. & O'Connor, K. F. (1982). Growth, phosphate absorption, and phosphorus chemical fractions in two *Chionochloa* species. *Journal of Ecology*, **70**, 305–21.

Chapin, F. S., III, Groves, R. H. & Evans, L. T. (1989). Physiological determinants of growth rate in response to phosphorus supply in wild and cultivated *Hordeum* species. *Oecologia*, **79**, 96–105.

Chapin, F. S., III & Kedrowski, R. A. (1983). Seasonal changes in nitrogen and phosphorus fractions and autumn retranslocation in evergreen and deciduous taiga trees. *Ecology*, **64**, 376–91.

Chapin, F. S., III, Moilanen, L. & Keilland, K. (1993). Preferential use of organic nitrogen for growth by a non-mycorrhizal arctic sedge. *Nature*, **361**, 150–3.

Chapin, F. S., III, Lubchenco, J. & Reynolds, H. R. (1995). Biodiversity effects on patterns and processes of communities and ecosystems. In *Global Biodiversity Assessment*, ed. UNEP, pp. 23–45. Cambridge: Cambridge University Press.

Chapin, F. S., III, Schulze, E.-D. & Mooney, H. A. (1990). The ecology and economics of storage in plants. *Annual Review of Ecology and Systematics*, **21**, 423–48.

Chapin, F. S., III & Shaver, G. R. (1985). Individualistic growth response of tundra plant species to environmental manipulations in the field. *Ecology*, **66**, 564–76.

Chapin, F. S., III, Shaver, G. R., Giblin, A. E., Nadelhoffer, K. G. & Laundre, J. A. (1995). Response of arctic tundra to experimental and observed changes in climate. *Ecology*, **76**, 694–711.

Chapin, F. S., III & Tryon, P. R. (1982). Phosphate absorption and root respiration of different plant growth forms from northern Alaska. *Holarctic Ecology*, **5**, 164–71.

Chapin, F. S., III, Van Cleve, K. & Tryon, P. R. (1983). Influence of phosphorus on the growth and biomass allocation of Alaskan taiga tree seedlings. *Canadian Journal of Forest Research*, **13**, 1092–8.

Charney, J. G., Quirk, W. J., Chow, S.-H. & Kornfield, J. (1977). A comparative study of effects of albedo change on drought in semiarid regions. *Journal of Atmospheric Science*, **34**, 1366–85.

Cody, M. L. (1986). Structural niches in plant communities. In *Community Ecology*, ed. J. Diamond & T. Case, pp. 381–405. San Francisco: Harper and Row.

Coley, P. D., Bryant, J. P. & Chapin, F. S., III (1985). Resource availability and plant anti-herbivore defense. *Science*, **230**, 895–9.

D'Antonio, C. M. & Vitousek, P. M. (1992). Biological invasions by exotic grasses, the grass-fire cycle, and global change. *Annual Review of Ecology and Systematics*, **23**, 63–87.

Davis, M. B., Sugita, S., Calcote, R. R. & Frelich, L. (1992). Invasion of forests by hemlock coincided with change in disturbance regime. *Bulletin of the Ecological Society of America*, **73**, 155.

Dodd, A. P. (1959). Biological control of prickly pear in Australia. In *Biogeography and Ecology of Australia*. Monographs in Biology, vol. 8, ed. A. Keast, R. L. Crocker & C. S. Christian, pp. 565–77. The Hague: Junk.

Ehrlich, P. R. & Ehrlich, A. H. (1981). *Extinction. The Causes and Consequences of the Disappearance of Species*. New York: Random House.

Estes, J. A. & Palmisano, J. F. (1974). Sea otters: their role in structuring nearshore communities. *Science*, **185**, 1058–60.

Ewel, J. J., Mazzarino, M. J. & Berish, C. W. (1991). Tropical soil fertility changes under monocultures and successional communities of different structure. *Ecological Applications*, **1**, 289–302.

Field, C. & Mooney, H. A. (1986). The photosynthesis–nitrogen relationship in wild plants. In *On the Economy of Plant Form and Function*, ed. T. J. Givnish, pp. 25–55. Cambridge: Cambridge University Press.

Flanagan, P. W. & Van Cleve, K. (1983). Nutrient cycling in relation to decomposition and organic matter quality in taiga ecosystems. *Canadian Journal of Forest Research*, **13**, 795–817.

Foley, J. A., Kutzbach, J. E., Coe, M. T. & Levis, S. (1994). Feedbacks between climate and boreal forests during the Holocene epoch. *Nature*, **371**, 52–4.

Grime, J. P. (1977). Evidence for the existence of three primary strategies in plants and its relevance to ecological and evolutionary theory. *American Naturalist*, **111**, 1169–94.

Grime, J. P. & Hunt, R. (1975). Relative growth rate: Its range and adaptive significance in a local flora. *Journal of Ecology*, **63**, 393–422.

Hairston, N. G., Smith, F. E. & Slobodkin, L. B. (1960). Community structure, population control and competition. *American Naturalist*, **94**, 421–5.

Heath, B., Sollins, P., Perry, D. & Cromack, J. K. (1988). Asymbiotic N fixation in litter from Pacific northwest forests. *Canadian Journal of Forest Research*, **18**, 68–74.

Hobbie, S. E. (1992). Effects of plant species on nutrient cycling. *Trends in Ecology and Evolution*, **7**, 336–9.

Hobbie, S. E. (1995). Direct and indirect species effects on biogeochemical processes in arctic ecosystems. In *Arctic and Alpine Biodiversity: Patterns, Causes and Ecosystem Consequences*, ed. F. S. Chapin III & C. Körner, pp. 213–24. Heidelberg: Springer.

Hobbs, R. J. & Mooney, H. A. (1991). Effects of rainfall variability and gopher disturbance on serpentine annual grassland dynamics. *Ecology*, **72**, 59–68.

Holdridge, L. R. (1947). Determination of world plant formations from simple climatic data. *Science*, **105**, 367–8.

Holling, C. S. (1986). Resilience of ecosystems: Local surprise and global change. In *Sustainable Development and the Biosphere*, ed. W. C. Clark & R. E. Munn, pp. 292–317. Cambridge: Cambridge University Press.

Hughes, F., Vitousek, P. M. & Tunison, T. (1991). Alien grass invasion and fire in the seasonal submontane zone of Hawai'i. *Ecology*, **72**, 743–7.

Jackson, L. E., Strauss, R. B., Firestone, M. K. & Bartolome, J. W. (1988). Plant and soil nitrogen dynamics in California annual grassland. *Plant and Soil*, **110**, 9–17.

Jenny, H. (1980). *The Soil Resources: Origin and Behavior*. New York: Springer.

Kielland, K. & Chapin, F. S., III (1994). Phosphate uptake in arctic plants in relation to phosphate supply: the role of spatial and temporal variability. *Oikos*, **70**, 443–8.

Kotanen, P. M. (1995). Responses of vegetation to a changing regime of disturbance: effects of feral pigs on a California coastal prairie. *Ecography*, **18**, 190–9.

Lambers, H. & Poorter, H. (1992). Inherent variation in growth rate between higher plants: a search for physiological causes and ecological consequences. *Advances in Ecological Research*, **23**, 187–261.

Lauenroth, W. K., Dodd, J. L. & Simms, P. L. (1978). The effects of water- and nitrogen-induced stresses on plant community structure in a semiarid grassland. *Oecologia*, **36**, 211–22.

Lawton, J. H. & Brown, V. K. (1993). Redundancy in ecosystems. In *Biodiversity and Ecosystem Function*, ed. E.-D. Schulze & H. A. Mooney, pp. 255-70. Berlin: Springer.

MacArthur, R. H. (1965). Patterns of species diversity. *Biological Reviews*, **40**, 510–33.

McNaughton, K. G. & Jarvis, P. G. (1991). Effects of spatial scale on stomatal control of transpiration. *Agricultural and Forest Meteorology*, **54**, 279–302.

McNaughton, S. J. (1977). Diversity and stability of ecological communities: A comment on the role of empiricism in ecology. *American Naturalist*, **111**, 515–25.

McNaughton, S. J. (1985). Ecology of a grazing ecosystem: the Serengeti. *Ecological Monographs*, **53**, 259–94.

McNaughton, S. J. (1988). Mineral nutrition and spatial concentrations of African ungulates. *Nature*, **334**, 343–5.

Melillo, J. M., Aber, J. D. & Muratore, J. F. (1982). Nitrogen and lignin control of hardwood leaf litter decomposition dynamics. *Ecology*, **63**, 621-6.

Menge, B. A., Berlow, E. L., Blanchette, C. A., Navarrete, S. A. & Yamada, S. B. (1994). The keystone species concept: variation in interaction strength in a rocky intertidal habitat. *Ecological Monographs*, **64**, 249–86.

Meyer, O. (1993). Functional groups of microorganisms. In *Biodiversity and Ecosystem Function*, ed. E.-D. Schulze & H. A. Mooney, pp. 67–96. Berlin: Springer.

Mills, L. S., Soule, M. E. & Doak, D. F. (1993). The keystone-species concept in ecology and conservation. *BioScience*, **43**, 219–24.

Naeem, S., Thompson, L. J., Lawler, S. P., Lawton, J. H. & Woodfin, R. M. (1994). Declining biodiversity can alter the performance of ecosystems. *Nature*, **368**, 734–7.

NRC (1994). *The Role of Terrestrial Ecosystems in Global Change: A Plan for Action*. Washington: National Academy Press.

Olson, A. M. & Lubchenco, J. (1990). Competition in seaweeds: linking plant traits to competitive outcomes. *Journal of Phycology*, **26**, 1–6.

Owen-Smith, R. N. (1988). *Megaherbivores: the Influence of Very Large Body Size on Ecology*. Cambridge: Cambridge University Press.

Pastor, J. (1995). Diversity of biomass and

nitrogen distribution among plant species in arctic and alpine ecosystems. In *Arctic and Alpine Biodiversity: Patterns, Causes and Ecosystem Consequences*, ed. F. S. Chapin III & C. Körner, pp. 225–69. Heidelberg: Springer.

Poorter, H. & Remkes, C. (1990). Leaf area ratio and net assimilation rate of 24 wild species differing in relative growth rate. *Oecologia*, **83**, 553–9.

Power, M. E. (1992). Hydrologic and trophic controls of seasonal algal blooms in northern California rivers. *Archiv für Hydrobiologie*, **125**, 385–410.

Prentice, I. C., Cramer, W., Harrison, S. P., Leemans, R., Monserud, R. A. & Solomon, A. M. (1992). A global biome model based on plant physiology and dominance, soil properties and climate. *Journal of Biogeography*, **19**, 117–34.

Radke, J. K., Andrews, R. W., Janke, R. R. & Peters, S. E. (1988). Low-input cropping systems and efficiency of water and nitrogen use. In *Cropping Strategies for Efficient Use of Water and Nitrogen*, ed. W. L. Hargrove, pp. 193–218. Madison: American Society of Agronomy.

Rasmussen, D. I. (1941). Biotic communities of Kaibab Plateau, Arizona. *Ecological Monographs*, **3**, 229–75.

Read, D. J. (1991). Mycorrhizas in ecosystems. *Experientia*, **47**, 376–91.

Robles, M. & Chapin, F. S., III (1995). Comparison of the influence of two exotic species on ecosystem processes in the Berkeley Hills. *Madroño*, **42**, 349–57.

Rodin, L. E. & Bazilevich, N. I. (1967). *Production and Mineral Cycling in Terrestrial Vegetation*. Edinburgh: Oliver and Boyd.

Running, S. W. & Coughlan, J. C. (1988). A general model of forest ecosystem processes for regional applications. I. Hydrologic balance, canopy gas exchange and primary production processes. *Ecological Modeling*, **42**, 125–54.

Schimel, J. (1995). Ecosystem consequences of microbial diversity and community structure. In *Arctic and Alpine Biodiversity: Patterns, Causes and Ecosystem Consequences*, ed. F. S. Chapin III & C. Körner, pp. 239–54. Heidelberg: Springer.

Schlesinger, W. H. (1991). *Biogeochemistry: An Analysis of Global Change*. San Diego: Academic Press.

Schlesinger, W. H., Reynolds, J. F., Cunningham, G. L., Huenneke, L. F., Jarrell, W. M., Virginia, R. A. & Whitford, W. G. (1990). Biological feedbacks in global desertification. *Science*, **247**, 1043–8.

Schulze, E.-D. & Chapin, F. S., III. (1987). Plant specialization to environments of different resource availability. In *Potentials and Limitations in Ecosystem Analysis*, ed. E.-D. Schulze & H. Zwolfer, pp. 120–48. Berlin: Springer.

Schulze, E.-D., Chapin, F. S., III & Gebauer, G. (1995). Nitrogen nutrition and isotope differences among life forms at the northern treeline in Alaska. *Oecologia*, **100**, 406–12.

Schulze, E.-D., Fuchs, M. & Fuchs, M. I. (1977). Spacial distribution of photosynthetic capacity and performance in a mountain spruce forest of northern Germany. III. The significance of the evergreen habit. *Oecologia*, **30**, 239–48.

Shaver, G. R. (1995). Plant functional diversity and resource control of primary production in Alaskan arctic tundras. In *Arctic and Alpine Biodiversity: Patterns, Causes and Ecosystem Consequences*, ed. F. S. Chapin III & C. Körner, pp. 199–211. Heidelberg: Springer-Verlag.

Shaver, G. R., Giblin, A. E., Nadelhoffer, K. J. & Rastetter, E. B. (In press). Plant functional types and ecosystem change in arctic tundras. In *Plant Functional Types*, ed. T. Smith, H. H. Shugart & F. I. Woodward. Cambridge: Cambridge University Press.

Shukla, J., Nobre, C. & Sellers, P. (1990).

Amazon deforestation and climate change. *Science*, **247**, 1322–5.

Singer, F., Swank, W. & Clebsch, E. (1984). Effects of wild pig rooting in a deciduous forest. *Journal of Wildlife Management*, **48**, 464–73.

Slobodchikoff, F. S. & Doyen, J. T. (1977). Effects of *Ammophila arenaria* on sand dune arthropod communities. *Ecology*, **58**, 1171–5.

Snow, A. A. & Whigham, D. F. (1989). Costs of flower and fruit production in *Tipularia discolor* (Orchidaceae). *Ecology*, **70**, 1286–93.

Solbrig, O. T. (ed.) (1991). *From Genes to Ecosystems: A Research Agenda for Biodiversity*. Cambridge, Massachusetts: International Union of Biological Sciences.

Stark, N. M. & Steele, R. (1977). Nutrient content of forest shrubs following burning. *American Journal of Botany*, **64**, 1218–24.

Strong, D. R. (1992). Are trophic cascades all wet? Differentiation and donor-control in speciose ecosystems. *Ecology*, **73**, 747–54.

Swift, M. J. & Anderson, J. M. (1993). Biodiversity and ecosystem function in agricultural systems. In *Biodiversity and Ecosystem Function*, ed. E.-D. Schulze & H. A. Mooney, pp. 15–41. Berlin: Springer.

Tilman, D. (1988). *Plant Strategies and the Dynamics and Function of Plant Communities*. Princeton: Princeton University Press.

Tilman, D. & Downing, J. A. (1994). Biodiversity and stability in grasslands. *Nature*, **367**, 363–5.

van Berkum, P. & Gohlool, B. B. (1980). Evolution of N fixation by bacteria in association with roots of tropical grasses. *Microbiological Reviews*, **980**, 491–517.

van Breemen, N. (1993). Soils as biotic constructs favouring net primary productivity. *Geoderma*, **57**, 183–211.

Van Cleve, K., Chapin, F. S., III, Dryness, C. T. & Viereck, L. A. (1991). Element cycling in taiga forest: State-factor control. *BioScience*, **41**, 78–88.

Van Cleve, K., Viereck, L. A. & Schlentner, R. L. (1971). Accumulation of nitrogen in alder (*Alnus*) ecosystems near Fairbanks, Alaska. *Arctic and Alpine Research*, **3**, 101–14.

van Hylckama, T. E. A. (1974). Water use by salt cedar as measured by the water budget method. *Geological Survey Professional Papers*, 491-E.

Vitousek, P. M. (1982). Nutrient cycling and nutrient use efficiency. *American Naturalist*, **119**, 553–72.

Vitousek, P. M. (1990). Biological invasions and ecosystem processes: Towards an integration of population biology and ecosystem studies. *Oikos*, **57**, 7–13.

Vitousek, P. M. & Hooper, D. U. (1993). Biological diversity and terrestrial ecosystem biogeochemistry. In *Biodiversity and Ecosystem Function*, ed. E.-D. Schulze & H. A. Mooney, pp. 3–14. Berlin: Springer.

Vitousek, P. M. & Reiners, W. A. (1975). Ecosystem succession and nutrient retention: a hypothesis. *BioScience*, **25**, 376–81.

Vitousek, P. M. & Walker, L. R. (1989). Biological invasion by *Myrica faya* in Hawai'i: plant demography, nitrogen fixation, ecosystem effects. *Ecological Monographs*, **59**, 247–65.

Vitousek, P. M., Walker, L. R., Whiteacre, L. D., Mueller-Dombois, D. & Matson, P. A. (1987). Biological invasion by *Myrica faya* alters ecosystem development in Hawaii. *Science*, **238**, 802–4.

Vrba, E. S. & Gould, S. J. (1986). The hierarchical expansion of sorting and selection: sorting and selection cannot be equated. *Paleobiology*, **12**, 217–28.

Walker, B. (1995). Conserving biological diversity through ecosystem resilience. *Conservation Biology*, **9**, 747–52.

Whittaker, R. H. (1953). A consideration of climax theory: The climax as a population and pattern. *Ecological Monographs*, **23**, 41–78.

Whittaker, R. H. (1972). Evolution and measurement of species diversity. *Taxon*, **21**, 213–51.

Whittaker, R. H. (1975). *Communities and Ecosystems*. New York: Macmillan.

Wilson, J. B. & Agnew, D. Q. (1992). Positive-feedback switches in plant communities. *Advances in Ecological Research*, **23**, 263–336.

Woodward, F. I. (1987). *Climate and Plant Distribution*. Cambridge: Cambridge University Press.

Part six
GCTE and Earth system science

㉒ Carbon and nitrogen interactions in the terrestrial biosphere: anthropogenic effects

J. M. Melillo

Introduction

Humans have dramatically altered the global cycles of carbon and nitrogen. Through fossil fuel combustion and biomass burning, we have added an average of about 7 PgC of carbon to the atmosphere each year during the 1980s. Of this amount, slightly less than half has been stored in the atmosphere. In the most recent Intergovernmental Panel on Climate Change (IPCC) analysis of the carbon cycle Schimel and his colleagues (1994) indicate that about half of the remainder is stored in the ocean each year and the other half is stored on land. It is likely that this terrestrial carbon storage is the result of several mechanisms including ones associated with our alteration of the nitrogen cycle.

The global nitrogen budget is being directly altered by us through industrial and agricultural activities. We are converting unreactive nitrogen to reactive nitrogen and adding it to both terrestrial and marine ecosystems. We are also causing the redistribution of reactive nitrogen among ecosystems.

We also have the potential to indirectly affect the global nitrogen budget. Through fossil fuel use and changes in land cover and land use, we are increasing the atmosphere's content of carbon dioxide, methane and nitrous oxide. The loading of the atmosphere with these greenhouse gases is expected to result in a warming of the Earth. The warming is predicted to accelerate soil organic matter decay and make more reactive nitrogen available to plants.

This chapter will explore how an increase in the availability of reactive nitrogen affects carbon storage in terrestrial ecosystems. The chapter has four parts. The first part reviews how humans are altering the global nitrogen budget. Next the direct effects of alterations of the global nitrogen budget on carbon storage in terrestrial ecosystems are discussed. The third part of the chapter considers the indirect effects of climate change on the global nitrogen budget and consequences for the global carbon budget. And

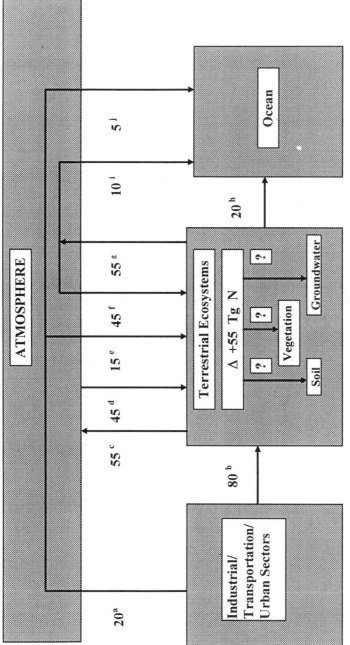

Figure 22.1 The 'anthropogenic' component of the global nitrogen budget: human influences on major nitrogen fluxes and the increment in the terrestrial nitrogen pool. Units are Tg N/y. All values rounded to nearest 5 Tg. The fluxes are as follows: a, NO$_x$ released in fossil fuel burning; b, fertilizer nitrogen applied to land; c, denitrification and other gaseous N losses from land; d, nitrogen fixation associated with agriculture; e, deposition on land of NO$_x$ derived from fossil fuel burning; f, deposition on land of NO$_x$ and NH$_x$ derived from a variety of human activities including crop fertilization, animal husbandry and biomass burning; g, loss from land of NO$_x$ and NH$_x$ associated with human activities listed in f above; h, flux of nitrogen from the land to the ocean via rivers; i, deposition on the ocean of NO$_x$ and NH$_x$ derived from human activities listed in f above; j, deposition on the ocean of NO$_x$ derived from fossil fuel burning. See the text for sources of the estimates.

finally, a blueprint is set out for future research designed to explore more completely how alterations in the global nitrogen budget affect the global carbon budget. The proposed research includes a range of activities from whole-ecosystem ^{15}N labelling experiments to regional and global terrestrial ecosystem model simulations.

Human influences on the global nitrogen budget: direct effects

Three processes, fossil fuel combustion, fertilizer production, and legume and rice cultivation, are responsible for converting unreactive nitrogen to various forms of reactive nitrogen. In recent years, this conversion rate annually has been about 145 Tg of N: 20 Tg N associated with fossil fuel combustion (Levy & Moxim, 1989), 80 Tg N in fertilizer production (FAO, 1989), and 45 Tg N resulting from legume and rice cultivation (Burns & Hardy, 1975). Throughout this section of the chapter all nitrogen fluxes and pools are rounded to the nearest 5 Tg.

The nitrogen from fertilizer production and legume and rice cultivation enters terrestrial ecosystems each year. The nitrogen from fossil fuel combustion first enters the atmosphere, mostly in oxidized forms, NO_x. Based on my interpretation of recent model calculations of atmospheric residence times of NO_x (e.g. Levy & Moxim, 1989; Penner et al., 1991), about 75% of the oxidized nitrogen or about 15 Tg N is deposited on the land and the remaining 25% or about 5 Tg N is deposited in the ocean. In summary, 140 of the 145 Tg N of the newly reactive nitrogen enters terrestrial ecosystems each year (Fig. 22.1).

Annually, terrestrial ecosystems lose a large fraction of this 140 Tg N to the atmosphere and the ocean. Galloway and colleagues (1994) estimate that about 50 Tg N of the 140 Tg N are transferred from the land to the atmosphere as N_2 in the process of denitrification. The estimate of N_2 flux to the atmosphere is probably the most speculative value in the global nitrogen budget presented in Fig. 22.1. Another 5 Tg N is transferred from the land to the atmosphere as N_2O in a variety of processes associated with human activities including biomass burning. Both N_2 and N_2O are generally unreactive and so remain in the atmosphere for long periods.

Human activities also cause the redistribution of nitrogen among the Earth's ecosystems. Each year about 55 Tg N enter the atmosphere from the land in forms that are reactive such as ammonia (NH_3) and nitric oxide (NO). Animal husbandry and biomass burning are the major sources of NH_3, while biomass burning is the major source of NO (Table 22.1).

Table 22.1 *Redistribution of reactive nitrogen as a result of human activities*

Nitrogen form (s)	Activity	Annual flux to the atmosphere (Tg N)
NH$_3$	Fertilization of fields	10
	Animal husbandry	32
	Biomass burning	5
NO	Biomass burning	8
NH$_3$ + NO		55

The NH$_3$ data is from Schlesinger and Hartley (1992) and the NO data is from Levy *et al.* (1991).

Ammonia has shorter residence time in the atmosphere than NO. Because of this, I estimate that about 85% of the NH$_3$ injected into the atmosphere as a consequence of human activities returns to the land, as compared with 75% for NO. Based on these assumptions, it appears that of the 55 Tg of the reactive N that enters the atmosphere from human activities including animal husbandry, biomass burning and soil fertilization, 45 Tg N returns to the land. The remaining 10 Tg N are transferred to the ocean.

Besides receiving a nitrogen input through the atmosphere, the ocean receives an input through rivers. About 20 Tg N move from the land to the ocean each year through rivers (Duce *et al.*, 1991).

Using a simple mass balance calculation, I estimate that there is an annual net accumulation of about 55 Tg N in the globe's terrestrial ecosystems. Galloway and colleagues (1994), who have done a similar analysis after which I patterned mine, have concluded that there is a net annual accumulation of 50 Tg N.

A critical question to ask is: Where is this 'new' nitrogen stored in terrestrial ecosystems? There are three places in terrestrial ecosystems where the 'new' nitrogen may be stored: groundwater, vegetation and soils. In the past decades, there has been a dramatic accumulation of nitrogen as nitrate in groundwater in many parts of the world. Recently Spalding and Exner (1993) reviewed a number of site-specific studies of groundwater pollution and concluded that many of the most severe problems are found in agricultural areas where nitrogen is added to croplands as industrially produced fertilizer or as animal wastes.

Vegetation and soils can also be sites of nitrogen accumulation. Recently, our group has conducted several studies of the fate of nitrogen inputs to forest ecosystems of the northeastern United States. We found that while both vegetation and soils do accumulate added nitrogen, the soil is the major nitrogen sink (Aber *et al.*, 1993; Nadelhoffer *et al.*, 1994).

How do these site-specific results translate to regional and global estimates of the fate of 'new' nitrogen added to terrestrial ecosystems? The answer is, we do not know for sure. We do not yet have a carefully validated georeferenced data set on nitrogen inputs to terrestrial ecosystems and we do not know, for various ecosystems, how much of the nitrogen will be retained in the ecosystems and where it will be stored. This kind of information is important to our understanding not only of the global N budget, but also of the global C budget. This is so because in many ecosystems of the world, the storage of carbon is thought to be limited by a lack of nitrogen.

Nitrogen-stimulated carbon storage in terrestrial ecosystems: our current understanding

Carbon storage in many temperate and boreal forests of the northern hemisphere is nitrogen limited, (e.g. Aber *et al.*, 1982; Miller, 1981; Stafford & Filip, 1974; Tamm, 1990). Many of these forests are receiving substantial inputs of nitrogen each year in precipitation. Some areas of western and central Europe have annual nitrogen inputs of more than 50 kg/ha each year. In 1989, I used the summaries of nitrogen deposition prepared by Barrie and Hales (1984), Buijsman and Erisman (1988), and Bonis *et al.* (1980), in conjunction with a global vegetation map, to produce estimates of the spatial patterns of wet plus dry nitrogen deposition in the temperate and boreal zones of North America and Europe (Melillo *et al.*, 1989). I assumed dry deposition to be equivalent in magnitude to wet deposition. Europe was broadly defined to include all of the temperate and boreal areas in western and eastern Europe, and all of the temperate and boreal areas in the former Soviet Union; that is, the vast forested landscape east of the Ural Mountains as well as 'European Russia' west of the Urals.

My estimate of nitrogen deposition in precipitation for these temperate and boreal zones of the Northern Hemisphere was about 18 Tg N per year for the period between 1975 and 1980, with North America receiving about 6 Tg N annually and Europe receiving about 12 Tg N annually (Melillo *et al.*, 1989). I further divided the deposition between temperate and boreal zones on the two 'continents'. In North America, the temperate forests received about 5 Tg N annually and the boreal forests about 2 Tg N.

For this study, I have used my 1989 estimates of nitrogen deposition to the temperate and boreal forest regions of the Northern Hemisphere with the methodology described by Melillo and Gosz (1983) and Peterson and Melillo

(1985), to estimate the magnitude of nitrogen-stimulated storage of carbon in these regions. I have also used a model of the global carbon and nitrogen cycles, the Terrestrial Ecosystem Model (TEM, Melillo *et al.*, 1993), to estimate whole-ecosystem C : N ratios for the temperate and boreal forests. The whole-ecosystem C : N ratios were weighted by area to account for the differences in C : N ratios among the deciduous, coniferous and mixed forests of the temperate zone, and between the boreal forests and boreal woodlands. In this study, I have assumed that nitrogen entering an ecosystem is stored in the vegetation and soil in the same proportions as it is found in mature forests. This assumption is consistent with our recent research on the fate of nitrogen added to forests (Aber *et al.*, 1993; Nadelhoffer *et al.*, 1994).

The maximum annual nitrogen-stimulated carbon storage I estimate for the temperate and boreal zones of the Northern Hemisphere is about 0.9 Pg C (Table 22.2). In making this estimate, I have assumed that all of the nitrogen that enters these ecosystems is retained within them; that is, none of the nitrogen entering the temperate and boreal forests in precipitation is lost through processes such as denitrification or leaching: a nitrogen retention efficiency of 100%. I have previously argued that this is a questionable assumption, and that it is more realistic to use a 60% retention efficiency for these calculations (Melillo & Gosz, 1983, Peterson & Melillo, 1985). Doing this, I estimate a nitrogen-stimulated carbon storage of 0.6 Pg annually in the forests of the temperate and boreal zones of the Northern Hemisphere (Table 22.2).

Recent estimates of nitrogen-stimulated carbon storage range from 0.2 to 6 Pg C annually (Table 22.3). The analyses producing this range differ in many ways including: (1) portion of the terrestrial biosphere considered; (2) magnitude and pattern of the nitrogen input; (3) nitrogen retention efficiency; (4) allocation of the nitrogen among ecosystem storage pools (e.g. vegetation, surface litter, soil); and (5) the C : N ratio of the storage pools. The more complete of the analyses (Kohlmaier *et al.*, 1988; Hudson *et al.*, 1994; Townsend, 1994) give estimates of between 0.2 and 1.7 Pg C for annual nitrogen-stimulated carbon storage in terrestrial ecosystems in recent years. My 'best estimate' from this study is towards the lower end of this range.

How important is nitrogen-stimulated carbon storage to the global carbon budget? Recent analyses which balance CO_2 accumulation in the atmosphere against fossil fuel release, the effects of changing land use, and ocean uptake, calculate a substantial residual term that is often called the 'missing sink'. The IPCC has recently argued that the missing sink is mostly on land and estimated that its magnitude, expressed as an average annual value for the 1980s, is 1.4 ± 1.5 Pg C (Table 22.4; Schimel *et al.*, 1994). In this context, my estimate of an annual storage of between 0.6 and 0.9 Pg C in

Table 22.2 *Estimates of nitrogen-stimulated carbon storage in the forests and wood-lands of North America and Europe (this study)*

Biome type	C : N ratio	Nitrogen input (Tg N)	Maximum[a] estimate of C storage (Pg C)	Likely[b] C storage (Pg C)
Boreal forests and woodlands	45	3	0.13	0.08
Temperate forests	54	15	0.81	0.49
Total		18	0.94	0.57

[a]Assumes 100% retention efficiency.
[b]Assumes 60% retention efficiency.

Table 22.3 *Recent estimates of nitrogen-stimulated carbon storage in terrestrial ecosystems: a literature survey*

Reference	Region(s) considered	N input (Tg N^{-1} year^{-1})	C store (Pg C^{-1} year^{-1})
Kohlmaier *et al.* (1988)	30–60° N	21 + 7	up to 0.7
Schindler & Bayley (1993)	Globe	13	inf 0.7 −2.0
Hudson *et al.* (1994)	Northern hemisphere	30	inf 1.0
Galloway *et al.* (1994)	Globe	50	2 − 6[a]
Townsend *et al.* (1994)	Global (excluding culti-vated areas)	7[b]	0.2 − 1.2[c]
Melillo (this study)	Temperate and boreal forests of North America and Europe	18	0.6 − 0.9[d]

[a]maximum range; [b]NO$_y$ only; [c]best estimate 0.3–0.6; [d]best estimate 0.6.

terrestrial ecosystems that is stimulated by nitrogen input is important in the global carbon budget.

Direct confirmation of nitrogen-stimulated carbon storage is difficult for two reasons. First, we are looking for a small change in a large pool; an annual change in the range of one part per thousand. Second, and equally important, several mechanisms are probably operating simultaneously and interactively to account for terrestrial carbon storage in the Northern Hemi-sphere ecosystem including: forest regrowth following harvest (e.g. Dixon

Table 22.4 *Average annual budget of CO_2 perturbations for 1980–1989*

	Pg C^{-1} year^{-1}
CO_2 sources	
Emissions from fossil fuel combustion and cement production	5.5 ± 0.5
Net emissions from changes in tropical land use	1.6 ± 1.0
Total anthropogenic emissions	7.1 ± 1.1
CO_2 sinks	
Storage in the atmosphere	3.2 ± 0.2
Oceanic uptake	2.0 ± 0.8
Uptake by Northern Hemisphere forest regrowth	0.5 ± 0.5
Additional terrestrial sinks (CO_2 fertilization, nitrogen fertilization, climatic effects)	1.4 ± 1.5

Fluxes and reservoir changes of carbon expressed in Pg C^{-1} year^{-1}, error limits correspond to an estimated 90% confidence interval. (Adapted from Schimel *et al.*, 1994).

et al., 1994), fertilization by carbon dioxide (Gifford, 1994), climate variability (Dai & Fung, 1993) and, of course, nitrogen stimulation. Kauppi and his colleagues (1992) have suggested that nitrogen-stimulated carbon storage may be the cause of the increases in European forest resources observed between 1950 and 1980. Houghton (1993) challenged this interpretation and argued that it is difficult to separate nitrogen stimulation from forest regrowth. Hudson and his colleagues (1994) have reviewed both positions and have concluded that the nitrogen stimulation argument cannot be dismissed for the reasons set forth by Houghton (1993). Clearly, additional research is needed to document the magnitude of the nitrogen-stimulated carbon storage.

If, as I have argued, nitrogen-stimulated carbon storage does account for about half of the 'missing sink', then it is important that projections of future atmospheric CO_2 levels take this into account. Environmental protection policies and the evolution of the world economy in the twenty-first century may reduce nitrogen-stimulated carbon storage and so lead to an increased rate of CO_2 accumulation in the atmosphere if fossil fuel use remains at current levels or increases.

Scientists in both Europe and North America have warned that chronic nitrogen inputs can adversely affect terrestrial, freshwater and coastal marine ecosystems (e.g. Aber *et al.*, 1989; Fisher *et al.*, 1988; Galloway *et al.*, 1987; Johnson *et al.*, 1991; Melillo *et al.*, 1989; Schindler & Bayley, 1993; Schulze, 1989; Tamm, 1990). The effects include forest dieback, lake and stream acidification and eutrophication of coastal estuaries and bays. In

Europe and North America, there may be government-mandated regulation of fossil fuel emissions to reduce nitrogen inputs to ecosystems. If the regulatory efforts are successful, nitrogen-stimulated carbon storage in these regions will be reduced.

Fossil fuel emissions in the developing countries, including many in the tropics, are likely to increase as their economies develop in the twenty-first century (IPCC, 1992). In many of the ecosystems in the tropics, including forests, carbon storage appears to be phosphorus-limited rather than nitrogen-limited (Sanchez et al., 1982). Therefore nitrogen-stimulated carbon storage per unit of fossil fuel combustion in the developing countries of the tropics will be less than it is today in the mid latitude region of the Northern Hemisphere.

Climate change, the global nitrogen budget and carbon storage in terrestrial ecosystems

The atmospheric concentrations of the major long-lived greenhouse gases continue to increase because of human activity (Watson et al., 1992). Current models of climate change predict that as part of the greenhouse effect, the global mean temperature will increase between 1.5 °C and 4.5 °C during the next century (IPCC, 1992). One result of global warming may be changes in the rates of temperature-dependent soil processes such as decomposition (Swift et al., 1979) and net nitrogen mineralization and nitrification (Focht & Verstraete, 1977). In recent years, my research group has conducted a series of field, laboratory and simulation modelling studies to explore, in a quantitative way, how predicted temperature increases will affect the nitrogen cycle, interactions between the nitrogen and carbon cycles, and carbon storage in terrestrial ecosystems. In this section I highlight some of the important results from these studies.

Field study

For the past 3 years we have been conducting a soil warming experiment in a deciduous forest stand at the Harvard Forest in central Massachusetts. At the research site we established 18 6 × 6 m plots in April, 1991. The plots are grouped into six blocks and the three plots within each block were randomly assigned to one of three treatments. The treatments are: (1) heated plots in which the average soil temperature is elevated 5 °C above ambient using buried heating cables; (2) disturbance-control plots that are identical to heated plots except they receive no electrical power; and (3) undisturbed control plots that have been left in their natural state. We are measuring a variety of parameters at the plots including CO_2 emissions from the soil and indexes of N availability. Details of the experimental design and the field and laboratory methods have recently been reported by Peterjohn *et al.* (1994).

During the first year of study (July 1991 to June 1992) we observed that heating increased the emission of CO_2 and the rate of nitrogen mineralization. We estimated CO_2 fluxes to be approximately 7100 kg C ha^{-1} year^{-1}, 7900 kg C ha^{-1} year^{-1} and 11 100 kg C ha^{-1} year^{-1} in the control, disturbance control and heated plots, respectively; an increase due to warming of approximately 3200 kg C ha^{-1} year^{-1}. Nitrogen mineralization was doubled in the surface soil which we defined as the forest floor plus the top 10 cm of the mineral soil. During the first growing season, we observed that in the forest floor, heating increased the average net mineralization rates from 1.02 to 2.47 mg N kg^{-1} day^{-1}. In the mineral soil, heating increased the average net mineralization rates from 0.09 to 0.19 mg N kg^{-1} day^{-1}. No nitrification was observed in either the forest floor or the mineral soil in any of the treatments.

In the second year of study (July 1992 to June 1993) we found that warming had a much less dramatic effect on CO_2 flux, but a sustained large effect on nitrogen mineralization. We estimated CO_2 fluxes in the second year to be approximately 6800 kg C ha^{-1}, 7600 kg C ha^{-1} and 8700 kg C ha^{-1} for control and disturbed control and heated treatments. Nitrogen mineralization rates remained doubled. Again, no nitrification was observed in either the forest floor or the mineral soil.

These results can be interpreted in several ways. One interpretation of the differences in the CO_2 response between the first and second years versus the constant response in nitrogen mineralization is as follows. There are two major soil organic matter pools; a fast turnover pool and a slow turnover pool. The fast pool contains high C : N ratio material (e.g. recent litter) and the decay of fast-pool material results in large CO_2 losses per unit of material processed and small net nitrogen release in the mineralization process. In contrast to the fast pool, the slow pool contains low C : N ratio material (e.g. meta-stable humus) and the decay of material in the slow pool results in a smaller loss of CO_2 per unit of material processed and a large net nitrogen release in the mineralization process. If this is an appropriate interpretation, it suggests that the slow pool is key to ecosystem response to climate change over the long term. As elevated soil temperature increases the rate of decay in the slow pool, both carbon (as CO_2) and nitrogen (as inorganic N) are released. The nitrogen then becomes available to be taken up by and stored in plants. Since the C : N ratio of plant material is substantially larger than the C : N ratio of soil organic matter in the slow pool, warming may lead to increased carbon storage in the ecosystem. The magnitude of such an increase would, of course, depend on how plant carbon balance is affected by other components of climate change including the availability of water and the effect of increased temperature on both photosynthesis and respiration.

A coupled plant–soil model

To evaluate how simultaneous changes in plant and soil processes affect ecosystem carbon balance and other terrestrial ecosystem responses to global change, my research group has developed a coupled plant–soil model that considers both carbon and nitrogen cycling (Raich et al., 1991; McGuire et al., 1992; Melillo et al., 1993). We call the model TEM, an acronym for the Terrestrial Ecosystem Model. It uses information on soils, vegetation, and climate variables such as temperature, precipitation, cloudiness and solar radiation input to make monthly estimates of important transfers of carbon and nitrogen in terrestrial ecosystems. The data sets used in TEM are currently organized into 0.5° latitude by 0.5° longitude units or grid cells that cover the vegetated surface of the Earth.

We applied the model to more than 56 000 grid cells in order to develop estimates of terrestrial plant and soil carbon stocks for current climate conditions and atmospheric CO_2 levels. Our model-based estimate of the total terrestrial carbon stocks is 2279 Pg, of which 977 Pg C is in the vegetation and 1301 Pg C is in soil (Table 22.5).

We did not include land-use change in this version of the model, and so the estimates are for potential plant and soil carbon stocks, not for present-day stocks. The conversion of natural ecosystems that have large woody plant components into agricultural systems has certainly reduced the Earth's plant carbon pool (e.g. Houghton et al., 1983; Melillo et al., 1988). This pool has also been reduced by firewood gathering, especially in poor tropical countries. In addition, the present-day soil carbon stock may be less than the potential stock because of the initial effects of converting natural ecosystems to agricultural systems that are tilled.

When both land area and carbon density are considered, tropical ecosystems account for about 47% of the total terrestrial carbon pool: 56% of the vegetation carbon and 41% of the soil carbon. With most of the world's population growth expected to occur in the tropics over the next century, these large carbon stocks are likely to be reduced through land-use change activities, and perhaps other aspects of global change.

I used the output from three climate models to run TEM (Melillo et al., 1993) to simulate changes in terrestrial carbon stocks under various climatic

Table 22.5. *Response of carbon storage (Pg C) by region for experiments involving two levels of atmospheric CO_2 and four levels of climate*

Climates	Contemporary[a]	CO_2 concentration (312.5 ppmv)			CO_2 concentration (625.0 ppmv)			
		GFDL	GISS	OSU	Contemporary	GFDL	GISS	OSU
Vegetation carbon								
Northern ecosystems[b]	140	+43	+32	+31	+4	+71	+42	+40
Temperate ecosystems[c]	290	−1	+24	+13	+33	+67	+80	+63
Tropical ecosystems[d]	547	−37	−50	−74	+108	+116	+110	+68
Total vegetation carbon	977	+5	+6	−30	+145	+254	+232	+171
Soil carbon								
Northern ecosystems	327	−42	−17	−8	+11	−7	+2	+8
Temperate ecosystems	441	−66	−47	−34	+73	+28	+36	+48
Tropical ecosystems	533	−103	−140	−138	+89	+8	−31	−37
Total soil carbon	1301	−211	−204	−180	+173	+29	+7	+19
Total carbon								
Northern ecosystems	466	+1	+15	+22	+15	+64	+44	+48
Temperate ecosystems	733	−67	−23	−20	+105	+95	+118	+112
Tropical ecosystems	1080	−140	−190	−212	+198	+124	+78	+30
Total terrestrial carbon	2279	−206	−198	−210	+318	+283	+240	+190

GFDL, Geophysical Fluid Dynamics Laboratory; GISS, Goddard Institute of Space Studies; OSU, Oregon State University.

[a] Pool size that is used as the baseline for responses.

[b] Northern ecosystems include polar desert, alpine tundra, wet/moist tundra, boreal woodland, and boreal forest. These ecosystems occupy a combined area of 28.2 million km^2.

[c] Temperate ecosystems include forests, savannas, grasslands, shrublands and deserts. These ecosystems occupy a combined area of 56.7 million km^2.

[d] Tropical ecosystems include forests, savannas and shrublands. These ecosystems occupy a combined area of 42.5 million km^2.

conditions. The three climate models, more correctly referred to as general circulation models (GCMs), were: the Goddard Institute of Space Studies (GISS) GCM; the Oregon State University (OSU) GCM; and the Geophysical Fluid Dynamics Laboratory (GFDL) GCM. These GCMs predict that with a CO_2 doubling: (1) mean global temperature will increase between 2.8 °C and 4.2 °C; (2) global precipitation will increase between 7.8% and 11.0%; and (3) cloudiness will decrease between 0.4% and 3.4%.

Changes in climate with no change in CO_2 concentration are predicted by TEM to result in substantial losses of carbon from terrestrial ecosystems (Table 22.5). For all three GCM climates the predicted carbon storage reductions are about 200 Pg and the losses come mostly from soils. This represents a 15% loss from the soil carbon pool.

Tropical ecosystems account for much of the global response. With the GFDL climates they account for about 75% of the carbon loss, and with the GISS and OSU climates they account for virtually all of the loss. In contrast to tropical systems, TEM predicted that northern ecosystems respond to climate change alone by increasing their carbon storage, albeit by relatively small amounts (Table 22.5). The increase in ecosystem-wide carbon storage occurred because the increase in vegetation carbon was larger than the decrease in soil carbon. The sequence of process-level changes leading to this increased carbon storage in the vegetation at the expense of the soil is thought to be as follows. As temperature increases, the soil loses carbon, and nitrogen mineralization is accelerated (see the earlier discussion of soil warming). The increased availability of inorganic nitrogen results in a plant fertilization response and an increase in plant productivity and plant carbon storage. This sequence represents an important carbon–nitrogen interaction at the ecosystem level that has global-scale consequences.

For doubled CO_2 with no climate change, TEM predicts an increase in global terrestrial carbon storage of 318 Pg, which is a 14% increase in terrestrial carbon storage relative to the estimated contemporary condition (Table 22:5). In absolute terms, tropical ecosystems are the most responsive to a CO_2 doubling alone. The model predicts that they will account for almost two-thirds of the increased carbon storage with doubled CO_2, but no climate change. The carbon storage in tropical ecosystems under this scenario is predicted to be almost equally divided between vegetation and soils. Northern ecosystems, including tundra, boreal woodlands and boreal forests, are predicted to be the least responsive to doubled CO_2 without climate change. The small response occurs because TEM predicts that productivity in nitrogen-limited ecosystems does not substantially respond to elevated CO_2 alone.

In response to changes in both CO_2 and climate, TEM predicts substantial increases in terrestrial carbon stocks, although the magnitude of the increases vary according to GCM climate (Table 22.5). The OSU climate combined with CO_2 doubling resulted in the smallest increase in terrestrial carbon storage globally, 190 Pg (an 8% increase), and the GFDL climate combined with a CO_2 doubling yielded the largest increase, about 280 Pg C (a 12% increase). Changes in vegetation carbon stocks accounted for about 90% of the increased storage. The vegetation of northern and temperate ecosystems is very responsive to the combination of CO_2 doubling and climate change because the nitrogen limitation is lessened by warming, through an increase in soil nitrogen mineralization (McGuire *et al.*, 1992).

Blueprint for future research

New research is needed to improve our understanding of how human activities affect terrestrial carbon storage through direct and indirect alterations of the global nitrogen budget. This new research should include both field research and modelling studies.

Research on direct effects

To better estimate the magnitude of nitrogen-stimulated carbon storage through the addition of 'new' nitrogen to terrestrial ecosystems, we need to improve our knowledge in five areas: (1) nitrogen deposition; (2) ecosystem nitrogen retention efficiency; (3) distribution of the retained nitrogen among ecosystem components (e.g. vegetation, litter, soils); (4) C : N ratios in ecosystems at various stages of maturity; and (5) land-use history.

To begin with, we need georeferenced estimates of nitrogen deposition for the world's terrestrial ecosystems. Models that estimate NO_y and NH_x deposition exist (NO_y, Levy & Moxim, 1989; Penner *et al.*, 1991; NH_y, Crutzen, personal communication). They use information on georeferenced sources, atmospheric residence times and atmospheric transport models to estimate deposition fields. With NH_x it will be important to separate natural from anthropogenic sources and keep track of them for the deposition calculations. Estimates of past and future deposition rates, as well as current deposition rates, will eventually be needed. Deposition estimates from past and current measurements should be used to validate the deposition models. Future deposition rates will have to be estimated from projected energy and fertilizer use.

Previously, I used the input−output budgets from small watershed studies to estimate nitrogen retention efficiencies for forest ecosystems (Melillo & Gosz, 1983). This kind of analysis will have to be done in a more rigorous way so as to include: (1) finer stratification by ecosystem type; (2) differences resulting from nitrate versus ammonium input because the former is more mobile in soils and therefore more likely to be lost from the system; and (3) possible differences associated with land-use history since 'ecological memory' may influence retention capacity.

Once we have determined the nitrogen retention efficiencies of various ecosystems, we will need to know how the 'retained' nitrogen is distributed among the ecosystems' components (i.e. vegetation, litter, soils). The distribution will be affected by a variety of factors including nitrogen form (i.e. nitrate versus ammonium), maturity of the ecosystem, and land-use history. The chronic nitrogen input study that we are conducting at the Harvard Forest (Aber et al., 1993) and the ^{15}N addition studies that we are carrying out at the Harvard Forest and in Maine (Nadelhoffer et al., 1994) are providing us with this information. In the ^{15}N studies we introduce a small amount of nitrogen that is highly enriched in ^{15}N to a forest stand and then follow its movement through time among the ecosystem's nitrogen pools (e.g. soil, litter and plant).

A number of literature reviews of C : N ratios in plants, litter and soils are available (e.g. Schindler & Bayley, 1993). These can be compared to model-derived estimates like the ones we used from TEM for this study (Fig. 22.2). A comprehensive data base on C : N ratios should be developed. It must consider how the C : N ratio changes in plants, litter, and soil with stages of ecosystem maturity and land-use history. Also, it is critically important to explore the plasticity of C : N ratios in plant tissues of various species. If the C : N ratio can be widened only a small amount, this would result in substantial carbon storage.

As noted above, a good knowledge of land-use history is essential for our understanding of nitrogen-stimulated carbon storage because ecosystems have 'memory', that is, past events affect current and future behaviour. Land-use history reconstructions at regional and global scales have been developed, albeit at relatively coarse scales (Melillo et al., 1988; Houghton et al., 1991). These early efforts used tabular records of land cover and land use. More recently, time series of satellite-acquired data on land-cover change have been used to calculate major shifts such as forests to pastures in tropical ecosystems (Skole & Tucker, 1993). This work is essential for many scientific purposes and should be encouraged. In addition, work to refine 'pre-satellite' land-use change estimates from tabular records should continue. This will involve an interdisciplinary effort that includes geographers and historians as well as ecologists.

All of this data, georeferenced when appropriate, should be drawn together, organized, and made available to the research community for use

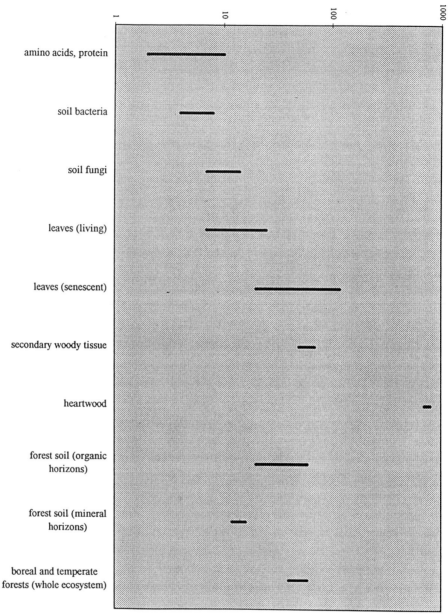

Figure 22.2   *The C : N ratio (weight basis).*

in model studies to estimate nitrogen-stimulated carbon storage. At least one study has been made using 'high-resolution' geographically referenced data on NO_y deposition with a dynamic ecosystem model for this purpose (Townsend *et al.*, 1994). While Hudson and his colleagues (1994) did not use 'high resolution' georeferenced data on nitrogen input, they did include considerations of retention efficiency, variable distribution of 'new nitrogen' among ecosystem compartments depending upon ecosystem maturity, and C : N ratios that varied among ecosystems. With this much progress in synthesis efforts already, I expect rapid advances in our understanding of nitrogen-stimulated carbon storage once the data sets I have called for become available to the research community.

Research on indirect effects

The research on this issue of indirect effects requires both biogeochemistry models such as TEM (Melillo *et al.*, 1993) and vegetation redistribution models such as BIOME (Prentice *et al.*, 1992) that operate at the global scale. Ideally, these models would be transient models, capable of running with time series data on climate change and changes in nitrogen deposition rate. Several models of each type already exist with transient versions, although they have not been coupled and run specifically to address this 'indirect effects' issue. This work should be encouraged.

Laboratory and process-level field studies on the 'indirect effects' issue should also be encouraged. These studies will be useful for developing modelling concepts and for providing parameters for models. Whole-ecosystem manipulations will also be useful for model parameterization and model validation. Here I am thinking about manipulations like the soil warming experiment I described earlier or like the free air circulation experiment (FACE) currently being conducted at the Duke Forest in North Carolina (W. H. Schlesinger, personal communication).

Prospects for action

The blueprint for future research proposed above is totally consistent with the International Geosphere-Biosphere Programme (IGBP). Several of the core projects and task groups of the IGBP are undertaking the data collection, laboratory and field experiments, and modelling activities required to further our understanding of how human activities affect terrestrial carbon storage through alterations of the global nitrogen budget. This work is needed by us to develop a scheme for managing the land systems of the Earth in a sustainable way.

Acknowledgements

This research is being supported by funds from the United States Environmental Protection Agency (Cooperative agreement CR817734-010, Athens Environmental Research Laboratory), the Department of Energy's National Institute for Global Change (Northeast Regional Center, subagreement

901214-HAR), and the National Aeronautics and Space Administration (contract NAGW-1825, subcontract 91-14). This chapter, somewhat modified, was presented at the Eighth Toyota Conference in November 1994.

References

Aber, J. D., Melillo, J. M. & Federer, C. A. (1982). Predicting the effects of rotation length, harvest intensity, and fertilization on fiber yield from northern hardwood forests in New England. *Forest Science*, **28**, 31–48.

Aber, J. D., Nadelhoffer, K. J., Steudler, P. A. & Melillo, J. M. (1989). Nitrogen saturation in northern forest ecosystems – hypotheses and implications. *BioScience*, **39**, 378–86.

Barrie, L. A. & Hales, J. M. (1984). The spatial distributions of precipitation acidity and major ion wet deposition in North America during 1980. *Tellus*, **36B**, 333–5.

Bónis, K., Mészáros, E. & Putsay, M. (1980). On the atmospheric budget of nitrogen compounds over Europe. *Idójárás* (periodical of the Hungarian Meteorological Service), **84**, 57–68.

Buijsman, E. & Erisman, J. W. (1988). Wet deposition of ammonium in Europe. *Journal of Atmospheric Chemistry*, **6**, 265–80.

Burns, R. C. & Hardy, R. W. F. (1975). *Nitrogen Fixation in Bacteria and Higher Plants*, pp. 54–8. New York: Springer-Verlag.

Dai, A. & Fung, I. Y. (1993). Can climate variability explain the 'missing' CO_2 sink? *Global Biogeochemical Cycles*, **7**, 599–610.

Deevey, E. S., Jr. (1970). Mineral cycles. *Scientific American*, **223**, 148–58.

Duce, R. A., Liss, P. S., Merrill, J. T., Atlas, E. L., Buat-Menard, P., Hicks, B. B., Miller, J. M., Prospero, J. M., Arimoto, R., Church, T. M., Ellis, W., Galloway, J. N., Hansen, L., Jickells, T. D., Knap, A. H., Reinhardt, K. H., Schneider, B., Soudine, A., Tokos, J. J., Tsunogai, S., Wollast, R. & Zhou, M. (1991). The atmospheric input of trace species to the world ocean. *Global Biogeochemical Cycles*, **5**, 193–259.

Fisher, D., Ceraso, J., Mathew, T. & Oppenheimer, M. (1988). *Polluted Coastal Waters: the Role of Acid Rain*. New York: Environmental Defense Fund.

Focht, D. D. & Verstraete, W. (1977). Biochemical ecology of nitrification and denitrification. In *Advances in Microbial Ecology*, ed. M. Alexander, pp. 135–214. New York: Plenum Press.

Food and Agriculture Organization (FAO) (1989). *FAO Fertilizer Yearbook*, vol. 43, Rome: FAO.

Galloway, J. N., Hendrey, G. R., Schofield, D. L., Peters, N. E. & Johannes, A. H. (1987). Processes and causes of lake acidification during spring snowmelt in the west-central Adirondack Mountains, New York. *Canadian Journal of Fisheries & Aquatic Sciences*, **44**, 1595–602.

Galloway, J. N., Schlesinger, W. H., Levy II, H., Michaels, A. & Schnoor, J. L. (1994). Nitrogen fixation: anthropogenic enhancement-environmental response. *Global Biogeochemical Cycles* (submitted).

Gifford, R. M. (1994). The global carbon cycle: a viewpoint on the missing sink. *Australian Journal of Plant Physiology*, **21**, 1–15.

Houghton, R. A., Hobbie, J. E., Melillo, J. M., Moore, B., Peterson, B. J., Shaver, G. R. & Woodwell, G. M. (1983). Changes in the carbon content of terrestrial biota and soils between 1860 and 1980: a net release of CO_2 to the atmosphere. *Ecological Monographs*, **53**, 235–62.

Houghton, R. A. (1991). Tropical deforestation and atmospheric carbon dioxide. *Climate Change*, **19**, 99–118.

Houghton, R. A. (1993). Is carbon accumulating in the northern temperate zone? *Global Biogeochemical Cycles*, **7**, 611–18.

Hudson, R. J. M., Gherini, S. A. & Goldstein, R. A. (1994). Modeling the global carbon cycle: nitrogen fertilization of the terrestrial biosphere and the 'missing' CO_2 sink. *Global Biogeochemical Cycles*, **8**, 307–33.

IPCC (1992). *Climate Change 1992: The Supplementary Report to the IPCC Scientific Assessment*. Cambridge: Cambridge University Press.

Johnson, D. W., Cresser, M. S., Nilsson, S. I., Turner, J., Ulrich, B., Binkley, D. & Cole, D. W. (1991). Soil changes in forest ecosystems: evidence for and probable causes. *Proceedings of the Royal Society (Edinburgh)*, **97B**, 81–116.

Kauppi, P. E., Mieliainen, K. & Kuusela, K. (1992). Biomass and carbon budget of European forests. *Science*, **256**, 70–4.

Kohlmaier, G. H., Janecek, A. & Plöchl, M. (1988). Modelling response of vegetation to both excess CO_2 and airborne nitrogen compounds within a global carbon cycle model. In *Advances in Environmental Modelling*, ed. A. Marani, pp. 207–34. New York: Elsevier.

Levy II, H. & Moxim, W. J. (1989). Simulated global distribution and deposition of reactive nitrogen emitted by fossil fuel combustion. *Tellus*, **41B**, 256–71.

Levy II, H., Kasibhatla, P. S., Moxim, W. J. & Logan, J. A. (1991). The global impact of biomass burning on tropospheric reactive nitrogen. In *Proceedings of the Chapman Conference on Global Biomass Burning*, ed. J. Levine. Cambridge, MA: MIT Press.

McGuire, A. D., Melillo, J. M., Joyce, L. A., Kicklighter, D. W., Grace, A. L., Moore III, B. & Vorosmarty, C. J. (1992). Interactions between carbon and nitrogen dynamics in estimating net primary productivity for potential vegetation in North America. *Global Biogeochemical Cycles*, **6**, 101–24.

Melillo, J. M. & Gosz, J. R. (1983). Interactions of biogeochemical cycles in forest ecosystems. In *The Major Biogeochemical Cycles and their Interactions*, ed. B. Bolin & R. B. Cook, pp. 177–222. New York: Wiley.

Melillo, J. M., Fruci, J. R., Houghton, R. A., Moore III, B. & Skole, D. L. (1988). Land-use change in the Soviet Union between 1850 and 1980: causes of a net release of CO_2 to the atmosphere. *Tellus*, **50B**, 116–28.

Mellilo, J. M., Steudler, P. A., Aber, J. D. & Bowden, R. D. (1989). Atmospheric deposition and nutrient cycling. In *Exchange of Trace Gases between Terrestrial Ecosystems and the Atmosphere*, ed. M. O. Andreae & D. S. Schimel, pp. 263–80. New York: Wiley.

Melillo, J. M., McGuire, A. D., Kicklighter, D. W., Moore III, B., Vorosmarty, C. J. & Schloss, A. L. (1993). Global climate change and terrestrial net primary production. *Nature*, **363**, 234–40.

Miller, H. G. (1981). Forest fertilization: some guiding concepts. *Forestry*, **54**, 157–67.

Nadelhoffer, K. J., Downs, M. R., Fry, B., Aber, J. D., Magill, A. H. & Melillo, J. M. (1994). The fate of ^{15}N labeled nitrate additions to a northern hardwood forest in eastern Maine, USA. *Oecologia* (submitted).

Penner, J. E., Atherton, C. S., Dignon, J., Ghan, S. J., Walton, J. J. & Hameed, S. (1991). Tropospheric nitrogen: a three-dimensional study of sources, distributions, and deposition. *Journal of Geophysical Research*, **96**, 959–90.

Peterjohn, W. T., Melillo, J. M., Steudler, P. A., Newkirk, K. M., Bowles, F. P. & Aber, J. D. (1994). Responses of trace gas fluxes and N availability to experimentally elevated soil temperatures. *Ecological Applications*, **4**, 617–25.

Peterson, B. J. & Melillo, J. M. (1985). The potential storage of carbon caused by eutrophication of the biosphere. *Tellus*, **37B**, 117–27.

Prentice, I. C., Cramer, W., Harrison, S. P., Leemans, R., Monserud, R. A. & Solomon, A. M. (1992). Predicting global vegetation patterns from plant physiology and dominance, soil properties and climate. *Journal of Biogeography*, **19**, 117–34.

Raich, J. W., Rastetter, E. B., Melillo, J. M., Kicklighter, D. W., Steudler, P. A., Peterson, B. J., Grace, A. L., Moore III, B. & Vorosmarty, C. J. (1991). Potential net primary productivity in South America: application of a global model. *Ecological Applications*, **1**, 399–429.

Sanchez, P. A., Bandy, D. E., Villachica, H. & Nicholaides, J. J. (1982). Amazon basin soils: management for continuous crop production. *Science*, **216**, 821–7.

Schimel, D., Enting, I., Heimann, M., Wigley, T., Raynaud, D., Alves, D. & Siegenthaler, U. (1994). *The Carbon Cycle*. IPCC interim assessment.

Schindler, D. W. & Bayley, S. E. (1993). The biosphere as an increasing sink for atmospheric carbon: estimates from increased nitrogen deposition. *Global Biogeochemical Cycles*, **7**, 717–34.

Schulze, E.-D. (1989). Air pollution and forest decline in a spruce (*Picea abies*) forest. *Science*, **244**, 776–83.

Skole, D. L. & Tucker, C. (1993). Tropical deforestation and habitat fragmentation in the Amazon: satellite data from 1978–1988. *Science*, **260**, 1905–10.

Spalding, R. F. & Exner, M. E. (1993). Occurrence of nitrate in groundwater – a review. *Journal of Environmental Quality*, **22**, 392–402.

Stafford, L. O. & Filip, S. M. (1974). Biomass and nutrient content of 4-year-old fertilized and unfertilized hardwood stands. *Canadian Journal of Forest Research*, **4**, 549–54.

Swift, M. J., Heal, O. W. & Anderson, J. M. (1979). *Decomposition in Terrestrial Ecosystems*. Berkeley, CA: University of California Press.

Tamm, C. O. (1990). *Nitrogen in Terrestrial Ecosystems*. New York: Springer-Verlag.

Townsend, A. R., Braswell, B. H., Holland, E. A. & Penner, J. E. (1994). Nitrogen deposition and terrestrial carbon storage: linking atmospheric chemistry and the global carbon budget. *Geophysical Research Letters* (submitted).

Watson, R. T., Meira Filho, L. G., Sanhueza, E. & Janetos, A. (1992). Sources and sinks. In *Climate Change 1992: The Supplementary Report to the IPCC Scientific Assessment*, ed. J. T. Houghton, B. A. Callander & S. K. Varney, pp. 25–46. Cambridge: Cambridge University Press.

Global dynamic vegetation modelling: coupling biogeochemistry and biogeography models

R. P. Neilson and S. W. Running

Introduction

Issues of global environmental change, involving both climate and land-use change, require the rapid development of process-based, predictive tools of the global land surface and its dynamics. Biophysical feedbacks, for example, between the biosphere and the atmosphere could act to either enhance or diminish the greenhouse effect (Martin, 1993). In addition, the impacts of global change on the biosphere could have profound effects on global resources of wood, food, fibre, biological diversity and water (Bolin, 1991). Quantification of these feedbacks and impacts is important for helping to shape the policy responses to global warming and land use. One step in the development of such predictive capability is the creation of a dynamic global vegetation model (DGVM) that can simulate vegetation redistribution as fully coupled with the biogeochemistry of carbon, water and energy. Currently, no such model exists, but the fundamental components have been developed among different classes of ecological models.

Two general classes of ecosystem models have emerged for studying the large-scale impacts of climatic and land-use change on ecosystems of the world. Ecosystem Function (biogeochemistry) models (EF) were designed for vegetation productivity estimates and for simulating carbon, nitrogen and water cycles. However, EF models are not able to simulate the location of different ecosystem types, nor their recruitment, disturbance and mortality dynamics, and require vegetation type as an input. By contrast, ecosystem dynamics (biogeography) models (ED) are able to predict the distribution of different vegetation types under any specified climate, but were not originally designed to grow vegetation, nor to calculate carbon and nutrient fluxes. Ideally, the two modelling strategies should be combined into one and, in fact, the distinction between the two classes is already becoming arbitrary. This discussion does not include ecosystem models originally designed for small spatial extents or single types of systems, such as stand succession or 'gap' models, although some coupled EF-ED models at the stand level are now appearing (Friend *et al.*, 1993).

A first step in defining an EF-ED model linkage is being undertaken by a group of scientists representing three EF models, CENTURY (Parton, 1992), TEM (Melillo *et al.*, 1993), Biome-BGC (Running & Hunt, 1993) and three ED models, BIOME (Prentice *et al.*, 1992), DOLY (Woodward, 1987; Woodward & McKee, 1991; Woodward *et al.*, 1995) and MAPSS (Neilson, 1993; Neilson, 1995), under VEMAP, a multi-US agency (USFS, NASA, NSF) and privately funded (Electric Power Research Institute) project. Under VEMAP (Vegetation/Ecosystem Modelling and Analysis Project), models of like type are being intercompared and linked across types for a coupled assessment of climate change impacts over the conterminous USA. The linkage will be rudimentary and one way, consisting of the ED generation of altered vegetation distributions, which will be passed to the EF models for prediction of altered productivity, water yield and nutrient cycling.

The next step in the process, a full, integrated linkage in code of the two classes of models, is the primary focus of this discussion. This paper is only intended to be an exploration of ideas in order to present some of the challenges and possible solutions to building a fully dynamic global vegetation model.

Background

The development of two classes of models, Ecosystem Function (EF, biogeochemistry) and Ecosystem Dynamics (ED, biogeography), arises from two historically separate subdisciplines within ecology, biogeochemical processes and population processes, respectively (O'Neill *et al.*, 1986). The ED models address the question of what can establish and survive where and in what mixtures; while the EF models address the question of how well the vegetation can grow and persist, given knowledge of what lives there.

The EF models operate by simulating the processes of carbon assimilation, autotrophic and heterotrophic respiration and biomass turnover (decomposition), based on the primary inputs of radiation, temperature, precipitation and nutrient availability. Assimilated carbon is partitioned into various components of respiration and above and below ground growth. Each year, litter is produced and, through decomposition, returns carbon and nutrients to the soil and atmosphere. Decomposition is usually a function of lignin content, soil temperature and moisture conditions. CENTURY and Biome-BGC were designed to simulate transient productivity, given a set of initial conditions; whereas TEM analytically solves the production and decomposition equations to arrive at a final 'steady-state' solution of productivity and carbon stocks in 'mature' systems (perhaps 100+ years in age). The main outputs of EF models important here are net primary productivity (NPP) and carbon content in above and below ground vegetation parts (Plate 3, between pp. 204 and 205).

The ED models operate from a fundamentally different basis, being more concerned with the abiotic constraints on what a system could potentially become, rather than how well it can grow. Two primary constraints are well recognized, temperature limitations (primarily cold) and moisture limitations. All of the ED models simulate the temperature constraints in a similar fashion, being based on degree-day sums or thermal thresholds, calibrated to represent physiological cold-hardiness limits. The thermal constraints on vegetation result in roughly parallel latitudinal belts of vegetation. Within any given thermal zone there could exist a range of moisture conditions from very wet to quite dry.

The moisture constraints on different vegetation lifeforms are more difficult to simulate, given the different rooting depths, leaf forms, leaf areas and stomatal properties. In fact, the complete water balance at a site must be simulated in order to determine the limitations of moisture on vegetation distribution. There is a certain circularity in this, in that the water balance of a site is, in part, a function of the vegetation present at the site, which is, of course, the main endpoint to be determined. Modellers have approached this circularity from two perspectives: (1) find the simultaneous solution to the site water balance and site vegetation type through a process of trial-and-error iteration (MAPSS, DOLY) or by pre-specification of the vegetation type (Biome-BGC, CENTURY); or (2) uncouple the water balance and site vegetation by estimating transpiration as a function of soil moisture (TEM, BIOME).

There are many different ways to simulate the site water balance, all of which require some assumptions and methodological choices. Water must be brought into the system and partitioned between runoff and evapotranspiration. Snow dynamics, and soil water infiltration and percolation must be considered. Each model has a unique implementation of these processes with many details left implicit by the model structure. Differences occur among models in the number and thickness of soil layers, the use of soil texture information, the definition of rooting depths, the formulations (implied or explicit) for saturated and unsaturated infiltration and percolation, and the simulation timestep and many other components of the site water balance.

Actual evapotranspiration (AET) requires simulating the demand for water from atmospheric inputs of air temperature, humidity, solar radiation and wind. Potential evapotranspiration (PET) is converted to AET through consideration of rooting depth, soil moisture, stomatal conductance, leaf area index (LAI) and vegetation roughness length. There is no consensus on the best method to calculate PET and there are significant differences in the sensitivities of the various available methods (Mckenney & Rosenberg, 1993). The different EF and ED models use different approaches to AET calculation with varying levels of coupling between biological, meteorological and hydrological processes.

Once a site water balance is calculated, the ED models either generate a water balance index that is used to partition vegetation into different lifeforms or biomes (BIOME), or use the site water balance to directly estimate the amount of vegetation that can be supported at the site (DOLY, MAPSS). MAPSS, for example, calculates the maximum LAI that can be supported at a site while just using up the available soil moisture during the growing season (Neilson, 1995). The calculated LAI then becomes the implicit water balance index for the model. Various levels of LAI define different vegetation densities and stature, information used for vegetation classification.

The ED models also employ rules based on the seasonal patterns of temperature and moisture to determine if the vegetation will be evergreen or deciduous and broadleaf or needleleaf (Woodward, 1987; Neilson *et al.*, 1992; Prentice *et al.*, 1992; Neilson, 1995). Some vegetation types are mixtures of different vegetation lifeforms, such as tree or shrub savannas. Competition between the two contrasting lifeforms for light, water and nutrients must be considered; however, competition is handled quite differently among the different models. Information of leaf form and phenology, vegetation stature and density, relative grass-woody mixtures and thermal zones is combined in a rule-base to form a vegetation classification (Plate 4, between pp. 204 and 205; Table 23.1).

The EF models (with exceptions) were primarily constructed to simulate the transient growth of actual ecosystems under current climate, land use and successional status, given specified vegetation characteristics. The ED models, however, simulate potential 'climax' vegetation at steady-state under any climate, past, present or future. Since few ecosystems are actually at 'climax', the relation between model output and current ecosystem conditions is problematical. Thus, the ED models are currently quite different than the EF models, specifically with regard to the ability to simulate transient vs. steady-state dynamics. Ultimately, simulation of transient ecosystem dynamics will be of considerably greater use than steady-state estimations of 'potential climax vegetation'.

Table 23.1 *Preliminary classification of vegetation types within MAPSS.*

	LAI	Phenology	Leaf	Zone
1. Ice	N/A			
2. Tundra	N/A			Tundra
3. Taiga/tundra	N/A			Taiga-Tundra
4. Forest evergreen needle (taiga)	>3.75	E	M	Boreal
5. Forest evergreen needle (temperate)	>3.75	E	M	Temperate
6. Forest mixed cool (temperate)	>9	D/E	B/M	Temperate
7. Forest deciduous broadleaf	3.75–9	D	B	Temperate
8. Forest mixed warm (EN)	>3.75	D/E	B/M	Subtropical
9. Forest mixed warm (DEB)	>3.75	D/E	B	Subtropical
10. Tree savanna mixed cool (EN)	2–3.75	D/E	B/M	Boreal
11. Tree savanna (evergreen needle)	2–3.75	E	M	Temperate
12. Tree savanna (deciduous broadleaf)	2–3.75	D	B	Temperate
13. Tree savanna mixed warm (EN)	2–3.75	D/E	B/M	Subtropical
14. Tree savanna mixed warm (DEB)	2–3.75	D/E	B	Subtropical
15. Tropical forest evergreen broadleaf	>3.75	E	B	Tropical
16. Tropical seasonal forest (moist)	2–3.75	D/E	B/M	Tropical
17. Tropical dry forest/savanna	0.65–2	D/E	B/M	Tropical
18. Chaparral	2.1–3.5	D/E	B/M	All
19. Open shrubland	<2.1	D/E	B/M	All
20. Shrub savanna mixed cool (EN)	<2.1	D/E	B/M	Boreal
21. Shrub Savanna evergreen microphyllous	<2.1	E	M	Temperate
22. Shrub savanna deciduous broadleaf	<2.1	D	B	Temperate
23. Shrub savanna mixed warm (EN)	<2.1	D/E	B/M	Subtropical
24. Shrub savanna mixed warm (DEB)	<2.1	D/E	B	Subtropical
25. Shrub savanna tropical	<2.1	E	B	Tropical
26. Tall grass prairie	2–6			All
27. Mixed grass prairie	1.15–2			All
28. Short grass prairie	0.45–1.15			All
29. Semi-desert grassland	0.01–0.45			All
30. Tropical desert	<0.1			Tropical
31. Subtropical desert	<0.1			Subtropical
32. Temperate desert	<0.1			Temperate
33. Boreal desert	N/A			Boreal
34. Extreme desert	N/A			All

D, deciduous; E, evergreen; B, broadleaf; M, microphyllous. Parenthetical abbreviations in 'Mixed' vegetation types indicate likely climax physiognomy; EN, evergreen needleleaf; DEB, deciduous/evergreen broadleaf. LAI, all-sided leaf area index. Savanna types encompass the full gradient from almost-closed woodland to sparse-tree savanna. Reprinted with permission from the *Journal of Vegetation Science*, Neilson and Marks, 1994.

Models capable of fully transient vegetation dynamics must account for vegetation establishment, growth, succession, decline, mortality, disturbance, dispersal and land use. The EF models are, at present, capable of growth and decline; while the ED models variously handle disturbance, but primarily to shift from potential 'climatic climax' to 'fire climax' (Clements, 1936). Even if the two classes of models were fully linked at the current stage of development, they would still not adequately simulate succession, mortality, catastrophic disturbance, dispersal or land use changes. Also, none of these models currently handles horizontal transport dynamics, which will be required for dispersal and disturbance (at some scales) and will require an explicit, georeferenced operating platform.

Model linkage

Steady-state models

Both CENTURY and Biome-BGC were designed to simulate transient dynamics, given a specified set of initial conditions. This ability is the strength of the models for transient simulations, but presents difficulties when attempting to simulate systems that have equilibrated to a $2 \times CO_2$ climate at some unknown point in the future. For example, what should be the 'initial' values for soil carbon pools or vegetation types? Soil carbon pools can take hundreds to thousands of years to accumulate; yet are important in that they represent a reservoir of raw material for N mineralization and heterotrophic respiration and also affect soil water balance. Typically, such transient models would be run for up to one or two thousand years under new, steady-state conditions in order to stabilize. However, such simulations could be prohibitive at a daily timestep for large areas.

The TEM model, by contrast, was constructed to simulate steady-state conditions directly and is probably the easiest model for linkage to steady-state ED models under such conditions. However, even TEM must go through a 'spin-up' stage before a solution is reached. Another possible source of difficulty is that the ED models, by simulating potential climatic climax vegetation, are implicitly simulating very old systems; while models like TEM were calibrated to systems perhaps 100 years in age; mature, but not really old (Fig. 23.1). Carbon stocks and many other system attributes could be quite different between mature and old systems.

Linkage between steady-state EF and ED models would be more easily accomplished than for transient versions of these models. The two classes of models could remain generally as 'stand-alone' models, with explicit

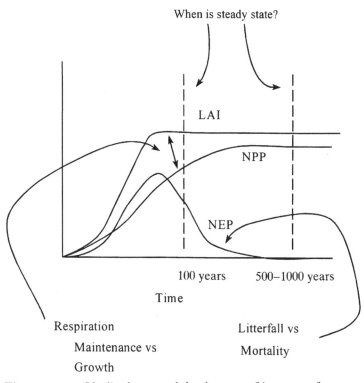

Figure 23.1 *Idealized temporal development of important forest ecosystem variables. Steady-state biogeochemistry models are usually calibrated using recent observations from ecosystems that are mature (100+ years old), but not ancient. Steady-state biogeography models imply 'potential' or ancient vegetation. Key issues with respect to transient dynamics involve, among others: (1) partitioning of respiration and its relation to the age and allometry of the 'stand', and (2) partitioning of ecosystem processes, such as litterfall, from population processes, such as mortality and disturbance. LAI, leaf area index; NPP, net primary productivity; NEP, net ecosystem productivity.*

linkages at two points. The ED (biogeography) model would be run first and would provide the vegetation type, soil moisture conditions and AET information required by the EF (biogeochemistry) model (for example, TEM). After completion of the EF simulation, the above-ground biomass, produced by the EF model, would be compared with the LAI (or foliar projective cover) produced by the ED model, to test for allometric consistency between the two simulations. If LAI and biomass results are not consistent between the models, then one or the other of the two models would be constrained to the lowest common metric of biomass and the simulations would be re-run until convergence is obtained.

An unresolved issue in these linkages is the role of nitrogen (Vitousek & Howarth, 1991). One philosophy, often taken by the biogeography modellers, is that, given sufficient time to attain 'climatic climax', nitrogen will be internally generated and will not be a limitation on the carrying capacity of the system. This situation is implied by most current and earlier generation biogeography models, by the lack of any nitrogen processes in the models. However, most of the ecosystem function (EF) models were constructed under a different philosophy, that even after very long times, nitrogen can still be a limitation on the ultimate carrying capacity of the system. The gist of the argument revolves around the rate of fixation of new nitrogen into the system relative to the rate of demand for nitrogen during the growing season. For example, in cold, high latitude systems temperature may limit both the supply of nitrogen through constraints on non-symbiotic nitrogen fixation and through constraints on N mineralization from decomposition. If sufficient N cannot be supplied through new fixation or mineralization, then potential production may not be realized due to constraints on nitrogen use efficiency (carbon fixed per nitrogen required). Under such circumstances, the system may never reach a steady state. Soils may continue to build as litterfall continuously exceeds decomposition and the total live biomass may never achieve its potential capacity under unlimited nitrogen. There is substantial debate on the global applicability of this logic.

Transient models

Consider a situation where a biome is being gradually changed from one type to another; yet the biomes are individually parameterized in the model. How does one alter the parameters to reflect the gradual transition? If the parameter space is a strict function of the abiotic environment, such as carrying capacity might be, then one could gradually change the parameters over time as climatic forcing changed. Still, there is the difficulty of knowing which direction and how much to change the parameters, which implies some foreknowledge of the shifting potential climatic climax condition. If, on the other hand, the parameters reflect intrinsic genetic characters, such as might be the case with maximum stomatal conductance, then the parameters should remain unchanged until the organism (or, lifeform) is replaced by a different kind through the process of succession. The issue of changing parameters over time is less of a difficulty with models that were constructed for transient operation. However, parameterization will still probably need to be constrained by thermal zone and life form.

Biome-BGC and MAPSS will be the primary focus for the discussion of transient modelling, but the issues are generic and could apply to any combination of models. The issues are very similar to those for steady-state linkage, such as sharing a common water balance simulation, structuring the soil layering for internal consistency, checking for allometric consistency between the two models and meshing different timesteps.

A schematic of the linkage planned between MAPSS and Biome-BGC is presented in Fig. 23.2. The linked model, MAPSS-BGC, will require daily and monthly aggregated inputs of air temperature, solar radiation, humidity, windspeed (depending on the potential transpiration algorithm), and precipitation (Fig. 23.2). The daily climate will be used within the Biome-BGC module for the calculations of water and carbon balance and the nitrogen budget. The monthly climate data will be used for the calculation of climate indices for cold stress thresholds, growing degree days, aridity and continentality. The monthly indices will be used by both the MAPSS and Biome-BGC components.

MAPSS and Biome-BGC were independently constructed under a similar philosophy, being structured around core, process-based hydrologic models (Running & Hunt, 1993; Neilson, in press). Canopy transpiration in both models is simultaneously limited by stomatal conductance and leaf area and will form a basis for model linkage. Both models simulate leaf area index (LAI), which will be the pivotal variable for integration.

Biome-BGC calculates water balance at a daily timestep (compared with monthly for MAPSS). Photosynthesis, dark respiration and maintenance

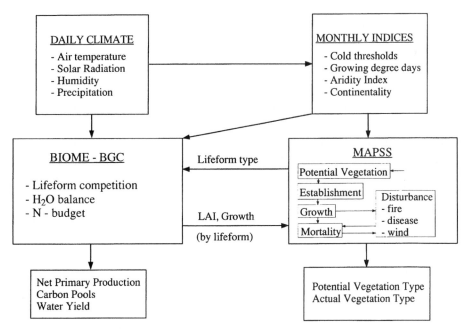

Figure 23.2 *Schematic representation of the planned MAPSS-BGC model linkage.*

respiration are calculated daily. The carbon fixed is accumulated over a full year and partitioned to roots and shoots with a calculation of growth respiration at that time for a final calculation of NPP (Plate 3). Litterfall and decomposition are calculated annually.

Initial linkage will require a common hydrologic model. Both Biome-BGC and MAPSS contain internal hydrologic models, which will necessitate a significant restructuring of both models. Some difficulties exist; for example, the two models use a different number of soil layers with different thicknesses and they operate at different timesteps. Both models calculate transpiration as a function of leaf area index and stomatal conductance, but use different stomatal conductance algorithms. It may be possible to modularize the stomatal conductance algorithms and test them independently.

Allometric consistency between LAI and biomass will be an important feature of the linked models and will have to be variable with ecosystem age. For example, in forested biomes LAI reaches a maximum value before total system carbon peaks (Fig. 23.1). Dense young stands attain high LAI at an early age and retain high LAI through the process of self-thinning and growth toward fewer, but larger trees.

The issues of allometric consistency can be illustrated by a comparison of global LAI maps produced by the two models (Plate 5, between pp. **204** and **205**). A simplified Biome-BGC representation was used for the global potential LAI calculation (Plate 5a), degraded to a monthly time-step for use with global long-term mean monthly climatology (Nemani & Running, 1995). Hydrologic equilibrium logic (similar in concept to MAPSS) was used to estimate a climatic potential for LAI rather than parameterizing LAI with NDVI, as is normally done for global Biome-BGC simulations. The difference in pixel level variability between the two simulations is caused primarily by the use of a variable soils data layer in the Biome-BGC simulation (Plate 5a), but not in the MAPSS simulation (Plate 5b). However, MAPSS can also calculate variable soil water holding capacity, given soil texture information. The Biome-BGC LAI calculations were constrained to integer values, so that areas such as the Sahara Desert were rounded up to unity. By contrast, MAPSS calculates fractional LAI values which were rounded to zero over much of the Sahara Desert. Thus, some of the discrepancy between the two models is a simple labelling difference.

The two maps are highly correlated ($P<0.0001$), but clearly display several interesting differences. MAPSS uses prescribed LAI values for energy-limited zones (primarily affecting the high latitude zones) to allow calculations of site water balance for hydrologic assessments. The prescribed LAI values have no effect on the simulated biogeography, which is under specific thermal constraints. The prescribed values were set too high in comparison with the dynamic calculations of energy-limited LAI from Biome-BGC. This serves to demonstrate one of the values of a linkage between MAPSS and Biome-BGC, which will effectively constrain high latitude LAI to the minimum sustainable under energy and nitrogen constraints as well as the water-limited constraints currently present in MAPSS.

Another area of discrepancy in the LAI simulations (Plate 5) is in the tropical deciduous forests (Type 16 in Table 23.1 and Plate 4). This proto-type version of MAPSS forces all tropical LAI to be evergreen and simply uses the LAI values as a water-balance index to 'window' the tropical deciduous vegetation between two 'evergreen' LAI thresholds (Table 23.1, Type 16). When a tropical deciduous 'rule', based on drought, is imple-mented in MAPSS, the water-limited LAI will reach a higher value in order to utilize the same amount of water, as described below. Thereafter, the two models are expected to simulate similar tropical deciduous LAI values. Other differences between the two LAI simulations are minor, but will be more fully explored in a separate manuscript. These two LAI plots (Plate 5) serve to demonstrate the overall comparability between the two models and to demonstrate the necessity of allometric consistency for such a critical, integrative variable.

MAPSS was primarily designed as a biogeographic tool with the funda-mental objective of producing a vegetation classification of potential vegeta-tion on any given site (Plate 4, Table 23.1). The classification is determined through a physiologically conceived rule-base that utilizes LAI, leaf form (broadleaf, needleleaf), leaf phenology (evergreen, deciduous) and thermal zone (tundra, taiga-tundra, boreal, temperate, subtropical, tropical) (Neilson, 1993, 1995). The LAI calculation and some of the leaf phenology rules require a process-based simulation of site water balance in order to deter-mine the maximum LAI that can be supported at the site with respect to water limitation. Various indices of productivity and water use efficiency are constructed by MAPSS to assist in the rule-based determination of leaf form and vegetation type. In a linked model framework the productivity, water use efficiency and LAI calculations would be handled by Biome-BGC' while, the rule-based portion of MAPSS would provide the determination of lifeforms supported at the site and the vegetation classifi-cation.

Biome-BGC will calculate lifeform competition between woody and grass vegetation based on light, water and nitrogen and will require knowledge of the lifeform types present on the site, as provided either by the MAPSS model or as initial conditions (Running et al., 1994). Competition between lifeforms is not currently included in Biome-BGC, but is a component of MAPSS. The competition module from MAPSS will be adapted for inclusion in the Biome-BGC model. Water competition in MAPSS occurs through different canopy conductance characteristics and variable rooting depths between the woody and grass lifeforms. Grasses are relegated to the surface 0.5 metre and the woody roots occupy the top 1.5 m of soil. Water is withdrawn from the top half-metre of soil for transpiration by the two lifeforms (woody and grass) as a direct proportion of their LAIs and stomatal conductances (canopy conductance). After depletion of the surface soil layer, woody plants are capable of extracting water from the second, 1 m thick soil layer. A third soil layer is included for accurate simulation of base flow runoff, but contains no roots. Nitrogen competition will be similarly structured.

Competition for light will be determined by the height and LAI of the respective grass and woody lifeforms using Beer's law for vertical light extinction through the canopy (Cannell & Grace, 1993). Carbon balance simulation will produce growth and LAI calculations by lifeform, which will then be passed to the MAPSS model. The upper limits or carrying capacity of LAI, as calculated by Biome-BGC, will be constrained by either water, energy (via growth rate) or nitrogen.

The initial vegetation conditions, required as inputs for MAPSS-BGC, will specify the lifeforms present and the leaf characteristics. Mortality will be a function of disturbance, physiological drought stress (lifeform specific stomatal responses) or a lack of growth. Establishment will be handled by MAPSS, which will constrain the lifeform types that can be supported at a site, specifically with respect to leaf characteristics, light and thermal tolerances. Succession will occur through differential mortality of different lifeforms as the light, soil water and nutrient conditions change due to vegetation growth and climate change.

The coupling of ED and EF models in the transient context has been discussed as if the two models will retain their comparatively unique characteristics. However, the coupling described here represents a very fundamental linkage at numerous locations in both models. Each model will require significant restructuring for a very intimate linkage of algorithms, essentially becoming one model.

Conclusions

Construction of a fully transient general vegetation model will require coupling the processes that control biogeochemical cycling with those that control biogeographic processes. The latter processes are largely in the realms of population biology and physiology; while, the former processes are in the realm of ecosystem science. Coupling of these two schools of science will open numerous opportunities for both applied and basic science.

With respect to basic science, issues of the linkage between or constraints upon population biology and ecosystem processes can be more fully explored. The relative importance, for example, of processes controlling recruitment and establishment versus those controlling growth, disturbance and mortality have long been debated. Many theoretical experiments will be possible with respect to coupled processes of hydrology, population biology, nutrient cycling, atmospheric forcing, fires, pests and wildlife. These model experiments will be useful as detailed hypotheses amenable to field testing.

Coupled, transient biogeochemistry and biogeography (EF-ED) will allow, with respect to applied science, the development of linked atmosphere–biosphere models for simulations of the whole Earth system. Such global simulations can be used with advanced remote sensing platforms to address the complex issues of global warming and land use, mitigation of global stressors and our adaptation to the ensuing global impacts (Running *et al.*, 1994). At regional and landscape scales, coupled EF-ED models will allow managers to explore alternate land-use prescriptions. Policymakers will be able to explore the potential impacts of a rapidly changing planet through the linkages between land use, disturbance (such as fire), stream flow and stream chemistry (relevant for fish habitat), forest productivity and forest resistance to drought and infestations. Multiple use management will be made more robust through deterministic simulation of carbon stores (global carbon cycle), wildlife habitat (biodiversity), forest and grassland distribution and productivity and disturbance. The initial EF-ED model coupling described here will only be the beginning of a powerful new generation of ecosystem simulation models, applicable to the general question of the ecological sustainability of spaceship Earth.

Acknowledgements

We want to thank E. R. Hunt Jr and R. R. Nemani for the global net primary production and global LAI images. SWR was supported by NASA grant # NAGW-3151 and the NASA contribution to the VEMAP project. RPN was supported by the USDA Forest Service and its contribution to the VEMAP project.

References

Bolin, B. (1991). The intergovernmental panel on climate change (IPCC). In *Climate Change: Science, Impacts, and Policy*, ed. J. Jager & H. L. Ferguson, pp. 19–22. Cambridge: Cambridge University Press.

Cannell, M. G. R. & Grace, J. (1993). Competition for light: detection, measurement, and quantification. *Canadian Journal of Forest Research*, **23**, 1969–79.

Clements, F. E. (1936). Nature and structure of the climax. *Journal of Ecology*, **24**, 252–84.

Friend, A. D., Shugart, H. H. & Running, S. W. (1993). A physiology-based gap model of forest dynamics. *Ecology*, **74**, 792–7.

Martin, P. (1993). Vegetation responses and feedbacks to climate: A review of models and processes. *Climate Dynamics*, **8**, 201–10.

Mckenney, M. S. & Rosenberg, N. J. (1993). Sensitivity of some potential evapotranspiration estimation methods to climate change. *Agricultural and Forest Meteorology*, **64**, 81–110.

Melillo, J. M., McGuire, A. D., Kicklighter, D. W., Moore, B., Vorosmarty, C. J. & Schloss, A. L. (1993). Global climate change and terrestrial net primary production. *Nature*, **363**, 234–40.

Neilson, R. P. (1993). Vegetation redistribution: A possible biosphere source of CO_2 during climatic change. *Water, Air and Soil Pollution*, **70**, 659–73.

Neilson, R. P. (1995). A model for predicting continental-scale vegetation distribution and water balance. *Ecological Applications*, **5**, 362–85.

Neilson, R. P., King, G. A. & Koerper, G. (1992). Toward a rule-based biome model. *Landscape Ecology*, **7**, 27–43.

Neilson, R. P. & Marks, D. (1994). A gobal perspective of regional vegetation and hydrologic sensitivities and risks from climatic change. *Journal of Vegetation Science*, **5**, 715–30.

Nemani, R. & Running, S. (1995). Satellite monitoring of global land cover changes and their impact on climate. *Climatic Change*, **31**, 395–413.

O'Neill, R. V., DeAngelis, D. L., Waide, J. B. & Allen, T. F. H. (1986). *A Hierarchical Concept of Ecosystems*. Princeton: Princeton University Press.

Parton, W. J. (1992). Development of simplified ecosystem models for applications in earth system studies: The CENTURY experience. In *Modeling the Earth System*, ed. D. Ojima, pp. 281–302. Boulder: UCAR.

Prentice, I. C., Cramer, W., Harrison, S. P., Leemans, R., Monserud, R. A. & Solomon, A. M. (1992). A global biome model based on plant physiology and dominance, soil properties and climate. *Journal of Biogeography*, **19**, 117–34.

Running, S. W. & Hunt, E. R. (1993). Generalization of a forest ecosystem process model for other biomes, BIOME-BGC, and an application for global-scale models. In *Scaling Processes between Leaf and Landscape Levels*, ed. J. R. Ehleringer & C. Field, pp. 141–58. San Diego: Academic Press.

Running, S. W., Loveland, T. R. & Pierce, L. L. (1994). A remote sensing based vegetation classification logic for use in global biogeochemical models. *Ambio*, **23**, 77–81.

Running, S. W., Justice, C. O., Salomonson, V., Hall, D., Barker, J., Kaufmann, Y. J., Strahler, A. H., Huete, A. R., Muller, J.-P., Vanderbilt, V., Wan, Z. M., Teillet, P. & Carneggie, D. (1994). Terrestrial remote sensing science and algorithms planned for EOS/MODIS. *International Journal of Remote Sensing*, **15**, 3587–620.

Vitousek, P. M. & Howarth, R. W. (1991). Nitrogen limitation on land and in the sea: How can it occur? *Biogeochemistry*, **13**, 87–115.

Woodward, F. I. (1987). *Climate and Plant Distribution*. Cambridge: Cambridge University Press.

Woodward, F. I. & McKee, I. F. (1991). Vegetation and climate. *Environmental International*, **17**, 535–46.

Woodward, F. I., Smith, T. M. & Emanuel, W. R. (1995). A global land primary productivity and phytogeography model. *Global Biogeochemical Cycles*, **9**, 471–90.

24 Global and regional land-use responses to climate change

M. L. Parry, J. E. Hossell, R. Bunce, P. J. Jones, T. Rehman, R. B. Tranter, J. S. Marsh, C. Rosenzweig, and G. Fischer

Introduction

The hypothesis posed here is that the potential effects of climate change on agricultural land use in any given region are likely to stem both from the response of the world food production and trade system to global climate change and from specific changes in the climate of that region. In any given region we might expect that changes in global production due to climate change would lead to differential changes in prices for different crops and livestock products, and would affect the demand for different crops and thus what crops would be grown in what proportions under the postulated economic and environmental conditions at that time. It is, at the same time reasonable to expect that changes in climate which, for a given region, are likely to lead to changes in the yields of different crops, would lead to changes in the comparative advantage of different regions as areas for growing various food crops and livestock fodder.

If these two types of effects could be estimated, the relative weight of effects on land use that might stem directly from regional-level changes in climate *per se* and those that may stem indirectly from global climate-induced changes in agricultural demand and prices could be inferred. To achieve this level of analysis requires evaluation of agricultural inputs at the regional level but within a global price environment and the development of a hierarchy of land-use and climate models. In this chapter, the results of using the outputs of a global analysis of the sensitivity of world food prices and demand to climate change as inputs to a regional land-use allocation model are presented. Research on the former has been published in full elsewhere (Rosenzweig & Parry, 1994); the latter is the product of a recent, previously unpublished, research project relating to land-use and climate change in England and Wales (Parry, 1996., forthcoming). A schema of the research approach followed for this work is given in Fig. 24.1.

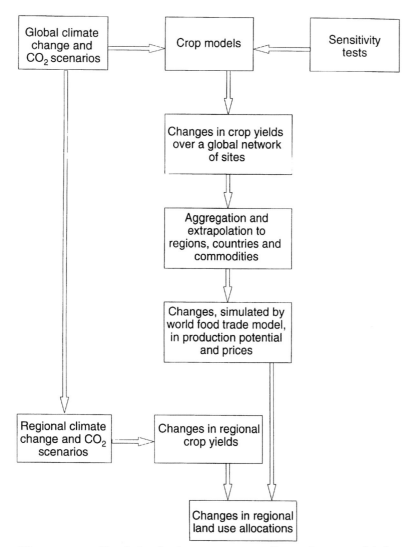

Figure 24.1 *Simulating land-use responses to climate change at global and regional levels.*

Effects of climate change on regional competitiveness and global prices

The first part of this paper summarizes current knowledge concerning the potential effects of climate change on the production potential of global agriculture. This is followed by a discussion of effects on prices of agricultural products and the altered competitiveness (or comparative advantage) of different agricultural regions with respect to the world market. There are two components to this global estimation. First, potential changes in national grain crop yields were estimated using crop models and a decision support system developed by the US Agency for International Development's International Benchmark Sites Network for Agrotechnology Transfer (IBSNAT, 1989). The crops modelled were wheat, rice, maize and soybeans. These crops account for more than 85% of the world trade in grains and legumes (for details, see Rosenzweig & Parry, 1994 and Fischer *et al.*, 1994). Second, the yield changes were used as inputs to a world food trade model, the Basic Linked System (BLS), a full description of which is given in Fischer *et al.* (1988). Outputs from simulations from the BLS provided information on global food production potential and food prices. These two research components are described in the following section.

Modelling effects on crop yields

Crop models

The IBSNAT crop models were used to estimate how climate change and the associated increased level of carbon dioxide could alter yields of the selected crops over the major production areas of the world and the major worldwide vulnerable regions at low, middle and high latitudes. These models employ simplified functions to predict the growth of crops as influenced by the major factors that affect yields, i.e. genetics, climate (daily solar radiation, maximum and minimum temperatures and precipitation), soils and management practices. The models used were: wheat (Ritchie & Otter, 1985; Godwin *et al.*, 1989); maize (Jones & Kiniry, 1986; Ritchie *et al.*, 1989); paddy and upland rice (Godwin *et al.*, 1989); and soybean (Jones *et al.*, 1989).

The IBSNAT models were selected for use in this study because they have been validated over a wide range of environments (see, for example, Otter-Nacke *et al.*, 1986) and are not specific to any particular location or soil type. They are better suited for large-area studies in which crop growing and soil conditions differ greatly than more detailed physiological models that have not been as widely tested. The validation of the crop models over different environments also improves the ability to estimate effects of changes in climate. Because the crop models have been tested over essentially the full range of temperature and precipitation regimes where crops are grown in today's climate, and to the extent that future climate changes bring temperature and precipitation regimes within these ranges, the models may be considered useful tools for assessment of potential climate change impacts. Furthermore, because management practice, such as the choice of varieties, the planting date, fertilizer application, and irrigation, may be varied in the models, these permit experiments that simulate adaptation by farmers to climate change.

The above crop models account for the beneficial physiological effects of increased atmospheric CO_2 concentrations on crop growth and water use through the use of ratios between measured daily photosynthesis and evapotranspiration rates for a canopy exposed to high CO_2 values. This was based on published experimental results (Kimball, 1983; Cure & Acock, 1986; Allen *et al.*, 1987). Crop modelling simulation experiments were performed at 112 sites in 18 countries for the baseline climate period (1951–80) and for General Circulation Model (GCM) doubled CO_2 climate change scenarios with or without the physiological effects of CO_2. The sites represent both major and minor production areas at low, mid and high latitudes (Fig. 24.2).

Aggregation of the yield effects to the national level

Estimates of altered yields were aggregated by weighting regional yield changes (based on current production) to estimate changes in national yields. This assumed the current mix of rain fed and irrigated production, and the current crop varieties, nitrogen management and soils. Changes in yield of crops that were not simulated were estimated on the basis of similarities between the assumptions of crop models used and the actual growing conditions, and on a wide search of previously published and unpublished climate change impact studies.

The primary source of uncertainty in the estimates thus obtained lies in the relatively few crop modelling sites which may not adequately represent the variability of agricultural regions within countries and the variability of agricultural systems within similar agro-ecological zones. However, since the site results relate to regions that account for about 70% of world grain production, both the method adopted and the conclusions for the scale of the study adopted here, are considered to be appropriate (Rosenzweig & Parry, 1994).

Modelling worldwide effects of climate change on food supply and prices

The estimates of climate-induced changes in yields were used as inputs to a dynamic model of the world food system, the BLS, in order to assess possible impacts on future levels of global food production and food prices.

The BLS is a modelling system for general equilibrium economic analysis which incorporates 20 national models that cover about 80% of the world food production and trade system. The remaining 20% is covered by 14 regional models for the countries which have broadly similar attributes (e.g. African oil exporting countries, Latin American high income countries,

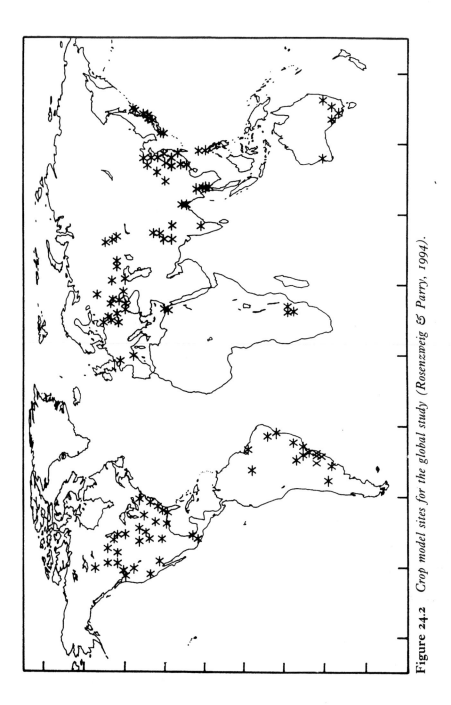

Figure 24.2 Crop model sites for the global study (Rosenzweig & Parry, 1994).

Asian low income countries, etc.). A full description is given in Fischer
et al (1988). Various countries are linked through trade, world market prices
and financial flows. The BLS represents a recursively dynamic production
and trade system: a first round of exports from all countries is calculated
for an assumed set of world prices, and international market clearance is
checked for each commodity. World prices are then revised, using an optimizing algorithm and again transmitted to the national level. Next, these
generate new domestic equilibria and adjust net exports. This process is
repeated until the world markets are cleared in all commodities.

At each stage of such reiterations domestic markets are in equilibrium
and the analyses make assumptions about the land that is available for
different agricultural uses. Information on land suitability is based on the
FAO's database on agro-ecological zones (FAO, 1991). This process yields
international prices as influenced by governmental and intergovernmental
agreements in a given year. The system is solved on a year-to-year basis,
simultaneously for all countries, with changes (largely increases) in demand
being met both by increases in land productivity and by increases in the
farmed area. Effects of changes in climate are introduced to the model as
changes in the average national or regional commodity yields.

Assumptions about the future

The reference scenario
The reference scenario involves projection of the world agricultural system
to the year 2060 assuming no climate change. It is assumed that there will
be some changes in the political or economic context of world food trade,
namely:

- medium rates of population growth based on UN estimates, giving a
  world total of 10.2 billion by 2060 (UN, 1989; IBRD/World Bank,
  1990)
- 50% trade liberalization in agriculture introduced gradually by 2020
- Moderate economic growth based on World Bank estimates, (ranging
  from 3.0% per year in 1980–2000 to 1.1% per year between 2040 and
  2060) (IBRD/World Bank, 1990)
- Increases in crop yields over time, based on FAO estimations. (Cereal
  yields for the world as a whole, developing countries and developed
  countries are assumed to increase annually by 0.7%, 0.9% and 0.6%,
  respectively; FAO, 1991.) According to FAO data, yields have been
  growing at an average of around 2% annually during the period
  1961–90, both for developed and developing (excluding China) countries (FAO, 1991). Recent increases (1965–85) in annual productivity

for less-developed countries average about 1.5%/year. In the 1980s, however, yields grew globally at an average yield increase of only 1.3%, implying a falling trend in yield growth rates.

Cereal prices (see Table 24.1) are estimated at an index of 121 (1990 equal to 100) for the year 2060, reversing the trend of falling cereal prices of the last 100 years. There are two phases of price change, these being strongly affected by the assumptions relating to tariff liberalization as a result of international agreements.

Scenarios of CO_2 and climate change

These are projections of the world food system including effects of climate on agricultural yields under three different climate scenarios. The scenarios for this study were constructed by changing observed data on current climate (1951–80) according to doubled CO_2 simulations of three GCMs. The GCMs used are those from the Goddard Institute for Space Studies (GISS), Geophysical Fluid Dynamics Laboratory (GFDL) and United Kingdom Meteorological Office (UKMO). The temperature changes of these GCM scenarios (4.0 – 5.2 °C) are nearer to the upper end of the range (1.5–4.5 °C) projected for doubled CO_2 warming by the IPCC (1990 and 1992) (Table 24.2). Mean monthly changes in climate variables from the appropriate gridbox were applied to observed daily climate records to create climate change scenarios for each site.

CO_2 concentrations were estimated to be 555 ppm in 2060 (based on Hansen *et al.*, 1988). The effective CO_2 doubling will have occurred round the year 2030, if current emission trends continue. The climate change caused by an effective doubling of CO_2 may be delayed by a further 30 to 40 years or longer, hence the projections for 2060 used in these studies.

Table 24.1. *Index of world food prices simulated by the Basic Linked system reference case (i.e. without climate change)*

	1980	2000	2020	2060
Cereals	102	125	126	121
Other crops	110	118	110	94
All crops	108	120	115	102
Livestock	105	131	135	119
Agriculture	107	123	121	107

1970 = 100
From Rosenzweig and Parry (1994).

Table 24.2 *General Circulation Model (GCM) doubled CO₂ climate change scenarios*

GCM	Year[a]	Resolution lat. × long.	CO_2 ppm	Change in average global temperature °C	precipitation %
GISS	1982	7.83° × 10°	630	4.2	11
GFDL	1988	4.4° × 7.5°	600	4.0	8
UKMO	1986	5.0° × 7.5°	640	5.2	15

[a]When calculated.
GISS, Goddard Institute for Space Studies; GFDL, Geophysical Fluid Dynamics Laboratory; UKMO, United Kingdom Meteorological Office.
From Rosenzweig and Parry (1994).

Adaptation

Two adaptation levels to cope with potential effects of climate change on yield and agriculture were considered: (a) those adaptations at the farm level that would not involve any major changes in agricultural practices, including changes in planting date, in amounts of irrigation, and in choice of crop varieties that are currently available; and (b) major changes in agricultural practices (e.g. large shifts of planting date, the availability of new cultivars, widespread expansion of irrigation and increased fertilizer application) which imply policy changes at the national and international level and significant impact on costs of agricultural production.

Altered potential cereal production and food prices

Fig. 24.3 shows the estimated changes in world cereal production for the GISS, GFDL, and low resolution UKMO $2 \times CO_2$ climate change scenarios allowing for the direct effects of CO_2 on plant growth. These yield estimations are used as inputs to the dynamic simulations by the BLS to indicate how the world food system might respond to climate-induced supply shortfalls of cereals and consequently higher commodity prices through dynamic increases in production factors (cultivated land, labour and capital, and inputs such as fertilizer).

World cereal production is estimated to decrease by between 1% and 7%, depending on the climate scenario. Under the UKMO scenario, global production is estimated to decrease by more than 7%, while under the GISS scenario (which assumes smaller temperature increases) cereal

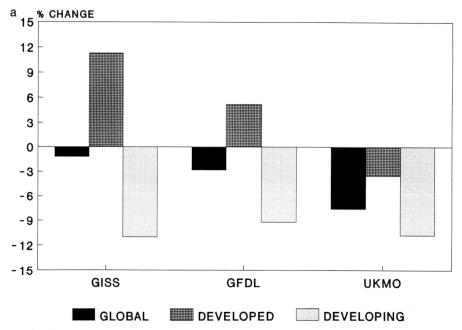

production is estimated to decrease by just over 1%. The largest negative changes occur in developing countries averaging between 9% to 11%. By contrast, in developed countries production is expected to increase under all but the UKMO scenario (+11% to −3%). Thus, existing disparities in crop production between developed and developing countries are estimated to increase.

Effects on world prices are shown in Table 24.3. These are estimated to be 24% to 145% higher than those estimated for the reference case (i.e. the future in 2060 without climate change, the REF-2060 scenario). Both minor and major levels of adaptation help restore world production levels (Fig. 24.3), but quite large price differences remain, particularly under the more extreme climatic scenarios (Fig. 24.4). These climate-affected prices are then used as inputs to the model of regional land use.

Effects on regional level land use

Changes of climate are likely to affect the use of agricultural land at a given place in two ways: first, through changes in yield potential at that place and, second, through effects on the competitive position of activities on that land *vis-à-vis* land elsewhere. The aim of the second part of this paper is to illustrate how an analysis of these two elements can be made in order to evaluate their integrated effect. A case study will be used to illustrate the process; in this instance, for England and Wales using the Climate Land

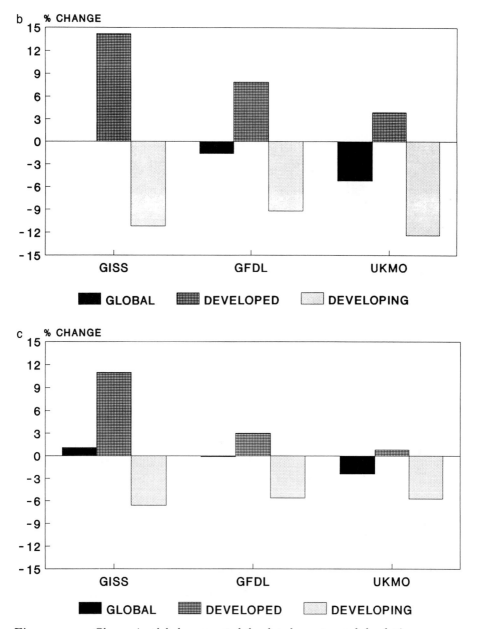

Figure 24.3 *Change in global, aggregated developed country and developing country cereal production in 2060 projected by the BLS under climate change scenarios under (a) no adaptation, (b) adaptation level 1 and (c) adaptation level 2. Reference scenario for 2060 (0 on graphs) assumes no climate change (world 3286 mmt, developed 1449 mmt, developing 1836 mmt. GISS, Goddard Institute for Space Studies; GFDL, Geophysical Fluid Dynamics Laboratory; UKMO, United Kingdom Meteorological Office. (Rosenzweig & Parry, 1994.)*

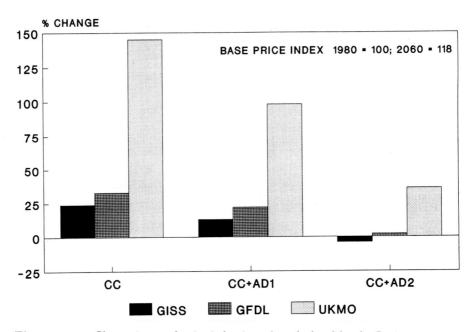

Figure 24.4 *Change in cereal price index in 2060 calculated by the Basic Linked System under climate change scenarios. Reference scenario for 2060 assumes no climate change (price index is 18% above 1980 levels). (Rosenzweig & Parry, 1994.)*

Use Allocation Model (CLUAM) and manipulating it in the context of (a) global price changes due to climate and (b) changes in yield potential in England and Wales due to climate. The combination of price and yield potential is used to calculate the land use providing highest returns for each of 155 235 1 km² cells of land in England and Wales (Barr *et al.*, 1994). Each cell is characterized by given physical qualities such as climate, soils, elevation, etc., termed Land Classes. There are 15 such Land Classes in England and Wales.

The Climate Land Use Allocation Model

The CLUAM was derived from an earlier model which had been originally designed and built to analyse the environmental and ecological consequences of changes in agricultural practices in response to adjustments in agricultural policies and market conditions (Harvey *et al.*, 1992). In constructing the CLUAM, data on the various factors that are important in determining agricultural land use are structured within the unified framework of a linear programming model that treats the agriculture of England and Wales as a single 'farm'. On this 'farm' a variety of production activities (or land uses) are employed to produce various agricultural outputs using a range of inputs and different types of land. The CLUAM allocates

Table 24.3 *Change in cereal production and prices in 2060 under GCM 2 × CO$_2$ climate change scenarios*

Region	Reference scenario	GISS %	GFDL %	UKMO %
Change in cereal production				
Global	3286	−1.2	−2.8	−7.6
Developed	1449	11.3	5.2	−3.5
Developing	1836	−11.0	−9.2	−10.8
Change in cereal price index (1970 = 100)				
Cereal prices	121	24	33	145

From Rosenzweig and Parry (1994).

land of various types to these different land-using activities subject to three basic types of constraint:

- the availability of land in each Land Class
- the total amount of production and the quantity of inputs used
- 'policy' constraints that restrict production activities or impose specific land utilization patterns in certain areas to conform to environmental or other objectives.

Within these constraints land utilization is determined according to the economic margins that can be earned by the various possible activities. Four types of use were identified (arable, ley, permanent pasture and rough grazing), with yield categories reflecting the range of production potential (three yield categories for arable land uses and five for each of the grassland types). The CLUAM allocates each cell of land of varying agricultural potential to its highest value use and its highest achievable yield.

In the base case (i.e. under current climate, the 'mid-1980s optimal') areas are calculated of the various crops and grass/herbage types. When the model is run for both the altered climatic or economic conditions in 2060, the economic optimum mix of enterprises that is chosen by the model depends on the altered yield potentials of the land combined with the various altered prices of inputs and outputs.

Effects of climate change on agricultural potential in England and Wales

Changes of climate may have two types of effect on agricultural potential in England and Wales: (a) The distribution of crop potential could be affected across the country by climate change. Local conditions would determine the degree to which yield changes reflect this response. (b) Crop yields could be

Table 24.4 *The difference between BLS (see text) estimated yield, price & demand changes for agricultural
products in England and Wales between the mid 1980s
optimal and GISS 2 × CO₂ climate change scenario*

Enterprise	Yield change	Price change	Demand change
Wheat	1.43	0.87	1
Barley	1.33	0.87	1
Coarse grains	1.49	0.91	1.38
Sugar beet	1.22	0.69	1.03
Potatoes	1.39	0.69	1.03
Peas & beans	1.39	0.69	1.03
OSR	1.39	0.69	1.03
Dairy			
Milk	1.45	0.86	0.96
calf	1.45	0.9	–
cull	1.45	0.9	–
Beef (meat)	1.45	0.9	0.93
Sheep			
lamb	1.45	0.9	0.93
wool	1.45	0.59	0.87
cull	1.45	0.9	–

Expressed as a proportion: mid 1980s optimal = 1.
OSR, oil seed rape.

directly affected by changes in ambient levels of the atmospheric carbon
dioxide.

Estimates of the combined effects of (a) and (b) were calculated using
results from crop model simulations described previously in this chapter
(Table 24.4) and the percentage changes applied to the yield categories in
England and Wales. For all crops climate information was used to define
criteria beyond which a given crop would not grow. Given the coarse resolution of climate data available, and the large number of crops included in
the model, these criteria were defined in terms of simple thermal limits,
with a measure of soil moisture conditions being used to assess the ability
to work the land sufficiently to both sow and harvest the crop. Similar
assumptions have been used to assess the potential suitability of land in
Great Britain and Europe for a number of crops under present and future
climate conditions (Carter *et al.*, 1991; Rowntree *et al.*, 1991). Production
of a crop in an area was prohibited if either the thermal or moisture

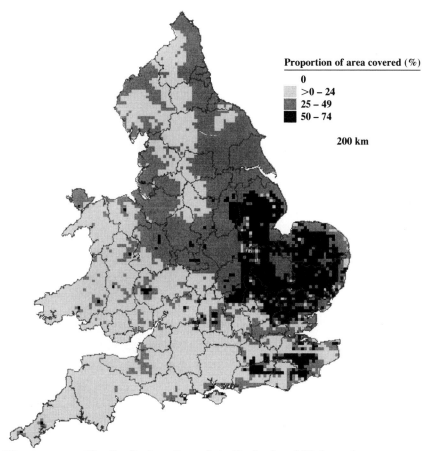

Figure 24.5 *The distribution of cereals in England and Wales under current climate (the mid 1980s optimal run) shown as the proportion of agricultural land covered.*

conditions severely limited growth. Moisture conditions were assumed to be limiting only if the levels were high enough to prevent use of machinery during either the entire preparation and sowing period or the whole harvesting period. Further details are given in Parry *et al.* (1996).

Land use in England and Wales in a future without climate change (REF-2060)

The projection of land-use changes under the REF-2060 scenario enables the effects due to economic and technological changes to be separated from the response due to climate change alone.

Figs. 24.5 and 24.6 show the national area under cereals for the present (mid 1980s) and for 2060. It is projected to decline in a future without climate change from 3.5 Mha to 2.1 Mha by 2060, a fall of some

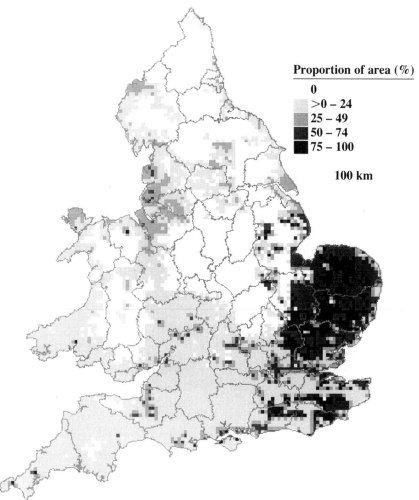

Figure 24.6   *The distribution of cereals in England and Wales under the REF-2060 run shown as the proportion of agricultural land covered.*

40%, with the main production tending to concentrate in Southeast England, the area most favourable to production. Elsewhere, large areas of the country fall out of cereal production altogether, largely due to increasing competition from other cereal producing regions in a less tariff-restricted trading environment.

Effects of climate changes on land use in England and Wales

As with the REF-2060 scenario, the BLS adopts a range of assumptions about the changes in yields, demand and prices under the climate change scenarios for which the CLUAM is run. To illustrate, Table 24.4 gives the yield, price and demand changes estimated for agricultural products in

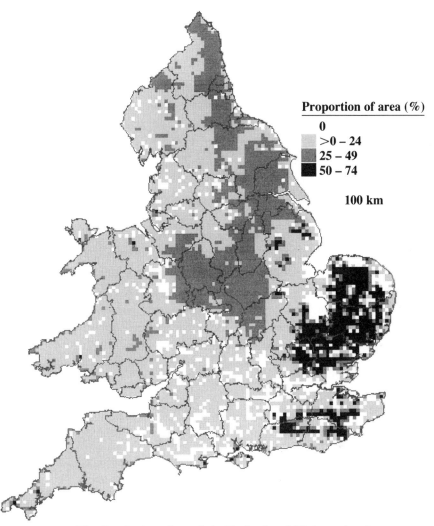

Figure 24.7   *The distribution of cereals in England and Wales under the*
GISS 2×CO₂ GCM run shown as the proportion of agricultural land covered.

England and Wales under the GISS 2 × CO$_2$ scenario. For reasons of
brevity only the estimations of land-use change under the GISS 2 × CO$_2$
scenario are mapped and described in this paper.

Fig. 24.7 shows that, while there is very little difference in the total
national cereals *area* under the REF-2060 and the GISS 2 × CO$_2$ scenarios,
there occurs a major change in the *location* of the area northwards. The
explanation for this is the projected reduction of soil moisture in traditional
cereal production areas in S.E. England. Cereal production is essentially
'seeking out' those Land Classes where moisture and temperature are at
their optimum levels for output. These tend to be further north and further

west under the GISS 2× CO_2 scenario than under current climatic conditions.

Conclusion

The purpose of this chapter has not been simply to report estimates of land-use change resulting from specified changes of climate. Its objective has been to describe means by which this can be done and demonstrate the use of those means. Regional land-use models can be developed which capture some of the broad-scale changes in the global agricultural environment that may occur as a result of climate change. These models can then be used to estimate the finer-scale land-use response. As in any exercise such as this, the sensitivity of the models and the use of them in combination needs to be explored in relation to the whole array of assumptions which are economic, technological and physical. There has not been the space to explore these tests fully in this paper; and not until they have been fully examined can the effects of climate change be properly evaluated. But the means exist by which this can be done.

References

Allen, L. H., Jr, Boote, K. J., Jones, J. W., Jones, P. H., Vale, R. R., Acock, B., Rogers, H. H. & Dahlman, R. C. (1987). Response of vegetation to rising carbon dioxide: Photosynthesis, biomass and seed yield of soyabean. *Global Biogeochemical Cycles*, **1**, 14.

Barr, C. J., Bunce, R. G. H., Clarke, R. T., Fuller, R. M., Furse, M. T., Gillespie, M. K., Groom, G. B., Hallam, C. J., Hornung, M., Howard, D. C. & Ness, M. J. (1994). *Countryside Survey 1990 Main Report*. London: Department of the Environment.

Carter, T. R., Parry, M. L. & Porter, J. H. (1991). Climatic change and future agroclimatic potential in Europe. *International Journal of Climatology*, **11**, 251–69.

Cure, J. D. & Acock, B. (1986). Crop responses to carbon dioxide doubling: a literature survey. *Agricultural and Forest Meteorology*, **38**, 127–45.

FAO (1991). *Agrostat/PC*. Rome: United Nations.

Fischer, G., Frohberg, K., Keyzer, M. A. & Parikh, K. S. (1988). *Linked National Models: a Tool for International Food Policy Analysis*. Dordrecht: Kluwer.

Fischer, G., Frohberg, K., Parry, M. L. & Rosenzweig, C. (1994). Climate change and world food supply, demand and trade: Who benefits, who loses? *Global Environmental Change*, **4**, 7–23.

Godwin, D., Ritchie, J. T., Singh, U. & Hunt, L. (1988). *A User's Guide to CERES-Wheat V2.10*. Muscle Shoals, Alabama: International Fertilizer Development Center.

Hansen, J., Russel, G., Rind, D., Stone, P., Lacis, A., Lebedeff, S., Ruedy, R. & Travia, L. (1983). Efficient three-dimensional global models for climate studies: Models I and II. *Monthly Weather Review*, **111**, 609–62.

Harvey, D. R., Rehman, T., Jones, P. &

Upton, M. (1992). The Centre for Agricultural Strategy's Land Use Allocation Model. Paper presented at the Agricultural Economics Society Annual Conference, Aberdeen, April, 1992.

International Bank for Reconstruction and Development (IBRD)/World Bank. (1990). *World Population Projections*. Baltimore: Johns Hopkins University Press.

International Benchmark Sites Network for Agrotechnology Transfer (IBSNAT) (1989). *Decision Support System for Agrotechnology Transfer version 2.1 (DSSAT V2.1)*. Honolulu: Department of Agronomy and Soil Science, College of Tropical Agriculture and Human Resources, University of Hawaii.

Intergovernmental Panel on Climate Change (IPCC) (1990). *Climate Change: The IPCC Assessment*, ed. W. J. McG. Tegart, G. W. Sheldon & D. C. Griffiths. Canberra: Australian Government Publishing Service.

Intergovernmental Panel on Climate Change (IPCC) (1992). *Climate Change 1992*, ed. J. T. Houghton, B. A. Callender & S. K. Varney. Cambridge: Cambridge University Press.

Jones, C. A. & Kiniry, J. R. (1986). *CERES-Maize: a Simulation Model of Maize Growth and Development*. College Station: Texas A&M Press.

Jones, J. W., Boote, K. J., Hoogenboom, G., Jagtap, S. S. & Wilkerson, G. G. (1989). *SOYGRO V5.42: Soyabean Crop Growth Simulation Model. Users' Guide*. Gainesville, Florida: Department of Agricultural Engineering and Department of Agronomy, University of Florida.

Kimball, B. A. (1983). Carbon dioxide and agricultural yield. An assemblage and analysis of 430 prior observations. *Agronomy Journal*, **75**, 779–88.

Parry, M. L., Hossell, J. E., Jones, P. J., Rehman, T., Tranter, R. B., Marsh, J. S. & Carson, I. (1996). Integrating global and regional analyses of the effects of climate change: a case study of land use in England and Wales. *Climatic Change* 1: 185–198.

Ritchie, J. T. & Otter, S. (1985). Description and performance of CERES-Wheat: a user-orientated wheat yield model. In *ARS Wheat Yield Project*, ed. W. O. Willis, pp. 159–75. Washington, DC: Department of Agriculture, Agricultural Research Service, AARS-38.

Ritchie, J. T., Singh, U., Godwin, D. & Hunt, L. (1989). *A User's Guide to CERES-Maize V2.10*. Muscle Shoals, Alabama: International Fertilizer Development Center.

Rosenzweig, C., Parry, M. L., Fischer, G. & Frohberg, K. (1993). *Climate Change and World Food Supply*. Research report no. 3. Oxford: Environmental Change Unit.

Rosenzweig, C. & Parry, M. L. (1994). Potential impact of climate change on world food supply. *Nature*, **367**, 133–8.

Rowntree, R. R., Callender, B. A. & Cochrane, J. (1989). Modelling climate change and some potential effects on agriculture in the UK. *Journal of the Royal Society of England*, **149**, 120–6.

United Nations (1989). *World Population Prospects 1988*. New York: United Nations.

25 Incorporating land-use change in Earth system models illustrated by IMAGE 2

R. Leemans

Introduction

The Global Change and Terrestrial Ecosystems (GCTE) core project of the International Geosphere-Biosphere Programme (IGBP) aims to analyse the consequences for natural and agro-ecosystems of changes in climate, atmospheric composition and land use. It further emphasizes the interactions of all those ecosystems with, and feedbacks to, the atmosphere and the physical climate system (Steffen *et al.*, 1992). Land-use change is thus a central theme within GCTE, which should be considered by most of the GCTE core research projects. However, most GCTE research has taken a passive attitude towards land-use change; it is listed as one of the important factors, but rarely effectively included in experimental designs or model developments. If included, the analysis is not directed towards land-use change, but only focused at major biogeochemical impacts. Land-use change is often restricted entirely to tropical deforestation.

It is, of course, envisioned that the Land-Use/Land-Cover Change core project (LUCC; Turner *et al.*, 1993) developed by both IGBP and the International Human Dimension Programme (IHDP) could lead to an improved understanding of global and regional land-use change, which can directly be integrated with the GCTE core research projects. This will probably be difficult in the short term, due to a strong emphasis of natural sciences by GCTE, while LUCC will consist of an integration of both natural and social sciences. The problems involved in such multidisciplinary integration will only yield results on a longer time scale. Therefore, GCTE clearly should address its land-use change requirements explicitly and comprehensively, should actively pursue the involvement of dynamic land-use change research and communicate openly its findings, results, additional requirements and understanding to LUCC and other core projects.

In this chapter, I present first the significance of land-use and land-cover change for influencing major processes of the global Earth system. I will illustrate this emphasizing the links between land-use and land-cover change and the sources, sinks and fluxes of different Greenhouse Gases

(GHGs), and their interactions with the global Earth system. This will be followed by a short review of different approaches to include land-use and land-cover change in different global biogeochemical models, their objectives and applications. This type of model is generally used to simulate the characteristic fluxes of GHGs.

The discussion will not be limited to past and current condition of the global Earth system. Those states could adequately be used, for example, to validate current understanding and to quantify past and current processes that characterize major biogeochemical and physical climate processes. However, the challenge now facing the research community is not only to describe and explain those past and current processes, but also to address their plausible future dynamics. Policy makers and their advisors (often well-known scientists) want to implement (cost-)effective policies to address climate change issues. This is most clearly manifested in the Climate Treaty, which was signed by c. 150 countries at the UNCED in 1992. The objective (Article 2) of this treaty is:

> To achieve stabilization of greenhouse gas concentrations in the atmosphere at a level that would prevent dangerous anthropogenic interference with the climate system. Such a level should be achieved within a time-frame sufficient to allow ecosystems to adapt naturally to climate change, to ensure that food production is not threatened and to enable economic development to proceed in a sustainable manner.

Satisfying this objective demands close collaboration between many disciplines and covers several sectors. Innovative modelling techniques that can integrate the understanding of those disciplines and sectors will be among the necessary tools to evaluate possible policy options. I will, therefore, present recent integrated modelling developments in which socio-economic and natural science aspects of the global system are incorporated. Such developments should ultimately result in improved assessments (Dowlatabadi & Morgan, 1993).

I will follow the accepted terminology of FAO (1993) and IGBP/IHDP (Turner et al., 1993) in defining land, land use and land cover. Land refers to the relatively stable solid surface of the Earth, including the soil with its underlying geology and hydrology, biosphere and atmosphere. Land cover refers to the actual appearance of land, while land use refers to the purpose for which humans exploit and alter land and land cover. Land use as defined here is thus of lesser importance for GCTE research, but land use and land-use change lead to significant changes in cover. Therefore land use is more than just an environmental issue. It has profound impacts on biogeochemical and physical climate processes.

Typologies of land-use and land-cover change

Anthropogenic land-cover change is traditionally classified by distinguishing two types of change: conversion and modification (Fig. 25.1). Land-cover conversion refers to the complete replacement of one cover type by another (Turner *et al.*, 1993). Major land-cover conversions have occurred as a consequence of, for example, the increased demand for crops, livestock and wood for fuel and timber. Land-cover conversions usually lead to a decrease in forests and other natural covers, while areas for agriculture, habitation, infrastructure, industry and mineral extraction increase. The major conversions (mainly deforestation) were already accounted for in the earliest global climate change studies, albeit poorly quantified.

Land-cover modification refers to the more subtle changes that affect the character of a land cover without changing its overall classification (Turner *et al.*, 1993). Land-cover change and especially degradation through erosion, overgrazing, forest dieback, nutrient depletion, desertification, salinization

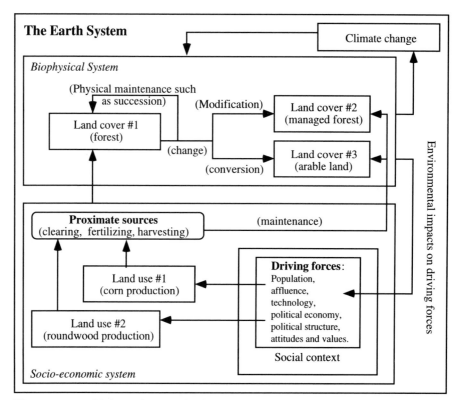

Figure 25.1   *Linkages between land use and land cover and defining different types of change. (Modified from Turner* et al., *1993.)*

and/or acidification is currently considered as a major environmental problem. Although the impacts of land-cover modifications are not very well documented, they could be considerable (Houghton, 1991). Unfortunately they are rarely considered in global climate change studies, but they should be, because they are a major influence of the global biogeochemical processes and required for estimating future concentrations of atmospheric GHGs.

Recently, a different typology has been defined (Turner *et al.*, 1990). This typology distinguishes between systemic and cumulative change. Systemic change involves the explicit modification of global properties of the Earth system, such as atmospheric composition and broad land-cover or climate patterns. The earliest global climate change research efforts have strongly emphasized this type of change by comprehensively focusing on global biogeochemical cycles and atmospheric and oceanic circulation.

Cumulative change has no immediate effects globally and is therefore much more difficult to perceive. For example, small local changes in land use resulting in minor modifications of land-cover seem to be negligible for the major properties of the Earth system. However, such changes often do not occur in isolation. One of the main features of current human activities is that they are repeated frequently, both in time and in space. Their intensity could further increase with an expanding population and further (lack of) economic development. The influence of all these activities becomes globally significant through their cumulative effects. Changes in forest cover, applications of fertilizers and increased urbanization are important examples of cumulative change. Cumulative change is not (yet) adequately covered in any global climate change analysis. Only some specific GHG emissions from modifications in land cover (Houghton, 1991) or land uses (Matthews & Fung, 1989; Crutzen & Andreae, 1990) have been estimated. Recently, cumulative change has obtained more attention with the wider application of geographic information systems, availability of global data sets and more advanced models.

The importance of land use for sources and sinks of GHGs

The atmospheric concentrations of many GHGs are closely linked to each other through their involvement with and interactions in chemical processes in the atmosphere (Prinn, 1994). When compiling a list of the sources and sinks of GHGs (Table 25.1) it is apparent that the terrestrial biosphere plays a major role in determining their actual emissions and thus final atmospheric GHG concentrations. Here I will shortly present the different GHGs and discuss their emissions in relation to land-use and land-cover

Table 25.1 *Sources and sinks of the major greenhouse gases*

	Sink	Source
Carbon dioxide (CO_2)	Oceans	Natural: Soils, ecosystems
	Terrestrial biosphere Atmospheric increase	Human: Fossil fuel combustion, land use changes
Methane (CH_4)	Atmospheric removal (through OH radical) Removal by soils Atmospheric increase	Natural: Wetlands, termites, oceans and freshwaters, CH_4-hydrates Human: Mining, rice paddies, enteric fermentation, animal waste, sewage treatment and landfills, biomass burning
Nitrous Oxide (N_2O)	Removal by soils Photolysis in the stratosphere Atmospheric increase	Natural: Oceans, soils, forests and grasslands Human: Cultivated soils, biomass burning, fossil fuel combustion, industrial production, fertilizer use
Halocarbons (e.g. CFCs, CCl_4, CH_3Cll_3)	Photolysis in the stratosphere Atmospheric increase	Natural: None Human: Aerosol propellants and blowing agents, industrial solvents, refrigeration
Ozone (O_3) and its tropospheric precursors (carbon monoxide, hydrocarbons and nitrogen oxides)	Reduction of stratospheric ozone by halocarbons Tropospheric increase	Natural: Ultraviolet radiation, lightning, soils Human: Fossil fuel combustion, biomass burning, aircraft emissions
Aerosol sulphates (H_2SO_4, an 'anti' greenhouse gas)	Precipitation from the atmosphere	Natural: Oceans (DMS), soils and vegetation, volcanic emissions Human: Fossil fuel combustion, biomass burning, aircraft emissions

Adapted from Houghton *et al.* (1990, 1992).

change. (Although water vapour is one of the major GHGs, I will not discuss it here. From general understanding of its role in the global hydrological cycle and climate system, it should be clear that land use and land-use change strongly interact on local, regional and global levels.)

Carbon dioxide

The globally averaged concentration of atmospheric carbon dioxide (CO_2) increased from its preindustrial values of *c*. 280 ppmv to 353 ppmv in 1990 and has been increasing since, although in the early nineties with a somewhat slower rate (Sarmiento, 1993). This increase is the direct result of emissions from the burning of fossil fuels and changes in land use, and affected by the uptake by the oceans and terrestrial biosphere. The magnitude of the latter sink is strongly influenced by ecological and physiological feedbacks and land cover. Many of these feedbacks have been analysed and modelled in a systemic way, but small-scale cumulative physiological processes have not been addressed.

Emissions from fossil fuel use are well established and are estimated to be 6.0 ± 0.5 GtC per year (Watson *et al.*, 1992). Emissions from land-use and land-cover change are less well known. Generally they are estimated to be between 1.1 and 3.6 GtC per year (Houghton, 1991). The wide range of these estimates is due to unreliable estimates of tropical deforestation and other land-cover characteristics. Actual deforestation rates appear to be lower than those estimates and many deforested areas show considerable regrowth (Skole & Tucker, 1993). In most global studies little attention is paid to the actual patterns of land use, the management activities involved and the consequences for the resulting land cover (Fig. 25.1). Most emphasis has been directed towards the large-scale conversion of land-cover types or ecosystems, while little or no attention has been paid to the finer-scale conversions and modifications. With adequate remote sensing techniques with a wide regional and continental cover, it is hoped that these limitations will be overcome shortly.

A major sink of CO_2 is the ocean. The total annual net flux from the atmosphere to the oceans is estimated to be 2.0 ± 0.8 GtC (Watson *et al.*, 1992). The magnitude of this flux is relatively well established, although nowadays more attention is drawn towards interannual, spatial and seasonal variation and links with oceanic circulation and the climate system (e.g. Winguth *et al.*, 1994). The terrestrial biosphere is the other important sink of CO_2, where it is absorbed through photosynthesis by plants. If a global C budget is compiled from the atmospheric increase, the emissions from combustion of fossil fuels and land-use and land-cover change, the oceanic uptake and the assumed terrestrial sink, then a significant amount of unaccounted C still remains.

This 'missing sink' is believed to be located in the terrestrial biosphere, because all other segments of the global C budget are supposed to be well understood or their uncertainties are supposed to be smaller. Several studies (e.g. Tans *et al.*, 1990; Enting & Mansbridge, 1991; Kauppi *et al.*, 1992) have indicated that especially biospheric processes in northern latitudes could be responsible for such a missing sink. Many of these processes are cumulative and not systemic, and therefore difficult to quantify. The processes involved are thought to be N-fertilization through air pollution and increased fertilizer use in agriculture, forest regrowth, climatic warming and CO_2-fertilization. However, the actual magnitude (and often also the net direction) of these processes are not well understood or documented, because heterogeneous spatial and temporal patterns are involved and most of the relevant processes are (again) strongly influenced by land-use and land-cover change. Up to now no comprehensive regional or continental evaluation of the importance of these processes exists. Without a better documentation and understanding, the discussion on balancing the global C budget will probably continue for some time.

Methane

Methane (CH_4) is another important greenhouse gas. CH_4 is strongly involved in atmospheric chemistry and tropospheric ozone production (Prinn, 1994). Its oxidation in the stratosphere leads to water vapour and CO_2, both significant GHGs (Watson *et al.*, 1992). Atmospheric CH_4 concentrations have more than doubled since preindustrial times and are still increasing, although with a somewhat decreasing rate, due to unknown combinations of different processes. Anthropogenic sources now dominate over natural sources. About 20% of the total emissions are related to human energy production or derived from activities such as mining, distribution and use of fossil fuels. Other important contributions are landfills and waste-treatment plants. There is a clear, but globally less significant, land-use component in all these human activities.

Emissions from ecosystems are by far the largest natural source of methane. Termites, tropical swamps and boreal peat bogs all produce significant amounts of CH_4. Future emissions from wetlands sources can be strongly influenced by a changing climate. The total extent of wetlands has decreased over the last century, due to reclamation for agricultural use. Such changing land cover could limit CH_4 emissions somewhat, but current trends seem to protect and restore large wetlands systems for their biodiversity, coastal protection or recreational value. Future policies concerning wetlands are thus important in estimating CH_4 emission trends.

Agricultural activities are a major anthropogenic source. Rice paddies and man-made wetlands contribute significantly globally. Regionally there are large differences due to management practices (such as fertilization and irrigation), species selection, soil characteristics and climate. Another significant agricultural source is animal husbandry. Especially the CH_4 emissions from enteric fermentation in cattle could be as large as those from rice paddies. All emissions from agricultural activities are logically closely related to land use. Their patterns are locally and regionally defined and strongly influenced by management practices, soil and climate. They are typical examples of cumulative contributions to atmospheric concentrations.

Soils are currently a significant sink of CH_4, but changes in land use through enhanced nitrogen input (either by fertilization or pollution) are decreasing this uptake (Mosier et al., 1991). This reduction is currently probably small, but this could change under a changing climate. The local and regional patterns are thus important in characterizing the future trends in CH_4 fluxes.

Nitrous oxide

Nitrous oxide (N_2O) is an important long-lived GHG and is also the primary source of stratospheric N, which controls the abundance and distribution of ozone. N_2O has increased by about 8% since pre-industrial times.

Ecosystems are important sources of N_2O. Denitrification and nitrification are the main processes in which N_2O is released. The release rates are strongly dependent on the aerobic, moisture and other environmental conditions within the soil. Quantification of global N_2O emissions from soils are difficult because of the large local and regional variability in these environmental conditions. Application of N fertilizer influences emission levels as it alters initial soil N levels. Land-use change, such as deforestation, could enhance N_2O emissions temporarily, but depending on the succeeding land management could return to earlier, or even lower levels (Keller et al., 1993). Much depends on the new environmental conditions. Land-use change is thus an important determinant of past, current and future N_2O emission levels and severe changes in emission patterns could occur under a changing climate.

Unfortunately, few studies on N_2O-emissions over wider regions are available to estimate comprehensively the total emissions. Up to now data are still too fragmented and limited to only a few ecosystems and environmental conditions.

Halocarbons

The halocarbons (e.g. CFCs and CCl$_4$), which are among the strongest greenhouse gases, all derive from human sources. They are used in technical processes, such as refrigeration and cleaning. The worldwide consumption of these gases has decreased significantly after the Montreal protocol and its subsequent amendments, because of the availability of effective substitutes. No land use or land-use change processes are involved in the emissions of halocarbons.

Ozone and its precursors

Ozone (O$_3$) is an effective GHG in the upper troposphere and lower stratosphere, and also plays a key role in the absorption of ultraviolet radiation, which can damage essential functions of organisms. A downward trend of ozone concentrations in high latitudes has been observed (Watson *et al.*, 1992). The main sink is the reduction of stratospheric ozone by halocarbons, while the main sources are stratospheric photolysis and tropospheric production through chemical reactions involving carbon monoxide, hydrocarbons and several oxides of nitrogen (Prinn, 1994). These gases are the most important ozone-precursors and must therefore be included in any integrated assessment of global climate change.

Land use or land-use change are not directly involved in the ozone cycle, but important determinants of the concentrations of ozone precursors. Ozone precursors derive from processes like fossil fuel combustion and biomass burning. The latter is strongly land use-defined and contributes significantly. The regional and temporal patterns of biomass burning and its related emissions are not very well described because of their heterogeneous character. Only coarse estimates of the total global emissions of ozone precursors exist. Biomass burning is a good example of cumulative change, while only temporally modifying land cover.

Others

Recently, sulphur-containing gases and related aerosols have received more attention. Besides their importance in atmospheric chemical processes, they scatter solar radiation directly back to space (Taylor & Penner, 1994) and influence cloud characteristics (Jones *et al.*, 1994). The total impact has been evaluated as a negative radiate forcing, thus neutralizing the increased greenhouse effect of GHGs. Defining a globally average radiative forcing is impossible, due to strongly regionalized sources and their short atmospheric lifetimes. The influence of these gases is therefore strongly regional and much more complex than the other GHGs. The major sources of sulphur dioxide (SO_2) are the combustion of fossil fuels and several industrial processes. These sources are located mainly in North America and Europe, but are nowadays significantly reduced through pollution control measures. In the near future a shift in emission patterns can therefore be expected towards the other, rapidly developing parts of the world, such as South-East Asia. Emissions from volcanos, oceans (S in dimethyl sulphide, DMS), plants, soils (DMS and H_2S) and biomass burning also add to the atmospheric sulphur budget. The emissions of the latter two sources are again strongly influenced by land use.

Earth system models

A short review of different approaches

Many global climate change assessments have emphasized the importance of atmospheric CO_2 since it possesses a strong radiative forcing, has a long atmospheric life time and is relatively abundant. Small changes in the global C budget can considerably influence the atmospheric CO_2 concentration. The global C cycle has therefore received most attention and, initially, other GHGs were sometimes neglected in defining atmospheric radiative forcing.

This interest has resulted in a wealth of different global C cycle models. Most models distinguished the main compartments: atmosphere, ocean and terrestrial biosphere (cf. Bolin, 1981). The earliest models were simple, just accounting for the budgets between these compartments. Each compartment was functionally divided into subsystems. Early models (e.g. Emanuel et al., 1984; Goudriaan & Ketner, 1984) distinguished a few ecosystem types and land-use and land-cover change was neglected or assumed to be identical to an assumed deforestation rate in the tropics. Only the impact of these prescribed conversions between ecosystems was modelled. Ecosystems were characterized by their extent and productivity and this approach allowed for straightforward terrestrial and global C budget calculations.

A next step was to increase the number of modelled ecosystems, to use more realistic global ecosystem patterns and to enhance the productivity calculations. Initially this was done by using bioclimatic classifications, such as the Holdridge Life Zones (Tans et al., 1990) or BIOME (Prentice et al., 1994). Global climate databases were used to delineate ecosystem patterns and characteristic local productivity figures were assigned to calculate the global terrestrial C storage. An advantage of this approach was that the climatic feedback on ecosystem distribution could be considered. The main conclusion of these studies was that the C content in the biosphere has been increasing since the last Glacial Maximum (18 000 BP) and could potentially continue to do so in the near future if climate warms. The main limitations of these approaches are that only potential or natural vegetation is considered and that important human influences and vegetation dynamics are neglected.

The modelling was further elaborated upon by Raich *et al.* (1991). They use a gridded database for current land cover, climate and soil to implement a regional and global C cycle model, the Terrestrial Ecosystem Model (TEM). TEM is a highly aggregated, process-based model, developed to investigate interactions among terrestrial ecosystems and environmental variables. It includes an advanced environmental and water balance module that controls C and N fluxes in and out of soil and vegetation, thereby dynamically determining and linking the C and N budgets in each grid's ecosystem. The direct responses to changing atmospheric CO_2 concentrations and climate change are simulated as a result of ecophysiological processes. Simulations are run until equilibrium states occur between N, C and hydrological cycles. Recent versions of TEM are able to include land-use and vegetation change, but a dynamically changing land cover is not yet envisioned (D. Kicklighter, personal communication). Although the model has been very useful in defining biospheric responses to a changing environment (Melillo *et al.*, 1993), the omission of dynamic changes in natural ecosystems and land use makes it less appropriate to be used in an integrated model for global climate change assessments.

Among the few global C cycle models that include land use in a more satisfactory way is the Osnabrück Biosphere Model (HRBM; Esser, 1991). This model is implemented on a grid and initialized with data on soil, climate, land cover and land-use change. Grid cells are modelled as heterogeneous units: their land cover consists of different assemblages of several agricultural and natural vegetation classes (Esser, 1984). Agricultural yield and ecological productivity is determined from a regression of production and climate. Changes in agricultural area (cf. changes in land use) are determined from historic country-based yield and population figures. Future land use is determined by logistic extrapolations of these trends. Localities where land-use change actually occurs are estimated from a simple ranking of all cells according to agricultural suitability (Esser, 1989). CO_2 fertilization, climate and other feedbacks within the terrestrial biosphere are included.

The advantage of this approach is that the consequences for the global C cycle can easily and directly be determined. However, the inclusion of land-cover change is still limited. Land-cover change is mainly defined as an increase of agricultural area and therefore only land-cover conversions are considered. The global significance of anthropogenic land-cover modification through land-use changes is still neglected. Further, only ecophysiological feedbacks on the major processes are considered. In all these models land-use and land-cover change do not lead to changes in emissions of other GHGs. For global climate change assessments, however, these emissions are an important contribution.

There has been a series of different global climate change assessments, but all were characterized by focusing on either the causes and resulting emissions or the impacts. These studies have always strongly emphasized energy and industrial activities and these sectors were often assessed with economic models (e.g. Darmstadter & Edmonds, 1989). The impacts assessments have traditionally been focused on specific topics, such as agriculture (Parry *et al.*, 1990), hydrology (Kwadijk, 1993), ecosystems (Leemans, 1992), and biodiversity (Peters & Lovejoy, 1992). The disciplinary focus of all these studies restricted their applicability for policy support. There have been several studies that link the impacts of climate change across sectors (e.g. Scott, 1991; McKenny *et al.*, 1992; Rosenzweig & Parry, 1994) and these, often called integrated assessments, only integrate series of impacts and possible responses, but neglect feedbacks and interactions with other parts of the Earth system.

The only true integrated assessments are the disciplinary expert judgements in 1990 and 1992 by IPCC (Houghton *et al.*, 1990; Izrael *et al.*, 1990; Bernthal, 1990; and their 1992 supplements). Several assessment models have been used to assist in different segments of these reports, but no globally comprehensive integrated assessment model was available then. Land-use and land-cover change has been discussed in detail in these reports, but deforestation and biomass burning were the only land-cover changes used for the identification of their emission scenarios.

During the last few years a suite of more advanced integrated assessment models have been developed for different regions (Dowlatabadi & Morgan, 1993). Most of these models emphasize the importance of the energy supply sector and are derived from different economic models and approaches. Although they calculate emissions from different sources and include atmospheric processes and simple functions for climate change and impacts (e.g. Rotmans, 1990; Dowlatabadi & Morgan, 1993; Edmonds *et al.*, 1994), land-use and land-cover change was not considered. Recently several groups (e.g. Batelle PNL, USA: Edmonds *et al.*, 1994; CGER: Morita *et al.*, 1994; CGER, Japan; MIT, USA: Prinn, 1994; and RIVM, The Netherlands: Alcamo, 1994) have started to develop more realistic global climate change models. Only one of those models, the IMAGE 2 model, has been fully documented to date. I will therefore limit the following discussion to this model, presenting the scope of the model and some of its potential applications. The other models differ in details, but the overall rationale is similar: to obtain an adequate simulation framework of global climate change by simulating major sources and sinks of GHGs, biogeochemical, atmospheric and climatic processes, and the linkages and feedbacks between different components of the system.

The IMAGE 2 Model

We (Alcamo, 1994) started to develop an integrated assessment model that should overcome many of the limitations of earlier assessments. This model, IMAGE 2 (Integrated Model to Assess the Greenhouse Effect) should ideally represent the major aspects of the global society-biosphere-climate system. Land-cover conversion and modification are both modelled, so that the most dominant characteristics of land use-related GHG emissions are simulated.

Objectives

Earlier versions of IMAGE (Rotmans, 1990; Rotmans *et al.*, 1990) were globally aggregated models to estimate GHG emissions from energy use, industrial activities and land-cover change (read tropical deforestation). These emissions were amalgamated in the atmosphere, while accounting for simple atmospheric chemistry and global carbon cycling. The final radiative forcing of the final atmospheric concentrations of GHGs was determined and used by the model to compute global annual mean temperature changes and sea-level rises. Little or no societal or ecological impacts of such changes were considered.

Later IMAGE was included in the ESCAPE framework (Evaluation of Strategies to Address Climate Change by Adaptation to and Preventing Emissions: Rotmans *et al.*, 1994). ESCAPE was an improvement of IMAGE, in that climate change was regionalized and its impacts were simulated for several sectors such as agriculture, natural ecosystems and transport. ESCAPE thus gave a more complete picture of many aspects within global climate change. However, these models were strongly based on a cause–effect modelling approach, where the impacts (climate change and subsequent) do not influence the causes. From many global climate change analyses, one knows that this is not realistic and future emission projections with these models could therefore be misleading. For example, the polewards shift of many vegetation zones will influence global carbon dynamics

(Smith *et al.*, 1992) and changes in agricultural productivity will lead to different land-use patterns (Solomon *et al.*, 1993).

The experience with these earlier models has led to the development of the more innovative framework for modelling of global climate change, IMAGE 2 (Alcamo, 1994). IMAGE 2 is a multidisciplinary, integrated model and is designed to provide a dynamic and long-term overview of climate change issues to support a science-based evaluation of national and international policies concerning the buildup of GHGs in the atmosphere. To provide reliable estimates the model incorporates the different linkages and feedbacks and their relative importance of the society-biosphere-climate system. The model should therefore be able to estimate important sources of uncertainty in such linked modelling systems. The model could further help in identifying gaps in knowledge and data availability in order to help set the agenda for global climate change research.

These objectives are ambitious and require the development of detailed and demanding model approaches. However, to support evaluation of different sets of policy options, results from a simulation should be obtained quickly and be relatively straightforward to analyse. These requirements lead to a different approach than those from highly specialized, disciplinary research models. IMAGE 2 consists therefore of a set of linked and simplified disciplinary models. Each submodel has been tested and validated against available data sets and has ideally been accepted by its own scientific discipline. The linkages between the models consist of transparent information flows that allow for tests of uncertainties and implementation of different assumptions. Most socio-economic parameters that influence the system, such as wealth, population and technology assumptions, are provided externally in order to describe (objectively) the impact of changes therein.

Structure

IMAGE 2 consists of three fully linked components (Fig. 25.2): the Energy–Industry System (EIS); the Terrestrial Environment System (TES); and the Ocean–Atmosphere System (OAS). Dynamic calculations are performed for a 100 year time horizon. All socio-economic submodels are regionally explicit (i.e. country or continent), while all environmental and physical submodels are geographically explicit and implemented on a grid of 0.5° longitude and latitude. Each grid cell is characterized by a set of climate, vegetation and soil properties. Atmosphere simulations in OAS are calculated for latitudinal belts and several vertical layers. The ocean calculations are done for different compartments, reproducing major

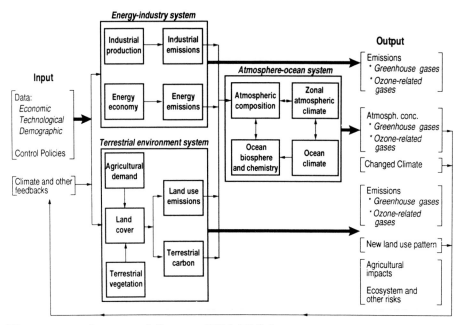

Figure 25.2   *A conceptual diagram of IMAGE 2.*

processes at different depths and surface layers. One of the most innovative accomplishments of the IMAGE 2 framework is an adequate linkage between the structurally different dimensions and resolutions of its submodels using heuristic models, similar to the cellular automata (Wolfram, 1984).

The objective of EIS is to compute the emissions of GHGs in each region as a function of energy consumption and industrial production. The models are designed especially for investigating the effectiveness of changed fuel mixes (including biomass), improved energy efficiency and technological development on regional GHG emissions. de Vries *et al.* (1994) give a complete description of this model, its inputs and underlying assumptions.

The objective of TES is to simulate global land use and land cover dynamically through time and employ these changes to determine GHG fluxes between the terrestrial biosphere and the atmosphere. The set of TES models can be used to evaluate the effectiveness of mitigation policies, such as changed agricultural practices or sequestration of carbon through forestation and the impact of climate change and changing atmospheric composition on ecosystems and agriculture. TES reconciles the demand for land with its potential (Zuidema *et al.*, 1994). Regional land-use demand is computed by calculating the demands for agricultural products (crops, livestock, lumber and fuelwood) using socio-economic and demographic input parameters and assuming some prescribed trade between regions. Land-use potential is determined from local potential vegetation and crop productivity using climate and soil characteristics (Leemans & van den Born, 1994). These two are linked through heuristic rules which reflect key driving factors in land-use change, such as proximity to current infrastructure, population and most productive lands. The result is a dynamically changing land cover. All land cover types with a potential agricultural production can be converted into new agricultural lands, so an increased agricultural area does not necessarily lead to deforestation as in many earlier models.

The heuristic rule-based model generates regionally different patterns of land-cover change. In several developed regions the extent of agricultural land contracts, while in others there is a large increase. These dynamics are a result of the regional differences in driving forces. The resulting land cover and agricultural activities serve as a basis for computing non-CO_2 GHG emissions (Kreileman & Bouwman, 1994). Each grid cell is characterized by its own C cycle, with its specific land use and environment-dependent carbon stocks and fluxes. Conversion from its original land cover to agriculture leads to a significant CO_2 emission from the total C stock. Conversions from agricultural land to a more natural state occur gradually, depending on the resulting land cover type. Feedback caused by changes in atmospheric CO_2 concentrations and climate are accounted for (Klein Goldewijk *et al.*, 1994).

The purpose of the OAS is to compute the build-up of GHGs in the atmosphere and the resulting change in temperature and precipitation patterns. Emissions from EIS and TES are combined and used to determine the physical and biological oceanic C uptake. The atmospheric composition model simulates globally averaged tropospheric and stratospheric chemistry, including some of the processes related to ozone production and destruction. The atmospheric energy balance submodel combines the resulting atmospheric level of GHGs with additional data to compute the Earth's energy balance and climate for latitudinal averaged zones. The longitudinal dimension is obtained by scaling these zonal averages to the terrestrial grid using results from global atmospheric circulation models and a high-resolution database of current climate.

TES is linked to EIS via the demand for fuelwood and land requirements for other biofuels. TES and OAS are directly linked through the greenhouse gas emissions, and indirectly through changes in soil moisture and albedo, caused by land-cover change. This feedback influences the global radiation balance. OAS is linked with TES through its climatic change patterns. The whole IMAGE 2 framework provides an integrated simulation of the Earth system, where influences in one section directly influence processes in other sections.

Applications

Several preliminary applications that capture some of the richness of information and results have been worked out. The heterogeneity and quantity of the results of a single IMAGE 2 simulation limit this presentation only to a few aspects. I will therefore only present a short summary of the model's capabilities and focus on the implications for land-use and land-cover change. Alcamo *et al.* (1994) provide more details and the scope of the different simulations. It is important to provide a baseline simulation with which others can be compared. We (Alcamo *et al.*, 1994) have defined the 'Conventional Wisdom Scenario' (CWS) as our reference or baseline scenario with which to compare the results of other simulations. IMAGE 2 thus aims to provide not only a plausible future state of the Earth system, but also allows comparison of the sensitivity of the system with a different set of complex assumptions.

The CWS makes 'conventional' assumptions about future demographic, economic and technological driving forces. No specific climate-related policies are implemented. Input data are, wherever possible, similar to the IS92a scenario of IPCC (Houghton *et al.*, 1992). Population will more than double by the year 2100, reaching 11.5 billion people, but regionally there are large differences. Economic growth assumptions range from 1.3 to 4.5% and decline with time in developed regions, while increasing significantly towards the end of next century for developing regions. The current per capita income gap between regions remains, nevertheless. The specific energy/industry-related assumptions are given in detail by de Vries *et al.* (1994) and Alcamo *et al.* (1994). In general, consumption increases proportionally with GNP, although the use of passenger vehicles shows a slower increase, based on the assumption that in many regions it is and will be severely constrained by available infrastructure. Through time fuel is converted more effectively into energy. Most GHG emission factors are held constant over time. Only N_2O emissions from the transport sector are assumed to increase due to a wider use of catalytic converters in vehicles.

Future agricultural demand strongly affects land-use patterns and leads to land-cover change and influences GHG fluxes. This demand depends not only on local demands, but also on demands created through possibilities of global trade. We have assumed that exports from the developed world increase by 50% from 1990 to 2050 and level off afterwards. Net exports from developing regions, if food demand from the local populations allows, double their 1990 level by 2100. Import is weighted according to regional consumption. The yield of a crop is assumed to increase based on increased fertilizer use and other technological inputs. These fertilizer-use figures are taken from the IS92a scenario.

These assumptions lead to a new future land-cover pattern (Plate 6, between pp. **204** and **205**). These patterns can easily be explained. In North America and Europe the consumption of many agricultural commodities reaches saturation levels early in the next century. This and continuing increases in agricultural productivity lead to a gradual decrease in the extent of agricultural land. We assume that the excess land is returned towards its climatically determined natural vegetation and thereby also gradually establishing its typical carbon storage. (Playing golf will thus not become a significant land use in these regions!) Latin America shows a somewhat different course. Current agricultural land and pasture increase for a few decades, but with declining rates. In the middle of next century a similar situation occurs as described above and the forest areas increase again.

The pattern is totally different in the regions of India, China and the rest of South-East Asia and, especially, Africa. The additional demands from an increased population and higher income levels reverse the increase in agricultural productivity, so the extent of agricultural land increases continuously during the next century and little natural vegetation cover remains. The global increase in agricultural land under this scenario during next century is not unrealistically large (Table 25.2) and consistent with the increase that we have seen during the last half of this century (Plucknett, 1994), but the simulated shift in deforestation from Latin America towards Africa and other regions illustrates the highly dynamic, regional and non-linear character of land-use and land-cover change.

To investigate the sensitivity for different energy assumptions on land use and land cover, the CWS assumptions have been changed slightly. Bio-fuels satisfy 75 EJ and 210 EJ in energy production in respectively 2050 and 2100. CWS has been based on the assumption that this requirement easily can be met by using crop residues and different kind of wastes (cf. Johannson *et al.*, 1993). Biofuels therefore do not require additional agricultural land. Two other scenarios were run with variations on this assumption. One, the biofuel crop scenario, assumes that residues cannot adequately satisfy this total biofuel demand and 60% of the energy comes from special biomass crops with short rotations, such as *Salix* spp. and *Miscanthus* spp. Additional agricultural land is needed. The third scenario just assumes that no biofuels are used and this additional energy demand is satisfied by fossil fuels. Table 25.2 lists the consequences for the extent of global agricultural land needed. The biofuel crop scenario requires the largest amounts of new agricultural land, which cannot be used for durable C sequestration. The atmospheric CO_2 concentration is therefore still considerably higher than the CWS and only a little lower than the fossil fuel scenario. Note, however, that the deforestation rates are only slightly

Table 25.2 *Difference in land cover patterns under different assumptions to satisfy future energy demand*

	Atmospheric CO_2 concentration (ppmv)	Agricultural area (10^3 km²)	Forest area (10^3 km²)
1990	492	26.7	47.2
2050			
Conventional wisdom	522	+9%	−26%
Biofuel crop	534	+30%	−32%
Fossil fuel	539	+9%	−26%
2100			
Conventional wisdom	777	+14%	−27%
Biofuel crops	821	+65%	−31%
Fossil fuel	857	+15%	−27%

Additional Energy needs: 75 EJ in 2050 and 210 EJ in 2100. Sources: Conventional Wisdom: 100% crop residues; biofuel crops: 60% crop residues and 40% biomass crops; fossil fuel: 100% oil and no biofuels.

higher in the biofuels scenario. The land-cover change model projects large biofuels plantations on grasslands and other non-forest land.

An unexpected result of this analysis is the unabated emissions of other GHGs, especially methane and the ozone precursors. CWS and the biofuels scenario gave higher concentrations of these gases, which lead to the conclusion that the total radiative forcing of the scenarios were not significantly different. Measures to reduce only CO_2 are therefore doomed to fail. The emissions of all GHGs have to be evaluated simultaneously in determining the effectiveness of greenhouse abatement policies. And above all: land use and its resulting land-cover patterns play a major role in such evaluation.

Another application of the IMAGE 2 model is the evaluation of the mitigation potential by forestation. New forests store C and if they are left undisturbed or their wood products are used in durable, long-term applications, considerable additional C stores can be achieved. Using forest products and their residues as biofuels could further decrease the dependence of fossil fuels. This greenhouse abatement policy has received considerable positive attention (e.g. Winjum *et al.*, 1993), but can not offset all CO_2 emissions from fossil fuels (Sedjo & Solomon, 1989). However, forest could satisfy many other functions, such as fuelwood supply, erosion management, recreation and biodiversity, so that forestation could be achieved effectively.

Table 25.3 *Land suitability and availability (not used for other purposes) for the forestation mitigation option, according to IMAGE 2 simulation for the conventional wisdom scenario*

Zone	Currently suitable (10^6 km^2)	Currently available (10^6 km^2)	Suitable in 2050 (10^6 km^2)	Available in 2050 (10^6 km^2)
North America	10.2	6.1	11.7	7.9
Latin America	11.1	6.2	14.3	6.5
Europe	3.9	1.3	4.2	1.4
Former Soviet Union	11.5	8.5	14.1	9.6
Africa	4.5	2.5	6.3	0.2
India	0.9	0.5	1.3	0.0
South East Asia	3.2	1.3	3.6	0.5
China	4.5	1.9	5.5	0.2

Growing forest as a mitigation option competes for land with agriculture and other land uses. Under a future, warmer climate it has been shown that forest could potentially cover a larger area (Leemans, 1992). The vegetation distribution model (BIOME; Prentice *et al.*, 1992), which is incorporated in the IMAGE 2 model, shows a similar pattern between now (1990) and the middle of next century (Table 25.3). There is thus a larger future potential to store C in future forested ecosystems. But, this potential must be weighted against the requirements for agricultural and other land uses. Table 25.3 lists the forest lands available now and in the future for different regions. The pattern becomes very clear: there is only a possible increase in forest extent in the developed, temperate regions. In most others, large parts of the potential forest areas will be occupied by agricultural land and little opportunity for forestation arises. Here, I show the broad picture and there will, of course, be many more opportunities on a smaller scale, such as planting trees in shelterbelts and agroforestry systems. But these modifications require major remodelling of modern agricultural landscapes, and the current trend is to move away from smaller scale landscape elements.

Future developments

The above examples present useful applications of the IMAGE 2 model. The linkages between the different submodels make it possible to evaluate system-wide responses. These integrated assessments have proved to be very useful for policy evaluations (Dowlatabadi & Morgan, 1993). The model will be more frequently used to assist in policy evaluation. An assessment of some of the implications of Article 2 of the Climate Treaty is an

ongoing activity and it is further being used to evaluate a series of mitigation options, proposed by Working Group II of IPCC in their forthcoming assessment.

The IMAGE model still has serious limitations and uncertainties. For
example, many economic, cultural and social processes are not (yet)
considered. Further, the potential rain-fed productivity of a series of crops
is based on the local climatic and soil conditions. In reality, such
productivity is never obtained and large gaps remain between the potential
and actual agricultural production (Plucknett, 1994). In IMAGE 2 the
productivity of current agricultural regions are calibrated using data from
FAO (1991). This is correct for the current situation, but probably not for
a changed future distribution, especially if production becomes more limited
due to land degradation. A major future research activity is to improve the
agricultural production models and include land degradation. Other scientific uncertainties involve the realism of the simulated vegetation response
and the magnitude of different feedback mechanisms.

IMAGE 2 is based on a series of global environmental databases. Many
of those are included on the Global Ecosystem CD-ROM (Kineman, 1992).
However, the quality, resolution and included variables are not always what
is required. We have therefore implemented the underlying data structure
of IMAGE 2 in such a way that it allows for inclusion of new and
improved data sets. This requires, however, different calibration of submodels and will lead to different results. A complete validation of such model
is difficult, maybe impossible (Oreskes *et al.*, 1994), but the reliability of the
results depends to a large extent on the quality of the available data sets for
development, calibration and validation. Here, a lot of progress still has to
be made to improve on these databases.

Conclusions

The importance of different aspects of land-use and land-cover change for
global climate change processes has been described. Land use, land-use
change and the resulting land-cover modifications and conversions are
strongly involved in determining the final levels of atmospheric GHGs.
Models that attempt to determine past, current and future trends of these
gases must, besides including all GHGs and their interactions, also emphasize land-use and land-cover changes. These aspects of the Earth system are
not yet very well understood and highly dependent on diverse human,
cultural and socio-economic factors. These factors are by definition heterogeneous and locally, and only sometimes regionally or globally, defined.

They interact with major biogeochemical cycles and the physical climate system and these interactions actually differentiate possible future levels of GHG emissions.

Earth system models that are designed to cope with global climate change issues must, therefore, adequately include land-use and land-cover change. Aggregating such changes globally, as is done in the early C cycle models and early integrated assessment models, is insufficient because the underlying causal factors with their spatial and temporal heterogeneity are completely neglected. A more satisfying approach to cope with the heterogeneity demands local and regional descriptions of human activities and a georeferenced approach to define the environment. This approach was already applied successfully in the eighties by the RAINS model for acidification, developed at IIASA (Alcamo et al., 1987). Recently, several other global climate change and integrated assessment models have also adopted such an approach, (MIT: Prinn, 1994; Batelle PNL; Edmonds et al., 1994; and CGER: Morita et al., 1994), but their scope is regional, or land-use change is not yet covered dynamically. Of these models IMAGE 2 (Alcamo, 1994) is currently the furthest developed, implemented and applied.

The development and implementation of such models is still limited by the available environmental and socio-economic data. Data on global climate patterns are available (e.g. Leemans & Cramer, 1991), but lacks information on the variability. Data on soils are available in several databases (e.g. Zobler, 1986), but are only a poor extract of the original FAO soil map. (Fortunately, FAO has just released a high-quality digitized version of this map (FAO, 1993)). Although several attempts have been made, reliable data on the current status of land (and its degradation) are generally missing. The GLASOD database (Oldeman, 1991–2) gives the current degradation status, but no characteristics of underlying processes. Global land cover data are readily available, but of a highly heterogeneous quality and classification. The data is often not time-referenced and based on miscellaneous categories (Leemans et al., 1996). Global socio-economic data is available from many sources, but only of an even lower quality than the environmental data (e.g. WRI, 1992). Many projects and organizations, such as IGBP-DIS and CIESIN, are in progress to improve on this situation, but they are limited not only by their amount of funding and other resources, but also by the lack of reliable data to be collected for and distributed to the global climate change research community. Many regional and (inter)-national organizations that collect data use their own methods and standards. Global harmonization of those data is not a trivial task, but essential for high-quality integrated global climate change models and assessments.

The experience obtained with such integrated assessment models is still scanty, but they appear to be very useful tools, not only for defining and evaluating scientific issues, but also to comprehensively assess the effectiveness of sets of different policy options. I am convinced that such an approach allows us to obliterate barriers between scientific disciplines, between social and natural sciences and between research and policy. Especially, if the model structure and linkages are transparent, if the underlying assumptions are clear, if scientists can communicate their own disciplines, and if users can easily comprehend the complexity, this can be accomplished, but there is still a long and exciting way to go.

Acknowledgements

I thank Kees Klein Goldewijk, Rob Swart, Jelle van Minnen and two anonymous reviewers for critical remarks on earlier drafts of this manuscript. The research presented in this paper is funded by the Dutch Ministry of Housing, Planning and the Environment under contract MAP410 to RIVM. The Terrestrial Environment System of the IMAGE 2 model is an official contribution to IGBP-GCTE core research.

References

Alcamo, J. (ed.) (1994). *IMAGE 2: Integrated Modeling of Global Climate Change*. Dordrecht: Kluwer.

Alcamo, J., Amann, M., Hettelingh, J.-P., Homberg, M., Hordijk, L., Kämäri, J., Kauppi, L., Kornai, G. & Mäkelä, A. (1987). Acidification in Europe: A simulation model for evaluating control strategies. *Ambio*, 16, 232–45.

Alcamo, J., van den Born, G. J., Bouwman, A. F., de Haan, B., Klein Goldewijk, K., Klepper, O., Leemans, R., Olivier, J. A., de Vries, B., van der Woerd, H. & van den Wijngaard, R. (1994). Modeling the global society-biosphere-climate system, part 2: computed scenarios. *Water, Air and Soil Pollution*, 76, 37–78.

Bernthal, F. M. (ed.) (1990). *Climate Change: The IPCC Response Strategies*. Geneva: World Meteorological Organization/United Nations Environment Program.

Bolin, B. (ed.) (1981). *Carbon Cycle Modelling*. New York: Wiley.

Crutzen, P. J. & Andreae, M. O. (1990). Biomass burning in the tropics: Impact on atmospheric and biochemical cycles. *Science*, 250, 1669–78.

Darmstadter, J. & Edmonds, J. (1989). Human development and carbon dioxide emissions: the current picture and the long-term prospects. In *Greenhouse Warming: Abatement and Adaptation*, ed. N. J. Rosenberg, W. E. Easterling, P. R. Crosson & J. Darmstadter, pp. 35–49. Washington, DC: Resources for the Future.

de Vries, B., van den Wijngaard, R., Kreileman, G. J. J., Olivier, J. A. & Toet, S. (1994). A model for calculating regional energy use and emissions for evaluating global climate scenarios. *Water, Air and Soil Pollution*, 76, 79–131.

Dowlatabadi, H. & Morgan, M. G. (1993).

Integrated assessment of climate change. *Science*, **259**, 1813, 1932.

Edmonds, J., Wise, M., & MacCracken, C. (1994). Advanced energy technologies and climate change: an analysis using the global change assessment model (GCAM). Report 1. Washington, DC: Pacific Northwest Laboratory.

Emanuel, W. R., Killough, G. G., Post, W. M. & Shugart, H. H. (1984). Modeling terrestrial ecosystems in the global carbon cycle with shifts in carbon storage capacity by land-use change. *Ecology*, **65**, 970–83.

Enting, I. G. & Mansbridge, J. V. (1991). Latitudinal distribution of sources and sinks of CO_2: results of an inversion study. *Tellus*, **43B**, 156–70.

Esser, G. (1984). The significance of biospheric carbon pools and fluxes for the atmospheric CO_2: a proposed model structure. *Progress in Biometeorology*, **3**, 253–94.

Esser, G. (1989). Global land-use changes from 1860 to 1980 and future projections to 2500. *Ecological Modelling*, **44**, 307–16.

Esser G. (1991). Osnabrück Biosphere model: structure, construction, results. In *Modern Ecology, Basic and Applied Aspects*, ed. G. Esser, & D. Overdieck, pp. 679–709. Amsterdam: Elsevier.

FAO (1991). Agrostat PC. FAO Database series 1. Rome: Food and Agriculture Organization of the United Nations.

FAO (1993). World Soil Resources: An Explanatory Note on the FAO World Soil Resources Map at 1 : 25 000 000 Scale. World Soil Resources Report 66 rev. 1. Rome: Food and Agriculture Organization of the United Nations.

Goudriaan, J. & Ketner, P. (1984). A simulation study for the global carbon cycle, including man's impact on the biosphere. *Climatic Change*, **6**, 167–92.

Houghton, J. T., Callander, B. A. & Varney, S. K. (1992). *Climate Change 1992: The Supplementary Report to the IPCC Scientific Assessment.* Cambridge: Cambridge University Press.

Houghton, J. T., Jenkins, G. J. & Ephraums, J. J. (eds.). (1990). *Climate Change: The IPCC Scientific Assessment.* Cambridge: Cambridge University Press.

Houghton, R. A. (1991). Tropical deforestation and atmospheric carbon dioxide. *Climatic Change*, **19**, 99–118.

Izrael, Y. A., Hashimoto, M. & Tegart, W. J. M. (eds.) (1990). *Climate Change: The IPCC Impact Assessment.* Canberra: Australian Government Publishing Service.

Johansson, T. B., Kelly, H., Reddy, A. K. N. & Williams, R. H. (eds.) (1993). *Renewable Energy: Sources for Fuels and Electricity.* Washington, DC: Island Press.

Jones, A., Roberts, D. L. & Slingo, A. (1994). A climate model study of indirect radiative forcing by anthropogenic aerosols. *Nature*, **370**, 450–3.

Kauppi, P., Mielikäinen, K. & Kuusela, K. (1992). Biomass and carbon budget of European forests, 1971 to 1990. *Science*, **256**, 70–4.

Keller, M., Veldkamp, E., Weitz, A. M. & Reiners, W. A. (1993). Effect of pasture age on soil trace-gas emissions from a deforested area of Costa-Rica. *Nature*, **365**, 244–6.

Kineman, J. J. (1992). Global Ecosystems Database Version 1.0 (on CD-ROM) User's Guide. Key to Geophysical Records Documentation no. 26. Boulder, Colorado: USDOC/NOAA National Oceanic and Atmospheric Administration.

Klein Goldewijk, K., van Minnen, J. G., Kreileman, G. J. J., Vloedbeld, M. & Leemans, R. (1994). Simulating the carbon flux between the terrestrial environment and the atmosphere. *Water, Air and Soil Pollution*, **76**, 199–230.

Kreileman, G. J. J. & Bouwman, A. F. (1994). Computing land use emissions of greenhouse gases. *Water, Air and Soil Pollution*, **76**, 231–58.

Kwadijk, J. (1993). The impact of climate change on the discharge of the River

Rhine. *Nederlandse Geografische Studies*, **171**, 1–201.

Leemans, R. (1992). Modelling ecological and agricultural impacts of global change on a global scale. *Journal of Scientific and Industrial Research*, **51**, 709–24.

Leemans, R. & Cramer, W. (1991). The IIASA database for mean monthly values of temperature, precipitation and cloudiness on a global terrestrial grid. Research Report RR-91-18. Laxenburg: International Institute of Applied Systems Analyses.

Leemans, R., Cramer, W. & van Minnen, J. G. (1996). Prediction of global biome distribution using bioclimatic equilibrium models. In *Effects of Global Change on Coniferous Forests and Grassland*, ed. J. M. Melillo & A. Breymeyer, (in press). New York: Wiley.

Leemans, R. & van den Born, G. J. (1994). Determining the potential global distribution of natural vegetation, crops and agricultural productivity. *Water, Air and Soil Pollution*, **76**, 133–61.

Matthews, E. & Fung, I. Y. (1989). Methane emission from natural wetlands: Global distribution, area, and environmental characteristics of sources. *Global Biogeochemical Cycles*, **1**, 61–8.

McKenny, M. S., Easterling, W. E. & Rosenberg, N. J. (1992). Simulation of crop productivity and responses to climate change in the year 2030: the role of future technologies, adjustments and adaptations. *Agricultural and Forest Meteorology*, **59**, 103–27.

Melillo, J. M., McGuire, A. D., Kicklighter, D. W., Moore, III, B., Vorosmarty, C. J. & Schloss, A. L. (1993). Global climate change and terrestrial net primary production. *Nature*, **363**, 234–9.

Morita, T., Matsuoka, Y., Kainuma, M., Kai, K., Harasawa, H. & Dong-Kun, L. (1994). Asian-Pacific Integrated Model to assess policy options for stabilizing global climate. AIM report 1.0. Tsukuba, Japan: National Institute for Environmental Studies.

Mosier, A., Schimel, D., Valentine, D., Bronson, K. & Parton, W. (1991). Methane and nitrous oxide fluxes in native, fertilized and cultivated grasslands. *Nature*, **350**, 330–2.

Oldeman, L. R. (1991–1992). Global extent of soil degradation. *ISRIC Bi-annual report*, 1991–2, 19–36.

Oreskes, N., Shraderfrechette, K. & Belitz, K. (1994). Verification, validation, and confirmation of numerical models in the earth sciences. *Science*, **263**, 641–6.

Parry, M. L., Porter, J. H. & Carter, T. R. (1990). Agriculture: climate change and its implications. *Trends in Ecology & Evolution*, **5**, 318–22.

Peters, R. L. & Lovejoy, T. E. (eds.) (1992). *Global Warming and Biological Diversity*. New Haven: Yale University Press.

Plucknett, D. L. (1994). Science and agricultural transformation. IFPRI Lecture Series. Washington, DC: International Food Policy Research Institute.

Prentice, I. C., Cramer, W., Harrison, S. P., Leemans, R., Monserud, R. A. & Solomon, A. M. (1992). A global biome model based on plant physiology and dominance, soil properties and climate. *Journal of Biogeography*, **19**, 117–34.

Prentice, I. C., Sykes, M. T., Lautenschlager, M., Harrison, S. P., Denissenko, O. & Bartlein, P. J. (1994). Modelling global vegetation patterns and terrestrial carbon storage at the last glacial maximum. *Global Ecology and Biogeography Letters*, **3**, 67–76.

Prinn, R. G. (1994). The interactive atmosphere: global atmospheric-biospheric chemistry. *Ambio*, **23**, 50–61.

Raich, J. W., Rastetter, E. B., Melillo, J. M., Kicklighter, D. W., Steudler, P. A., Peterson, B. J., Grace, A. L., Moore, III, B. & Vörösmarty, C. J. (1991). Potential net primary productivity in South America: application of a global model. *Ecological Applications*, **1**, 399–429.

Rosenzweig, C. & Parry, M. L. (1994). Potential impact of climate change on world food supply. *Nature*, **367**, 133–8.

Rotmans, J. (1990). *IMAGE: An Integrated Model to Assess the Greenhouse Effect*. Dordrecht: Kluwer.

Rotmans, J., de Boois, H. & Swart, R. J. (1990). An integrated model for the assessment of the greenhouse effect: the Dutch approach. *Climatic Change*, **16**, 331–56.

Rotmans, J., Hulme, M. & Downing, T. E. (1994). Climate change implications for Europe: An application of the ESCAPE model. *Global Environmental Change*, **4**, 97–124.

Sarmiento, J. L. (1993). Carbon cycle – atmospheric CO_2 stalled. *Nature*, **365**, 697–8.

Scott, M. J. (1991). The MINK study: the regional effects of changing CO_2 and climate. DOE Research Summary no. 12. Oak Ridge, Tennessee: Oak Ridge National Laboratory.

Sedjo, R. A. & Solomon, A. M. (1989). Climate and forests. In *Greenhouse Warming: Abatement and Adaptation*, ed. N. J. Rosenberg, W. E. Easterling, III, P. R. Crosson & J. Darmstadter, pp. 105–119. Washington, DC: Resources for the Future.

Skole, D. & Tucker, C. (1993). Tropical deforestation and habitat fragmentation in the Amazon: Satellite data from 1978 to 1988. *Science*, **260**, 1905–10.

Smith, T. M., Leemans, R. & Shugart, H. H. (1992). Sensitivity of terrestrial carbon storage to CO_2-induced climate change: comparison of four scenarios based on general circulation models. *Climatic Change*, **21**, 367–84.

Solomon, A. M., Prentice, I. C., Leemans, R. & Cramer, W. P. (1993). The interaction of climate and land use in future terrestrial carbon storage and release. *Water, Air and Soil Pollution*, **70**, 595–614.

Steffen, W. L., Walker, B. H., Ingram, J. S. & Koch, G. W. (1992). *Global Change and Terrestrial Ecosystems: The Operational Plan*. IGBP-Report no. 21.

Stockholm: International Geosphere-Biosphere Programme.

Tans, P. P., Fung, I. Y. & Takahashi, T. (1990). Observational constraints on the global atmospheric CO_2 budget. *Science*, **247**, 1431–8.

Taylor, K. E. & Penner, J. E. (1994). Response of the climatic system to atmospheric aerosols and greenhouse gases. *Nature*, **369**, 734–7.

Turner, B. L., II, Kasperson, R. E., Meyer, W. B., Dow, K. M., Golding, D., Kasperson, J. X., Mitchell, R. C. & Ratick, S. J. (1990). Two types of global environmental change: definitional and spatial-scale issues in their human dimensions. *Global Environmental Change*, **1**, 14–22.

Turner, B. L., Moss, R. H. & Skole, D. L. (1993). *Relating Land Use and Global Change: A Proposal for an IGBP-HDP Core Project*. IGBP Report no. 24 and HDP Report no. 5. Stockholm: International Geosphere-Biosphere Programme and the Human Dimensions of Global Environmental Change Programme.

Watson, R. T., Meira Filho, L. G., Sanheuza, E. & Janetos, A. (1992). Greenhouse gases: sources and sinks. In *Climate Change 1992. The Supplementary Report to the IPCC Scientific Assessment*, ed. J. T. Houghton, B. A. Callander & S. K. Varney, pp. 27–46. Cambridge: Cambridge University Press.

Winguth, A. M. E., Heimann, M., Kurz, K. D., Maierreimer, E., Mikolajewicz, U. & Segschneider, J. (1994). El Niño-Southern Oscillation related fluctuations of the marine carbon cycle. *Global Biogeochemical Cycles*, **8**, 39–63.

Winjum, J. K., Dixon, R. K. & Schroeder, P. E. (1993). Forest management and carbon storage: an analysis of 12 key forest nations. *Water, Air and Soil Pollution*, **70**, 239–57.

Wolfram, S. (1984). Cellular automata as models of complexity. *Nature*, **311**, 419–24.

WRI (1992). *World Resources, 1992–1993*, Washington, DC: World Resources Institute.

Zobler, L. (1986). *A World Soil File for Global Climate Modeling*. Technical Memorandum. New York: NASA.

Zuidema, G., van den Born, G. J., Alcamo, J. & Kreileman, G. J. J. (1994). Simulating changes in global land cover as affected by economic and climatic factors. *Water, Air and Soil Pollution*, **76**, 163–98.

Developing the potential for describing the terrestrial biosphere's response to a changing climate

F. I. Woodward

Introduction

Projecting the likely nature of the terrestrial biosphere into a future climate and atmospheric composition is a major aim of the IGBP core project Global Change and Terrestrial Ecosystems (GCTE; Steffen *et al.*, 1992). This projection will be achieved by the parallel use of experiments in simulated future environments and modelling, based on our current mechanistic understanding of ecosystem responses to a changing environment. Modelling developments will, of necessity, be dependent on new findings from experiments. Experimental designs and protocols should also be modified to incorporate the forward and usually up-scaled projections of models. A simple case to illustrate the latter point was the move to consider competitive interactions between plant species with C_3 and C_4 photosynthetic metabolisms, once it was realized that these two functional types of species possessed differing responses to CO_2 enrichment (Patterson & Flint, 1980; Zangerl & Bazzaz, 1984; Wray & Strain, 1987; Bazzaz & Garbutt, 1988; Curtis *et al.*, 1989; Bazzaz, 1990; Woodward *et al.*, 1991).

A greater problem for both modelling and experiment has emerged with longer-term experiments on the CO_2 enrichment of tree species. The first point is that no trees have been grown in CO_2-enriched atmospheres, yet in the modelling arena trees and forests receive the greatest attention globally (Shugart, 1984; Woodward, 1987). However only young saplings have been grown and raised for 3 to 4 years under CO_2 enrichment (Idso & Kimball, 1992; Norby *et al.*, 1992). The second point to emerge from the work on young tree saplings is that while photosynthesis can be stimulated in the long term under CO_2 enrichment, total plant growth may not (Norby *et al.*, 1992). The explanation for this lack of simple scaling-up lies in plant allocation patterns, and in particular with the responses of plant roots and mycorrhizas to the increased supplies of carbohydrates (Read, 1991; Woodward & Smith, 1994). These are murky areas of mechanistic understanding and severely limit the development of adequate global-scale models of tree and forest behaviour.

At the global-scale of plant and vegetation modelling there are many limitations to further developments. The first is the realization that there is very little, if any, information about the behaviour of the major plant species of the world, and even less information about their responses to environmental change. GCTE is attempting to circumvent this problem by defining plant functional types: agglomerations of species into types with very similar responses to the environment (Smith *et al.*, 1996).

The second major obstacle to the development of global vegetation models is the lack of a global database and unified modelling framework for including the effects of landscape structure and function. Two areas of IGBP (Townshend, 1992; Turner *et al.*, 1993) are beginning to address the issue, but this is a long way from providing generalized approaches for the global-scale modeller.

The aim of this chapter is to describe the approach being developed at the University of Sheffield with a number of collaborators around the world, towards deriving a model of the terrestrial biosphere's dynamic response to a changing climate. The aim is not to describe the model, as it is not yet complete, but to describe some of the current problems and limitations in the model development. The objective of the modelling is for a forward look, so that the description of the future must be derived from models. The global view is the focus, so that everything which follows must be top-down. However, our understanding of plant processes at the small and fine scales underpin the large scale, and so our understanding is more bottom-up.

A Dynamic Global Vegetation Model (DGVM)

Introduction

The objective of predicting a future view of the terrestrial biosphere will be achieved by developing a Dynamic Global Vegetation Model (DGVM). Once this is in place, then it becomes possible to delve into smaller-scale regions for more specific detail. A major objective of GCTE is to develop one or more DGVMs (Steffen *et al.*, 1992). A DGVM is a global-scale model which is capable of describing the functional responses (i.e. processes such as water and CO_2 exchange) and structural responses (i.e. life-form changes in abundance and distribution) of vegetation to a changing environment. The model should therefore be capable of changing the structural nature of vegetation in ecological step (i.e. realistically simulating migration and change of life-form composition) with changing climate and atmospheric CO_2 concentration.

The need for DGVM originates from a number of objectives, the first is as the vegetation component of climate within General Circulation Models (GCMs). In this case the model should provide the appropriate biophysical feedbacks on climate. The second requirement is more ecologically oriented to provide general predictions of regional and continental-scale changes in vegetation. Such projections would integrate small-scale patch and regional models of processes, based on extensive field campaigns. The third requirement is for a terrestrial-scale model to fit into a general Earth System Model. In this case the model would integrate with biogeochemical and ocean models, such as in the case of developing a global carbon model (Tans et al., 1990; Moore & Braswell, 1994), which is a major feature of the IGBP Task Force Global Analysis, Interpretation and Modelling (GAIM; Moore & Braswell, 1994).

Current approach and status in developing a DGVM

A DGVM is being developed at the University of Sheffield, in collaboration with other groups at the UK Hadley Centre of the Meteorological Office, the University of Cranfield, UK, the Institute of Terrestrial Ecology, UK and the University of Virginia, USA.

The general framework of the model (Fig. 26.1) indicates the major features which are outlined elsewhere (Woodward et al., 1995). It is important to note that predictions of net primary productivity (NPP) and leaf area index (LAI) are determined from well-characterized plant physiological responses to climate, CO_2 and soil nutrients and soil water. Other published global scale vegetation models (e.g. Neilson et al., 1992; Prentice et al., 1992) do not include this full range of mechanistic responses to the aerial and edaphic environments. As described (Fig. 26.1) the model is a technique of scaling-up from plant processes to vegetation processes. In fact all of these processes are currently calculated for a grid square – cell – of $0.5°$ of latitude by $0.5°$ of longitude. The cell size is therefore the resolution limit for the model runs and for the data necessary to drive the model. The world is divided into these cells, and a base set of data are required for each cell to run the model.

A critical feature of the model is that it can respond mechanistically to changes in CO_2 concentration, through changes in photosynthesis, stomatal conductance and NPP. In addition the cells are not initialized with vegetation characteristics. Therefore the model predicts the nature of the vegetation based only on climate, CO_2 and soils. So no current vegetation data, and this includes remotely sensed data, are used to initialize the model. This has two important advantages, the first is that the model can be run

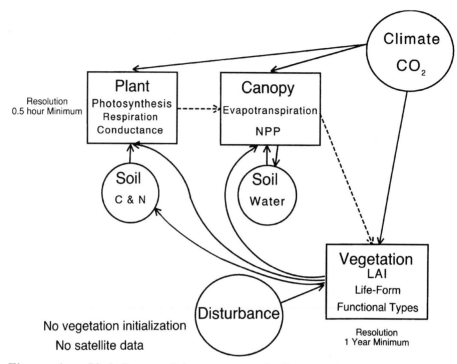

Figure 26.1   *Block diagram of the components of a Dynamic Global Vegetation Model.*

under any reasonable climatic and CO_2 scenarios, as initialization with vegetation is unnecessary. The second advantage is that, for current climate at least, the model predictions can be validated against current vegetation maps and data from satellites. The data components have been discussed elsewhere (Woodward & Smith, 1994), however it should be noted (Fig. 26.1) that the model is designed to operate from a time-scale of minutes, for processes such as gas exchange, to years for predicting changes in the structure and distribution of vegetation.

A typical product of the model (Plate 7 between pp. **204** and **205**) is a global-scale distribution of LAI, in this case a present-day climate and soil determined simulation (Woodward *et al.*, 1995). LAI is a vegetation characteristic which can be predicted and tested against satellite surface reflectance data (Running *et al.*, 1989; Potter *et al.*, 1993; Woodward & Smith, 1994). Maps of annual NPP can also be determined (Plate 8 between pp. **204** and **205**), based on well-characterized plant physiological processes (Fig. 26.1). This case is also for current climate and soils (Woodward *et al.*, 1995). Validating NPP is much more problematical than is the case for LAI. There is no global-scale validation set and it is clear that site-specific measurements of NPP have significant error components. So at present it is impossible to indicate whether NPP models are good, bad, or some intermediate state. However, it is likely that there will be a global-scale correlation between LAI and NPP (Potter *et al.*, 1993; Woodward *et al.*, 1995). Indeed models which are initialized with satellite reflectance data will have a significant proportion of the global-scale variance in NPP defined.

The model output, in this its simplest form, assumes that all vegetation is natural or potential vegetation. The production of a rule-based approach for predicting the distribution of areas dominated by human land-use, such as areas of agriculture, is currently under development by the University of Cranfield, UK. This map will be an overlay on the NPP and LAI maps and which will exclude potential vegetation from the model runs.

The model as described so far can therefore predict LAI and NPP for the world's terrestrial surface, and produces global-scale maps which agree closely with other models (e.g. Potter *et al.*, 1993). LAI and NPP under future climatically changed conditions, and with changes in CO_2, can also be determined.

Process to structure

The DGVM can track changes in climate and CO_2, changing its predictions of LAI and NPP. These changes are next used to determine the proportions of the major life-forms – trees, shrubs and grasses – and the area of bare ground, in each cell. The proportion of bare ground is a critical requirement of GCMs, influencing hydrological processes and energy transfer in general. The proportion of bare ground at vegetation equilibrium is determined as the fraction of irradiance not intercepted by the predicted LAI. The methods by which changes in processes translate to changes in structure are still under development, however general procedures have been established. In this case LAI and NPP are used together to calculate life-form specific rates of establishment and mortality (Woodward, 1987; Melillo *et al.*, 1993). The rate of change in the fraction of the three life forms is modelled in a matrix compartment model (Woodward & Lee, 1995). In general terms, low LAI and NPP maximize the growth and proportion of shrubs, increasing the NPP and LAI first favours grasses and then finally trees, which dominate and shade the more dwarf life forms. In addition, the lower the NPP the slower the rate of increase in vegetation cover. A sample output (Fig. 26.2) for a cell with a predicted LAI of 5 and NPP of tC ha^{-1} year^{-1} assumes that the cell starts from bare ground and simulates successional change to a forest.

The cell may, at any particular year be at some stage of succession, recovering from disturbance. If disturbance has occurred within a period of a few years, then the whole of the cell may be dominated by grasses (Fig. 26.2). It is unlikely that a whole cell will be subjected to a single and complete disturbance. Therefore, there will be a need to partition each cell into more realistic simulations of the spatial extents, types and times of disturbances. This major task has not been completed but is being achieved by a mixture of data from historical archives of different vegetation types, plus calculations from satellite-series data for vegetation types with no historical data archives. In addition, there is a need to apply a general disturbance generator within the model, based at least on the probabilities of fire and blow down. This has not yet been achieved, however the approach of Johnson and Gutsell (1994) seems potentially useful.

A typical range of successional responses for a cell with three initial states: disturbance to bare ground at year 0, disturbance at year 25 and no disturbance, shows (Fig. 26.3) clear parallel trajectories of change, as the NPP and LAI of the cell gradually increase. This basic approach will therefore be applied to all cells once the disturbance data set and simulators are completed.

When the succession model is calculated for a range of LAI and NPP, then the 'process space' of the three life-forms becomes clear (Fig. 26.4). Shrub dominance is predicted for cells with the lowest LAIs and NPPs, with the occurrence of some grass as a subdominant (the second highest cover). Increasing NPP at low LAI leads to a change in dominance from shrubs to grasses, with a subordinate shrub component. Once the LAI exceeds 3 then trees are expected to dominate the vegetation. However at low NPP the subdominant life-form is shrubs, with grasses predicted for high NPP. The vegetative cover is only defined in terms of LAI (Fig. 26.4). The general process space of life-forms conforms quite well with previous discussions (Woodward, 1987; Prentice *et al.*, 1992).

Life-forms to functional types

As stated by Running *et al.* (1994) 'For global modeling requirements, the development of realistic models of climate, carbon cycles, hydrology etc., all rely on an *unambiguous, repeatable definition of existing land-cover*' [my emphasis]. In essence this means that for most applications of a DGVM, defining life-form as only tree, shrub or grass is inadequate. Further details about the life-form are required, such as phenology and physiognomy (e.g. evergreen, deciduous, needleleaved or broadleaved).

Woodward (1987) described a scheme for defining some of these features based on a physiological survey of the responses of these different physiognomic types to climate, a particularly important controller is absolute

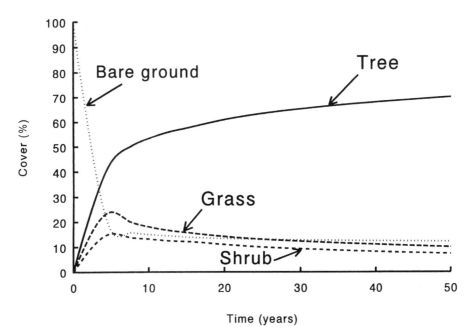

Figure 26.2 *Predicted change in life-form components of vegetation (cover %) with a projected LAI of 5 and NPP of 5 tC ha⁻¹ year⁻¹.*

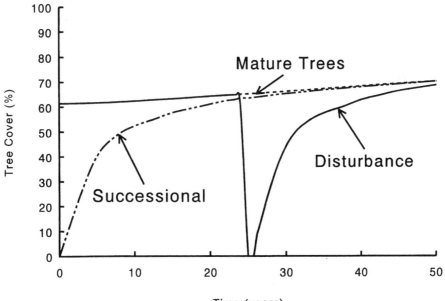

Figure 26.3 *Responses of tree cover to increases in NPP and LAI, and with initial disturbance ('successional'), disturbance at year 25 ('disturbance') and no disturbance ('mature trees'). Initial NPP = 5, LAI = 5; final NPP = 6, LAI = 6.*

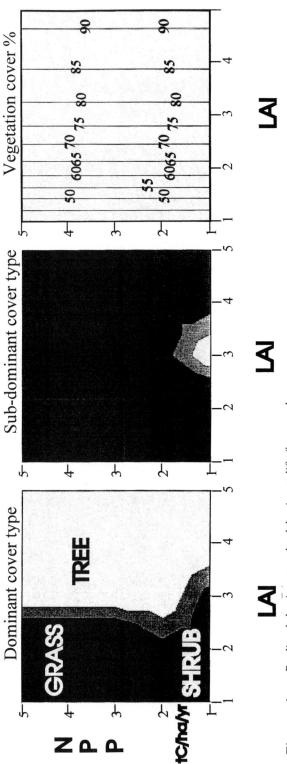

Figure 26.4 *Predicted dominant and subdominant life-forms and vegetation cover (%) for a range of LAI from 1 to 5 and a range of NPP from 1 to 5 tC ha⁻¹ year⁻¹. Shading key: black is grass; mid grey is shrub and white is tree. Dark and light grey indicate zones of overlap with shared dominance between adjacent life-forms.*

minimum temperature. A particularly clear example is shown by the ranges of frost resistance shown by species from boreal, broadleaf deciduous and broadleaf evergreen forests (Fig. 26.5). Species from the boreal forest can endure the lowest temperatures, whilst broadleaf evergreen species are the most frost sensitive. An important feature is the observation that species from all the three vegetation types can survive in the warmest conditions (minimum temperatures greater than -15 °C), i.e. this approach provides no explanation for the absence of boreal forest species from warm subtropical forests. Woodward (1987) suggested that competitive exclusion was the likely, but untested explanation. Prentice *et al.* (1992) explicitly included a competitive hierarchy to resolve the problem, while Neilson *et al.* (1992) applied a rule-based approach to predicting vegetation characteristics. These latter two approaches provide clearly defined (Fig. 26.6) climate spaces for species from the three vegetation types. The problem with these approaches is that the exclusion of a species or functional type is not mechanistic, just correlational. Therefore the approach may have limited applicability to future environments. What is needed is a mechanistic approach to defining functional type exclusion or inclusion.

The sharp transitions from one physiognomic class to another will not be seen naturally. A re-analysis of the data collected by Woodward (1987) shows smoother transitions (Fig. 26.7). In addition the ordinate is not simply a measure of survival; it also includes a measure of species abundance. Taken as a global average, i.e. including all potential species in a physiognomic class, then the responses can be normalized (Fig. 26.7). At a regional or continental scale it is quite possibly the case, as a consequence of various historical accidents, that the relative proportion of species in a particular physiognomic class is less than the global average (Fig. 26.8). In such a case, e.g. if the broadleaf evergreen species are very rare, it would be possible for the broadleaf deciduous species to extend their distribution and abundance into warmer climates (Fig. 26.8), a feature which cannot be concluded from the simple classification with implied competition (Fig. 26.6). Further analysis of Woodward's data on frost survival (Woodward, 1987) indicates the occurrence of a significant fraction of temperate conifers which have frost resistances very similar to broadleaf deciduous species. When the relative contributions of the two physiognomies are normalized (Fig. 26.9) it can be seen that the broadleaf deciduous climate space has two zones, rather than one, with significant overlap between the two physiognomies.

The implication of using the species compensation approach (Figs. 26.8 and 26.9) can be seen from a simple set of simulations. The first case (Fig. 26.10) describes the absolute minimum temperature isotherms for an

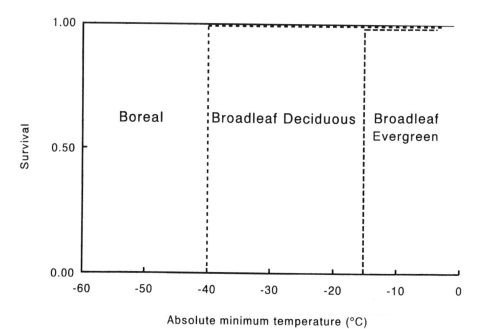

Figure 26.5   *Theoretical vegetation functional type definitions based on absolute minimum temperature survival.*

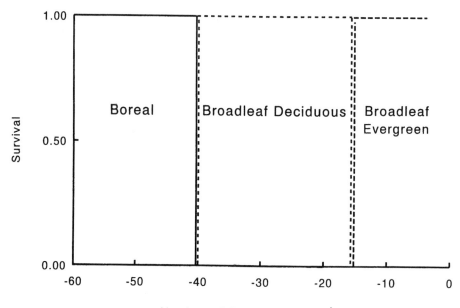

Figure 26.6   *Theoretical vegetation functional type definitions based on absolute minimum temperature survival and a simple competitive hierarchy.*

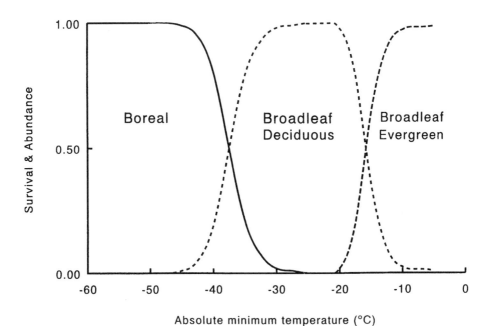

Figure 26.7 *Vegetation functional type definitions based on absolute minimum temperature survival and species abundance.*

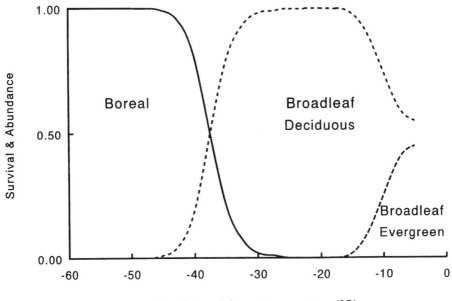

Figure 26.8 *Compensatory changes in broadleaf deciduous functional type abundance in response to low abundance of broadleaf evergreen functional type.*

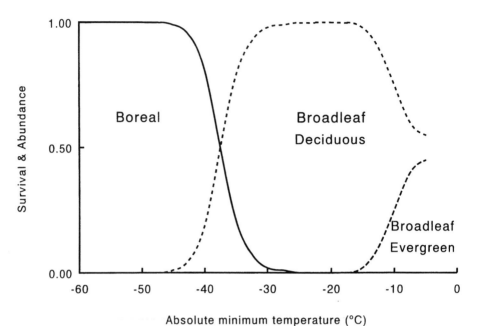

Figure 26.9 *Overlapping broadleaf deciduous and temperate conifer functional types and abundance.*

area and in which temperate conifers are assumed to be absent. The map of the broadleaf deciduous species indicates the highest probability of occurrence in the central region and the broadleaf evergreen at the base, in the warmest conditions. The boreal forest would dominate the coldest regions of the map, but this feature has been excluded from the map for simplicity.

In the second case (Fig. 26.11) both broadleaf deciduous and temperate conifers are assumed to occur in the abundances shown on Fig. 26.9. In this case there is no change in the response of the broadleaf evergreen types but the distribution of the broadleaf deciduous species is more strongly constrained to the central climatic zone, than in the base case (Fig. 26.10).

The final example (Fig. 26.12) shows the interaction between two types, broadleaf deciduous and broadleaf evergreen types. Fig. 26.12a repeats the earlier case (Fig. 26.10). In Fig. 26.12b it is assumed that the broadleaf evergreen species are absent from the lower right corner of the area. There is very clear and marked compensation of distribution and abundance by the broadleaf deciduous types.

The significance of these simulations is that when defining the distribution and abundance of functional types it is critical that there is geographical initialization, i.e. the actual and current distributions of functional types are important. There is a carry over of vegetational history which will influence the future responses of plants and vegetation to climate change.

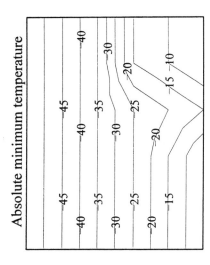

Absolute minimum temperature

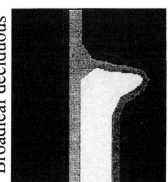

Temperate conifer

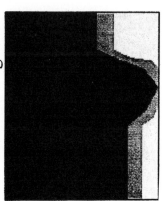

Broadleaf deciduous

Broadleaf evergreen

Figure 26.10 *Predicted abundances of functional types in response to a theoretical distribution of absolute minimum temperatures (isotherms in °C). Temperate conifers absent. For remaining functional types white is dominant, black is absent and increasing lightness of grey indicates increasing abundance.*

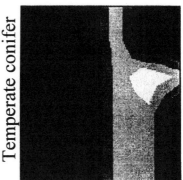

Temperate conifer

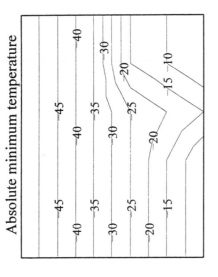

Absolute minimum temperature

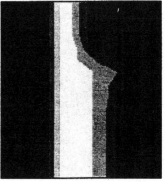

Broadleaf deciduous

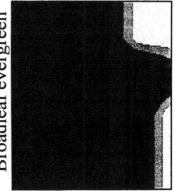

Broadleaf evergreen

Figure 26.11 *As Figure 26.10 but with the inclusion of temperate conifers.*

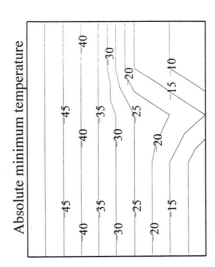

Absolute minimum temperature

Broadleaf evergreen

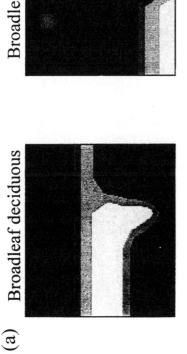

Broadleaf deciduous

(a)

Broadleaf evergreen

Broadleaf deciduous

(b)

Figure 26.12 *(a) Broadleaf deciduous and broadleaf evergreen functional types directly distributed in relation to absolute minimum temperatures (°C). (b) Broadleaf evergreen density reduced (Figure 26.8) in lower left corner of area and absent from lower right corner. Density scales as for Figure 26.10.*

Conclusions

The major conclusion is that we are not yet ready to describe a dynamic terrestrial biosphere in a future climate with much confidence; however, the development of a full DGVM is moving quickly, and this will greatly enhance future projections. Very general projections have been made (Smith *et al.*, 1992; Woodward, 1989) but these do not incorporate any natural dynamics of change, and so have limited value in predicting the course of future vegetational change. An important conclusion from modelling functional types is that when vegetation detail is required beyond, for example, evergreen forest, then it will be necessary to initialize models with actual and current distributions of functional types. The history of the vegetation is important and it will influence vegetation responses to environmental change.

Acknowledgements

The modelling work in Sheffield has been funded by the Natural Environment Research Council under the Terrestrial Initiative in Global Environmental Research (Grant Number GST/02/696). The project to develop a DGVM is a Core Research project of the IGBP core project Global Change and Terrestrial Ecosystems (GCTE). I am grateful for considerable assistance, from my collaborators in preparing aspects of this chapter and the associated, but different, oral presentation at the first GCTE Science Conference in Woods Hole, USA on 27 May 1994. The collaborators in this venture were David Beerling, Susan Lee, Susannah Diamond, John Sheehy and Peter Mitchell at Sheffield, Tom Smith at the University of Virginia, USA, Bill Emanuel at Oak Ridge National Laboratory, USA, Peter Cox at the Hadley Centre of the UK Meteorological Office, Steve Cousins and Mark Strathern at Cranfield University, UK, and Jim Eastwood and Barry Wyatt at the Institute of Terrestrial Ecology, UK.

References

Bazzaz, F. A. (1990). The response of natural ecosystems to the rising global CO_2 levels. *Annual Review of Ecology and Systematics*, **21**, 167–96.

Bazzaz, F. A. & Garbutt, K. (1988). The response of annuals in competitive neighborhoods: effects of elevated CO_2. *Ecology*, **69**, 937–46.

Curtis, P. S., Drake, B. G., Leadley, P. W., Arp, W. J. & Whigham, D. F. (1989). Growth and senescence in plant communities exposed to elevated CO_2 concentrations on an estuarine marsh. *Oecologia*, **78**, 20–6.

Idso, S. B. & Kimball, B. A. (1992). Seasonal fine-root biomass development of sour orange trees grown in atmospheres of ambient and elevated CO_2 concentration. *Plant, Cell and Environment*, **15**, 337–41.

Johnson, E. A. & Gutsell, S. L. (1994). Fire frequency models, methods and interpretations. *Advances in Ecological Research*, **25**, 239–87.

Melillo, J. M., McGuire, A. D., Kicklighter, D. W., Moore III, B., Vorosmarty, C. J. & Schloss, A. L. (1993). Global climate change and terrestrial net primary production. *Nature*, **363**, 234–40.

Moore, B. III & Braswell, B. H. Jr (1994). Planetary metabolism: understanding the carbon cycle. *Ambio*, **23**, 4–12.

Neilson, R. P., King, G. A. & Koerper, G. (1992). Towards a rule-based biome model. *Landscape Ecology*, **7**, 27–43.

Norby, R. J., Gunderson, C. A., Wullschleger, S. S., O'Neill, E. G. & McCracken, M. K. (1992). Productivity and compensatory responses of yellow-poplar trees in elevated CO_2. *Nature*, **357**, 322–4.

Patterson, D. T. & Flint, E. P. (1980). Potential effects of global atmospheric CO_2 enrichment on the growth and competitiveness of C_3 and C_4 weed and crop plants. *Weed Science*, **28**, 71–5.

Potter, C. S., Randerson, J. T., Field, C. B., Matson, P. A., Vitousek, P. M., Mooney, H. A. & Klooster, S. A. (1993). Terrestrial ecosystem production: a process model based on global satellite and surface data. *Global Biogeochemical Cycles*, **7**, 811–41.

Prentice, I. C., Cramer, W., Harrison, S. P., Leemans, R., Monserud, R. A. & Solomon, A. M. (1992). A global biome model based on plant physiology and dominance. *Journal of Biogeography*, **19**, 117–34.

Read, D. J. (1991). Mycorrhizas in ecosystems. *Experientia*, **47**, 376–91.

Running, S. W., Nemani, R. R., Peterson, D. L., Band, L. E., Potts, D. F., Pierce, L. L. & Spanner, M. A. (1989). Mapping regional forest evapotranspiration and photosynthesis by coupling satellite data with ecosystem simulation. *Ecology*, **70**, 1090–101.

Running, S. W., Loveland, T. R. & Pierce, L. L. (1994). A vegetation classification logic based on remote sensing for use in global biogeochemical models. *Ambio*, **23**, 77–81.

Shugart, H. H. (1984). *A Theory of Forest Dynamics. The Ecological Implications of Forest Succession Models*. New York: Springer.

Smith, T. M., Leemans, R. & Shugart, H. H. (1992). Sensitivity of terrestrial carbon storage to CO_2-induced climate change: a comparison of four scenarios based on general circulation models. *Climatic Change*, **21**, 367–84.

Smith, T. M., Shugart, H. H. & Woodward, F. I. (1996). *Plant Functional Types*, Cambridge: Cambridge University Press.

Steffen, W. L., Walker, B. H., Ingram, J. S. I. & Koch, G. W. (1992). *Global Change and Terrestrial Ecosystems. The Operational Plan*. IGBP Global Change Report No. 21. Stockholm: IGBP.

Tans, P. P., Fung, I. Y. & Takahashi, T. (1990). Observational constraints on the global atmospheric CO_2 budget. *Science*, **247**, 1431–8.

Townshend, J. R. (1992). *Improved Global Data for Land Applications*. IGBP Global Change Report No. 20. Stockholm: IGBP.

Turner, B. L., Moss, R. H. & Skole, D. L. (1993). *Relating Land Use and Global Land-Cover Change*. IGBP Global Change Report No. 24. Stockholm: IGBP.

Woodward, F. I. (1987). *Climate and Plant Distribution*. Cambridge: Cambridge University Press.

Woodward, F. I. (1989). Plants in a greenhouse world. *New Scientist*, **1663**, 1–4.

Woodward, F. I. & Lee, S. E. (1995). Global scale forest function and distribution. *Forestry*, **68**, 317–25.

Woodward, F. I. & Smith, T. M. (1994). Global photosynthesis and stomatal conductance: modelling the controls by soil and climate. *Advances in Botanical Research*, **20**, 1–41.

Woodward, F. I., Smith, T. M. & Emanuel, W. R. (1995). A global land primary productivity and phytogeography model. *Global Biogeochemical Cycles*, **9**, 471–90.

Woodward, F. I., Thompson, G. B. & McKee, I. F. (1991). The effects of elevated concentrations of CO_2 on indi-vidual plants, populations, communities and ecosystems. *Annals of Botany* (Supplement 1) **67**, 23–38.

Wray, S. M. & Strain, B. R. (1987). Competition in old field perennials under CO_2 enrichment. *Ecology*, **68**, 1116–20.

Zangerl, A. R. & Bazzaz, F. A. (1984). The response of plants to elevated CO_2. *Oecologia*, **62**, 412–17.

Data requirements for global terrestrial ecosystem modelling

W. Cramer and A. Fischer

Introduction[1]

Spatially comprehensive global biosphere models are a current focus of global change research because they are expected to help us understand some of the major issues of global environmental concern, such as the global carbon balance (Smith *et al.*, 1993; IGBP, 1994) or the potential shift of major bioclimatic zones (Cramer & Leemans, 1993). Two main groups of such models are the terrestrial biogeochemical models (TBMs)[1] and the biome models (Prentice, 1993). TBMs describe ecosystem processes such as photosynthesis, respiration, transpiration, nitrogen cycles and their interactions (for example, the Frankfurt Biosphere Model, Lüdeke *et al.*, 1994). Biome or biogeophysical models describe the shifts of the potential distribution of major ecosystem types (for example, BIOME 1, Prentice *et al.*, 1992). Because vegetation is a physically dominating component of most ecosystems, and because it is stationary in space, ecosystem models tend to be mainly vegetation models. Here, however, we use the term 'ecosystem models' to indicate that it is the behaviour of the entire biosphere that is the focus of the modelling activities. Ecosystem models differ from biosphere models in that they are not necessarily global in nature and, if they are, they consider different parts (areas) of the biosphere in different ways, depending on their structural characteristics. The first fundamental distinction between ecosystems on the globe is between marine and terrestrial ecosystems, and this paper is concerned with terrestrial ecosystems only.

Global ecosystem models require data to be provided for several purposes, most of which concern *calibration, validation* and *application* of the model. Generally, each of these steps requires independent data sets. There is only one globe available for providing data, and its surface is only very incompletely covered by measurements at some detail and resolution. Therefore, to adequately simulate its behaviour and specific response to changes in the major driving forces, we must deal with this data limitation.

1. Technical abbreviations and acronyms are defined in the Appendix at the end of this chapter.

This paper attempts to provide some general guidelines for overcoming this problem. The design of the underlying database for a simulation model has an important bearing on the set of processes that can be described. In a complex system such as the biosphere, the processes which are modelled are usually chosen on the basis of a compromise between a ruling scientific paradigm or theory and the availability of data. Observable (physiognomic) features of the system are a convenient starting point for the definition of what is to be simulated. For the global (terrestrial) biosphere, *pattern* exists in both the spatial and the temporal domain: the ideal simulation model should therefore describe both, the spatial distribution of major types of ecosystems and their temporal dynamics (such as their appearance, growth and decline). Ecosystem *processes* either control the observable features of the pattern directly (dispersal, establishment, growth, competition), or they may be less visible, underlying mechanisms, influenced by structure and composition, such as productivity, allocation of carbon, or trace gas fluxes.

The availability of primary data for either type of model feature is usually very limited. Data sources may be point-based inventories and observations of subsystems of the Earth system, such as atmosphere, land cover, or the pedosphere. They may also be observations of physical properties that are indications of structure or function of the biosphere (such as the radiances observed by space- or airborne sensors). Many of the ground-based data sets are collected at locations and/or times that are not selected with the primary goal of adequate spatial or temporal coverage of biosphere models. Statistical techniques or the use of intensive satellite missions may reduce this problem.

Frequently, data sets are also limited because the data have been collected by different organizations and/or different nations. This situation turns the data availability problem into an organizational or political issue with high relevance for model development. Some national and international activities are underway to produce Data and Information Systems (DISs) which make such data sets available to the public domain, while other sources are becoming less accessible due to commercialization or other limitations. Our capabilities of predicting the future state of the Earth system under changing boundary conditions will be seriously limited if free and well-managed access to those data archives is not ensured.

Ecosystem models and the variables they use

General considerations

Ecosystems are characterized by the relationships between their components (organisms) and their environment, including the modifying effects that ecosystem components have on their own and their neighbours' environment (Odum, 1973). The purpose of ecosystem modelling is to identify and describe those relationships between organisms and their environment that are of fundamental importance to the function of the organisms and/or the structure of the ecosystem as a whole. The problem of data requirements for global ecosystem models is therefore directly linked to the nature of the processes described. This has influenced the development of databases that have been tailored for the needs of ecosystem models (Leemans & Cramer, 1991). Model development, and particularly the selection of processes described by the models, has also been influenced by the availability of such databases. Most current global-scale ecosystem models can be classified by the distinctions between ecosystem structure or function models, equilibrium or transient models, and diagnostic or prognostic models (see definitions below).

Many biosphere modelling activities have been initiated by the need to understand the major environmental constraints on the distribution of major *structural* types of ecosystems and by interest in prediction of climate change-driven changes of this distribution (Emanuel *et al.*, 1985; Prentice *et al.*, 1992; Neilson, 1993). Other studies have been geared towards the understanding of the role of the biosphere in controlling amounts of trace gases in the atmosphere and by the question of how the current biosphere might change its *trace gas flux* pattern if climate or atmospheric chemistry would change (Raich *et al.*, 1991; Melillo *et al.*, 1993; Esser *et al.*, 1994; Lüdeke *et al.*, 1994). Traditionally, structural models have involved the implicit assumption of quasi-stable trace-gas fluxes (apart from diurnal and annual cycles), and trace-gas models have involved the implicit assumption of stable ecosystem structure. Several groups currently develop models that combine structural and trace gas flux descriptions (e.g. Plöchl & Cramer, 1995).

Another distinction between ecosystem models that has a bearing on their data requirement is between equilibrium and transient models. *Equilibrium* models assume that the biosphere is in equilibrium with the environment, and they therefore do not require that time-dependent processes are modelled explicitly. They may involve a description of the seasonal pattern throughout a year, but this pattern is usually oriented towards the 'average year' only. In contrast to this, a *transient* model must respond to changes in climate with time-dependent lags, and it therefore needs more explicit formulations of processes such as growth, mortality and dispersal. Some recent trace-gas flux models can be said to function also outside the immediate vicinity of the equilibrium (Lüdeke *et al.*, 1994), but none of them contains facilities to simulate longer-term change in structure.

Ecosystem models may also be either diagnostic or prognostic. *Diagnostic* models require their input variables to be based on real-time observations. Most frequently, they are used with remotely sensed input data (Heimann & Keeling, 1989; Running *et al.*, 1989). As parts of a monitoring strategy, diagnostic models allow the production of global maps of variables that cannot be recorded directly. An example for this is the seasonal and latitudinal features of CO_2 fluxes between biosphere and atmosphere (Maisongrande *et al.*, 1995). *Prognostic* ecosystem models are designed to explain the basic mechanisms in nature. Their driving variables are usually climatic only, and the climate data may come from simulated or steady-state climatologies. Hence, they can be used to predict possible future developments of the Earth system (Prentice *et al.*, 1989). In some cases, prognostic models have been used in a diagnostic mode. For example, Claußen (1993) used a prognostic ecosystem structure model (BIOME) to diagnose shifts in climatic zones from a general circulation model.

Data sets used in global ecosystem modelling can be classified into data sets for calibration, validation or application of the model, depending on the stage of model development they are required for. Biotic variables are characteristics of the ecosystem itself, for example land cover data sets or ecosystem properties such as leaf area index. Abiotic variables are included in climatological or soil data sets. Point-based variables are descriptions or measurements of ecosystem structure, ecosystem processes, or environmental conditions at known locations. Spatial variables are regular fields of data showing the pattern of ecosystems or their environment over large areas. A further distinction is between locally and remotely sensed data. Data that are based on measurements made with sensors such as thermometers or chemical analysers are usually seen as being measured 'directly' and therefore interpretable in a physical way. Remotely sensed data are valuable because their space and time coverage is much more complete than that from most other sources. Also, these observations are independent of ground-based measurements and can therefore be used for validation of models fitted to those measurements (Coakley *et al.*, 1990). In the following, some of the major data requirements of large-scale ecosystem models will be reviewed.

Topography and soils

The earth's surface is a complex landscape with variable elevation, coastlines and physical properties of the soils. Most local-scale features of ecosystems are determined by topography and soil heterogeneity, either directly or through the mediating effect topography has on climate or hydrology. Among the variables that are used from topographical databases are elevation and the descriptors of surface runoff (slope, aspect, and stream connectivity). The soil variables that are of primary interest to ecosystem modelling are those dealing with soil texture, such as soil water capacity, wilting point and others.

Some ecosystem models pay particular attention to the organic processes occurring in the pedosphere, namely those that control the availability of nutrients as a function of climate, site factors and parameters from the ecosystem (Parton *et al.*, 1993). The additional data requirements concern the existing and potential stores of soil nutrients, as well as the major characteristics of the ecosystem.

Climate and hydrology

For global and continent-wide ecosystem models, the dominating importance of climatic constraints to ecosystems was recognized before human-induced climate change became an issue of interest (von Humboldt, 1807; Köppen, 1884), and it has been acknowledged again in the context of global ecosystem research (Box, 1981; Woodward, 1987). Consequently, the primary focus of database development has been on climatic variables. The climate database for a global ecosystem model must be spatially explicit and geographically comprehensive. Rather than being an interpretation derived from ecosystem distribution, it should be based on measurements from weather stations or satellites.

Monthly means of temperature and precipitation have been the standard climatic variables used for global applications. Mean temperatures are clearly relevant to many ecosystem processes, such as plant growth or the risk of frost. In many cases, however, temperature extremes or temperature sums over certain thresholds (growing degree days) are more adequate because of their functional importance as enhancing or constraining factors of ecosystem processes.

Monthly precipitation sums are less directly correlated to ecosystem function because they do not translate easily into the availability of moisture to the plant. Moisture indices have been derived from precipitation and other climate data using empirical relationships and have been used to describe the general 'dryness' of a site (for example, Thornthwaite, 1948). Such information has been used to predict major distributional limits of ecosystem distribution on a global scale. The availability of moisture is directly related not only to precipitation but to all vertical exchanges of water and energy between the atmosphere and the ecosystem. This vertical component of the hydrological cycle is often modelled more mechanistically, using the fluxes of water between soil, vegetation and atmosphere. Required data sets are precipitation, air and soil humidity, soil texture, the radiation balance, snow cover, wind speed and turbulence (Monteith, 1965; Cramer & Prentice, 1988). Common to some of the current approaches is that the estimation of moisture availability is performed separately from the ecosystem model and therefore generates spatial databases of moisture availability that later can be used as input to the ecosystem model (Prentice *et al.*, 1993). In real ecosystems, however, the moisture balance is an internal process, because evapotranspiration and other fluxes directly depend on the structure of the ecosystem and the soil. Because of short-term processes, such as infiltration, this is often done using daily time steps, derived from observed time-series or from a weather generator. Several recent models use the climatic variables in conjunction with information about the state of the ecosystem in order to determine these features of the hydrological cycle as dynamical aspects of the ecosystem (Martin, 1992; Friend *et al.*, 1993; Vörösmarty *et al.*, in press).

The horizontal component of the hydrological cycle, such as the surface and subsurface routing of runoff, is more difficult to describe in ecosystem models. A spatially connected model of the hydrological cycle will require data for both the vertical fluxes and the horizontal flow of water. Topographical data, including elevation and river networks, are required for this, as well as spatial coverage of soil texture.

Ecosystem properties

Many ecosystem models require data about system-level properties of the land cover itself. Trace-gas flux models, for example, require information for the parameterization of the energy and water balance (for example, maximum leaf area index or its seasonality, roughness length of the canopy, albedo, leaf maximum conductance), endogenous growth processes (for example, the relation between live and dead plant tissue, partitioning between above-ground and below-ground biomass, or the amount of light competition from neighbouring individuals), or other ecophysiological phenomena (for example, the seasonal stress limitation to photosynthesis). The simplest approximation to a data set with these variables is a land cover map combined with parameterizations of each of the land cover types. The parameters usually come from empirical studies of typical sites (Körner, 1994; Schulze *et al.*, 1994). Spatial data sets of these variables can be derived from satellite imagery or may be output from other models (for example, from biome models which derive the major aspects of vegetation structure on the basis of climate data). Several recent experiments have aimed at providing both ground-based and remotely sensed data in connection with each other (such as the International Satellite Land Surface Climatology Programme campaigns, Sellers *et al.*, 1995). One of the major advantages of these experiments is that they enhance the usefulness of satellite data also from sites that have not had intensive ground truth.

Human land use is a major variable influencing the dynamics of most of the world's ecosystems, ranging from weak influences on remote areas (for example, tourism, extensive grazing) to the complete replacement of the natural ecosystem with a human-made ecosystem (for example, agricultural crops or pastures) or technical facilities (industrial and urban areas) (Simpson, 1993). Databases with human land-use information are required as a mask that constrains the potential development of ecosystems in a model, because, in an urban environment, it clearly does not make much sense to simulate the long-term dynamics of a natural landscape. This area of research is the most important issue for cooperation between ecosystem modellers and social scientists, because of the feedback processes between climate-driven changes in land use potential and the response of different land users to these changes (Turner *et al.*, 1993).

The problem of spatial and temporal resolution

A four-dimensional view of the biosphere

In broad-scale ecosystem models, the temporal and the spatial dimensions of the biosphere are closely connected. One 'snapshot' view of the three-dimensional (topographic) distribution of ecosystems, or a long-term average of sequential snapshots, is insufficient to provide understanding of the inner workings of the global ecosystem. On the other hand, a mechanistic four-dimensional model of every organism's relationship with the environment at each point in time is infeasible, not only due to computational but probably also due to conceptual problems. To avoid unnecessary detail, the development of a model and the underlying database require explicit statements to be made about the temporal *and* spatial aggregations of dynamical features of the system into well-defined time-slices and spatial units.

This problem can be dealt with in two ways. On the one hand, the sensitivity of a deterministic process description to the spatial and temporal resolution of a model can be assessed experimentally. This requires data about the process to be available at each time step and area unit. A soil–vegetation–atmosphere transfer (SVAT) model, for example, could be run at hourly time-steps, using the variable intensity of rainfall events and their

implications for infiltration into the soil. It could also be run daily or
weekly and test the assumption that the purpose of the model does not
require the level of precision in the process model that the hourly time-step
would allow. Most importantly, the data for calibration or validation of the
hourly routine may not exist, and the longer aggregation may therefore be
mandatory. In contrast, a stochastic method can be used to simulate a
number of patches being exposed to a range of differing conditions. In this
case, the data are used in the form of a statistical aggregation only, and one
does not need a one-to-one match with observations for every time step and
every area unit. An example of this is the stochastic simulation of gap-phase
dynamics in a forest, where only the summary statistics of a range of sites
are compared with the summary statistics of a range of model patches.
Whichever method is chosen, it remains to be an *ecological* (rather than
numerical) assessment of the method itself, whether or not a given spatial
or temporal aggregation remains inside the domain of applicability of the
model or not (Glassy & Running, 1994; Kicklighter *et al.*, 1994).

Spatial pattern

The spatial pattern of ecosystems is variable within comparatively narrow
units, ranging from only tens of metres up to tens of kilometres. A gently
sloping, forested area, for example, may be characterized by altitudinal belts
which reflect small changes in temperature at various elevations. Other
features may be running perpendicular to these belts, such as those
ecosystem types that follow water courses down the slope. Repeated
mapping of this ecosystem pattern would yield fairly similar results from
one year to the other, and sometimes over much longer periods as well.
The map would, however, show a rather great (spatial) diversity of
ecosystem types with different functional features within a few square
kilometres.

During the development of an ecosystem model, decisions need to be
made about the way the spatial heterogeneity of the landscape should be
captured by the model. Some pattern can be aggregated effectively by
lumping similar units at a coarser scale and simulating only one unit for
each class. This can be done if the fine-scale pattern repeats itself
throughout the region or if the diversity of ecosystem function and structure occurring within it is of no relevance to the general purpose of the
model. A south-facing slope on one mountain may look similar to another
south-facing slope on a neighbouring mountain – or it may be different,
depending, for example, on the bedrock geology of the area. The mosaic of
fields and woodlands around one village may look similar to such a mosaic

near another village – or it may not, depending, for example, on the socio-economic difference between the two villages. In any case, the decision needs to be made whether the overall function of the model in relation to its desired properties allows the aggregation of the mosaic into one homogeneous or repetitive spatial structure. Each of these methods has implications for the underlying environmental databases.

Lack of data makes it rarely possible to identify the appropriate level of spatial aggregation experimentally, because this would *a priori* require several different scales of resolution to be achievable. Modelling studies have shown, however, that the influence of spatial resolution on the output of the model may be substantial (Vörösmarty *et al.*, 1993). Fig. 27.1 shows how a simple water-balance model produces widely differing values for soil moisture for different resolutions of a hypothetical landscape.

Usually, a compromise is required between the requirements of the model and the availability data. Guidance can be found from an assessment

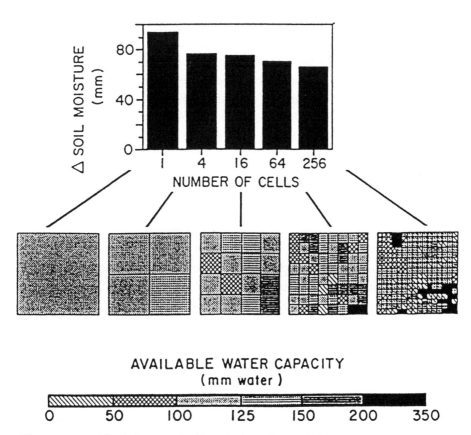

Figure 27.1 *The influence of grid resolution of water balance components during a dry to wet season transition using a simple water balance model. (From Vörösmarty et al., 1989.)*

of the types of landscape processes in an area. In mountainous areas, for example, major differences exist between conditions in lower and higher areas, such as in forests and in alpine vegetation, and these differences have major implications for most aspects of ecosystem dynamics (Körner, 1992). If the goal is the assessment of landscape fragmentation itself, such as that from human land use, then high resolution clearly is imperative for the study (Skole & Tucker, 1993). Appropriate representation of such areas in the database requires higher spatial resolution than is required for areas with less topographic or land use diversity.

Interpolation is the conventional technique for changing the spatial resolution of a database. For climatic data, it is often used to generate spatially regular coverages from irregularly spaced observations. Ecosystem model development has benefited from the recent methodological progress in this field. Examples are those interpolation methods that use circumstantial information, namely topography, as a further variable in addition to the two-dimensional spatial coordinates (Hutchinson & Bischof, 1983; Phillips *et al.*, 1992; Daly *et al.*, 1994; Hutchinson & Gessler, 1994). The problem of spatial resolution may also be addressed using nesting strategies. Some processes are described at finer spatial resolution than others, and aggregation and disaggregation are carried out where they are required. It is not always necessary, at a given level of resolution, that all detailed information from the next lower level is described in explicit coordinates. Such a fractional approach may be sufficient to parameterize models at the aggregated level (Vörösmarty *et al.*, 1993).

Temporal pattern

Ecosystem dynamics occur on a variety of time-scales. Most ecosystem simulation models cover only one or a few levels of temporal resolution, depending on the types of processes they describe. Photosynthetic and respiratory processes, for example, vary typically over a time-scale of a few minutes (involving even faster processes at the underlying biochemical level). These dynamics are driven by the rapidly changing conditions inside and near the individual leaf, such as the partial pressures of CO_2 and H_2O, and radiation. Many ecophysiological processes have a distinct diurnal cycle, driven by the changes in light intensity, temperature and humidity. Growth of most plants can be observed over periods of some days or weeks. Phenological processes, such as the development and death of leaves, normally follow the annual cycle. Longer-term life-cycle processes, such as germination, establishment, dispersal and mortality have even longer cycles. Some of them occur only at specific times of the year, and they may follow a pattern that is defined by a number of years.

A mechanistic ecosystem model might be expected to reflect all these processes. Because weather conditions also change rapidly, such a model would have to be driven by data with almost infinitely high temporal resolution. This represents not only a computational problem, due to the huge number of time steps that need to be resolved, but it is also a problem of data availability, because the temporal pattern of change for many driving variables may be incompletely known. The general solution usually is the temporal aggregation of process steps, based on the assumption of repeatability of conditions from one time step to the next. The criteria for this aggregation are rarely defined in the documentation of published models. An alternative way is the *a priori* assumption of a certain behaviour of the system over a limited number of time steps. For example, hourly time steps throughout a day can often be replaced by a mathematical integration along the typical curve of this diurnal cycle. The implication for the database is that we only need to know some parameters of this curve, assuming that we can ignore the specific conditions at every time step.

Here, the distinction between transient models and equilibrium models is crucial. Equilibrium models consider temporal trends only within the regular fluctuations of the environment, such as the seasonal or diurnal cycle, and the underlying database therefore only needs to contain data for these. To run equilibrium models in a transient mode requires assumptions about the lags between ecosystem processes and any changes in the environment. From a database point of view, this implies the need of sufficient resolution along the time axis so that disturbances may be 'seen' by the model at their true intensity. Monthly mean temperatures, for example, are probably too smooth to capture the single day with extreme frost that has killed a plant and therefore changed the ecosystem.

The determination of the appropriate temporal resolution for a transient model, like its spatial resolution, can only rarely be based on experiments with several levels of resolution, because these data usually do not exist. Understanding the nature of the ecological processes involved will usually be the main information for decisions on the design of both database and model resolution. Table 27.1 lists some classes of temporal aggregations for equilibrium and transient models, along with their respective implicit assumptions.

Table 27.1 *Temporal aggregation criteria used in equilibrium and transient ecosystem models*

Class of model	Shortest (aggregated) time steps	Criteria for temporal aggregation
Equilibrium	Minutes–hours	Arbitrary (computational resources)
	Daily min/max	Simplified diurnal cycle
	Days	Implicit diurnal cycle
	Weeks–months	Simplified annual cycle
	Years	Implicit annual cycle with static conditions
	Years with min/max	Implicit annual cycle with variability around mean
Transient	Minutes–hours	Arbitrary (computational resources)
	Days	Implicit diurnal cycle
	Weeks–months	Simplified annual cycle
	'Average' year + trend	Implicit annual cycle with semi-static conditions
	'Average' year + trend + variability	Implicit annual cycle with noisy trend
	Observed or simulated years	Implicit annual cycle with realistic trend

Data availability

In the following section, some of the available database products for ecosystem modelling will be described, along with some remarks on their usability. We do not attempt to provide a full catalogue here because the developments in this field are rapid. Rather, we focus on some aspects of these data sets that should be taken into account when they are used for the development of global ecosystem models.

Topography and soils

One global topography data set is used frequently (ETOPO5, from the US National Geophysical Data Center, 1988), but it contains several regions with unsatisfactory data quality and has a coarse vertical resolution only. It is primarily a digital elevation model, providing elevation data at a resolution of five arc-minutes. Improved data sets are to be expected in the near future, such as sets derived from the Digital Chart of the World, or such as satellite-based measurements using a synthetic aperture radar (SAR) system (Arnell, 1993). An initiative of the NASA EOS Land Processes DAAC is to generate a new regional and continental DEM (0.5 to 1 km) from remote sensing data (Gesch, 1994). From topographical databases, hill and channel slopes can be defined for the horizontal component of hydrological models (Morris & Heerdegen, 1988). If additional data about the stream network are available, then the accuracy of these derived terrain maps is much improved (Morris & Flavin, 1990).

Soil *texture* data sets have been derived from a global map of dominating soil types (FAO/UNESCO, 1974), which in itself is a compilation of many soil pedon descriptions from different sources that have been assigned to more or less large regions. The texture values are generated by mapping soil texture classes on the classification of soil types contained in the original map (Wilson & Henderson-Sellers, 1985; Staub & Rosenzweig, 1986; Prentice *et al.*, 1992). The principal difficulty of constructing a global, spatially comprehensive soil map is that spatial coverages need to be derived from point-based pedon data. These pedons are classified units within the soil taxonomy, rather than continuous variables, and therefore do not allow interpolation techniques to be applied. The major limitations to the current approaches are (i) inconsistencies in the soil type classifications used for the different regional inventories that are part of the global database, (ii) large gaps between the individual pedons studied, and (iii) the problem of 'spatialization' between classified data points. Another type of soil data is the content of major nutrients in soils world-wide (Zinke *et al.*, 1984), which are derived from point-based data in a similar way. Currently, a global soils data set is under development that holds promise for significant improvements (Scholes *et al.*, 1994). The construction of this database involves three steps:

1. Compilation of a new pedon database, containing profile descriptions and analytical results from five major international pedon databases. These databases contain many records that have not previously been used in the global data sets about soil parameters.
2. Development of statistically derived products, such as summaries by soil type.
3. Development of spatially derived products, essentially map databases, including the best possible approximations of the spatial coverage for each of the soil types.

Once established, this database will be a major improvement over most current data sets being used for global ecosystems modelling.

Climate and hydrology

Climate data sets should have a high enough spatial resolution for distinction of the major pattern of ecosystems, such as the latitudinal or altitudinal gradients. Inspection of existing maps of the global climate, however, shows that the majority of these maps, until recently, had only relatively coarse spatial resolution. Many of them appear to be based on observations of the major ecological zones, rather than on meteorological records. The information content of these climate maps is therefore not independent of the ecosystems, and they are therefore not usable as databases for ecosystem model development (Cramer & Leemans, 1993).

The large climate diagram world atlas by Walter & Lieth (1960–67) is one of the earliest, comprehensive databases for point-based climate observations worldwide. The atlas covers the globe with diagrams from an impressive number of weather stations, rather than with a continuous field. An interpolation derived from data in this atlas was used for the classical study by Emanuel *et al.* (1985). It was, however, not before the 1980s that globally comprehensive, measurement-based databases of climatic variables became more widely available (United Kingdom Meteorological Office, 1966, 1972, 1973, 1978, 1980, 1983; Willmott *et al.*, 1981a,b; Müller, 1982; Shea, 1986; Chadwyck-Healey, 1992; Vose *et al.*, 1992). Most of these data sets contain either time series or long-term averages of monthly mean temperature and precipitation, as well as of several other standard climatological variables.

In addition to these multi-variable data sets, some of the newer products focus on specific climate variables, such as precipitation (Cogley, 1991; Hulme, 1992), cloudiness (Mokhov & Schlesinger, 1993) or solar radiation (Stanhill & Moreshet, 1992). Many data sets are developed particularly for the detection of *temporal* signals rather than detailed spatial pattern (Jones & Wigley, 1980; Folland *et al.*, 1992; Vose *et al.*, 1992). Proxy data for historical climate reconstructions form a separate group (Huntley & Webb, 1988; Kong & Watts, 1992; Stocker & Mysak, 1992; Wing & Greenwood, 1993). The current network of weather observing stations, as well as the management of these observations, has been shown to be inadequate for climate change assessments because of inadequate spatial and temporal coverage, and other shortcomings (Karl *et al.*, 1993), and it is threatened by continuing decay.

For *spatial* coverage of climate data at high resolution, interpolation and gridding methods are indispensable (Hutchinson *et al.*, 1984; Legates & Willmott, 1990). Because ecosystem boundaries frequently occur on elevational gradients, it is important that the climate data account for this as

accurately as possible. In regional applications, it is possible to develop
a local atmospheric model that mechanistically describes air flow and
the associated weather variables, which can then be aggregated to provide
climate statistics at high spatial resolution. Although this might be the
preferred method, its use is practically impossible for global applications,
and we must use interpolation methods instead. Previously, some inter-
polations used a simple scheme of correcting temperatures for elevation
with standard lapse-rates (Leemans & Cramer, 1991). This method is unsatisfactory, however, because (i) it is applicable for temperature only (but
precipitation varies as well); (ii) its apparent physical background (humidity
changing systematically with changes in elevation) is valid only for specific
parcels of air rising or subsiding and breaks down when observations are
aggregated to monthly means; and (iii) it cannot account for inversions or
other local atmospheric irregularities.

Several new interpolation methods have been developed with the aim of
using circumstantial information about the landscape or the spatial correlation structure of the climate data fields. Combined with observations from
weather stations, these methods can be used to generate climate data for
points other than the weather stations themselves. Kriging (Hevesi *et al.*,
1992; Philips *et al.*, 1992) and splines (Hutchinson & Bischof, 1983; Hutchinson & Gessler, 1994) are popular methods because they permit variables
other than eastness and northness (for example, elevation, slope, aspect,
etc.) to be included into the interpolation.

Climate databases, which are gridded by interpolation, are useful for equilibrium models because they contain long-term monthly means for the most
important variables. They are, on the other hand, not suitable for investigations of transient trends in the Earth system. Climate databases for time-
series observations have been used for the detection of climatic trends
during this century (for example, Wigley & Barnett, 1990, or Schönwiese,
1994). Because these data sets are based only on stations with long observational series and high precision, they have so far only been gridded to very
coarse resolutions (4–5°), and they are therefore usually not appropriate as
drivers for terrestrial ecosystem models.

A separate independent source of climate drivers for ecosystem models
are output fields from atmospheric general circulation models (AGCMs).
Such summaries usually have global coverage and short time-intervals, e.g.
the solar radiation measurements by Planton *et al.* (1991). They can therefore be useful as alternatives to observed and interpolated data sets. The
spatial resolution of AGCMs, however, is coarse and their reliability for
regional patterns low. It is possible to use statistical techniques to convert
high-frequency, coarse-grid AGCM output to climatic variables with the

appropriate temporal and spatial scale of ecosystem models (Frey-Buness *et al.*, 1994; Gyalistras *et al.*, 1994), but such applications are obviously limited by the reliability of the climate model. They are useful for sensitivity studies of the climate situations, but not for predictive applications of ecosystem models (Claußen, 1993).

Meteorological satellites, which are mainly used for short-term weather observations, can be used for climatological purposes as well. They can provide useful data sets at high spatial resolution, including high-frequency time series. Recent reviews describe the satellite methods to derive surface *solar irradiance* (Pinker *et al.*, 1995) and *rainfall* rates (Petty, 1995). One such product is the set of satellite-based global radiation maps from the NASA Langley Research Center, covering selected months from 1985 to 1988 (Di Pasquale & Whitlock, 1993; Whitlock *et al.*, 1993). The current establishment of a SSM/I data set, from the joint NASA/NOAA Pathfinder project (Justice *et al.*, 1995), will probably include overland instantaneous rain rate estimation (Petty, 1995) for the period from April 1987 to November 1988. Zhao (1994) showed that the combination of infrared and thermal information from TOVS allows the monitoring of *atmospheric temperature*, *water vapour*, and *cloudiness*. A pathfinder data set is also currently designed for TOVS, with the objective of producing these variables on a 1° grid, the temporal resolution ranging from bi-daily to monthly (Maiden & Greco, 1994). In his critical review of the estimation of *longwave fluxes* from satellite data, Ellington (1995) concludes that there are, up to now, too many uncertainties to correctly determine the net radiative budget. Besides the NASA Pathfinder projects, the operational production at NOAA of global data sets for snow cover, precipitation, and snowpack parameters from SSM/I, as well as insolation and clear sky surface temperatures from GEOS I/M, is likely to be organized within the next 2 years (Justice *et al.*, 1995).

A further source of meteorological data is the ECMWF four-dimensional data assimilation (4DDA). Parameters such as wind components, surface and dew point temperature, pressure, albedo, net surface short-wave and long-wave radiations, will be collated, four times daily for the period January 1987 to December 1988, at 1.125° resolution on the ISLSCP CD-ROM (Sellers *et al.*, 1995).

A current trend appears to be that the existing meteorological station network is reduced, based on the expected increase in the capacity of satellites to replace them as data sources. In the interest of long-term monitoring of climatic trends, this decay must be seen as a serious threat to future investigations of possible impacts of climatic change.

Land cover and ecosystem properties

Land cover maps from inventories

The structure of the terrestrial biosphere is represented in several data sets
which are based on compilations of vegetation maps (Matthews, 1983;
Olson, et al., 1985; Wilson & Henderson-Sellers, 1985). These and several
other ecosystem maps (including some derived from models, Emanuel,
1985), have been compared with each other (Townshend et al., 1991;
Leemans et al., 1996), showing that the maps imply considerable variation
between the total areas and distribution of the various ecosystems. Incompatible classification systems are only part of the explanation for this problem.
Another is the basic shortage of spatially comprehensive national or continental inventories. The fundamental problem of land-use inventories is,
however, that, due to human land-use activities, land cover is fluctuating
more rapidly than the data can be compiled. The existing maps are therefore interpretations of the average condition of the terrestrial biosphere
during the last few decades. Because of the problems of global data collection, there is little reason to expect further developments from this
approach, and remotely sensed data products are likely to dominate in the
future.

Land cover maps from satellite data

Satellite measurements have been used for more than 20 years for mapping
and monitoring ecosystems. The photosynthetic activity of vegetation is
well captured by vegetation indices (VIs), which are a combination of
visible and near infrared signals. The most frequently used VI is the
Normalized Difference Vegetation Index (NDVI). At the global scale, the
daily overpass of coarse-resolution satellites like NOAA/AVHRR since 1982
provides the temporal signatures of NDVI for various ecosystems and gives
thereby a good representation of their phenological development throughout
the year (Goward et al., 1985; Justice et al., 1985). This has been used to
distinguish ecosystem types and their functioning, and to generate regional
or global land cover classifications from time series AVHRR data. The
AVHRR data products used for that purpose are the GVI (Tarpley et al.,
1984) at 15 km resolution, the GAC data at 4 km resolution, or the LAC
data at 1 km resolution. Fischer et al. (this volume) give a review of such
land cover classifications, and show the results of a regional multitemporal
classification, with 3 years of radiometrically corrected AVHRR data.
Running et al. (1995) define vegetation classes which can be directly trans-
lated into biophysical parameters for global climate and biogeochemical
models.

Because of high cloudiness or low illumination conditions, the use of the

NDVI may seem to be limited over the equatorial or high-latitude zones. Nevertheless, NOAA/AVHRR data have been successfully used for the determination of areas with particular disturbance patterns even in these zones. Kasischke *et al.* (1993), for example, have monitored the wildfires on boreal forests. Similar data also have provided accurate estimates of deforestation rates in the tropics (Malingreau & Tucker, 1988).

Investigations of the original GVI product have revealed significant shortcomings, and a comprehensive evaluation of it is given by Goward *et al.* (1993). This has resulted in the recent reprocessing of this data set (Goward *et al.*, 1994*a*). Gutman (1994) showed that reprocessed vegetation indices result in a signal that better represents the behaviour of vegetation. With the current establishment of global higher-level AVHRR databases and with the complete reprocessing of the raw data (Townshend, 1994), new global land cover classifications will become available soon. In the framework of IGBP-DIS, a 1 km data set derived from AVHRR LAC data for April 1992 to September 1994 is currently in development at the USGS EDC (Eidenshink & Faundeen, 1994). The availability of such a data set will allow the production of a Conventional Land Cover Stratification (Justice *et al.*, 1995). Other initiatives are the World Vegetation Map of Murai *et al.* (1990), giving a 3 year period classification of GVI.

Microwave data (active or passive) do not suffer from the same limitations that affect the measurements in the solar spectrum, and they can be used to detect other aspects of land cover (Choudhury, 1989). The SAR is sensitive to structural properties (roughness, canopy architecture). Dobson *et al.* (1995) showed that data from the ERS-1 and JERS-1 SARs yielded land cover categorizations at great accuracy. This is especially interesting for the distinction of forest-types (Ranson & Sun, 1994). Although not especially designed for vegetation characterization, the wind scatterometer onboard ERS-1 can be used to identify the main global vegetation types, especially in tropical regions (Mougin *et al.*, 1995). In a preliminary analysis for Africa, the authors detected the strong capability of the ERS-1 scatterometer to monitor vegetation changes at the global scale. For passive microwave sensors, Rock *et al.* (1993) state that the global data set from SMMR (difference of vertically and horizontally polarized brightness temperature at 37 GHz from January 1979 to December 1985 at $0.25° \times 0.25°$) could be a highly valuable tool to study seasonal and interannual vegetation change. Land surface type classifications are part of the SSM/I pathfinder product of NASA/NOAA (Maiden & Greco, 1994).

Ecosystem properties from inventories

Functional aspects of ecosystems are traditionally observed at relatively small sites with extensive measurement programmes, monitoring various ecosystem variables over shorter or longer periods. For the purpose of global ecosystem modelling, such observations are crucial because they link the broad-scale models to observable features of the system. The problem of extrapolation, however, exists both in time and in space. Ground-based measurements of LAI, canopy conductance, or trace gas fluxes can hardly be made over large areas or over long periods. Some models have therefore used a few well-studied sites for each ecosystem type only (Melillo *et al.*, 1993), thereby ignoring the possible variability of ecosystem properties throughout the range of that type. The documentation of some other models remains unclear about the origin of their calibration data. Recent work has shown that a combination of well-selected ground observations and specific models can significantly improve understanding of the global pattern of ecosystem properties, such as canopy conductance (Schulze *et al.*, 1994).

Ecosystem properties from satellite data

Several major vegetation properties can be derived from satellite measurements at various wavelengths. These include reflectances of sun radiation (short wavelengths), brightness temperatures (thermal infrared and passive microwaves) and backscatter coefficients (active microwaves). Theoretical models as well as empirical data have shown that these variables can be related to the fraction of Absorbed Photosynthetically Active Radiation (fPAR), LAI, fractional cover, biomass density, canopy roughness, canopy water content, evapotranspiration or soil moisture.

Hall *et al.* (1995) and Myneni *et al.* (1995) discuss both empirical and physically based algorithms which can be used in the short wavelengths to invert fPAR or LAI. Empirical algorithms use VIs, such as NDVI. Several studies have dealt with the definition of vegetation indices that are less sensitive to soil background effects (Huete, 1989; Qi *et al.*, 1994), canopy closure (Nemani *et al.*, 1993*b*), or atmospheric perturbations (Pinty & Verstraete, 1991; Kaufman & Tanré, 1992). Local calibration is generally provided by field experiments. Estimates of LAI or fPAR at the global scale have been used as input variables into global biogeochemical models (Heimann & Keeling, 1989; Ruimy *et al.*, 1994; Maisongrande *et al.*, 1995). The AVHRR Pathfinder product (James & Kalluri, 1994) will provide a long-term (from 1981 on) global data set at 4 km resolution with consistent geometric and radiometric processing. This will improve the derivation of biophysical properties and help to monitor their temporal change at regional or global scale. The production of a global land cover parameterization from the 1 km AVHRR data set currently produced by EDC will involve LAI, albedo, vegetation structure, fPAR, NPP (Townshend *et al.*, 1994). Physically based algorithms are currently not applicable to satellite data because they require multi-angle measurements in both visible and near-infrared channels (Pinty *et al.*, 1990; Verstraete *et al.*, 1990; Rahman *et al.*, 1993*a,b*). The new sensors to be launched in the coming decade (e.g. ADEOS/ POLDER, EOS/MODIS) are likely to provide the necessary measurements for this purpose. The multi-angle acquisitions, aerosol, ozone and vapour water content estimates will help parameterize atmospheric corrections models, as well as in the inversion of bi-directional reflectance using physical radiative transfer models.

Information on the surface energy budget (the partitioning between sensible and latent heat flux) can be derived from observations in the thermal infrared (Norman *et al.*, 1995). The daily integration of the instantaneous latent heat flux (evapotranspiration) is an important variable for global vegetation modelling. Seguin *et al.* (1994) have confirmed that regional scale estimation of evapotranspiration using NOAA/AVHRR or Meteosat data is possible.

The signature of the canopy in microwave measurements provides information about the structure (roughness, above-ground biomass) and the water content of vegetation. The highest wavelengths are particularly sensitive to vegetation characteristics. Dobson *et al.* (1995) describe how quantitative values of roughness and above-ground biomass can be estimated from ERS-1 and JERS-1 SAR data. The global microwave data sets which exist or are in preparation are those from SMMR or SSM/I (Maiden & Greco, 1994). From these data, land surface temperature, vegetation density (Choudhury, 1989), and vegetation water content (Calvert *et al.*, 1994) can be retrieved.

Soil *moisture* is a dynamic variable that is required for calibration or validation of spatially comprehensive models of the hydrological cycle. It is a difficult variable to measure on a consistent and spatially comprehensive basis, but can be estimated from remote sensing (Schmugge & Jackson, 1994; Engman & Chauhan, 1995; Hall *et al.*, 1995). Data from the SMMR onboard Nimbus-7, combined with information about the vegetation from NOAA/AVHRR, have allowed large-scale surface soil moisture estimation (Van de Griend & Owe, 1994). Although the existing global data set for SMMR data (Choudhury & Tucker, 1987) is not suitable for soil moisture estimation (Rock *et al.*, 1993), the new above-mentioned SMMR Pathfinder data set will be more appropriate (Maiden & Greco, 1994). Currently, the SAR's onboard ERS-1 and JERS-1 are the only operational microwave satellites with frequencies suitable for soil moisture (Engman & Chauhan, 1995), even though their coverage is not global. It has been suggested that a combination of optical and thermal wavelengths can be used to estimate soil moisture at the surface and in the root zone (Carlson *et al.*, 1990). Regional applications with NOAA/AVHRR have produced average soil moisture conditions (such as drought) throughout the growing period (Nemani *et al.*, 1993*a*; Goward *et al.*, 1994*b*).

A particular requirement which becomes more urgent with the development of global transient dynamic ecosystem models is for calibration or validation of *disturbance* processes, where information is required about the nature and intensity of disturbances in different ecosystems. Several projects are currently addressing this problem. An example is the IGBP-DIS working group that is being set up to generate algorithms for the detection of fire frequencies from the global 1 km AVHRR data base (Justice, 1994). Regional land cover classification from high spatial resolution (Landsat TM since 1974, SPOT HRV since 1986) have clearly displayed land use changes in sensitive areas. The Landsat/Landcover Pathfinder data set of NASA EOS will provide, among others, maps of deforestation for Amazon, Central Africa, South-East Asia, and of land cover changes in North America, Mexico, and the Caribbean (Justice *et al.*, 1995).

Combined data set packages

The need for global, integrated databases for climate assessments and, particularly, modelling of biosphere processes has been recognized (Hastings *et al.*, 1991; Townshend, 1992; Marshall, 1993), and this has resulted in several products covering compatible data sets for several variables being available now (Boden *et al.*, 1990; NOAA-EPA Global Ecosystems Database Project, 1992). The Global Ecosystems Database, for example, consists of several CD-ROM disks containing global data sets from many different sources. Most of the data sets are provided in their original format as well as in a format that has been regridded to a common global grid. In addition to some printed documentation and a users guide, the CD also contains scanned images of the original publications documenting the data sets.

Another package is the ISLSCP Initiative I CD-Rom which provides several data sets, such as global soil moisture estimates, on a $1° \times 1°$ grid, four times daily for the years 1987–8, produced by ECMWF four-dimensional data assimilation (Sellers *et al.*, 1995), as well as other hydro-meteorological parameters.

Organizational issues

Access to data may be limited for several reasons, such as technical limitations (Justice *et al.*, 1995) (now being slowly reduced by new technologies, such as CD-ROM devices or the worldwide web), lack of access to the scientific networks (which often may limit knowledge of existing data) and the increasing problem of data commercialization (Hulme, in press). The issue of data availability has been given high priority since the beginning of the International Geosphere-Biosphere Programme (IGBP). To achieve maximum coordination between the various core projects within IGBP, a separate core project DIS (Data and Information Systems) was established. In some cases, single large-scale research programmes have designed their own information systems, such as FIS (the FIFE Information System, Sellers *et al.*, 1992) or EOSDIS (NASA, 1993). In IGBP-DIS focus 1 (data set development; IGBP, 1994), nine main activities are listed:

1. Development of a global 1 km land-surface database from AVHRR and similar satellite sensors
2. Sample high-resolution satellite data
3. Global land cover database generation
4. Global soil pedon data set
5. Development of a global fire database
6. Estimation of the areal extent and characteristics of global wetlands, from remote sensing and other sources
7. Determination of terrestrial NPP by remote sensing and other sources
8. Global topographic data sets
9. Global land biomass database

All of these activities are directly relevant to the development of global ecosystem models, providing either initialization or calibration data. Some (like NPP) are also partly dependent on the development of appropriate diagnostic ecosystem models.

Another international data set production initiative is the Global Climate Observing System (GCOS), the concept of which was outlined at the Second World Climate Conference in 1990. Its goals are (Heal *et al.*, 1993):

1. Climate system monitoring, climate change detection and response monitoring, especially in terrestrial ecosystems
2. Data for application to national economic development
3. Data for research towards improved understanding, modelling and prediction of the climate system
4. Eventually, a comprehensive observing system for climate forecasting.

Similar to GCOS, a Global Terrestrial Observing System (GTOS) is

currently in preparation. GTOS will provide the link between various monitoring activities for the terrestrial biosphere, IGBP, and other international
research efforts that require efficient monitoring of the global state of terrestrial ecosystems (Heal *et al.*, 1993). Like GCOS, this system is planned to
provide monitoring schemes, using ground-based data collection and satellites that shall allow for detection of global change processes and impacts as
early as possible.

Calibration and validation

Data-based activities that initialize or parameterize an ecosystem model can
be summarized using the term *calibration*. If ecosystem models could be
built on first principles of the functioning of nature alone, then this data
requirement would be absent because the inherent structure of the
biosphere would simply emerge from the basic processes. In practice,
however, this level of development may not be achievable, and we must use
a range of data sources to determine the empirical relationships between the
environment (in time and space) and those components of the ecosystem
that we cannot simulate otherwise. For all models, an independent *validation* is required to ensure that the relationships described have validity also
beyond the domain given by the calibration data set. Validation of a calibrated simulation model can only be done if independent sets of observations under otherwise equivalent conditions exist. In a global ecosystem
model, we can validate only parts of the system, for example by comparing
the behaviour of a submodel with measured data from real locations. We
can also compare the spatial pattern of features that can be mapped globally
(ecosystem type, plant type composition, or leaf area index). Diagnostic
models, for example those derived from satellite data, can be calibrated
using some years of observations and validated with other years.

   In some cases, output from a detailed, mechanistic ecosystem model,
which has been parameterized for one or a few locations, can be used as
validation 'data' for a more general type of model with wide spatial
coverage. In other cases, the ability of a detailed model to perform inside
the ranges given by a general model may partly validate the performance of
the detailed model. Clearly, the uncertainties of either simulation will somewhat contaminate the other, but this procedure may be sufficient to achieve
a basic test that could not be achieved otherwise. The same data set may be
treated by one model as initialization or calibration data, and by another as
validation data. A leaf area index database, for example, may be required as
input into a biogeochemical model, but it could be an independent validation for the performance of an ecosystem structure model.

Calibration/validation using field data

Field surveys of ecosystem characteristics, such as biomass or trace-gas fluxes, have been used to calibrate various mechanisms in ecosystem models. This can be done either by simple parameterization of an empirical relationship, or through the determination of the value of parameters with direct biophysical meaning. In the latter case, data sets from different locations or different times provide the validation of the model. Problems occur as soon as a generalization across ecosystem types or large regions is expected, because the spatial variability of the parameters is not known *a priori*, and some major processes in the vegetation behaviour may also change between different biomes. Several intensive studies have been performed in recent years which aimed specifically at combining field measurements with diagnostic and prognostic models (Mascart *et al.*, 1991; Sellers & Hall, 1992; Waring & Peterson, 1994; Prince *et al.*, 1995).

Calibration/validation using satellite data

For the present time, remotely sensed data are the only appropriate tool to continuously monitor the entire Earth, and they allow therefore the validation of those variables in global ecosystem models that may be related to radiative measurements. LAI is a good example for such a variable, and it is difficult to estimate on the ground at a landscape or regional scale.

The *ecosystem LAI* is assumed to be in long-term equilibrium with soil, climate, and vegetation structure. An empirical model for the estimation of the LAI in mature coniferous forests has been proposed by Grier & Running (1977) for the northern Rocky Mountains. Nemani & Running (1989) have validated a similar model by comparison with the LAI derived from a NOAA/AVHRR satellite image. Once they are validated, satellite mapped LAI data allow the estimation of other model parameters the spatial variation of which is unknown. Also, with satellite data over natural reserves, Woodward & Smith (1994) have validated the continental-scale pattern of potential LAI following a more mechanistic concept of equilibrium between precipitation and canopy structure (Woodward, 1987).

Seasonality of the LAI

Prognostic ecosystem models should be capable of capturing the major aspects of phenology of deciduous ecosystems. The FBM (Frankfurt Biosphere Model; Lüdeke *et al.*, 1994), for example, describes photosynthesis, respiration, and the allocation of carbon to leaves, roots, and stems, using daily time steps. The climate data used to initialize and drive this model were derived from long-term monthly means, and the model therefore produces a mean temporal profile of LAI. Such a profile could be partly validated by the comparison with a mean NDVI profile over several years, or with the mean NDVI profile which is representative for that ecosystem (M. K. B. Lüdeke, et al., in press).

A more detailed discussion on the use of temporal NDVI profile to initialize/calibrate a model for the seasonal development of the canopy is given by Kergoat et al., 1995 and Fischer et al., this volume. Temporal satellite data are useful to estimate the values of model parameters with the following characteristics: (i) they have a strong influence on the satellite signal; (ii) they have a strong influence on output variables from the ecosystem model (trace-gas fluxes, structure development, biomass production); and (iii) they are poorly known, especially regarding their spatial pattern. The strategy described allows assessment of the values of site-dependent parameters, both between and within biomes. If one year of satellite data can provide the calibration of parameters controlling the canopy development, then another year (used as an independent data set) can be used for validation. This indirect way would be useful in the development of a dynamic global vegetation model, because it provides a simple (and cheap) way to access the spatial variations of key parameters.

Conclusion: Data needs for the development and application of a dynamic global vegetation model (DGVM)

Several current research initiatives are aiming towards the development of dynamic global vegetation models (Steffen et al., 1992). These models are designed to reflect the transient response of the biosphere to changes in the environment (climate and atmospheric composition). The current philosophy involves a multi-layer, nested approach along both the spatial and the temporal scales, ranging from minutes to years, and from individual leaves to grid cells of several tens of kilometres. Because of the extremely high resolution required by some compartments of the model, not all processes would be simulated with precise geographical coordinates. Instead, these components would be run for 'average' locations within the coarser pixels or longer time steps of the next higher hierarchical level. At all levels, however, the state descriptions of the grid cells would be partly controlled by the dynamics of the system itself, and partly by a calibration database. The dynamics of ecosystem development need to be described in terms of the concepts currently used in fine-scale vegetation dynamics models, such as establishment, growth, and decline. The DGVM, however, would require parameterization of these processes for functional types of plants, rather than species. It would also involve modules for the dispersal of such types, and therefore requires explicit treatment of landscape pattern with global coverage.

The requirements for the underlying database go beyond the capabilities of any currently existing product. For the climatic part, it must contain spatially and temporarily comprehensive time-series data. The site variables also need data at the highest available resolution for topography, soils, hydrological relationships (including the horizontal surface routing of water runoff), and current land use. It appears to be feasible to generate many of these data sets at spatial resolutions near to 1 km, although this will sometimes represent a smoothed representation of the real landscape.

Validation of the DGVM against data will occur in various ways. Being composed from several existing components, such as soil–vegetation–atmosphere–transfer (SVAT) models, or modified forest dynamics models, these components may be validated for specific sites. At a large spatial scale, the behaviour of the system as a whole can, in some of its key variables, be validated on two time-scales: (i) against paleoecological reconstructions of past distributions of ecosystems, or (ii) against short-term changes as they are observed by satellites. Both lines will be followed, and they require separate database developments. For the long-term historic reconstruction, it is essential that the driving variables of the DGVM are independent of the observation being used for validation. Therefore, the climatic driving variables must be derived from general circulation models or other, non-biological methods to reconstruct past climates (Prentice *et al.*, 1993). For the short-term reconstruction, data sets of optical reflectances and their derived indices must be matched as closely as possible with climatic data of similar spatial and temporal structure.

Applications to future climatic conditions, such as those that can be expected to occur at a doubled CO_2 content of the atmosphere, will be performed as part of the general activities towards better understanding of the sensitivities of the Earth system to global change. The predictive value of such analyses, however, will depend on the interactive coupling between general models of the atmospheric and ocean circulation and the DGVM. Until such coupling is achieved, the DGVM will require output from existing GCMs to be used as separate climatic driving forces. To provide these kind of data is not only an issue of archiving and transmitting near-surface climatic conditions as they are simulated in GCMs, but also of appropriate scale conversion techniques in the spatial and temporal domain.

Acknowledgements

We thank Will Steffen, Canberra (Australia) for reading an earlier draft of the chapter. The manuscript was much improved due to comments provided by the editors, John Townshend, College Park (Maryland, USA), and an anonymous reviewer. Parts of the work have been supported by the Norwegian Research Council (Oslo, Norway), the CSIRO Division of Wildlife and Ecology (Canberra, Australia) and the German IGBP secretariat (Berlin, Germany).

Appendix. Acronyms

4DDA	Four-Dimensional Data Assimilation
ADEOS	Advanced Earth Observation Satellite
AGCM	Atmospheric General Circulation Model
AVHRR	Advanced Very High Resolution Radiometer
DAAC	Distributed Active Archive Center
DEM	Digital Elevation Model
DGVM	Dynamic Global Vegetation Model
DIS	Data and Information Systems (IGBP core project)
ECMWF	European Center for Medium Range Weather Forecast
EDC	EROS Data Center
EOS	Earth Observing System
EOSDIS	Earth Observing System Data and Information System
EPA	US Environmental Protection Agency
EROS	Earth Resources Observation System
ERS-1	European Remote Sensing Satellite
FIFE	First ISLSCP Field Experiment
FIS	FIFE Information System
fPAR	fraction of Absorbed PAR
GAC	Global Area Coverage
GCOS	Global Climate Observing System
GEOS I/M	Geostationary Observational . . .
GTOS	Global Terrestrial Observing System
GVI	Global Vegetation Index
HRV	High Resolution Visible
IGBP	International Geosphere-Biosphere Programme
ISLSCP	International Satellite Land Surface Climatology Program
JERS-1	Japanese Earth Resource Satellite

LAC	Local Area Coverage
LAI	Leaf Area Index
MODIS	Moderate-resolution Imaging Spectrometer
NDVI	Normalized Difference Vegetation Index
NPP	Net Primary Productivity
PAR	Photosynthetically Active Radiation
POLDER	Polarization and Directionality of Earth Reflectances
SAR	Synthetic Aperture Radar
SMMR	Scanning Multichannel Microwave Radiometer
SPOT	Système Probatoire d'Observation de la Terre
SSM/I	Special Sensor Microwave Imager
SVAT	Soil-Vegetation-Atmosphere Transfer Model
TBM	Terrestrial Biogeochemical Model
TIROS	Television and Infrared Observation Satellite
TM	Thematic Mapper
TOVS	TIROS Operational Vertical Sounder
USGS	United States Geological Survey
VI	Vegetation Index

References

Arnell, N. W. (1993). Data requirements for macroscale modelling of the hydrosphere. In *Macroscale Modelling of the Hydrosphere*, pp. 139–49. Yokohama, Japan: IAHS.

Boden, T. A., Kanciruk, P. & Farrell, M. P. (1990). *Trends '90: A Compendium of Data on Global Change*. Carbon Dioxide Information Analysis Center. No. 36. ORNL/CDIAC-36.

Box, E. O. (1981). *Macroclimate and Plant Forms: An Introduction to Predictive Modeling in Phytogeography*. The Hague: Junk.

Calvet, J.-C., Wigneron, J.-P., Mougin, E., Kerr, Y. H. & Brito, J. L. S. (1994). Plant water content and temperature of the Amazon forest from satellite microwave radiometry. *IEEE Transactions on Geoscience and Remote Sensing*, **32**, 397–408.

Carlson, T. N., Perry, E. M. & Schmugge, T. J. (1990). Remote estimation of soil moisture availability and fractional vegetation cover for agricultural fields. *Agricultural and Forest Meteorology*, **52**, 45–69.

Chadwyck-Healey (1992). *World Climate Disk: Global Climate Change Data*. Cambridge: Chadwyck-Healey.

Choudhury, B. J. (1989). Monitoring global land surface using Nimbus-7 37 GHz data. Theory and examples. *International Journal of Remote Sensing*, **10**, 1579–605.

Choudhury, B. J. & Tucker, C. J. (1987). Monitoring global vegetation using NIMBUS-7 37 GHz data: some empirical relations. *International Journal of Remote Sensing*, **8**, 1085–90.

Claußen, M. (1993). *Shift of biome patterns due to simulated climate variability and climate change*. Report. 115. Hamburg: Max-Planck-Institut für Meteorologie.

Coakley, J. A. J., Pinker, R. T., Arkin, P. A., Gelman, M. E., Miller, A. J., Nagatani, R. M., Reynolds, R. W., Ropelewski, C. F., Gruber, A., Ohring,

G. & Stowe, L. L. (1990). Use of satellite data in climate analysis. In *Weather Satellites: Systems, Data, and Environmental Applications*, ed. P. K. Rao, S. J. Holmes, R. K. Anderson, J. S. Winston & P. E. Lehr, pp. 429–47. Boston: American Meteorological Society.

Cogley, J. G. (1991). *GGHYDRO – Global hydrographic data release 2.0*. Department of Geography, Trent University. Trent Climate Note. No. 91–1.

Cramer, W. & Leemans, R. (1993). Assessing impacts of climate change on vegetation using climate classification systems. In *Vegetation Dynamics and Global Change*, ed. A. M. Solomon & H. H. Shugart, pp. 190–217. New York: Chapman and Hall.

Cramer, W. & Prentice, I. C. (1988). Simulation of soil moisture deficits on a European scale. *Norsk Geografisk Tidskrift*, 42, 149–51.

Daly, C., Neilson, R. P. & Phillips, D. L. (1994). A statistical-topographic model for mapping climatological precipitation over mountainous terrain. *Journal of Applied Meteorology*, 33, 140–58.

Di Pasquale, R. C. & Whitlock, C. H. (1993). First WCRP long-term satellite estimates of surface solar flux for the globe and selected regions. In *25th International Symposium on Remote Sensing and Global Environmental Change, 4–8 April 1993*, Graz, Austria.

Dobson, M. C., Ulaby, F. T. & Pierce, L. E. (1995). Land-cover classification and estimation of terrain attributes using Synthetic Aperture Radar. *Remote Sensing of Environment*, 51, 199–214.

Eidenshink, J. C. & Faundeen, J. L. (1994). The 1 km AVHRR global land data set: first stages in implementation. *International Journal of Remote Sensing*, 15, 3443–62.

Ellington, R. G. (1995). Surface longwave fluxes from satellite observations: a critical review. *Remote Sensing of Environment*, 51, 89–97.

Emanuel, W. R., Shugart, H. H. &

Stevenson, M. P. (1985). Climatic change and the broad-scale distribution of terrestrial ecosystems complexes. *Climatic Change*, 7, 29–43.

Engman, E. T. & Chauhan, N. (1995). Status of microwave soil moisture measurements with remote sensing. *Remote Sensing of Environment*, 51, 189–98.

Esser, G., Hoffstadt, J., Mack, F. & Wittenberg, U. (1994). *Szenarienrechnungen: Landnutzungs- und Rodungsszenarien*. Mitteilungen aus dem Institut für Pflanzenökologie der Justus-Liebig-Universität Gießen. 1.

FAO/UNESCO (1974). *Soil map of the world, 1 : 5 000 000*. Paris: United Nations Food and Agriculture Organization.

Folland, C. K., Karl, T. R., Nicholls, N., Nyenzi, B. S., Parker, D. E. & Vinnikov, K. Y. (1992). Observed climate variability and change. In *Climate Change 1992. The Supplementary Report to the IPCC Scientific Assessment*, ed. J. T. Houghton, B. A. Callander & S. K. Varney, pp. 135–70. Cambridge: Cambridge University Press.

Frey-Buness, A., Heimann, D. & Sausen, R. (1994). A statistical-dynamical downscaling procedure for global climate simulations. *Theoretical and Applied Climatology*, 50, 117–31.

Friend, A. D., Shugart, H. H. & Running, S. W. (1993). A physiology-based gap model of forest dynamics. *Ecology*, 74, 792–7.

Gesch, D. B. (1994). *Topographic data requirements for EOS global change research*. US Geological Survey Open-File Report.

Glassy, J. M. & Running, S. W. (1994). Validating diurnal climatology logic of the MT-CLIM model across a climatic gradient in Oregon. *Ecological Applications*, 4, 248–57.

Goward, S. N., Dye, D. G., Turner, S. & Yang, J. (1993). Objective assessment of the NOAA global vegetation index data product. *International Journal of Remote Sensing*, **14**, 3365–94.

Goward, S. N., Tucker, C. J. & Dye, D. G. (1985). North American vegetation patterns observed with the NOAA-7 advanced very high resolution radiometer. *Vegetatio*, **64**, 3–14.

Goward, S. N., Turner, S., Dye, D. G. & Liang, S. (1994*a*). The University of Maryland improved Global Vegetation Index product. *International Journal of Remote Sensing*, **15**, 3365–95.

Goward, S. N., Waring, R. H., Dye, D. G. & Yang, J. (1994*b*). Ecological remote sensing at OTTER: satellite macroscale observations. *Ecological Applications*, **4**, 322–43.

Grier, C. C. & Running, S. W. (1977). Leaf area of mature Northwestern coniferous forests: relation to site water balance. *Ecology*, **58**, 893–9.

Gutman, G. G. (1994). Global data on land surface parameters from NOAA AVHRR for use in numerical climate models. *Journal of Climate*, **7**, 669–80.

Gyalistras, D., Von Storch, H., Fischlin, A. & Beniston, M. (1994). Linking GCM generated climate scenarios to ecosystems: case studies of statistical downscaling in the Alps. *Climate Research*, **4**, 167–89.

Hall, F. G., Townshend, J. R. & Engman, E. T. (1995). Status of remote sensing algorithms for estimation of land surface state parameters. *Remote Sensing of Environment*, **51**, 138–56.

Hastings, D. A., Kineman, J. J. & Clark, D. M. (1991). Development and application of global databases: considerable progress, but more collaboration needed. *International Journal of Geographical Information Systems*, **5**, 137–46.

Heal, O. W., Menaut, J.-C. & Steffen, W. (1993). *Towards a global terrestrial observing system (GTOS). Detecting and monitoring change in terrestrial ecosystems*, IGBP Global Change Report no. 26. Stockholm: International Geosphere-Biosphere Programme.

Heimann, M. & Keeling, C. D. (1989). A three-dimensional model of atmospheric CO_2 transport based on observed winds. 2. Model description and simulated tracer experiments. *Geophysical Monographs*, **55**, 237–75.

Hevesi, J. A., Istok, J. D. & Flint, A. L. (1992). Precipitation estimation in mountainous terrain using multivariate geostatistics. 1. Structural analysis. *Journal of Applied Meteorology*, **31**, 661–76.

Huete, A. R. (1989). Soil influences in remotely sensed vegetation-canopy spectra. In *Theory and Applications of Optical Remote Sensing*, ed. G. Aswar, pp. 107–41. New York: Wiley.

Hulme, M. (1992). A 1951–80 global land precipitation climatology for the evaluation of general circulation models. *Climate Dynamics*, **7**, 57–72.

Hulme, M. (in press). The cost of climate data – a European experience. *Weather*, **6**.

Huntley, B. & Webb III, T. (eds.) (1988). *Vegetation History*. Dordrecht: Kluwer.

Hutchinson, M. F. & Bischof, R. J. (1983). A new method for estimating the spatial distribution of mean seasonal and annual rainfall applied to the Hunter Valley, New South Wales. *Australian Meteorological Magazine*, **31**, 179–84.

Hutchinson, M. F., Booth, T. H., McMahon, J. P. & Nix, H. A. (1984). Estimating monthly mean values of daily total solar radiation for Australia. *Solar Energy*, **32**, 277–90.

Hutchinson, M. F. & Gessler, P. E. (1994). Splines – more than just a smooth interpolator. *Geoderma*, **62**, 45–67.

IGBP (1994). *IGBP Global Modelling and Data Activities 1994–1998*, Global Change Report no. 30. Stockholm: International Geosphere-Biosphere Programme.

James. M. E. & Kalluri, S. N. V. (1994). The Pathfinder AVHRR land data set: an improved coarse resolution data set for terrestrial monitoring. *International Journal of Remote Sensing*, **15**, 3347–64.

Jones, P. D. & Wigley, T. M. L. (1980).

Northern hemisphere temperatures, 1881–1979. *Climate Monitor*, **9**, 43–7.

Justice, C. O. (ed.) (1994). *IGBP-DIS Satellite Fire Algorithm Workshop Technical Report*. Paris: IGBP Data and Information Systems, Core Project Office.

Justice, C. O., Bailey, G. B., Maiden, M. E., Rasool, S. I., Strebel, D. E. & Tarpley, J. D. (1995). Recent data and information system initiatives for remotely sensed measurements of the land surface. *Remote Sensing of Environment*, **51**, 235–44.

Justice, C. O., Townshend, J. R. G., Holben, B. N. & Tucker, C. J. (1985). Analysis of the phenology of global vegetation using meteorological satellite data. *International Journal of Remote Sensing*, **6**, 1271–318.

Karl, T. R., Quayle, R. G. & Groisman, P. Y. (1993). Detecting climate variations and change – new challenges for observing and data management systems. *Journal of Climate*, **6**, 1481–94.

Kasischke, E. S., French, N. H. F., Harrel, P., Christensen Jr, N. L., Ustin, S. L. & Barry, D. (1993). Monitoring of wildfires in boreal forests using large area AVHRR NDVI composite image data. *Remote Sensing of Environment*, **45**, 61–71.

Kaufman, Y. & Tanré, D. (1992). Atmospherically resistant vegetation index (ARVI) for EOS-MODIS. *IEEE Transactions on Geoscience and Remote Sensing*, **30**, 261–70.

Kergoat, L., Fischer, A., Moulin, S. & Dedieu, G. (1995). Satellite measurements as a constraint on estimates of vegetation carbon budget. Tellus Series B. *Chemical and Physical Meteorology*, **47**, 251–63.

Kicklighter, D. W., Melillo, J. M., Peterjohn, W. T., Rastetter, E. B., McGuire, A. D., Steudler, P. & Aber, J. D. (1994). Aspects of spatial and temporal aggregation in estimating regional carbon dioxide fluxes from temperate forest soils. *Journal of Geophysical Research*, **99**, 1303–15.

Kong, W.-S. & Watts, D. (1992). A unique set of climatic data from Korea dating from 50 BC, and its vegetational implications. *Global Ecology and Biogeography Letters*, **2**, 133–8.

Köppen, W. (1884). Die Wärmezonen der Erde, nach der Dauer der heissen, gemässigten und kalten Zeit und nach der Wirkung der Wärme auf die organische Welt betrachtet. *Meteorologische Zeitschrift*, **1**, 215–26 (+ map).

Körner, C. (1992). Response of alpine vegetation to global climate change. *Catena*, **22**, 85–96.

Körner, C. (1994). Leaf diffusive conductances in the major vegetation types of the globe. In *Ecophysiology of Photosynthesis*, ed. E.-D. Schulze & M. M. Caldwell, pp. 463–90. Berlin, Heidelberg, New York: Springer.

Leemans, R. & Cramer, W. (1991). *The IIASA Database for Mean Monthly Values of Temperature, Precipitation and Cloudiness of a Global Terrestrial Grid*. RR-91-18. Laxenburg, Austria: International Institute for Applied Systems Analysis (IIASA).

Leemans, R., Cramer, W. & Van Minnen, J. G. (1996). Prediction of global biome distribution using bioclimatic equilibrium models. In *Carbon Cycling in Grassland and Forested Ecosystems*, ed. J. M. Melillo & A. Breymeyer. New York: Wiley (in press).

Legates, D. R. & Willmott, C. J. (1990). Mean seasonal and spatial variability in global surface air temperature. *Theoretical and Applied Climatology*, **41**, 11–21.

Lüdeke, M. K. B., Badeck, F.-W., Otto, R. D., Häger, C., Dönges, S., Kindermann, J., Würth, G., Lang, T., Jäkel, U., Klaudius, A., Ramge, P., Habermehl, S. & Kohlmaier, G. H. (1994). The Frankfurt Biosphere Model. A global process oriented model for the seasonal and long-term CO_2 exchange between terrestrial ecosystems and the atmosphere. I. Model description and illustrative results for cold deciduous and boreal forests. *Climate Research*, **4**, 143–66.

Lüdeke, M. K. B., Ramge, P. H. & Kohlmaier, G. H. (in press). The use of satellite-detected NDVI data for the validation of global vegetation phenology models and application to the Frankfurt Biosphere Model. *Ecological Modelling*.

Maiden, M. E. & Greco, S. (1994). NASA's Pathfinder data set programme: land surface parameters. *International Journal of Remote Sensing*, **15**, 3333–46.

Maisongrande, P., Ruimy, A., Dedieu, G. & Saugier, B. (1995). Monitoring seasonal and interannual variations of gross primary productivity, net primary productivity and net ecosystem productivity using a diagnostic model and remotely-sensed data. Tellus Series B. *Chemical and Physical Meteorology*, **47**, 178–90.

Malingreau, J. P. & Tucker, C. (1988). Large scale deforestation in the southeastern Amazon basin of Brazil. *Ambio*, **17**, 49–55.

Marshall, E. (1993). Global change – fitting planet earth into a user-friendly database. *Science*, **261**, 846.

Martin, P. (1992). EXE: a climatically sensitive model to study climate change and CO_2 enrichment effects on forests. *Australian Journal of Botany*, **40**, 717–35.

Mascart, P., Taconet, O., Pinty, J. P. & Mehrez, M. B. (1991). Canopy resistance formulation and its effect in mesoscale models – a HAPEX perspective. *Agricultural and Forest Meteorology*, **54**, 319–51.

Matthews, E. (1983). Global vegetation and land use: new high-resolution data bases for climate studies. *Journal of Climate and Applied Meteorology*, **22**, 474–87.

Melillo, J. M., McGuire, A. D., Kicklighter, D. W., Moore III, B., Vörösmarty, C. J. & Schloss, A. L. (1993). Global climate change and terrestrial net primary production. *Nature*, **363**, 234–40.

Mokhov, I. I. & Schlesinger, M. E. (1993). Analysis of global cloudiness. 1. Comparison of METEOR, NIMBUS-7, and International Satellite Cloud Climatology Project (ISCCP) satellite data. *Journal of Geophysical Research*, **98**, 12849–68.

Monteith, J. L. (1965). Radiation and crops. *Experimental Agriculture Review*, **1**, 241–51.

Morris, D. G. & Flavin, R. W. (1990). A digital terrain model for hydrology. In *4th International Symposium on Spatial Data Handling*, pp. 250–62. Zürich, Switzerland.

Morris, D. G. & Heerdegen, D. G. (1988). Automatically derived catchment boundaries and channel networks, and their hydrological applications. *Geomorphology*, **1**, 131–41.

Mougin, E., Lopes, A., Frison, P. L. & Proisy, C. (1995). Preliminary analysis of ERS-1 wind scatterometer data over land surfaces. *International Journal of Remote Sensing*, **16**, 391–8.

Müller, M. J. (1982). *Selected Climatic Data for a Global Set of Standard Stations for Vegetation Science*. The Hague: Junk.

Murai, S., Honda, Y., Asakura, K. & Goto, S. (1990). *An Analysis of Global Environment by Satellite Remote Sensing*. University of Tokyo, 7–22, Roppingi Minato-ku, Tokyo, Japan: Institute of Industrial Science.

Myneni, R. B., Maggion, S., Iaquinta, J., Privette, J. L., Gobron, N., Pinty, B., Kimes, D. S., Verstraete, M. M. & Williams, D. L. (1995). Optical remote sensing of vegetation: Modeling, caveats, and algorithms. *Remote Sensing of Environment*, **51**, 169–88.

NASA (1993). *EOS Reference Handbook*. NASA publication no. 202. Washington, DC: NASA Earth Science Support Office.

National Geophysical Data Center. (1988). *10-minute Topography Data Base*. Washington DC: US Department of Commerce.

Neilson, R. P. (1993). Vegetation redistribution: a possible biosphere source of CO_2 during climatic change. *Water, Air and Soil Pollution*, **70**, 659–73.

Nemani, R., Pierce, L., Running, S. & Goward, S. (1993*a*). Developing satellite-derived estimates of surface moisture status. *Journal of Applied Meteorology*, **32**, 548–57.

Nemani, R. P., Pierce, L., Running, S. W. & Band, L. (1993*b*). Forest ecosystem processes at the watershed scale – sensitivity to remotely sensed leaf area index estimates. *International Journal of Remote Sensing*, **14**, 2519–34.

Nemani, R. R. & Running, S. W. (1989). Testing a theoretical climate–soil–leaf area hydrological equilibrium of forests using satellite data and ecosystem simulation. *Agricultural and Forest Meteorology*, **44**, 245–60.

NOAA-EPA Global Ecosystems Database Project (1992). *Global Ecosystems Database Version 1.0.* User's Guide, Documentation, Reprints and Digital Data on CD-ROM. Boulder, Colorado: USDOC/NOAA National Geophysical Data Center.

Norman, J. M., Divakarla, M. & Goel, N. S. (1995). Algorithms for extracting information from remote thermal-IR observations of the Earth's surface. *Remote Sensing of Environment*, **51**, 157–68.

Odum, E. P. (1973). *Fundamentals of Ecology*, third edition. Philadelphia: Saunders.

Olson, J., Watts, J. A. & Allison, L. J. (1985). *Major World Ecosystem Complexes Ranked by Carbon in Live Vegetation: A Database.* Carbon Dioxide Information Center. NDP-017.

Parton, W. J., Scurlock, J. M. O., Ojima, D. S., Gilmanov, T. G., Scholes, R. J., Schimel, D. S., Kirchner, T., Menaut, J.-C., Seastedt, T. R., Moya, E. G., Kamnalrut, A. & Kinyamario, J. I. (1993). Observations and modeling of biomass and soil organic matter dynamics for the grassland biome worldwide. *Global Biogeochemical Cycles*, **7**, 785–809.

Petty, G. W. (1995). The status of satellite-based rainfall estimation over land. *Remote Sensing of Environment*, **51**, 125–37.

Phillips, D. L., Dolph, J. & Marks, D. (1992). A comparison of geostatistical procedures for spatial-analysis of precipitation in mountainous terrain. *Agricultural and Forest Meteorology*, **58**, 119–41.

Pinker, R. T., Frouin, R. & Li, Z. (1995). A review of satellite methods to derive surface shortwave irradiance. *Remote Sensing of Environment*, **51**, 108–24.

Pinty, B. & Verstraete, M. M. (1991). GEMI: A non-linear index to monitor global vegetation from satellites. *Vegetatio*, **111**, 15–20.

Pinty, B., Verstraete, M. M. & Dickinson, R. (1990). A physical model of the bidirectional reflectance of vegetation canopies. 2. Inversion and validation. *Journal of Geophysical Research*, **95**, 11767–75.

Planton, S., Déqué, M. & Bellevaux, C. (1991). Validation of an annual cycle simulation with a T42-L20 GCM. *Climate Dynamics*, **5**, 189–200.

Plöchl, M. & Cramer, W. (1995). Coupling global models of vegetation structure and ecosystem processes. An example from Arctic and Boreal ecosystems. Tellus Series B. *Chemical and Physical Meteorology*, **47**, 240–50.

Prentice, I. C. (1993). Process and production. *Nature*, **363**, 209–10.

Prentice, I. C., Cramer, W., Harrison, S. P., Leemans, R., Monserud, R. A. & Solomon, A. M. (1992). A global biome model based on plant physiology and dominance, soil properties and climate. *Journal of Biogeography*, **19**, 117–34.

Prentice, I. C., Sykes, M. T. & Cramer, W. (1993). A simulation model for the transient effects of climate change on forest landscapes. *Ecological Modelling*, **65**, 51–70.

Prentice, I. C., Sykes, M. T., Lautenschlager, M., Harrison, S. P., Denissenko, O. & Bartlein, P. J. (1993). Modelling global vegetation patterns and terrestrial carbon storage at the last glacial maximum. *Global Ecology and Biogeography Letters*, **3**, 67–76.

Prentice, I. C., Webb, R. S., Ter-Mikhaelian, M. T., Solomon, A. M., Smith, T. M., Pitovranov, S. E., Nikolov, N. T., Minin, A. A., Leemans, R., Lavorel, S., Korzuhin, M. D., Helmisaari, H. O., Hrabovszky, J. P., Harrison, S. P., Emanuel, W. R. & Bonon, G. B. (1989). *Developing a Global Vegetation Dynamics Model: Results of an IIASA Summer Workshop*. Laxenburg, Austria: International Institute for Applied Systems Analysis.

Prince, S. D., Kerr, Y. K., Goutorbe, J.-P., Lebel, T., Tinga, A., Bessemoulin, P., Brouwer, J., Dolman, A. J., Engman, E. T., Gash, J. H. C., Hoepffner, M., Kabat, P., Monteny, B., Said, F., Sellers, P. & Wallace, J. (1995). Geographical, biological and remote sensing aspects of the Hydrologic Atmospheric Pilot Experiment in the Sahel (HAPEX-Sahel). *Remote Sensing of Environment*, **51**, 215–34.

Qi, J., Chehbouni, A., Huete, A. R., Kerr, Y. H. & Sorooshian, S. (1994). A modified soil adjusted vegetation index. *Remote Sensing of Environment*, **48**, 119–26.

Rahman, H., Pinty, B. & Verstraete, M. M. (1993*a*). Coupled Surface-Atmosphere reflectance (CSAR) model. 2. Semiempirical surface model usable with NOAA advanced very high resolution radiometer data. *Journal of Geophysical Research – Atmosphere*, **98**, 20791–801.

Rahman, H., Verstraete, M. M. & Pinty, B. (1993*b*). Coupled Surface-Atmosphere reflectance (CSAR) model. 1. Model description and inversion on synthetic data. *Journal of Geophysical Research – Atmosphere*, **98**, 20779–89.

Raich, J. W., Rastetter, E. B., Melillo, J. M., Kicklighter, D. W., Steudler, P. A., Peterson, B. J., Grace, A. L., Moore III, B. & Vörösmarty, C. J. (1991). Potential net primary productivity in South America: application of a global model. *Ecological Applications*, **1**, 399–429.

Ranson, K. J. & Sun, G. (1994). Northern forest classification using temporal multifrequency and multipolarimetric SAR images. *Remote Sensing of Environment*, **47**, 142–53.

Rock, B. N., Skole, D. L. & Choudhury, B. J. (1993). Monitoring vegetation change using satellite data. In *Vegetation Dynamics and Global Change*, ed. A. M. Solomon & H. H. Shugart, pp. 153–67. New York: Chapman and Hall.

Ruimy, A., Saugier, B. & Dedieu, G. (1994). Methodology for the estimation of terrestrial net primary production from remotely sensed data. *Journal of Geophysical Research*, **99**, 5263–83.

Running, S. W., Loveland, T. R., Pierce, L. L., Nemani, R. R. & Hunt, J. E. R. (1995). A remote sensing based vegetation classification logic for global land cover analysis. *Remote Sensing of Environment*, **51**, 39–48.

Running, S. W., Nemani, R. R., Peterson, D. L., Band, L. E., Potts, D. F., Pierce, L. L. & Spanner, M. A. (1989). Mapping regional forest evapotranspiration and photosynthesis by coupling satellite data with ecosystem simulation. *Ecology*, **70**, 1090–101.

Schmugge, T. & Jackson, T. J. (1994). Mapping surface soil moisture with microwave radiometers. *Meteorology and Atmospheric Physics*, **54**, 213–23.

Scholes, R. J., Skole, D. & Ingram, J. S. (1994). *A Global Database of Soil Properties: Proposal for Implementation (draft)*. International Geosphere-Biosphere Program, Data and Information System. IGBP-DIS Working Paper. No. 10.

Schönwiese, C. D. (1994). Analysis and prediction of global climate temperature change based on multiforced observational statistics. *Environmental Pollution*, **83**, 149–54.

Schulze, E.-D., Kelliher, F. M., Körner, C., Lloyd, J. & Leuning, R. (1994). Relationships between plant nitrogen nutrition, carbon assimilation rate, and maximum stomatal and ecosystem surface conductances for evaporation: a global ecology scaling exercise. *Annual Review of Ecology and Systematics*, **25**, 629–60.

Seguin, B., Courault, D. & Guérif, M. (1994). Surface temperature and evapotranspiration: Application of local scale methods to regional scales using satellite data. *Remote Sensing of Environment*, **49**, 287–95.

Sellers, P. J. & Hall, F. G. (1992). FIFE in 1992: results, scientific gains, and future research directions. *Journal of Geophysical Research*, **97**, 19091–109.

Sellers, P. J., Hall, F. G., Asrar, G., Strebel, D. E. & Murphy, R. E. (1992). An overview of the First International Satellite Land Surface Climatology Project (ISLSCP) Field Experiment (FIFE). *Journal of Geophysical Research*, **97**, 18345–371.

Sellers, P. J., Meeson, B. W., Hall, F. G., Asrar, G., Murphy, R. E., Schiffer, R. A., Bretherton, F. P., Dickinson, R. E., Ellingson, R. G., Field, C. B., Huemmrich, K. F., Justice, C. O., Melack, J. M., Roulet, N. T., Schimel, D. S. & Try, P. D. (1995). Remote sensing of the land surface for studies of global change: Models-algorithms-experiments. *Remote Sensing of Environment*, **51**, 3–26.

Shea, D. J. (1986). *Climatological Atlas: 1950–1979. Surface Air Temperature, Precipitation, Sea-level Pressure, and Sea-Surface Temperature (45°S–90°N)*. Technical note no. 269. Boulder, Colorado: National Center for Atmospheric Research.

Simpson, J. R. (1993). Urbanization, agroecological zones and food production sustainability. *Outlook on Agriculture*, **22**, 233–9.

Skole, D. & Tucker, C. (1993). Tropical deforestation and habitat fragmentation in the Amazon: satellite data from 1978 to 1988. *Science*, **260**, 1905–10.

Smith, T. M., Cramer, W., Dixon, R. K., Leemans, R., Neilson, R. P. & Solomon, A. M. (1993). The global terrestrial carbon cycle. *Water, Air and Soil Pollution*, **70**, 19–37.

Stanhill, G. & Moreshet, S. (1992). Global radiation climate changes – the world network. *Climatic Change*, **21**, 57–75.

Staub, B. & Rosenzweig, C. (1986). *Global Digital Data Sets of Soil Type, Soil Texture, Surface Slope, and Other Properties: Documentation of Archived Tape Data*. NASA Technical Memorandum no. 100685. Greenbelt, Maryland: Goddard Space Flight Center.

Steffen, W. L., Walker, B. H., Ingram, J. S. & Koch, G. W. (1992). *Global Change and Terrestrial Ecosystems: The Operational Plan*. Global Change Report no. 21. Stockholm: International Geosphere-Biosphere Programme.

Stocker, T. F. & Mysak, L. A. (1992). Climatic fluctuations on the century time scale – a review of high-resolution proxy data and possible mechanisms. *Climatic Change*, **20**, 227–50.

Tarpley, J. D., Schneider, S. R. & Money, R. L. (1984). Global vegetation indices from the NOAA-7 meteorological satellite. *Journal of Climate and Applied Meteorology*, **23**, 491–4.

Thornthwaite, C. W. (1948). An approach toward a rational classification of climate. *The Geographical Review*, **38**, 55–94.

Townshend, J. R. G. (1992). *Improved Global Data for Land Application: A Proposal for a New High Resolution Data Set*. IGBP report no. 20. Stockholm: International Geosphere-Biosphere Programme.

Townshend, J. R. G. (1994). Global data sets for land applications from the Advanced Very High Resolution Radiometer: an introduction. *International Journal of Remote Sensing*, **15**, 3319–32.

Townshend, J. R. G., Justice, C. O., Li, W., Gurney, C. & McManus, J. (1991). Global land cover classification by remote sensing: present capabilities and future possibilities. *Remote Sensing of Environment*, **35**, 243–55.

Townshend, J. R. G., Justice, C. O., Skole, D., Malingreau, J.-P., Cihlar, J., Teillet, P., Sadowski, F. & Ruttenberg, S. (1994). The 1 km resolution global data set: needs of the International Geosphere Biosphere Programme. *International Journal of Remote Sensing*, **15**, 3417–41.

Turner, B. L., Moss, R. H. & Skole, D. L. (1993). *Relating Land Use and Global Change: A Proposal for an IGBP-HDP Core Project*. Global Change Report no. 24. HDP Report no. 5. Stockholm: International Geosphere-Biosphere Programme.

United Kingdom Meteorological Office (1966). *Tables of Temperature, Relative Humidity and Precipitation for the World. Part V. Asia*. London: HMSO.

United Kingdom Meteorological Office (1972). *Tables of Temperature, Relative Humidity, Precipitation and Sunshine for the World. Part III. Europe and the Azores*. London: HMSO.

United Kingdom Meteorological Office (1973). *Tables of Temperature, Relative Humidity and Precipitation for the World. Part VI. Australasia and Pacific Ocean*. London: HMSO.

United Kingdom Meteorological Office (1978). *Tables of Temperature, Relative Humidity and Precipitation for the World. Part II. Central and South America, the West Indies and Bermuda*. London: HMSO.

United Kingdom Meteorological Office (1980). *Tables of Temperature, Relative Humidity, Precipitation and Sunshine for the World. Part I. North America and Greenland (including Hawaii and Bermuda)*. London: HMSO.

United Kingdom Meteorological Office (1983). *Tables of Temperature, Relative Humidity, Precipitation and Sunshine for the World. Part IV. Africa, the Atlantic Ocean South 35°N and the Indian Ocean*. London: HMSO.

Van de Griend, A. A. & Owe, M. (1994). Microwave vegetation optical depth and inverse modelling of soil emissivity using Nimbus/SMMR satellite observations. *Meteorology and Atmospheric Physics*, **54**, 225–39.

Verstraete, M. M., Pinty, B. & Dickinson, R. E. (1990). A physical model of the biodirectional reflectance of vegetation canopies. I. Theory. *Journal of Geophysical Research*, **95**, 11755–65.

Von Humboldt, A. (1807). *Ideen zu einer Geographie der Pflanzen neben einem Naturgemälde der Tropenländer*. Tübingen, Germany: Lotta.

Vörösmarty, C. J., Grace, A. L., Moore III, B., Choudhury, B. J. & Willmott, C. J. (in press). A strategy to study regional hydrology and terrestrial ecosystem processes using satellite remote sensing, ground-based data and computer modeling. *Acta Astronautica*.

Vörösmarty, C. J., Gutowski, W. J., Person, M., Chen, T.-C. & Case, D. (1993). Linked atmosphere-hydrology models at the macroscale. In *Macroscale Modelling of the Hydrosphere*, pp. 3–27. Yokomaha, Japan: IAHS.

Vörösmarty, C. J., Moore III, B., Grace, A. L., Gildea, M. P., Melillo, J. M., Peterson, B. J., Rastetter, E. B. & Steudler, P. A. (1989). Continental scale models of water balance and fluvial transport: an application to South America. *Global Biogeochemical Cycles*, **3**, 241–65.

Vose, R. S., Schmoyer, R. L., Steurer, P. M., Peterson, T. C., Heim, R., Karl, T. R. & Eischeid, J. K. (1992). *The Global Historical Climatology Network: Long-term Monthly Temperature, Precipitation, Sea-level Pressure, and Station Pressure Data*. ORNL/CDIAC-53. Oak Ridge, Tennessee: Environmental Science Division, Oak Ridge National Laboratory.

Walter, H. & Lieth, H. (1960–67). *Klimadiagramm-Weltatlas*. Stuttgart: Fischer Verlag.

Waring, R. H. & Peterson, D. L. (1994). Oregon Transect Ecosystem Research (OTTER) Project. *Ecological Applications*, **4**, 210.

Whitlock, C. H., Charlock, W. F., Staylor, W. F., Pinker, R. T., Laszlo, I., Di Pasquale, R. C. & Ritchey, N. A. (1993). *WCRP surface radiation budget shortwave data product description-version 1.1.* NASA Tech. Mem. 107747. NASA.

Wigley, T. M. L. & Barnett, T. P. (1990). Detection of the greenhouse effect in the observations. In *Climate Change: The IPCC Scientific Assessment*, ed. J. T. Houghton, G. J. Jenkins & J. J. Ephraums, pp. 239–55. Cambridge: Cambridge University Press.

Willmott, C. J., Mather, J. R. & Rowe, C. M. (1981a). Average monthly and annual surface air temperature and precipitation data for the world. Part 1: The eastern hemisphere. *Publications in Climatology*, **34**, 1–395.

Willmott, C. J., Mather, J. R. & Rowe, C. M. (1981b). Average monthly and annual surface air temperature and precipitation data for the world. Part 2. The western hemisphere. *Publications in Climatology*, **35**, 1–378.

Wilson, M. F. & Henderson-Sellers, A. (1985). A global archive of land cover and soils data for use in general circulation models. *Journal of Climate*, **5**, 119–43.

Wing, S. L. & Greenwood, D. R. (1993). Fossils and fossil climate – the case for equable continental interiors in the Eocene. *Philosophical Transactions of the Royal Society of London, Series B*, **341**, 243–52.

Woodward, F. I. (1987). *Climate and Plant Distribution*. Cambridge: Cambridge University Press.

Woodward, F. I. & Smith, T. M. (1994). Global photosynthesis and stomatal conductance: modelling the controls by soil and climate. *Advances in Botanical Research*, **20**, 1–41.

Zhao, B. (1994). Study of TOVS applications in monitoring atmospheric temperature, water vapor, and cloudiness in East Asia. *Meteorology and Atmospheric Physics*, **54**, 261–70.

Zinke, P. J., Stangenberger, A. G., Post, W. M., Emanuel, W. R. & Olson, J. S. (1984). *Worldwide Organic Soil Carbon and Nitrogen Data*. ORNL/TM-8857. Oak Ridge, Tennessee: Oak Ridge National Laboratory.

28 Satellite data for monitoring, understanding and modelling of ecosystem functioning

A. Fischer, S. Louahala, P. Maisongrande, L. Kergoat
and G. Dedieu

Introduction

The prediction of possible future behaviour of terrestrial ecosystems under global change requires understanding of both ecosystem structure (*pattern*) and function (*process*), which together influence the interactions between vegetation and climate. A model of these interactions needs to be prognostic, but it must be validated for the current climatic conditions. The complexity and diversity of processes to account for is huge, and their description in a simulation model therefore needs to be based on observations of both detailed mechanisms and global trends. It is necessary to look for all possible data sources for modelling and validation of the various subsystems. Useful information can be derived from satellite observations directly. It can also be the result of the application of a diagnostic model using satellite observations as input. The use of satellite data to monitor various vegetation features and changes, from the local to the global scale, has been discussed several times (e.g. reviewed by Rock *et al.*, 1993). For example, satellite-derived maps of land cover can either be a validation of equilibrium biosphere models and socio-economic models of land use change, or they can simply prescribe the ecosystem type for the parameterization of biogeochemical models.

Time series (for one or several years) of satellite observations can be related to the temporal features of some variables required by diagnostic models. Such variables can be used, for example, to simulate the spatiotemporal variations of CO_2 fluxes at the global scale (Potter *et al.*, 1993; Esser *et al.*, 1994). These variations are important for the understanding of the actual trend of the global carbon cycle, and they therefore can give some answers to the question of whether the terrestrial biosphere acts as a source or a sink of carbon. Time series can also be used as an independent observation data set for the calibration and the validation of the seasonal and interannual features simulated by a vegetation model. For example, one year of satellite data can provide the calibration of site-dependent parameters of a process-based vegetation model, and another year can be used for the validation. This could be seen as a basic requirement for the assessing of the accuracy of prognostic vegetation models which are developed.

Radiative measurements in the solar spectrum are mostly related to the green biomass and to the photosynthetic activity of the vegetation. At the current stage, the Advanced Very High Resolution Radiometer (AVHRR) of the National Oceanic and Atmospheric Administration (NOAA) satellite is the only instrument which allows a daily global coverage of the Earth. In the near future, several new sensors will be available: the POLarization and Directionality of the Earth's Reflectances (POLDER) sensor onboard the ADvanced Earth Observing Satellite (ADEOS) in 1996, the *Végétation* sensor onboard the 4th Système Probatoire d'Observation de la Terre (SPOT4) in 1997, and the Moderate-Resolution Imaging Spectrometer (MODIS) sensor onboard the Earth Observing Satellite (EOS) in the late 1990s. In comparison to AVHRR, these will be improved instruments in terms of several spectral, radiometric and geometric properties. They will provide new capabilities for processing, atmospheric correction, and interpretation of the satellite signal (Townshend *et al.*, 1991; Achard *et al.*, 1992; Deschamps *et al.*, 1994). However, during the last decade, NOAA/ AVHRR data have been very useful for monitoring of vegetation at the regional or global scale, for example, for phenology (Goward *et al.*, 1985) and land cover (Tucker *et al.*, 1985). This was done mainly by using Vegetation Indices (VIs), which are combinations of visible and near infrared radiative measurements. They have also been used to estimate the spatial and temporal variations of characteristic canopy variables like the leaf area index (LAI), or the fraction of the photosynthetically active radiation (fPAR) that is absorbed by the canopy. These variables allow the estimation of net primary productivity (NPP) (Heimann & Keeling, 1989; Potter *et al.*, 1993; Ruimy *et al.*, 1994) or biogeochemical fluxes (Running *et al.*, 1989). The spatial features provided by an equilibrium LAI model (Nemani & Running, 1989; Woodward & Smith, 1994), or by a global terrestrial NPP model (Francois *et al.*, 1994), or the seasonal variations simulated by an ecosystem model of Sahelian grasslands (Lo Seen Chong *et al.*, 1995) have been validated using NOAA/AVHRR data.

Other wavelengths have also to be used to monitor vegetation. A variable such as the surface temperature could be derived from the thermal channels of AVHRR, in order to estimate the soil moisture status and the canopy hydric status (Nemani *et al.*, 1993). Roughness, vegetation structure and water content, and soil moisture are variables that can be derived from satellite passive and active microwave observations. This radiation is not absorbed by clouds, and a global data set of microwave vegetation index has been produced with the Scanning Multichannel Microwave Radiometer (SMMR, Choudhury & Tucker, 1987). This chapter especially focuses on the questions which can be addressed using multitemporal satellite data in the short wavelengths. We will describe the value and the limits of the various strategies of using temporal series of VIs.

We can distinguish the use of multitemporal satellite data (restricted here to solar spectrum) in three categories: (i) observation and description of ecosystem function, (ii) determination of driving variables for ecosystem models, (iii) understanding and modelling of vegetation processes, and therefore assessing and validating ecosystem models. Each of these are required for the description of biosphere behaviour. For each category, this chapter gives a brief review of typical applications. Then a more detailed example of recent work is used to discuss the importance of appropriate signal processing, as well as of the correct interpretation of the radiative observations. The first example shows a regional land cover classification that was derived from time series data. The second discusses how the interannual variability of global NPP may be estimated from satellite data using a diagnostic model. The third deals with the initialization of a crop production model using the seasonal profile of the satellite observations. To achieve a quantitative interpretation of the data, and to fully benefit from the multitemporal aspect, data are processed with particular attention to perturbating effects, e.g. clouds, atmosphere, sensor accuracy, and the sensitivity of the radiative measurements to both sun and satellite angles.

Analysis of ecosystem pattern

Review

When observed in the solar spectrum, most biomes or land cover types differ from each other by their phenology (Reed *et al.*, 1994). Phenology can be traced by VIs, which, therefore, are generally well suited to classify land cover types according to their temporal signatures. The first applications have concerned crops, using multitemporal Landsat data (Badhwar *et al.*, 1982; Lo *et al.*, 1986). At continental or global scale, phenology analysis has been used for a decade with coarse resolution satellites like NOAA/AVHRR (Goward *et al.*, 1985; Justice *et al.*, 1985; Goward & Dye, 1987), and is mainly based on the Global Vegetation Index (GVI) product. Tucker *et al.* (1985) first used the temporal signal of the Normalized Differential Vegetation Index (NDVI) to classify African land cover. Some classification algorithms are based on satellite data alone, while others use additional data sources (Brown *et al.*, 1993). Some earlier studies used principal component analysis (for example, Townshend *et al.*, 1987, for South America), or simply the integrated value of the NDVI during the growing season (Malingreau, 1986, for South-east Asia), or during the year (Koomanoff, 1989). Later, automatic classifications based on clustering algorithms (Viovy, 1990) successfully retrieved the main regional pattern of land cover of West Africa (Arino *et al.*, 1991) and France (Derrien *et al.*, 1992). Other approaches use additional data. Supervised classifications using ground observations have been applied at the global (Lloyd, 1990; DeFries & Townshend, 1994) and regional scale (Kremer & Running, 1993). Others have combined information about climate, elevation and eco-regions to refine the classification (Loveland *et al.*, 1991; Running *et al.*, 1994).

Some limitations of these land cover classifications from AVHRR data have been discussed in Townshend *et al.* (1991). Some of them concern the choice of the satellite product and its processing. Clouds which still remain in the Maximum Values Composite (MVC) (Holben, 1986) images, for example, strongly affect the result of the automatic classification procedure (Moody & Strahler, 1994). Goward *et al.* (1993) provided a complete assessment of the GVI product. The GVI conserves several intrinsic negative features due to its resampling strategy and its temporal aggregation algorithm (Goward *et al.*, 1991). Nevertheless, some improvements are still possible (Goward *et al.*, 1994) which allow significant quantitative relationships with land surface properties to be obtained. These approaches involve a more appropriate calibration of digital counts to radiances, e.g. Teillet & Holben (1993), their conversion into reflectances, filtering of clouds, e.g. Stowe *et al.* (1991), and possibly atmospheric corrections, e.g. Berthelot *et al.* (1994). The same radiometric corrections can also be applied on Global Area Coverage (GAC) or Local Area Coverage (LAC) data, and the analysis of the temporal signal provides tools which result in an improved normalization of the observations towards remaining perturbing effects (clouds, directional effects) compared to those that use simple MVC techniques (Leroy & Roujean, 1994; Sellers *et al.*, 1994).

After signal processing, the result of the multitemporal classifications are classes that are interpreted as signatures of ecosystem function. The same biome can be stratified into several classes, for example, when there is a soil or climatic gradient leading to different behaviour that is perceptible on the NDVI profile. However, a land-cover classification that is only based on one year of vegetation index data is obviously characterized by the specific meteorological conditions of that year, because of its influence on phenology and hence on the radiometric signal (Townshend *et al.*, 1991). Therefore, we present the result of a land cover classification from several years of NOAA/AVHRR LAC data. We also discuss the characteristics of the temporal profiles for some particular land cover classes, and their interannual variations.

Regional land cover classification using satellite observations from several years

Site description and ground truth

The study area is a region of approx. 400 km × 600 km in the southwestern part of France. Several different land cover types occur in this region: The Landes pine forests close to the Atlantic Ocean, coniferous forests in the mountains (Pyrénées and Massif Central), mixed with deciduous forests, mountain pastures, mixed farming in the countryside around Toulouse, vineyards and orchards in the Garonne valley, intensive maize crops in the Adour basin, and sparse vegetation in the Mediterranean area. In 1988, the relative occurrence of various crops in each administrative district (average area: $100-400$ km^2), as well as of forests and urban areas, were provided by administrative services. We have combined and gridded these data sets to produce a land use map. The land cover types were grouped into six major components: winter crops, spring crops, vineyards/orchards, grasslands, forests, and others (e.g. urban, bare soils). The administrative units were then aggregated according to the relative proportion of each of these six components. This led to 64 land use map units.

Satellite data processing

We have used NOAA/AVHRR data at 1 km resolution for the years 1987–9. The satellite data were processed (5 days MVC) with the aim of deriving a temporal profile of the NDVI as close as possible to its value measured at the surface with a radiometer in normalized angular conditions. The use of inflight calibration coefficients (Holben *et al.*, 1990) allows a correction of the calibration drift due to sensor ageing, and the computation of the top of atmosphere reflectances takes the variations in the illumination conditions into account. Atmospheric corrections are provided by an adaptation of SMAC (Rahman & Dedieu, 1994) for composite images. The resulting 'surface NDVI' signal shows remaining fluctuations because of clouds, strong directional effects and high content of aerosols. These are reduced by a dynamic filtering procedure, e.g. BISE, (Viovy *et al.*, 1992). Then, a Fourier filtering is applied to smooth the profiles (more details in Louahala *et al.*, 1991).

Automatic classification of the NDVI temporal profile

Finally, an automatic clustering was performed to collect pixels with similar NDVI temporal profiles in classes (Viovy, 1990). The classification (24 classes) of the separate NDVI profiles for the three years 1987–9 allows the retrieval of the main features of the landscape (Louahala *et al.*, 1991). However, because the NDVI temporal profiles also depend on the specific meteorological conditions of each year, there are significant variations between the three annual classifications, 1989 being a dry year. Therefore, we have applied the classification for the 3 years together, and we expected the resulting classes to be representative for the typical interactions between ecosystem structure and mean climate conditions. By using several years, more of the interannual variability is taken into account, and its spatial features should therefore be captured. Regions with a rather homogeneous vegetation, for example, may display the same mean NDVI profile over several years for the whole area, but one part may show a greater interannual variability due to a more fluctuating climate. In such a case, the classification assigns that part to a specific class, different from that of the rest of the area. This stratification is based on an implicit understanding of the spatial and temporal variations of the ecosystem functioning.

Result of the 3 years' classification

The stratification shown on Plate 9 (between pp. **204** and **205**) is in good agreement with the structure of the landscape in the southwestern part of France. The temporal profiles for six major classes A–F in Plate 9 are re-presented for 1988 and 1989 on Fig. 28.1. They were compared with the specific land use map units found in the same geographical area. Class A (orange in Plate 9) appears in the countryside around Toulouse, where mixed farming occurs. A representative land use map unit has 25% winter crops, 23% spring crops, and 24% pastures. The first peak of the annual profile shows the growth of the winter crops, and the second one the growth of the spring crops. These later crops were especially affected by the drought in 1989. Class B (yellow) appears especially in mixed forest pasture areas of the Massif Central (45% dominating pastures, 29% forests). The NDVI profiles mainly show the increasing activity of these pastures during spring, followed by a temporary decrease during summer in 1989 due to the effect of the drought on the pastures. Class C (green) appears in the Adour basin, and partially within the Landes pine forest, in some areas of intensive maize production (45% spring crops, 22% forests). The second mode of the NDVI profile shows the growth of maize alone, which was not immediately affected by the drought in 1989 because of the irrigation system in this region. Class D (blue) occurs in deciduous forests

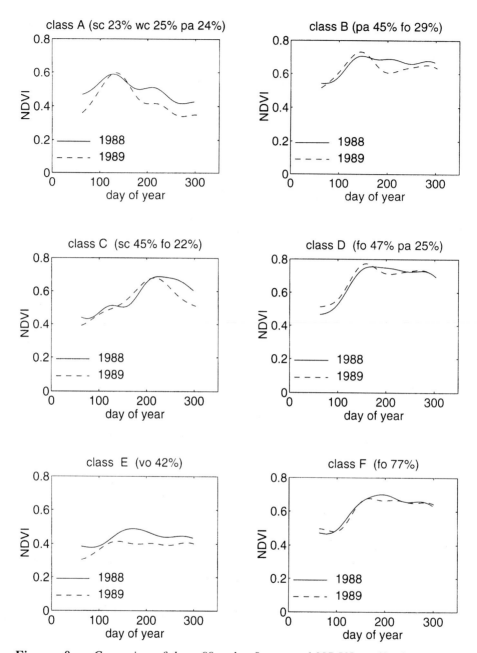

Figure 28.1   *Comparison of the 1988 and 1989 seasonal NDVI profiles for six land cover classes in southwestern France, with indication of land use according to the knowledge of land use map units occurring in the same geographical areas as the previous classes. Percentages refer to the area covered by each type in the representative area on the land use map. A, spring crops (23%), winter crops (25%) and pastures (24%); B, pastures (45%) and forests (29%); C, spring crops (45%) and forests (22%); D, forests (47%) and pastures (25%); E, vineyards/orchards (42%); F, forests (77%).*

(47% dominating forests: mainly oak in Massif Central, and beech in the Pyrenees, 25% pastures). In these ecosystems, the drought of 1989 was less visible than over crops or grasslands. Class E (red) appears in the Mediterranean area dominated by vineyards, orchards and scrubs (42% vineyard-orchards). Bare soil is always visible through the sparse canopies and the resulting signal shows a very low amplitude. Class F (yellow-orange) appears in the Landes pine forest (77% forests). This is not a closed forest, and the spring growth of the understorey is clearly perceptible. The proximity of ground water protects this area from being affected by drought. The structure of another class is clearly visible on Plate 9 for the Garonne valley (cyan), where the landscape is a complex mixture of major crops, gardens, woodlands and orchards.

Discussion

Compared with the knowledge of the landscape provided by ground information, the quality of the classification in terms of land use is very good. The use of 3 years allows the retrieval of very detailed structures, such as the Bouconne forest in the west of Toulouse, or the valleys coming from the Pyrénées through the Lanmezan Plateau, as indicated on Plate 9. At this step, the classification is mainly checked by geographical matching of the result with the topographic and land use map. Use of other wavelengths have been shown to be a successful alternative to classical NOAA products for land use classification (Choudhury & Tucker, 1987). Nevertheless, this example shows that the use of several years of NOAA/AVHRR optical data at full resolution, and processed with care, results in a good classification of the landscape related to the land use. We were mainly interested in the typical profiles of each class, which indicate how the seasonal features of the various ecosystems are perceived from space, and how the vegetation responds to variability in the climate. The spatial discrimination is therefore related to ecosystem functioning, and it could validate some prediction of ecosystem structure. These results are provided by radiative information only, and the next sections show that the combined use of vegetation models with satellite data will allow either the translation of this information into variables like CO_2 fluxes and net primary productivity, or the analysis of the seasonal variations simulated by a vegetation model.

Estimation of global net primary productivity

Diagnostic approach for the global terrestrial NPP

Several models for the terrestrial annual NPP have appeared these last years (Prince et al., 1994). Independently of the formulation of the processes of dry matter assimilation and growth, we call these models diagnostic when the temporal behaviour of the vegetation characteristics are given as input data from satellite observations. Such diagnostic models have been successfully applied at the regional scale (e.g. Lüdeke *et al.*, 1991), or at the continental scale (Goward & Dye, 1987). At the global scale, Warnant *et al.* (1994*a*) applied a relatively detailed photosynthesis model to a gridded map, where the percentage of green (photosynthetically active) vegetation along the year is derived from monthly NOAA/AVHRR NDVI values. A simple way to estimate NPP at large scale is to use the efficiency model of Kumar & Monteith (1981) with satellite inputs (Prince, 1991). In its simplest formulation, this model requires the temporal pattern of incoming solar photosynthetically active radiation (PAR), the temporal pattern of the fraction of absorbed PAR (fPAR), and the conversion efficiency (*e*) of radiation into dry matter. Heimann & Keeling (1989) first applied this at the global scale to estimate terrestrial NPP, with a constant *e* value everywhere, and using satellite data for the solar radiation as well as for the spatio-temporal variations of fPAR. Potter *et al.* (1993) have refined this approach with the use of reduction factors which account for the effects of temperature and water stress on the maximum potential efficiency e^*. In that case, e^* is calibrated with observed NPP values at specific sites. In contrast, Prince & Goward (1995) modelled the value of *e* according to environmental factors and physiological processes.

In their global NPP estimation, Ruimy *et al.* (1994) have improved the GVI data processing at various levels: by calibration (Holben *et al.*, 1990), atmospheric correction (Rahman & Dedieu, 1994), and by dynamic filtering of the temporal NDVI signal (Viovy *et al.*, 1992). They also discuss the use of a biome-dependent conversion efficiency. The importance of defining a validation strategy is emphasized by the diverging results of the various annual biome NPP, depending on the sources, such as diagnostic and statistical models (e.g. Box, 1988) or literature data (e.g. Lieth, 1975). Atmospheric CO_2 concentration fields can be used for validation, too, if a soil respiration model is added to the NPP model for computation of the net ecosystem productivity (NEP; Fung *et al.*, 1987). This information, combined with the CO_2 fluxes between atmosphere and the oceans, has been used as an input for an atmospheric CO_2 transport model (Heimann & Keeling, 1989). (For more information about global datasets for terrestrial ecosystem models see Cramer & Fischer, this volume.)

The major interest of a diagnostic biosphere model, based on satellite data, is to describe the spatial features and the temporal behaviour of terrestrial productivity, for example the zonal variations (Ruimy *et al.*, 1994). Observations of the trends through several years may be expected to give some answers to the unresolved questions of the contribution of the terrestrial biosphere to the global carbon cycle, and its relation to climatic events, such as the El Niño Southern Oscillation (ENSO).

A model for the interannual variations of productivity

Maisongrande *et al.* (1995) used 6 years of GVI data to simulate the spatio-temporal variations of the CO_2 fluxes over the terrestrial biosphere. To describe these CO_2 fluxes, which include biospheric productivity, NPP, and soil respiration (SR), the approach is the following: Kumar and Monteith's model is adapted to simulate gross primary productivity (GPP) from satellite data. Autotrophic respiration is assumed to be a temperature-dependent fraction of GPP. The range of the biome-dependent conversion efficiency found in the literature is mainly interpreted as an effect of temperature. Consequently, if this effect is taken into account by parameterization of the autotrophic respiration, the photosynthetic efficiency which is required by the GPP model can be kept constant for the whole globe (4.45 g CO_2 · MJ^{-1}). In theory, SR depends on temperature, precipitation and soil type (mainly litter structure). However, at this step, SR is modelled as a function of temperature only. To avoid unreasonable carbon stores in the soil, we chose a local equilibrium balance between CO_2 uptake and release over the year. The comparison of the temporal variations of NEP for each latitudinal band simulated by this model with the temporal variations of the atmospheric CO_2 concentrations measured at stations from the temperate northern hemisphere (larger vegetated areas, and reduced ocean–atmosphere interactions) gives a rough validation of the seasonal phases of the CO_2 exchanges, and of the combined NPP + SR model. This model of the terrestrial biosphere indicates that, in the northern hemisphere, the atmospheric CO_2 concentrations are well related to the terrestrial biospheric fluxes. In fact, as was expected, the seasonal variations simulated by the NPP model (and hence by the NEP model) strongly depend on latitude, and are related to the local climatic factors driving the phenology, such as precipitation, irradiance and temperature (Maisongrande *et al.*, 1995).

Observed interannual variations

The model described above uses a long-term mean climatology for solar radiation and temperature. Consequently, the simulated interannual variations come from the satellite data only. We assume that GVIs mainly carry information on green biomass and photosynthetic activity and that the observed interannual changes are due to the effect of meteorological events on productivity (ENSO events, strong in 1987, weaker in 1991, eruption of Mt Pinatubo in June 1991). Some problems appear when we want to analyse the interannual changes quantitatively because some variations are due to the satellite system itself. These are partly corrected during data processing. For example, the sensor ageing is taken into account using the estimated temporal drift of the calibration estimated by Kaufman & Holben (1993). The orbital drift of NOAA-9 resulted in a progressive delay in the time of equatorial crossing (16 : 10 at the end of its life, November 1988), which leads to different directional and atmospheric conditions. At this stage, the processing of the 6 years of GVI does not account for these effects. Because such changes in the measurement conditions are the same for the whole globe, the retrieval of typical spatial features resulting from events which have regional signatures in the NDVI signal should be possible.

Results and discussion

Plate 10 (between pp. **204** and **205**) shows the global NPP map derived from this model for the year 1990 (above), and the NPP difference between 1990 and 1987 (below). The global NPP map shows the same general spatial pattern as the output from similar diagnostic NPP models (Heimann & Keeling, 1989; Potter *et al.*, 1993; Ruimy *et al.*, 1994), or from other terrestrial NPP models like the Miami model (Lieth, 1975) or the CARAIB model (Warnant *et al.*, 1994*b*). As usual, the ENSO warm phase of 1987 was associated with drought in some tropical or subtropical regions, like southeastern Australia, southern Africa and southern Brazil (WMO, 1989). These areas are those which show the largest difference between the NPP of the 2 years (Plate 10), and it suggests that the lower NPP in 1987 was due to the limitation of rainfall. An ENSO warm phase is typically associated with increased precipitation in equatorial East Africa (WMO, 1989), and a corresponding increase in NPP is visible on our map for 1987.

There are several uncertainties in this simple model, mainly for the accuracy of fPAR, as it is derived from NDVI, but also for the conversion efficiency (see Ruimy *et al.* (1994) for a discussion of its sensitivity for global and zonal NPP results). Our results should therefore be interpreted with care. However, it appears as if the regions where the biospheric components of the carbon cycle have large interannual variations are clearly depicted by such a model. Such a simple diagnostic model for the global estimation of terrestrial GPP, NPP, and NEP, and the interannual change of their spatial pattern, is therefore useful. The quantitative interpretation of the results will become easier if the accuracy of both input satellite data and climate data are improved.

Satellite data

Using standardized principal components for the analysis of monthly raw NDVI time series for 1986–8 over Africa, Eastman & Fulk (1993) showed that the significant anomalies in the output of the sensor system itself were of greater magnitude than the signature of the (however strong) ENSO event of 1987. In the present study, the calibrated values lead to a NDVI that is more representative for vegetation behaviour than the NDVI computed from raw data. Goward *et al.* (1994) indicate that, with such processing, the questions concerning the interannual variations in biospheric activity could be explored, both globally and regionally. A quantitative comparison of the yearly NPP maps is possible if we assume that the atmospheric conditions encountered do not change too much from one year to another. The eruption of Mt Pinatubo (6 June 1991) provides an example that such an assumption may be wrong. The drastic increase in the amount of aerosols that were ejected into the atmosphere and were spreading out between 30°N and 20°S (McCormick *et al.*, 1995) lead to a clear reduction of the top of atmosphere NDVI signal that no composite method can avoid. For this particular case, Vermote *et al.* (1994) have shown a way to correct the composite NDVI. It would certainly be interesting to apply atmospheric corrections to the reflectances (Rahman & Dedieu, 1994), given that the required information about aerosol contents is available. In any case, we must determine the sensitivity of the resulting spatial features of GPP, NPP, or NEP to the accuracy of the signal.

Climate data

Maps of the yearly climate variations are needed, rather than maps of the mean climate. A result of the high aerosol content in tropical regions after the eruption of Mt Pinatubo was the decrease in temperatures (Dutton & Christy, 1992), which might have large-scale consequences on respiration, as was suggested by Sarmiento (1993). This phenomenon can only be taken into account in the NEP diagnostic model if the yearly temperatures are available.

Validation and calibration of the seasonal features simulated by a process model: assimilation strategy

General considerations

The previous examples show how satellite data can be used to detect land cover and to monitor how vegetation behaviour responds to the annual climatic conditions. Many improvements can be added to the NPP model, for both the satellite processing and the derivation of fPAR, as well as for the estimation of the conversion efficiency (as discussed for forests by Landsberg *et al.*, 1995), in order to provide improved quantitative results. However, the spatial and temporal trends of the terrestrial CO_2 global cycle which are depicted add to our understanding. Besides these diagnostic models, another important point is the development of prognostic models, which are able to simulate the seasonal and interannual behaviour of the ecosystems on a physiological and ecological basis, driven by soils and climate, without satellite input. Because they are based on understanding of ecophysiological mechanisms, these models should be usable in a predictive mode under different scenarios of climatic change (Lüdeke *et al.*, 1995; McGuire *et al.*, 1993; Melillo *et al.*, 1993), although further developments are needed to better understand and describe the feedbacks between processes (Prentice, 1993).

Some models are developed for specific ecosystems at the regional scale (e.g. Mougin *et al.*, 1995), others are more generic and work at the global scale (e.g. TEM; Raich *et al.*, 1991). At the present stage, the Frankfurt Biosphere Model (Lüdeke *et al.*, 1994) is one of the most advanced in terms of the simulation of the internal dynamics of carbon in the living biomass. Particularly important is the understanding of the seasonal features, which drive the trace gas and energy fluxes, and which are key elements for the prediction of the structure of ecosystems as well as for their transient behaviour. Changes in phenological events may signal important year-to-year climatic variations or even global environmental change (Reed *et al.*, 1994). If such models are designed to simulate the response of the ecosystems to climate change, they first must correctly describe the interannual and spatial variability of the vegetation behaviour due to variations in driving variables or environmental conditions, like the occurrence of El Niño. In terms of seasonal dynamics this implies, for example, that a model for shortgrass prairie and annual grassland should respond to rainfall events (Rauzi & Dobrenz, 1970), and a model for temperate forests should respond to spring warming, winter chilling, and the photoperiod (Hunter & Lechowicz, 1992). To do so, the calibration of those processes that are empirically described in such models, or which may be difficult to generalize (in space, in time or under new environmental conditions), must be improved. We show here that satellite time series can be used for this purpose. First of all, they provide a validation of the seasonal pattern simulated by the models, given that some of the variables which are simulated by the model can be related to satellite measurements. Using observations in the solar spectrum, Lüdeke *et al.* (in press) showed how phenological events can be validated. Furthermore, a control strategy (assimilation of satellite time series within a functional model), allows for improvement of the modelling, as we attempt to demonstrate in the next section.

Description of the assimilation strategy

The assimilation of satellite time series data is based on the idea of modelling the temporal behaviour of the radiative signal, as it can be measured from the satellite. Its principle is described following the schematic example of Fig. 28.2. We chose a specific kind of functional model: a process-based crop production model for winter wheat (AFRCWHEAT2, Porter, 1993), applied to a temperate region (meteorological conditions of Beauce, southwest of Paris, France). For the description of the mass and energy fluxes, the structure of this model is close to that of the biogeochemical models developed for natural vegetation, like FOREST-BGC (Running & Coughlan, 1988). Driven by soil and climate variables, the model simulates photosynthesis, maintenance and growth respiration, evapotranspiration processes, and nitrogen uptake using daily (or shorter) time steps. The phenological development and its interaction with the ecophysiological processes are based on data available for the various crops. The accuracy of the temporal simulation of the LAI is of major importance: the simulation of the fluxes, and therefore of the productivity, strongly depends on the LAI profile through the growing season (Weir et al., 1984). In the bottom part of Fig. 28.2, the temporal profiles of three variables (LAI, CO_2 fluxes and accumulated NPP) simulated by the crop model are plotted for two different parameter sets, as explained below. The resulting yields are also indicated.

Two parameters differ between the simulations. One key parameter (an initial condition, *sensu stricto*) for any crop model is the sowing date (sd). For crops with a vegetative period during winter, another key parameter is related to the photoperiod saturation sensitivity (psat), which has an impact on the phenological submodel. The output variables of such a crop model are biological or ecophysiological variables, and they cannot be directly compared to satellite observations, which are radiative measurements. The relations between biophysical canopy variables and radiative properties are described using a directional radiative transfer model. The directional radiative transfer model SAIL, (Verhoef, 1984) was therefore coupled to the daily LAI to simulate the visible and near infrared reflectances. Apart from LAI, the input variables of this radiative model are canopy characteristics, such as leaf and soil optical properties, leaf orientation distribution, eventually fractional cover (such variables could be outputs of the biological model), and acquisition characteristics, such as sun and satellite angles. The

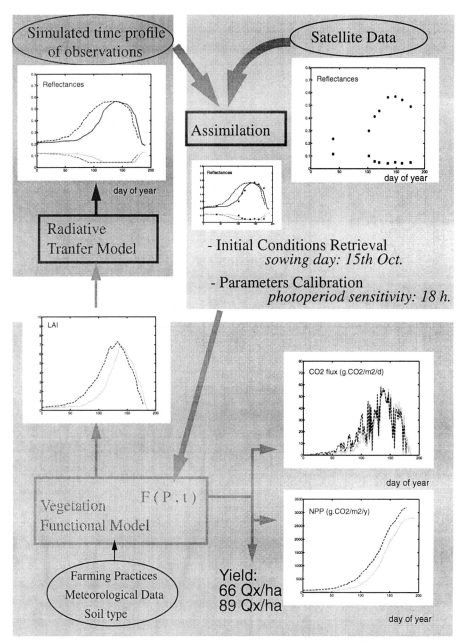

Figure 28.2 *Diagram describing the assimilation strategy. Bottom frame: output from the vegetation model. Upper left frame: coupling with a radiative transfer model. Upper right frame: assimilation of observations to calibrate and initialize the vegetation model.*

reflectance temporal profiles are then simulated for the conditions of observation. The satellite observations provide an irregular data set of radiative measurements over the field (similar to high spatial resolution in that case), and they are corrected for atmospheric perturbations. The assimilation procedure consists of the adjustment of the key parameters of the biological model in order to improve the coherency between simulation and observations. The values sd = October 15, and psat = 18 hours produced reflectance profiles close to the ones observed, and we therefore assume that they correspond to the optimal parameterization of the model.

Applications

Crops

The feasibility of such a strategy has been first demonstrated by Bouman (1991) using ground radiometric measurements over sugar beet fields. For crops, a study with SPOT data over winter wheat fields in Beauce estimated the sowing date only because of the limited number of images available during the growing season. Four observations of the near infrared reflectances during spring 1992 lead to the average estimation of 14 October as sowing date for the area round Chartres, which agrees well with data from the agricultural services (Kergoat *et al.*, 1995).

Natural vegetation

Lo Seen Chong *et al.* (1995) have shown that NOAA/AVHRR GVI can be used to validate the radiative temporal profile simulated by a model for tropical grasslands. In that case, ancillary data provided the knowledge required for the correct parameterization of various relations within the model, and especially for the representation of the seasonal pattern. Such behaviour is related to phenology and allocation processes, and generally, great uncertainty exists about their parameterization as well as about the spatial variability of the parameters. The control of the temporal profile from satellite data might provide calibration of those processes which describe the seasonal pattern of the canopy for tropical grasslands or deciduous forests. In a simulation study for a temperature deciduous forest, Kergoat *et al.* (1995) showed that the assimilation of a temporal sample of noisy model-generated reflectances allows the correct retrieval of the three parameters that mainly determine canopy development.

Effects of atmospheric perturbations and directional radiative properties

Experience with temporal series of satellite data in a quantitative way emphasizes the fact that all perturbing effects must be taken into account, as described by Townshend *et al.* (1994), for NOAA/AVHRR LAC data. First, for calibration, the top of atmosphere reflectances must be computed to normalize the variations of illumination, and an atmospherical radiative transfer model needs to be applied. Second, when NOAA/AVHRR data are used, a vegetation index should be used with a dynamic filtering process for the elimination of perturbed data (clouds, high aerosol content, high viewing angles). It seems preferable to keep as much clear data as possible to have a better sampling, and then to discard the maximum values using composite techniques. Beside the signal processing, the use of a directional radiative transfer model accounts to some extent for the influence of sun and satellite angles on the radiative signal (non-lambertian properties of the surface). The simulated reflectances are then assumed to be comparable to the observed ones.

A test is performed to determine the sensitivity of the results to the data processing. Nadir, near infrared and visible reflectances are simulated at the canopy level (full curves in Fig. 28.3). Some noise is added to simulate 'plausible' random directional and atmospheric effects. The experiment was performed with SPOT data, and hence the viewing angles follow a uniform distribution between $+30°$ and $-30°$. The reflectance for each possible viewing angle is simply derived from the nadir reflectance according to the results given by Cabot *et al.* (1994). Reflectances at the top of the atmosphere are computed with SMAC (Rahman & Dedieu, 1994), with a Gaussian distribution for water vapour, and a Poisson distribution for the aerosol content. One example of such 'noisy' reflectances is shown on Fig. 28.3 for eight dates sampled every 10 days. As commonly observed, the main effect of the atmosphere is to increase the visible reflectance, and to decrease the near infrared reflectance. In our case, because of the association of simulated directional and atmospheric effects, the resulting near infrared reflectance might be increased for some specific surface directional conditions. In this example, most of the 'observations' take place during the ascending part of the LAI curve (decrease of visible reflectance, increase of near infrared reflectance). As a consequence, when these data are used to adjust the initial conditions of the model, the noise of the 'observations' shifts the estimation of the sowing date to a later period in order to constrain the model simulations to follow the 'observations'. When averages over 100 runs with random noise are made as described above, this shift is between 10 and 15 days.

Discussion

The methodology described above is especially useful when no ground truth is available that could provide information about the unknown

Simulated data

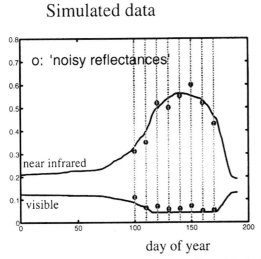

day of year

Figure 28.3 *Simulation of a sampling of 'noisy' visible and near infrared reflectances. This 'noise' is assumed to be representative of the directional and atmospheric effects.*

parameters or about the biological variables that describe the canopy development. If the temporal radiative features simulated by the coupling of a vegetation model with a radiative transfer model agree with the observation, then this should allow for higher confidence in the processes simulated by the vegetation model. We want to emphasize this by underlining two points:

1. CO_2 fluxes and NPP values are not directly available from radiative measurements, and field campaigns occur only locally, usually during a short period. This strategy therefore provides an indirect way to constrain the simulation.

2. In contrast, LAI is a variable which has been often derived directly from satellite data, mainly by applying empirical relationships to vegetation indices. The use of a radiative transfer model in the direct mode to simulate the reflectances which are then compared to the satellite observations avoids having to invoke such empirical relations, which need to be locally calibrated (Running *et al.*, 1989). Over forests, research has been carried out on the inversion of geometric-optical reflectance models (Wu & Strahler, 1994), but such models also involve considerable local parameterization. More physical radiative transfer models have been developed (Verstraete *et al.*, 1990), which can be coupled atmospheric-directional models (Rahman *et al.*, 1993). Their inversion requires multiangular measurements (Pinty *et al.*, 1990) and is expected to be possible with future observations,

such as from POLDER; it should theoretically be possible also with
NOAA/AVHRR, assuming that no LAI change takes place during
successive days with various viewing angles. During active growth,
however, we attempt to detect the periods of rapid changes (for
example, during bud-burst) over seasonal ecosystems (crops, tropical
grassland, deciduous forests) at times where such an assumption about
LAI appears to be doubtful. Furthermore, for most agricultural landscapes, the problem of spatial heterogeneity within the coarse pixel
makes the interpretation even more difficult and requires the
combined use of coarse and high resolution data (Fischer, 1994,
Moulin *et al.*, 1995). The alternative that we propose here, using the
assimilation strategy, bases the estimation of the LAI temporal profile
on the biological model first, while the satellite data are used to
constrain the model and to determine the values of uncertain
parameters.

Because this method involves a radiative transfer model, the consequences of the uncertainties on the parameters describing the optical properties, or on the influence of the canopy structure on the radiative behaviour,
need to be understood. The choice of the radiative transfer model must be
considered, as well as the sensitivity of the assimilation result to the
parameterization. For many ecosystem types, particularly in the case of
natural vegetation, the calibration of the radiative transfer model is missing.
High spatial resolution data can be used, and they will also provide information about the radiative behaviour of heterogeneous areas.

Conclusion

This chapter has presented three different ways to use satellite time series
data which help to understand ecosystem behaviour. There is no new
concept in the regional land cover classification described in the first
section, but the need for special care in data processing has been emphasized. The use of data at 1 km spatial resolution leads to the retrieval of
fine structures which display typical functioning during the year, as well as
interannual variations related to meteorological events.

In the diagnostic NPP model of the second section, satellite data provide
a simple way to investigate photosynthetic activity during the year for
various ecosystems at the global scale. The parametric *e* model has been
adapted to simulate GPP, NPP, and NEP, as well as their interannual variations. It therefore gives a clear image of the spatial and temporal features,
and of their changes, occurring in the major terrestrial components of the

global carbon cycle, related to the climate fluctuations. We have mainly shown the potential use of such an approach, because many important factors were not considered at this step. Improved data processing (atmospheric effects, dynamic filtering) is required, but the calibration of the relation between NDVI and fPAR could also be done locally to account for the influence of the ecosystem or soil structure.

The calibration of ecosystem models with satellite data as discussed in the last section is a promising way to access the spatial variability of biological parameters, or to initialize conditions that are difficult to estimate otherwise. Long-time series of coarse resolution satellite data appear to be a means of understanding the functioning of vegetation in areas where processes driving the seasonal behaviour of the canopy are poorly known. When a dynamic vegetation model will account for the mechanisms of disturbance, competition and dispersal, then the local and instantaneous zoom possibilities provided by high spatial resolution satellite data will provide accurate data on regional features, such as the regular trend in the tree/herbaceous ratio in savannas.

High as well as coarse spatial resolution satellite data are therefore needed for development and validation of ecosystem models attempting to work in a prognostic mode at the global scale. The use of other spectral domains (mean infrared, thermal infrared, passive and active microwaves) has not been discussed in this paper, but they also provide important information related to soil moisture properties, water cycle, surface temperatures, and roughness (stand structure). This paper has mainly focused on monitoring and modelling of vegetation seasonal dynamics. The understanding of the phenological mechanism is weaker than the understanding of the instantaneous trace gas and energy fluxes, but the description of the seasonal features of terrestrial biosphere is a major requirement for climate models (Henderson-Sellers *et al.*, 1986; Gutman, 1994) or dynamic global vegetation models. The long-term pattern of the ecosystems results from the seasonal processes as well as from their interactions. Our final conclusion is that satellite observations have appeared as a source of data allowing development, calibration and validation of global vegetation models, provided that the information contained in the radiative measurements is not misinterpreted.

Acknowledgements

Parts of this work have been supported by the Commission of the European Union (project 'The Global Carbon Cycle and its Perturbation by Man and Climate' of the Environment programme, and contract with the Joint

Research Center at Ispra: 'Pilot Project for Application of Remote Sensing to Agricultural Statistics'), the French PNTS and the Region Midi-Pyrénées. We especially thank the people of the Bioclimatological station of INRA (Avignon), Bernard Seguin, Richard Delécolle, Martine Guérif, for various discussions on multitemporal satellite data and crop models and Gilbert Saint and many GCTE scientists for their work towards a better link between ecologists and remote sensing specialists. The manuscript has been improved due to valuable comments given by Wolfgang Cramer and Will Steffen.

References

Achard, F., Malingreau, J.-P., Phulpin, T., Saint, G., Saugier, B., Seguin, B. & Vidal-Madjar, D. (1992). *The 'Végétation' instrument onboard SPOT-4: a mission for global monitoring of the continental biosphere.* CESBIO, 18 av. E. Belin, bpi 2801, F-31055 Toulouse Cedex.

Arino, O., Viovy, N. & Belward, A. S. (1991). Vegetation dynamics of West Africa classified using AVHRR NDVI time-series. In *5th AVHRR data users' meeting* Tromsø, Norway. 25–28 June 1991.

Badhwar, G. D., Carnes, J. G. & Austin, W. W. (1982). Use of Landsat derived temporal profiles for corn-soybean feature extraction and classification. *Remote Sensing of the Environment*, **12**, 57–79.

Berthelot, B., Dedieu, G., Cabot, F. & Adam, S. (1994). Estimation of surface reflectances and vegetation index using NOAA/AVHRR: methods and results at global scale. In *6th International Symposium on Physical Measurements and Signatures in Remote Sensing*, pp. 33–40. Val d'Isère, France, 17–21 January 1994: CNES.

Bouman, B. A. M. (1991). The linking of crop growth models and multi-sensor remote sensing data. In *5th International Colloquium on Physical Measurements and Signatures in Remote Sensing*, pp. 583–8. Courchevel, France, 14–18 January, 1991: ESA.

Box, E. O. (1988). Estimating the seasonal carbon source-sink geography of a natural, steady-state terrestrial biosphere. *Applied Meteorology*, **27**, 1109.

Brown, J. F., Loveland, T. R., Merchant, J. W., Reed, B. C. & Ohlen, D. O. (1993). Using multisource data in global land-cover characterization: concepts, requirements, and methods. *Photogrammetric Engineering and Remote Sensing*, **59**, 977–87.

Cabot, F., Qi, J., Moran, M. S. & Dedieu, G. (1994). Test of surface bidirectional reflectance models with surface measurements: results and consequences for the use of remotely sensed data. In *6th Int. Symposium on Physical Measurements and Signatures in Remote Sensing*. Val d'Isère, France, 17–21 January 1994: CNES.

Choudhury, B. J. & Tucker, C. J. (1987). Monitoring global vegetation using NIMBUS-7 37 GHz data: some empirical relations. *International Journal of Remote Sensing*, **8**, 1085–90.

DeFries, R. S. & Townshend, J. R. G. (1994). NDVI-derived land cover classifications at a global scale. *International Journal of Remote Sensing*, **15**, 3567–86.

Derrien, M., Farki, B., Legléau, H. & Sairouni, A. (1992). Vegetation cover mapping over France using NOAA-11/AVHRR. *International Journal of Remote Sensing*, **13**, 1787–95.

Deschamps, P.-Y., Bréon, F.-M., Leroy, M., Podaire, A., Bricaud, A., Buriez, J.-C. & Sèze, G. (1994). The POLDER mission: instrument characteristics and scientific objectives. *IEEE Transactions on Geoscience and Remote Sensing*, **32**, 598–615.

Dutton, E. G. & Christy, J. R. (1992). Solar radiation forcing at selected locations and evidence for global lower tropospheric cooling following the eruption of Eı Chichòn and Pinatubo. *Geophysical Research Letters*, **19**, 2313–16.

Eastman, J. R. & Fulk, M. (1993). Long sequence time series evaluation using standardized principal components. *Photogrammetric Engineering and Remote Sensing*, **59**, 991–6.

Esser, G., Hoffstadt, J., Mack, F. & Wittenberg, U. (1994). High Resolution Biosphere Model: Documentation Model version 3.00.00. *Institut für Pflanzenökologie, Justus-Liebig-Universität Gießen. Mitteilungen.* No. 2.

Fischer, A. (1994). A model for the seasonal variations of vegetation indices in coarse resolution data and its inversion to extract crop parameters. *Remote Sensing of the Environment*, **48**, 220–30.

Francois, L., Gérard, J.-C. & Warnant, P. (1994). Global carbon cycle modelling and remote sensing. In *Belgian Scientific Space Research*. Belgian Space Science Office.

Fung, I. Y., Tucker, C. J. & Prentice, K. C. (1987). Application of advanced very high resolution radiometer vegetation index to study atmosphere biosphere exchange of CO_2. *Journal of Geophysical Research*, **92**, 2999–3015.

Goward, S. N. & Dye, D. G. (1987). Evaluating North American net primary productivity with satellite observations. *Advances in Space Research*, **7**, 165–74.

Goward, S. N., Dye, D. G., Turner, S. & Yang, J. (1993). Objective assessment of the NOAA global vegetation index data product. *International Journal of Remote Sensing*, **14**, 3365–94.

Goward, S. N., Markham, B., Dye, D. G., Dulaney, W. & Yang, J. (1991). Normalized difference vegetation index measurements from the Advanced Very High Resolution Radiometer. *Remote Sensing of the Environment*, **35**, 257–77.

Goward, S. N., Tucker, C. J. & Dye, D. G. (1985). North American vegetation patterns observed with the NOAA-7 advanced very high resolution radiometer. *Vegetatio*, **64**, 3–14.

Goward, S. N., Turner, S., Dye, D. G. & Liang, S. (1994). The University of Maryland improved Global Vegetation Index product. *International Journal of Remote Sensing*, **15**, 3365–95.

Gutman, G. G. (1994). Global data on land surface parameters from NOAA AVHRR for use in numerical climate models. *Journal of Climate*, **7**, 669–80.

Heimann, M. & Keeling, C. D. (1989). A three-dimensional model of atmospheric CO_2 transport based on observed winds. 2. Model description and simulated tracer experiments. *Geophysical Monographs*, **55**, 237–75.

Henderson-Sellers, A., Wilson, M. F., Thomas, G. & Dickinson, R. E. (1986). *Current, global land-surface data sets for use in climate-related studies.* pp 272 STR. Boulder, Colorado: National Center for Atmospheric Research.

Holben, B. N. (1986). Characteristics of maximum-value composite images from temporal AVHRR data. *International Journal of Remote Sensing*, **7**, 1417–34.

Holben, B. N., Kaufman, Y. J. & Kendall, J. D. (1990). NOAA-11 AVHRR visible and near-IR inflight calibration. *International Journal of Remote Sensing*, **11**, 1511–19.

Hunter, A. F. & Lechowicz, J. (1992). Predicting the timing of budburst in temperate trees. *Journal of Applied Ecology*, **29**, 597–604.

Justice, C. O., Townshend, J. R. G., Holben, B. N. & Tucker, C. J. (1985). Analysis of the phenology of global vegetation using meteorological satellite data. *International Journal of Remote Sensing*, **6**, 1271–318.

Kaufman, Y. G. & Holben, B. N. (1993). Calibration of the AVHRR visible and near-IR bands by atmospheric scattering, ocean glint and desert reflection. *International Journal of Remote Sensing*, **14**, 21–52.

Kergoat, L., Fischer, A., Moulin, S. & Dedieu, G. (1995). Satellite measurements as a constraint on estimates of vegetation carbon budget. *Tellus*, **47B**, 251–63.

Koomanoff, V. A. (1989). *Analysis of Global Vegetation Patterns: a Comparison Between Remotely Sensed Data and a Conventional Map.* Biogeography research series, 890201. College Park, Maryland: Dept of Geography, Univ. of Maryland.

Kremer, R. G. & Running, S. W. (1993). Community type differentiation using NOAA/AVHRR data within a sagebrush-steppe ecosystem. *Remote Sensing of the Environment*, **45**, 311–18.

Kumar, M. & Monteith, J. L. (1981). Remote sensing of crop growth. In *Plants and the Daylight Spectrum*, ed. H. Smith, pp. 133–44. New York: Academic Press.

Landsberg, J. J., Prince, S. D. Jarvis, P. G., McMurtrie, R. E., Luxmoore, R. & Medlyn, B. E. (1995). Energy conversion and use in forests: the analysis of forest production in terms of radiation utilisation efficiency (e). In *The Use of Remote Sensing in Modelling Forest Productivity at Scales from the Stand to the Globe*, ed. H. Gholz, K. Nakane & H. Shimoda. Amsterdam: Kluwer.

Leroy, M. & Roujean, J. L. (1994). Sun and View angle corrections on reflectances derived from NOAA/AVHRR data. *IEEE Transactions on Geoscience and Remote Sensing*, **32**, 684–97.

Lieth, H. (1975). Primary productivity in ecosystems: comparative analysis of global patterns. In *Unifying Concepts in Ecology*, ed. H. Lieth & R. H. Whitaker, pp. 300–21. Berlin: Springer.

Lloyd, D. (1990). A phenological classification of terrestrial vegetation cover using shortwave vegetation index imagery. *International Journal of Remote Sensing*, **11**, 2269–79.

Lo Seen Chong, D., Mougin, E., Rambal, S., Gaston, A. & Hiernaux, P. (1995). A regional Sahelian grassland model to be coupled with multispectral satellite data. II. Toward the control of its simulations by remotely sensed indices. *Remote Sensing of the Environment*, **52**, 194–206.

Lo, T. H. C., Scarpace, F. L. & Lillesand, T. M. (1986). Use of multitemporal spectral profiles in agricultural land-cover classification. *Photogrammetric Engineering and Remote Sensing*, **52**, 535–44.

Louahala, S., Fischer, A., Podaire, A. & Viovy, N. (1991). Classification de profils temporels de NDVI AVHRR/NOAA et sensibilité a l'occupation du sol et aux conditions climatiques en région temperée. In *5th International Symposium on Physical Measurements and Signatures in Remote Sensing*, pp. 647–50. Courchevel, France, 14–18 January 1991: ESA.

Loveland, T. R., Merchant, J. W., Ohlen, J. F. & Brown, J. F. (1991). Development of a land-cover characteristics database for the coterminous US. *Photogrammetric Engineering and Remote Sensing*, **57**, 1453–63.

Lüdeke, M., Janecek, A. & Kohlmaier, G. H. (1991). Modelling the seasonal CO_2 uptake by land vegetation using the global vegetation index. *Tellus*, **43B**, 188–96.

Lüdeke, M. K. B., Badeck, F.-W., Otto, R. D., Häger, C., Dönges, S., Kindermann, J., Würth, G., Lang, T., Jäkel, U., Klaudius, A., Ramge, P., Habermehl, S. & Kohlmaier, G. H. (1994). The Frankfurt Biosphere Model. A global process oriented model for the seasonal and long-term CO_2 exchange between terrestrial ecosystems and the atmosphere. I. Model description and illustrative results for cold deciduous and boreal forests. *Climate Research*, **4**, 143–66.

Lüdeke, M. K. B., Dönges, S., Otto, R. D., Kindermann, J., Badeck, F.-W., Ramge, P., Jäkel, U. & Kohlmaier, G. H. (1995). Responses in NPP and carbon stores of the northern biomes to a CO₂ induced climatic change, as evaluated by the Frankfurt Biosphere Model (FBM). *Tellus*, **47B**, 191–205.

Lüdeke, M. K. B., Ramge, P. H. & Kohlmaier, G. H. (in press). The use of satellite-detected NDVI data for the validation of global vegetation phenology models: application to the Frankfurt Biosphere Model. *Ecological Modelling*.

Maisongrande, P., Ruimy, A., Dedieu, G. & Saugier, B. (1995). Monitoring seasonal and interannual variations of Gross Primary Productivity, Net Primary Productivity using a diognostic model and remotely sensed data. *Tellus* **47B**, 178–90.

Malingreau, J. P. (1986). Global vegetation dynamics: satellite observations over Asia. *International Journal of Remote Sensing*, **7**, 1121–46.

McCormick, M. P., Thomason, L. W. & Trepte, C. R. (1995). Atmospheric effects of the Mt Pinatubo eruption. *Nature*, **373**, 399–404.

McGuire, A. D., Joyce, L. A., Kicklighter, D. W., Melillo, J. M., Esser, G. & Vörösmarty, C. J. (1993). Productivity response of climax temperate forests to elevated temperature and carbon dioxide: a North American comparison between two global models. *Climatic Change*, **24**, 287–310.

Melillo, J. M., McGuire, A. D., Kicklighter, D. W., Moore III, B., Vörösmarty, C. J. & Schloss, A. L. (1993). Global climate change and terrestrial net primary production. *Nature*, **363**, 234–40.

Moody, A. & Strahler, A. H. (1994). Characteristics of composited AVHRR data and problems in their classification. *International Journal of Remote Sensing*, **15**, 3473–91.

Mougin, E., Lo Seen Chong, D., Rambal, S., Gaston, A. & Hiernaux, P. (1995). A regional Sahelian grassland model to be coupled with multispectral satellite data. I. Model description and validation. *Remote Sensing of the Environment*, **52**, 181–93.

Moulin, S., Fischer, A., Dedieu, G. & Delécolle, R. (1995). Temporal variations in satellite reflectances at field and regional scales compared with values simulated by linking crop growth and SAIL models. *Remote Sensing of the Environment*, **54**, 261–72.

Nemani, R., Pierce, L., Running, S. & Goward, S. (1993). Developing satellite-derived estimates of surface moisture status. *Journal of Applied Meteorology*, **32**, 548–57.

Nemani, R. R. & Running, S. W. (1989). Testing a theoretical climate–soil–leaf area hydrological equilibrium of forests using satellite data and ecosystem simulation. *Agricultural and Forest Meteorology*, **44**, 245–60.

Pinty, B., Verstraete, M. M. & Dickinson, R. (1990). A physical model of the bidirectional reflectance of vegetation canopies. 2. Inversion and validation. *Journal of Geophysical Research*, **95**, 11767–75.

Porter, G. R. (1993). AFRCWHEAT2: A model of the growth and development in wheat incorporating responses to water and nitrogen. *European Journal of Agronomy*, **2**, 69–82.

Potter, C. S., Randerson, J. T., Field, C. B., Matson, P. A., Vitousek, P. M., Mooney, H. A. & Klooster, S. A. (1993). Terrestrial ecosystem production: a process model based on global satellite and surface data. *Global Biogeochemical Cycles*, **7**, 811–41.

Prentice, I. C. (1993). Process and production. *Nature*, **363**, 209–10.

Prince, S. D. (1991). A model of regional primary production for use with coarse-resolution satellite data. *International Journal of Remote Sensing*, **12**, 1313–30.

Prince, S. D. & Goward, S. N. (1995).

Global net primary production: the remote sensing approach. *Journal of Biogeography*, 22 (in press).

Prince, S. D., Justice, C. O. & Moore III, B. (1994). *Monitoring and Modelling of Terrestrial Net and Gross Primary Production*. IGBP-DIS Working Paper 8.

Rahman, H. & Dedieu, G. (1994). SMAC: a simplified method for the atmospheric correction of satellite measurements in the solar spectrum. *International Journal of Remote Sensing*, 15, 123–43.

Rahman, H., Verstraete, M. M. & Pinty, B. (1993). Coupled Surface-Atmosphere Reflectance (CSAR) model. I. Model description and inversion on synthetic data. *Journal of Geophysical Research: Atmosphere*, 98, 20779–89.

Raich, J. W., Rastetter, E. B., Melillo, J. M., Kicklighter, D. W., Steudler, P. A., Peterson, B. J., Grace, A. L., Moore III, B. & Vörösmarty, C. J. (1991). Potential net primary productivity in South America: application of a global model. *Ecological Applications*, 1, 399–429.

Rauzi, F. & Dobrenz, A. K. (1970). Seasonal variation of chlorophyll in western wheatgrass and blue gamma. *Journal of Range Management*, 23, 372–3.

Reed, B. C., Brown, J. F., Van der Zee, D., Loveland, T. R., Merchant, J. W. & Ohlen, D. O. (1994). Measuring phenological variability from satellite imagery. *Journal of Vegetation Science*, 5, 703–14.

Rock, B. N., Skole, D. L. & Choudhury, B. J. (1993). Monitoring vegetation change using satellite data. In *Vegetation Dynamics and Global Change*, ed. A. M. Solomon & H. H. Shugart, pp. 153–67. New York: Chapman and Hall.

Ruimy, A., Saugier, B. & Dedieu, G. (1994). Methodology for the estimation of terrestrial net primary production from remotely sensed data. *Journal of Geophysical Research*, 99, 5263–83.

Running, S. W. & Coughlan, J. C. (1988). A general model of forest ecosystems processes for regional applications. I. Hydrological balance, canopy gas exchange and primary production processes. *Ecological Modelling*, 42, 125–54.

Running, S. W., Loveland, T. R. & Pierce, L. L. (1994). A vegetation classification logic based on remote sensing for use in global biogeochemical models. *Ambio*, 23, 77–81.

Running, S. W., Nemani, R. R., Peterson, D. L., Band, L. E., Potts, D. F., Pierce, L. L. & Spanner, M. A. (1989). Mapping regional forest evapotranspiration and photosynthesis by coupling satellite data with ecosystem simulation. *Ecology*, 70, 1090–101.

Sarmiento, J. L. (1993). Carbon cycle: atmospheric CO_2 stalled. *Nature*, 365, 697–8.

Sellers, P. J., Tucker, C. J., Collatz, C. J., Los, S. O., Justice, C. O., Dazlich, D. A. & Randall, D. A. (1994). A global 1 by 1 NDVI data set for climate studies. Part 2. The generation of global fields of terrestrial biophysical parameters from the NDVI. *International Journal of Remote Sensing*, 15, 3519–45.

Stowe, L. L., McClain, E. P., Carey, R., Pellegrino, P., Gutman, G. G., Davis, P., Long, C. & Hart, S. (1991). Global distribution of cloud cover derived from NOAA/AVHRR operational satellite data. *Advances in Space Research*, 3, 51–4.

Teillet, P. M. & Holben, B. N. (1993). Toward operational radiometric calibration of NOAA AVHRR imagery in visible and near-infrared channels. *Canadian Journal of Remote Sensing*, 15, 20–32.

Townshend, J. R. G., Justice, C. O. & Kalb, V. (1987). Characterization and classification of South American land cover types using satellite data. *International Journal of Remote Sensing*, 8, 1189–207.

Townshend, J. R. G., Justice, C. O., Li,

W., Gurney, C. & McManus, J. (1991). Global land cover classification by remote sensing: present capabilities and future possibilities. *Remote Sensing of the Environment*, **35**, 243–55.

Townshend, J. R. G., Justice, C. O., Skole, D., Malingreau, J.-P., Cihlar, J., Teillet, P., Sadowski, F. & Ruttenberg, S. (1994). The 1 km resolution global data set: needs of the International Geosphere Biosphere Programme. *International Journal of Remote Sensing*, **15**, 3417–41.

Tucker, C. J., Townshend, J. R. & Goff, T. E. (1985). African land-cover classification using satellite data. *Science*, **227**, 369–75.

Verhoef, W. (1984). Light scattering by leaf layers with application to canopy reflectance modeling: the SAIL model. *Remote Sensing of the Environment*, **16**, 125–41.

Vermote, E., El Saleous, N., Kaufman, Y. J. & Dutton, E. (1994). Data pre-processing: stratospheric aerosol perturbing effect on the remote sensing of vegetation: correction method for the composite NDVI after the Pinatubo eruption. In *6th International Symposium on Physical Measurements and Signatures in Remote Sensing*, pp. 151–8. Val d'Isère, France, 17–21 January 1994: CNES.

Verstraete, M. M., Pinty, B. & Dickinson, R. E. (1990). A physical model of the bidirectional reflectance of vegetation canopies. I. Theory. *Journal of Geophysical Research*, **95**, 11755–65.

Viovy, N. (1990). Etude spatiale de la biosphère terrestre: intégration de modèles écologiques et de mesures de télédétection. Institut National Polytechnique de Toulouse.

Viovy, N., Arino, O. & Belward, A. S. (1992). The Best Index Slope Extraction (BISE): a method for reducing noise in NDVI time-series. *International Journal of Remote Sensing*. **13**, 1585–90.

Warnant, P., Francois, L. D. S. & Robinet, F. (1994*a*). Forcing of a global model of plant productivity with climatic and remote sensing data. In *Vegetation, Modelling and Climatic Change Effects*, ed. F. Veroustraete et al., pp. 179–86. The Hague: SPB Academic Publishing.

Warnant, P., Francois, L., Strivay, D. & Gerard, J. C. (1994*b*). CARAIB: A global model of terrestrial biological productivity. *Global Biogeochemical Cycles*, **8**, 255–70.

Weir, A. H., Bragg, P. L., Porter, J. R. & Rayner, J. H. (1984). A winter wheat crop simulation model without water or nutrient limitations. *Journal of Agricultural Science*, **102**, 371–82.

WMO. (1989). *The Global Climate System: Climate System Monitoring, June 1986–1988*. Geneva: World Meteorological Organization.

Woodward, F. I. & Smith, T. M. (1994). Global photosynthesis and stomatal conductance: modelling the controls by soil and climate. *Advances in Botanical Research*, **20**, 1–41.

Wu, Y. & Strahler, A. H. (1994). Remote estimation of crown size, stand density, and biomass on the Oregon transect. *Ecological Applications*, **4**, 299–312.

Part seven

Conclusion

Predicting a future terrestrial biosphere: challenges to GCTE science

B. H. Walker

The success of a scientific project is judged against the degree to which it meets its objectives, and so it is appropriate to start an analysis of GCTE's future directions by recalling that the objectives (described in the Introduction to this book) are two-fold: predicting the effects of global change on ecosystems, and determining how these effects feed back to the atmosphere and the climate system. To begin the analysis it is helpful to put the project into some overall context; a larger framework that puts GCTE into perspective. An appropriate framework in this case is the well-known earth system wiring diagram (Anon., 1988), developed by Frances Bretherton and his colleagues on NASA's Earth System Sciences Committee.

Fig. 29.1 shows the terrestrial part of the wiring diagram with GCTE's research at the Activity level superimposed. At the bottom is a list of Activities that are not catered for by the diagram. Though it was familiar to all in a general way, this earth system 'model' was not used in the development of GCTE's Science plan and it is therefore interesting to note the almost one-to-one way in which the GCTE Operational Plan can be mapped onto it. In terms of what is needed for a global analysis, GCTE is at least addressing all that the terrestrial part of the wiring diagram calls for, with some minor differences. The question we need to ask is how well the various components are being addressed and the directions in which the research is going. Is GCTE likely to achieve its goals and what are the major hurdles still to be overcome?

The list at the bottom of Fig. 29.1 identifies research considered important in GCTE, but not in the earth system model. Mostly this reflects the fact that the earth system approach focuses on the ways in which induced changes in terrestrial ecosystems will feed back to further influence atmospheric composition and climate. This is one of GCTE's two objectives. What the earth system diagram does not include is the other objective: the changes in ecosystems that are more of local and regional concern, important because humans live in and depend on these ecosystems even though some of the changes may not have significant global feedback consequences.

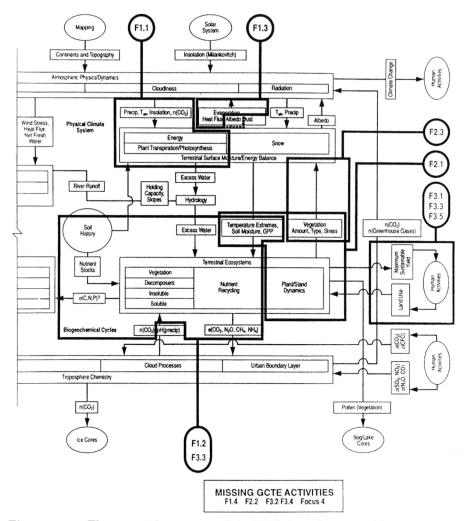

Figure 29.1   *The terrestrial component of the NASA earth system wiring diagram (Anon., 1988) with the GCTE research Activities superimposed.*

These sorts of changes are mostly encompassed within Foci 3 and 4 (agriculture and forestry and ecological complexity/biodiversity, respectively).

One particular activity in the list at the bottom of Fig. 29.1 that does not fit in to the earth system diagram is landscape dynamics (F2.2). No doubt the proponents of the diagram would claim that these are subsumed as part of the 'plant/stand' dynamics. However, it is an important, unresolved question whether or not stand (or patch) scale dynamics can be agglomerated to regional and global responses without taking into account landscape-scale processes and phenomena. The working hypothesis in GCTE is that there are scale-dependent processes (for example fire, grazing, water redistribution down catenas) at the landscape scale that significantly modify individual patch functions and which therefore cannot be omitted. It would be a lot easier for dynamic global vegetation models (DGVMs) if this hypothesis could be justifiably rejected but, given the sort of evidence provided at this meeting, we cannot yet do so. Unfortunately, it is also an area in ecological science that lags behind the developments at the patch and regional scales. It is one of the hurdles we need to overcome, and a challenge to aspiring young ecologists looking for a scientific niche to make their mark is to try to crack the problem of the significance of landscape-scale processes. Contributions in this area will be much appreciated.

A second Activity that stands out even more (F3.2) is the one on agricultural pests and diseases. In the context of the diagram it may be seen as a secondary mechanism, internal to the agriculture box. But from GCTE's perspective its effects will not be taken into account unless the ecology of weeds, pests and diseases under global change are explicitly addressed. It is an area of potentially major agricultural and forestry impact; but unfortunately one which has as yet not progressed very much.

Two other GCTE activities that don't fit neatly into the wiring diagram are the integrating Activities in ecosystem physiology (F1.4) and in agricultural systems (F3.4). They are designed to bring together all the factors that influence the interactive effects of increased CO_2 with changes in biogeochemistry and the water/energy balance on ecosystem physiology, and on the productivity and viability of complex agro-ecosystems, respectively. They are particular examples of an emerging way of doing science, almost a new kind of science, that IGBP is helping to foster: an integrating earth system science. Fig. 29.1 needs to be drawn in three dimensions to illustrate the active research links between the various earth system components, but such a figure would resemble the proverbial spaghetti diagram. As a single illustration, the individual objectives within the research areas being undertaken by BAHC, IGAC and GCTE are being drawn together explicitly, in a co-ordinated research effort, via a combined working group of the three projects (known as the BIG committee). This is no mean feat. Scientists, by nature and tradition, tend to work alone or in small, tight groups. On a personal basis, this new kind of science is less efficient, more frustrating and time-consuming, and more threatening in terms of the traditional scientific rewards of peer recognition and contribution. The new technologies are opening up exciting possibilities. Interdisciplinary science is coming of age and, as this volume has demonstrated, GCTE's research requires collaboration amongst chemists, physiologists, demographers, remote sensers, soil scientists, agronomists and many others.

Where are we at and where are we going?

It would be tedious to attempt a summary of all that is in this book, but a short statement on some of the big remaining questions is appropriate.

The ecological significance of elevated CO_2

The CO_2 research network established to achieve the objectives of Activity 1.1 (see GCTE Report No. 1) has made good progress in quantifying the net effects on photosynthesis in natural communities. The projects involved have shown that the major effects are below ground and they have highlighted the importance of compensatory mechanisms that reduce initially observed increases in productivity and carbon sequestration. But the scientists concerned are not satisfied with their understanding of the net effects on growth and productivity. Four questions, in particular, stand out:

1. How to measure the effects on soil C and N pools and fluxes. The conclusion that the answer lies in the soil raises the question of how to detect small changes, relative to very big pools and fluxes. The approach is becoming clear: a variety of techniques with ultimate testing of well-founded hypotheses in FACE systems to test the interactive effects under the limitations imposed by natural conditions.

2. The so-called 'biotic growth factor', β, (Gates 1985) an empirical coefficient used to describe the relative response of net primary production (or net ecosystem production) to an increase in CO_2 concentration. It amounts to the CO_2 fertilization effect. The variation in ecosystem production response to elevated CO_2 is poorly known, and badly (or not at all) represented in global models of the carbon cycle. The set of experiments and regional measurements proposed by the CO_2 network is aimed at moving from the empirical estimates of β to a more mechanistic understanding of CO_2 effects on vegetation function.

3. How species composition will change as a consequence of altered competitive relationships, both through the direct (differential) effects of elevated CO_2 on species, and through net changes in the environment resulting from interactive effects of changes with CO_2, water, temperature and nutrients.

4. The longer-term effects on communities via effects on reproduction and through evolutionary responses.

Biogeochemistry

The biogeochemistry objectives which led to the development of the IGBP transects remain as a major hurdle: the capacity to predict changes in the spatial distribution and in the net pools and fluxes of C, N and other elements at regional scales, in response to changes in land use (on its own) and to the interactive effects of changes in land use, climate and atmospheric composition. As one example, the Terrestrial Ecosystem Model (Melillo *et al.*, 1993) does as good a job as can be expected with current data sets and understanding, but net changes in pools and fluxes due to the interactive effects can not yet be adequately assessed.

Predicting changes in ecosystem structure

The patch-scale modelling has achieved many of its aims but there are two areas, in particular, that still need further development at this scale: (i) Multiple-life form models (e.g. trees and grasses) in which competition and perhaps other factors lead to mixed life form communities, and (ii) better linkage of physiological models with composition models in which physiology takes spatial heterogeneity into account. Apart from these two patch-scale developments, the future emphasis in modelling is now in two directions: first, the development of dynamic global vegetation models (DGVMs); second, the kinds and significance of landscape dynamics, including the landscape-scale interactive effects of spatial heterogeneity.

A key issue in all these modelling efforts is how to include the effects of spatial heterogeneity at sub-grid cell scales in regional or continental/global models. Massive parallel processing to model the world at patch scales is not feasible and the most promising direction is to use some sort of distribution function, as suggested by Emanuel and Drake (1995), which allows the effects to be parameterized. Their approach is aimed at a more realistic terrestrial carbon cycle model that deals with fine-scale dynamics by means of a function that expresses the area of a land unit as a function of the density per unit area of carbon in vegetation. An important feature of this approach is that it does not require specifying the exact locations of the different sub-grid classes (in their case, carbon densities). The dynamics of the area distribution function in response to plant growth and land use is described by a first-order, partial differential equation. This sort of approach would be appropriate for various kinds of area distribution models, allowing for changes in proportions of land classes, vegetation types or conditions, etc., without the need for explicitly identifying their exact locations.

The issue of functional types

In order to achieve the DGVMs and the generic landscape-scale models for each biome, the Focus 2 task groups have set as a high future priority the resolution of the functional types problem. The task is to avoid the use of the phylogenetic classification of plants (which has no predictive capacity) and to develop a minimum set of plant functional types that can be used in all biomes around the globe. This will allow for prediction of changes in vegetation structure as a consequence of functional responses to changes in environment. It is another area requiring some innovative thinking (aspiring young graduates take note), building on the various schemes developed so far (for non-GCTE purposes) and summarized in the initial GCTE work-shop on this topic (Smith *et al.*, 1996). Whether or not a useful universal FT classification can be developed remains to be seen. Are we looking at the development of a new and ecologically more meaningful taxonomy, or merely minor, useful constructs that will come and go? It is likely that different schemes will be needed for different scales of resolution. At the finer (patch or community) scales, at least, there is a need for a concerted set of field experiments to determine and test the validity of the proposed FTs, for example, such as the reciprocal manipulations of composition and function described in the Operational Plan (Task 2.1.2, pp. 34–5).

Empirical data requirements for vegetation dynamics models

In addition to the rather obvious set of improved site data needed for acceptable model predictions (soil depth, soil fertility, present composition, etc.) there are at least two major questions that need to be answered by experiments and observations:

1. What are the nature and magnitudes of key events (extreme conditions, sequences of events, etc.), in each major biome, that bring about significant changes in ecosystem structure? 'Significant' here needs to be defined either in terms of the consequences for feedback effects on the climate and atmosphere, or for the value of the ecosystem to humans.

2. What are the maximum rates of dispersal and migration of the major species or FTs? This information is important at two scales. First, at continental scales, for the DGVMs, will differences in migration rates lead to changes in the composition of biomes as they move in response to shifts in climatic zones? Second, at landscape and regional scales, how will differences in dispersal rates affect the viability of frag-mented meta-populations? The relative rates of recolonization and local extinction will determine the long-term persistence of such popu-

lations (e.g. Blaustein *et al.*, 1994). Analysis of meta-population dynamics tells us that fragmented populations are very likely to be doomed to extinction before all the patches that contain subpopulations are destroyed. Being able to predict the critical proportions of the total population required to prevent eventual extinction of that population depends on knowing the rates of subpopulation loss and of recolonization. Lack of knowledge about dispersal rates is presently a major stumbling block in applied conservation biology, and it is looming as an equally important problem in predicting the consequences of global change.

The consequences for agriculture and forestry

The crop networks and the forests measurement and modelling programs are major efforts that will take GCTE into the next century. The big hurdle will be bringing together the CO_2 network findings and the crop and forest models, the work on pests and diseases, and the work on soil effects, to give the net consequences of global changes for the complex agricultural and agro-forestry systems. The assumption that multi-species cropping systems embedded in mixed livestock–crop enterprises are more robust and likely to be more resilient in the face of global change, is being questioned (e.g. Ramakrishnan, 1992). The analogy with the 'eggs-in-one-basket' paradigm is too simplistic, and the interdependencies in these complex agro-ecosystems may in fact, in a number of cases, make them more vulnerable.

Ecological complexity

The fundamental question for GCTE concerns the ways in which complexity influences ecosystem functions, and, consequently, how these 'ways' are likely to be influenced by global change. The first phase of the research is therefore to work with other programmes (especially DIVERSITAS) to answer the first part of the question. Only then can we address the issue of how global climate and atmospheric changes may interact with changes in ecological complexity to affect ecosystem function.

Changes in ecological complexity (including but not restricted to reductions in biotic diversity) are being, and will continue to be, brought about primarily by land use. Predicting future losses calls for an improved theory of driven extinctions in declining populations of species that were originally very abundant, as opposed to the better studied stochastic extinction of small, capped populations (cf. Caughley, 1994).

Estimating ecosystem response rates

Thus far I've dealt with the purely scientific challenges we face if we are to achieve the predictive capacity that is expected of us. They comprise a complex and difficult set of tasks, but the scientific community is prepared, and knows how, to deal with them. We are already well involved and for the most part relish the prospect of working through them. There remains, however, a somewhat different sort of challenge that scientists are less comfortable about addressing. It has to do with being asked to make predictions before we're ready for them; the best guesstimates problem that could have far reaching implications. It is best summed up in the words of Article 2 of the United Nations Framework Convention on Climate Change, 1992:

> *The ultimate objective of this Convention and any related legal instruments that the Conference of the Parties may adopt is to achieve, in accordance with the relevant provisions of the Convention, stabilization of greenhouse gas concentrations in the atmosphere at a level that would prevent dangerous anthropogenic interference with the climate system. Such a level should be achieved within a time frame sufficient to allow ecosystems to adapt naturally to climate change, to ensure that food production is not threatened and to enable economic development to proceed in a sustainable manner.*

We are not able to answer the questions in this Article with any scientific assurance; yet we are obliged to try. The Intergovernmental Panel on Climate Change (IPCC) is attempting to address it and has asked for help, since they are bound to come up with some estimates to guide policy decisions. The issue is all about rates; the relative rates of climatic/atmospheric change and of ecosystem responses.

I suggest that a way forward is along two paths, adopting a successive approximation approach. First, using the available global models of vegetation distribution and change, such as BIOME (Prentice *et al.*, 1992), DOLY (Woodward, *et al.*, 1995) and others, identify regions of the world where, according to the mechanisms and assumptions in these models (reflecting our current understanding) vegetation (= ecosystems) changes, relatively, to a greater extent in terms of composition and/or structure in response to a selected set of environmental changes. This would involve, for example, examining the separate and interactive effects of different levels of precipitation and temperature change. By comparing the model results, and by combining them with other sources of information about vegetation stability and responses to environmental disturbance, we would be able to build up a picture of global 'hot-spots'. Based on this picture we need then to apply more detailed patch- and landscape-models to these identified regions, to sharpen up the estimates of magnitudes of response to nominated degrees of environmental change.

The second approach is to develop for each of a representative set of ecosystem types, selected independently of the first approach, an expert system or decision support system (DSS) using the knowledge of experts on those ecosystems. A comparison of the outcomes of the two approaches would enable us to produce a 'best guesstimate' to the questions in Article 2. The problem we face is that even this unnerving short-cut will take too long to meet the needs of the present IPCC assessment, unless significant funding was made available. The challenge nevertheless remains; and it is worth addressing despite the time constraint.

The future of the world's ecosystems

The IPCC challenge regarding assessment of ecosystem response rates leads to the question of what *will* happen to the world's ecosystems. Without attempting a map of the world's future vegetation, a few observations are appropriate:

- Virtually all of the global vegetation models and many of the local-scale ones, including those that are species-based (e.g. Box *et al.*, 1993), identify either absolute minimum temperature or annual winter temperatures as a very important if not the most important climatic factor determining vegetation distribution. It is the minimum temperatures that are predicted by GCMs to change most in response to CO_2 increase, and so refining the regional- and local-scale predictions of minimum temperature changes, and the predictions of vegetation sensitivity to these changes, are essential steps before we can expect good predictions of ecosystem change.
- Changes in precipitation and temperature will influence the ratio of actual to potential evapotranspiration (AET/PET), and this ratio is critically important in determining vegetation structure (e.g. Prentice *et al.*, 1991). Precipitation is poorly predicted by GCMs and reliable regional- and local-scale predictions via a combination of GCMs and weather generators are required before reliable prediction of vegetation change can be achieved.
- The existing global-scale models of vegetation change assume no lag effects, imposed either by dispersal rates (yet we know this to be untrue even under the climate change since the last ice age), or through inertia by existing communities 'hanging on' under sub-

optimal conditions. Relaxing these two assumptions will reduce the sorts of biome shifts and vegetation changes predicted by equilibrium models, or even some of the transient models.

▨ Most of the global scale models are driven by mean climate data and hence are not capable of handling the ecologically important extreme weather events. Furthermore, many of the local scale models that incorporate stochastic weather generators are based on assumptions (e.g. use of monthly weather data, normal distribution of monthly temperature) that conceal extreme events. For some parts of the world this may capture reality sufficiently well, but in others it is the occurrence of extreme events (e.g. exceptional high-intensity storms) that brings about change in ecosystem composition; and including the effects of these events and forecast changes in their frequencies will probably exacerbate the sorts of vegetation changes currently predicted. What we need is a synthesis of knowledge about disturbance regimes in major ecosystem types, and the ways in which these disturbance regimes structure vegetation, and determine its composition.

▨ The predicted climate-driven changes in ecosystems will be further modified by the degree to which resources (other than water) determine ecosystem structure. In their excellent review Field *et al.* (1992) highlight the significance of compensation mechanisms (carbon and nutrient allocation, rates of uptake, etc.) that are likely to dampen expected responses to purely climate and CO_2 change.

Drawing these sorts of qualifying observations together suggests that the magnitude of changes in the world's ecosystems which will occur over the next 50 years or so will be less uniform than some of the current model predictions and, in some places at least, may well exhibit strong lag effects – which means that the future vegetation will be even further from equilibrium with the climate than it is now. But (and this is important) significant changes will occur if the GCM predictions of climate change hold up. Suggestions that global warming and its effects on the world's ecosystems is a trivial issue and can be ignored are both dangerous and mischievous. It is a difficult and time-consuming task to resolve these questions, but only by doing so will we be able to prepare and adapt in time to avoid serious adverse effects on the people living in these ecosystems.

Conclusion

It is not GCTE's task to develop global databases, for example for topography, soil water holding capacity, etc., but without them the results of the best DGVMs will be severely limited, and the feedback effects of the terrestrial biosphere in GCMs will be inadequately treated. The problem is one of pulling together the vast amounts of disparate data in many different agencies and coupling these with new ways of measuring the earth from space. And one of the main limiting factors is available expertise. It is one of the major difficulties facing GCTE. The Data and Information Systems Framework Activity of IGBP is attempting to develop the soils database, as an initial high priority. One of their problems (the same as ours) is availability of suitably qualified scientists. The same problem faces those responsible for developing the Global Terrestrial Observing System (GTOS). Unfortunately, both these groups call on the same limited pool of people who are involved in GCTE Activities and other core projects, highlighting a serious issue in global change research: there are probably sufficient excellent scientists in the world to deal with the problem but there are definitely too few of them available, without direct salary compensation, to spend the necessary time on international as opposed to home-based projects. It is a frustrating and continuing concern of those involved in trying to put together globally directed international research.

If we are to achieve internationally cohesive and integrated research projects at the level which is needed, then national research agencies will have to allow and encourage more of their scientists to participate in the international projects, and provide supporting resources. This is the only 'political', non-scientific statement I make. It arises from a conviction that this is the major research obstacle blocking the resolution of global environmental problems. I can only hope the message gets through to, and is taken on board by, those responsible for national funding policies.

There are 37 Tasks in the GCTE Core Project, each under the leadership of one or more Task Leaders. A few have still to become fully developed, but all of them are on track to achieve substantial results and there is little doubt they will make valuable contributions to the global change issue. It is, therefore, difficult to pick out particular contributions, since those omitted will have been unjustifiably overlooked. Nevertheless, it may be useful to conclude with a very reduced list, combining many tasks, that at this stage of GCTE's development appear likely to be lasting contributions and which, over the next several years, will involve the participation of many of the world's ecologists:

- The IGBP transects, integrating biogeochemistry, the hydrological cycle and vegetation dynamics
- The Long-term Ecological Modelling Activity (LEMA) and the development of (i) DGVMs and (ii) landscape-scale impact models
- The research networks
  The elevated CO_2 network
  The crop networks and generic crop models
  The rangelands and pastures network and models
  The forest network and forest production models
- An integrated farming system model, including soil effects and pests
- Resolution of the ecological complexity issue (still to be defined operationally).

GCTE does not have a definite lifetime. The current set of core projects and those still to be developed in order to meet the objectives of the Operational Plan will take us through the turn of the century. A thorough review at that time will determine whether GCTE should continue as it is, be transformed into a different kind of project, or terminate.

References

Anon. (1988). *Earth System Science*. Washington, DC: Earth System Sciences Committee, NASA Advisory Council, NASA.

Blaustein, A. R., Wake, D. B. & Sonsa, W. P. (1994). Amphibian declines: judging stability, persistence, and susceptibility of populations to local and global extinctions. *Conservation Biology*, 8, 60–71.

Box, E. O., Crumpacker, D. W. & Hardin, E. D. (1993). A climatic model for location of plant species in Florida, USA. *Journal of Biogeography*, 20, 629–44.

Caughley, G. C. (1994). Directions in conservation biology. *Journal of Animal Ecology*, 6, 215–44.

Emanuel, W. R. & Drake, J. B. (1995). Modelling carbon cycling on disturbed landscapes. *Ecological Modelling* (in press).

Field, C. B., Chapin III, F. S., Matson, P. A. & Mooney, H. A. (1992). Responses of terrestrial ecosystems to the changing atmosphere: a resource-based approach. *Annual Review of Ecological Systematics*, 23, 201–35.

Gates, D. M. (1985). Global biospheric response to increasing atmospheric carbon dioxide concentration. In *Direct Effects of Increasing Carbon Dioxide on Vegetation*, ed. B. R. Strain & D. Cure, pp. 171–84. Washington, DC: US Department of Energy, DO/ER-0238.

Melillo, J. M., McGuire, A. D., Kicklighter, D. W., Moore III, B., Vorosmarty, C. J. & Schloss, A. L. (1993). Global climate change and terrestrial net primary production. *Nature*, 363, 234–40.

Prentice, I. C., Sykes, M. T. & Cramer, W. (1991). The possible dynamic response of northern forests to global warming. *Global Ecology and Biogeography Letters*, 1, 129–35.

Prentice, I. C., Cramer, W., Harrison, S. P., Leemans, R., Monserud, R. A. & Solomon, A. M. (1992). A global biome model based on plant physiology and dominance, soil properties and climate. *Journal of Biogeography*, **19**, 117–34.

Ramakrishnan, P. S. (1992). *Shifting Agriculture and Sustainable Development: an Interdisciplinary Study from Northeastern India*. UNESCO-MAB series. Paris: Parthenon (republished by Oxford University Press, New Delhi, 1993).

Smith, T. M., Shugart, H. H. & Woodward, F. I. (1995). *Plant Functional Types: Their Relevance to Ecosystem Properties and Global Change*. IGBP series no. 1. Cambridge: Cambridge University Press.

Woodward, F. I., Smith, T. M. & Emanuel, W. R. (1995). A global primary productivity and phytogeography model. *Global Biogeochemical Cycles*, **9**, 471–90.

Index

Folio numbers in bold denote diagrams and tables